1/25/02

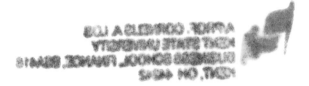

Springer Finance

Springer
*Berlin
Heidelberg
New York
Barcelona
Hong Kong
London
Milan
Paris
Singapore
Tokyo*

Springer Finance

Springer Finance is a new programme of books aimed at students, academics and practitioners working on increasingly technical approaches to the analysis of financial markets. It aims to cover a variety of topics, not only mathematical finance but foreign exchanges, term structure, risk management, portfolio theory, equity derivatives, and financial economics.

Credit Risk: Modelling, Valuation and Hedging
T.R. Bielecki and M. Rutkowski
ISBN 3-540-67593-0 (2001)

Risk-Neutral Valuation: Pricing and Hedging of Finance Derivatives
N.H. Bingham and R. Kiesel
ISBN 1-85233-001-5 (1998)

Visual Explorations in Finance with Self-Organizing Maps
G. Deboeck and T. Kohonen (Editors)
ISBN 3-540-76266-3 (1998)

Mathematics of Financial Markets
R. J. Elliott and P. E. Kopp
ISBN 0-387-98533-0 (1999)

Mathematical Finance - Bachelier Congress 2000 • Selected Papers from the First World Congress of the Bachelier Finance Society, held in Paris, June 29-July 1, 2000
H. Geman, D. Madan, S.R. Pliska and T. Vorst (Editors)
ISBN 3-540-67781-X (2001)

Mathematical Models of Financial Derivatives
Y.-K. Kwok
ISBN 981-3083-25-5 (1998)

Efficient Methods for Valuing Interest Rate Derivatives
A. Pelsser
ISBN 1-85233-304-9 (2000)

Exponential Functionals of Brownian Motion and Related Processes
M. Yor
ISBN 3-540-65943-9 (2001)

Damiano Brigo   Fabio Mercurio

Interest Rate Models Theory and Practice

Springer

Damiano Brigo
Fabio Mercurio
Banca IMI
San Paolo IMI Group
Corso Matteotti 6
20121 Milan
Italy
e-mail: brigo@bancaimi.it
          fmercurio@bancaimi.it

Mathematics Subject Classification (2000): 60H10, 60H35, 62P05, 65C05, 65C20, 90A09
JEL Classification: G12, G13, E43

Cataloging-in-Publication Data applied for

Die Deutsche Bibliothek - CIP-Einheitsaufnahme
Brigo, Damiano:
Interest rate models theory and practice / Damiano Brigo ; Fabio
Mercurio. - Berlin ; Heidelberg ; New York ; Barcelona ; Hong Kong ;
London ; Milan ; Paris ; Singapore ; Tokyo : Springer, 2001
  (Springer finance)
  ISBN 3-540-41772-9

ISBN 3-540-41772-9 Springer-Verlag Berlin Heidelberg New York

This work is subject to copyright. All rights are reserved, whether the whole or part of the material is concerned, specifically the rights of translation, reprinting, reuse of illustrations, recitation, broadcasting, reproduction on microfilm or in any other way, and storage in data banks. Duplication of this publication or parts thereof is permitted only under the provisions of the German Copyright Law of September 9, 1965, in its current version, and permission for use must always be obtained from Springer-Verlag. Violations are liable for prosecution under the German Copyright Law.

Springer-Verlag Berlin Heidelberg New York
a member of BertelsmannSpringer Science+Business Media GmbH

http://www.springer.de

© Springer-Verlag Berlin Heidelberg 2001
Printed in Germany

The use of general descriptive names, registered names, trademarks etc. in this publication does not imply, even in the absence of a specific statement, that such names are exempt from the relevant protective laws and regulations and therefore free for general use.

Cover design: *design & production*, Heidelberg
Typesetting by the authors using a Springer LaTeX macro package
Printed on acid-free paper   SPIN 10755152   41/3142db-5 4 3 2 1 0

To Our Families

Preface

*"We don't do it for the glory. We don't do it for the recognition...
We do it because it needs to be done. Because if we don't, no one else will.
And we do it even if no one knows what we've done.
Even if no one knows we exist. Even if no one remembers we ever existed."*

Kara, "Christmas with the Superheroes" 2,
"Should Auld Acquaintance Be Forgot", 1989, DC Comics

MOTIVATION

The idea of writing a book on interest-rate modeling crossed our minds a couple of years ago, after a few months of work at Banca IMI, where, as quantitative analysts, we have been supporting traders, dealing with mathematical modeling for financial engineering. We had reached this position after a look at the academic environments and after a subsequent first period as quantitative analysts in a different Italian bank. In Banca IMI we were given the task of studying and developing financial models for the pricing and hedging of a broad range of derivatives, and we have been involved in medium/long-term projects. The farsighted policy of Aleardo Adotti, Head of the Product and Business Development Group, allowed us to work in a serene and stimulating environment, also benefiting from the continuous and valuable feedback of several of our bank's traders.

We began writing technical reports on what we were studying and implementing, and soon we realized that our material was covering some gaps in the existing financial literature. We looked at the material collected and wondered whether it be possible and worthwhile to reorganize it in a coherent structure with the ultimate aim of achieving some circulation. "Why not?", we answered ourselves, and our book-writing project started to be outlined.

We had to tread carefully in this new "book-writing territory", since our writing experience had been limited in the main to academic papers for journals and our Ph.D. theses.

We also asked for the advice of our Head, who immediately gave us his enthusiastic approval and granted us enough time and freedom to devote to our writing.

A year and a half later we had reached this final version and eventually decided to stop thinking of further modifications. Hopefully, we will have the chance of implementing any suggestion for improvements in later editions of the book.

In conclusion of this short initial chronicle, let us say a final word about the language used. We would venture to say it looks like American English, if we could be sure that no one would turn his/her nose up. Indeed, our text editor featured an American spell-checking program.

We are quite aware that many sentences could have been written in a more stylish and classical English, yet we resort to the excuse that our English reflects our Latin background, and at times the temptation to transliterate from Italian has simply been irresistible. So we hope you will find our *Italian English* intelligible if amusing!

Let us now describe our motivations and the content of the book in a more detailed manner, starting by answering the following fundamental, inevitable and obligatory question.

Why a New Book?

In years where every month a new book on financial modeling or on mathematical finance comes out, one of the first questions inevitably is: why one more, and why one on interest-rate modeling in particular?

The answer springs directly from our job experience as quantitative analysts in financial institutions. Indeed, one of the major challenges any financial engineer has to cope with is the practical implementation of mathematical models for pricing derivative securities.

When pricing market financial products, one has to address a number of theoretical and practical issues that are often neglected in the classical, general basic theory: the choice of a satisfactory model, the derivation of specific analytical formulas and approximations, the calibration of the selected model to a set of market data, the implementation of efficient routines for speeding up the whole calibration procedure, and so on. In other words, the general understanding of the theoretical paradigms in which specific models operate does not lead to their complete understanding and immediate implementation and use for concrete pricing. This is an area that is rarely covered by books on mathematical finance.

Undoubtedly, there exist excellent books covering the basic theoretical paradigms, but they do not provide enough instructions and insights for tackling concrete pricing problems. We have therefore thought of writing this book in order to cover this gap between theory and practice.

The Gap between Theory and Practice

A gap, indeed. And a fundamental one. The interplay between theory and practice has proved to be an extremely fecund ingredient in the progress of science and modeling in particular. We believe that practice can help to appreciate theory, thus generating a feedback that is one of the most important and intriguing aspects of modeling and more generally of scientific investigation.

If theory becomes deaf to the feedback of practice or vice versa, great opportunities can be missed. It may be a pity to restrict one's interest only to extremely abstract problems that have little relevance for the scientists or quantitative analysts working in "real life".

Now, it is obvious that everyone working in the field owes a lot to the basic fundamental theory from which such extremely abstract problems stem. It would be foolish to deny the importance of a well developed and consistent theory as a fundamental support for any practice involving mathematical models. Indeed, practice that is deaf to theory or that employs a sloppy mathematical apparatus is quite dangerous.

However, besides the extremely abstract refinement of the basic paradigms, which are certainly worth studying but that interest mostly an academic audience, there are other fundamental and more specific aspects of the theory that are often neglected in books and in the literature, and that interest a larger public.

Is This Book about Theory? What kind of Theory?

In the book, we are not dealing with the fundamental no-arbitrage paradigms with great detail. We resume and adopt the basic well-established theory of Harrison and Pliska, and avoid the debate on the several possible definitions of no-arbitrage and on their mutual relationships. Indeed, we will raise problems that can be faced in the basic framework above. Insisting on the subtle aspects and developments of no-arbitrage theory more than is necessary would take space from the other theory we need to address in the book and that is more important for our purposes.

Besides, there already exist several books dealing with the most abstract theory of no-arbitrage. On the theory that we deal with, on the contrary, there exist only few books. What is this theory? To give a flavor of it, let us select a few questions at random:

- How can the market interest-rate curves be defined in mathematical terms?
- What kind of interest rates does one select when writing the dynamics? Instantaneous spot rates? Forward rates? Forward swap rates? Are there models for each such basic variable?
- What is a sufficiently general framework for expressing no-arbitrage in interest-rate modeling?
- Is there a definition of cap or caplet volatility (and of their term structures) in terms of interest-rate dynamics that is consistent with market practice?
- What kinds of diffusion coefficients in the rate dynamics are compatible with different qualitative evolutions of the term structure of volatilities over time?

- How is "humped volatility shape" translated in mathematical terms and what kind of mathematical models allow for it?
- What is the most convenient probability measure under which one can price a specific product, and how can one derive concretely the related interest-rate dynamics?
- Are different market models of interest-rate dynamics mutually compatible?
- Is it possible to restore some compatibility between incompatible market models via some kind of approximation? How is this done precisely? What are the related formulas?
- What does it mean to calibrate a model to the market in terms of the chosen mathematical model? Is this always possible? Or is there a degree of approximation involved?
- Does terminal correlation among rates depend on instantaneous volatilities or only on instantaneous correlations? Can we analyze this dependence?
- Can terminal correlations among rates be computed with an approximation independent of the particular probability measure chosen? Under which kind of models is this possible? If so, what are the related formulas?
- What is the volatility smile, how can it be expressed in terms of mathematical models and of forward-rate dynamics in particular?
- What is the link between dynamics of rates and their distributions?
- What kind of model is more apt to model correlated interest-rate curves of different currencies, and how does one compute the related dynamics under the relevant probability measures?
- When does a model imply the Markov property for the short rate and why is this important?
- ......

We could go on for a while with questions of this kind. Our point is, however, that the theory dealt with in a book on interest-rate models should consider this kind of question.

We sympathize with anyone who has gone to a bookstore (or perhaps to a library) looking for answers to some of the above questions with little fortune. We have done the same, several times, and we were able to find only limited material and few reference works. We hope this book is a successful step forward towards addressing such questions.

We also sympathize with the reader who has just finished his studies or with the academic who is trying a life-change to work in industry or who is considering some close cooperation with market participants. Being used to precise statements and rigorous theory, this person might find answers to the above questions expressed in contradictory or unclear mathematical language. This is something else we too have been through, and we are trying not to be disappointing on this side either.

Preface   XI

Is This Book about Practice? What kind of Practice?

We try and answer some questions on practice that are again overlooked in most of the existing books in mathematical finance, and on interest-rate models in particular. Again, here are some typical questions selected at random:

- What are accrual conventions and how do they impact on the definition of rates?
- Can you give a few examples of how time is measured in connection with some aspects of contracts? What are "day-count conventions"?
- What is the interpretation of most liquid market contracts such as caps and swaptions? What is their main purpose?
- What kind of data structures are observed in the market? Are all data equally significant?
- How is a specific model calibrated to market data in practice? Is a joint calibration to different market structures always possible or even desirable?
- What are the dangers of calibrating a model to data that are not equally important, or reliable, or updated with poor frequency?
- What are the requirements of a trader as far as a calibration results are concerned?
- How can one handle path-dependent or early-exercise products numerically? And products with both features simultaneously?
- What numerical methods can be used for implementing a model that is not analytically tractable? How are trees built for specific models? Can instantaneous correlation be a problem when building a tree in practice?
- What kind of products are suited to evaluation through Monte Carlo simulation? How can Monte Carlo simulation be applied in practice? Under which probability measure is it convenient to simulate?
- Is there a model flexible enough to be calibrated to the market smile for caps?
- What typical qualitative shapes of the volatility term structure are observed in the market?
- What is the impact of the parameters of a chosen model on the market volatility structures that are relevant to the trader?
- What is the accuracy of analytical approximations derived for swaptions volatilities and terminal correlations?
- Does there exist an interest-rate model that can be considered "central" nowadays, in practice? What do traders think about it?
- Can you present some concrete examples of calibration to market data?
- How can we express mathematically the payoffs of some typical market products?
- How do you handle in practice products depending on more than one interest-rate curve at the same time?
- ......

Again, we could go on for a while, and it is hard to find a single book answering these questions with a rigorous theoretical background. Also, answering some of these questions (and others that are similar in spirit) motivates new theoretical developments, maintaining the fundamental feedback between theory and practice we hinted at above.

AIMS, READERSHIP AND BOOK STRUCTURE

Contrary to the equity-derivatives area, interest-rate modeling is a branch of mathematical finance where no general model has been yet accepted as "standard" for the whole sector, although the LIBOR market model is emerging as a possible candidate for this role. Indeed, there exist market standard models for both main interest-rate derivatives "sub-markets", namely the caps and swaptions markets. However, such models are theoretically incompatible and cannot be used jointly to price other interest-rate derivatives.

Because of this lack of a standard, the choice of a model for pricing and hedging interest-rate derivatives has to be carefully dealt with. In this book, therefore, we do not just concentrate on a specific model leaving all implementation issues aside. We instead develop several types of models and show how to use them in practice for pricing a number of specific products.

The main models are illustrated in different aspects ranging from theoretical formulation to a possible implementation on a computer, always keeping in mind the concrete questions one has to cope with. We also stress that different models are suited to different situations and products, pointing out that there does not exist a single model that is uniformly better than all the others.

Thus our aim in writing this book is two-fold. First, we would like to help quantitative analysts and advanced traders handle interest-rate derivatives with a sound theoretical apparatus. We try explicitly to explain which models can be used in practice for some major concrete problems. Secondly, we would also like to help academics develop a feeling for the practical problems in the market that can be solved with the use of relatively advanced tools of mathematics and stochastic calculus in particular. Advanced undergraduate students, graduate students and researchers should benefit as well, from seeing how some sophisticated mathematics can be used in concrete financial problems.

The Prerequisites

The prerequisites are some basic knowledge of stochastic calculus and the theory of stochastic differential equations in particular. The main tools from stochastic calculus are Ito's formula and Girsanov's theorem, which are, however, briefly reviewed in an appendix.

The Book is Structured in Two Parts

The first part is more academic and develops the theoretical basis that is needed for tackling the concrete pricing problems we want to solve.

We start by reviewing some basic concepts and definitions and briefly explain the fundamental theory of no-arbitrage and its implications as far as pricing derivatives is concerned.

We then review some of the basic short-rate models, forward-rate models and market models describing their distributional properties, discussing their analytical tractability and proposing numerical procedures for approximating the interest-rate dynamics. We will make extensive use of the "change-of-numeraire" technique, which is explained in details in a initial section.

The second part in contrast is devoted to concrete applications. We in fact list a series of market financial products that are usually traded over the counter and for which there exists no uniquely consolidated pricing model. We consider some typical interest-rate derivatives dividing them into two classes: i) derivatives depending on a single interest-rate curve; ii) derivatives depending on two interest-rate curves. We also propose a section where we hint at pricing equity derivatives under stochastic interest rates.

Appendices

It is sometimes said that no one ever reads appendices. This book ends with four appendices, and the fourth one is an interview with a quantitative trader, which should be interesting enough to convince the reader to have a look at the appendices, for a change.

The first appendix briefly reviews some basic results from stochastic calculus that are mentioned and applied in the book. The second appendix reports a useful calculation, whereas the third one deals with a general approximation of a diffusion process with a tree.

FINAL WORD AND ACKNOWLEDGMENTS

Whether our treatment of the theory fulfills the targets we have set ourselves, is up to the reader to judge. A disclaimer is necessary though. Assembling a book in the middle of the "battlefield" that is any trading room, while quite stimulating, leaves however little space for planned organization. Indeed, the book is not homogeneous, some topics are more developed than others.

We have tried to follow a logical path in assembling the final manuscript, but we are aware that the book is not optimal in respect of homogeneity and linearity of exposition. Hopefully, the explicit contribution of our work will emerge over these inevitable little misalignments.

Acknowledgments

A book is always the product not only of its authors, but also of their colleagues, of the environment where the authors work, of the encouragements and critics gathered from conferences, referee reports for journal publications, conversations after seminars, and many analogous events. While we cannot do justice to all the above, we thank explicitly our colleagues Gianvittorio "Tree and Optimization Master" Mauri and Francesco "Monte Carlo" Rapisarda, for their help and continuous interaction concerning both modeling and concrete implementations on computers. Francesco also helped by proofreading the manuscript and by suggesting modifications.

We express a thought of gratitude for our Head Aleardo Adotti in Banca IMI, for his farsightedness in allowing us to write this book and more generally to work on the frontiers of mathematical finance inside a bank.

The feedback from the interest-rate-derivatives desk has been fundamental, in the figures of Luca Mengoni and later on also of Antonio Castagna, who have stimulated many developments with their objections, requirements and discussions. Their feeling for market behavior has guided us in cases where mere mathematics and textbook finance could not help us that much.

Two trainees, Giulio Sartorelli and Cristina Capitani, have also contributed to the manuscript: both helped in developing numerical tests and Cristina also proofread part of the book manuscript.

The staff at Springer-Verlag has been active and supportive in the figures of Catriona Byrne and Susanne Denskus, whose help in our first "non-thesis" book has been very valuable.

All mistakes are, needless to say, ours.

Last but not least, we are grateful to our families and friends. Also, Damiano is grateful to the "historical" comics/anime/role-playing/motorcycle friends in Venice, and Fabio to his girlfriend, for their supportive enthusiasm.

Finally, our ultimate gratitude is towards transcendence and is always impossible to express with words. We just say that we are grateful for the Word of the Gospel and the Silence of Zen.

A Special Final Word for Young Readers and Beginners

We close this long preface with a particular thought and encouragement for young readers. Clearly, if you are a professional or academic experienced in interest-rate modeling, we believe you will not be scared by a first quick look at the table of contents and at the chapters.

However, even at a first glance when flipping through the book, some young readers might feel discouraged by the variety of models, by the difference in approaches, and might indeed acquire the impression of a chaotic sequence of models that arose in mathematical finance without a particular order or purpose. Yet, we assure you that this subject is interesting, relevant, and that it can (and should) be fun, however "cliched" this may sound to you. We have tried at times to be colloquial in the book, in an attempt to avoid writing a book on formal mathematical finance from A to Zzzzzzzzzz... (where have you heard this one before?).

We are trying to avoid the two apparent extremes of either scaring or boring our readers. Thus you will find at times opinions from market participants, guided tours, intuition and discussion on things as they are seen in the market. We would like you to give it at least a try. So, if you are one of the above young readers, and be you a student or a practitioner, we suggest you take it easy. This book might be able to help you a little in entering this exciting field of research. This is why we close this preface with the by-now classic recommendations...

"May fear and dread not conquer me"
Majjhima Nikaya VIII.6

"Do not let your hearts be troubled and do not be afraid"
St. John XIV.27

Indeed, *Don't Panic!*

Venice and Milan, April 1, 2001

Damiano Brigo and Fabio Mercurio

DESCRIPTION OF CONTENTS BY CHAPTER

We herewith provide a detailed description of the contents of each chapter.

Part I: MODELS: THEORY AND IMPLEMENTATION

Chapter 1: Definitions and Notation. The chapter is devoted to standard definitions and concepts in the interest-rate world, mainly from a static point of view. We define several interest-rate curves, such as the LIBOR, swap, forward-LIBOR and forward-swap curves, and the zero-coupon curve.

We explain the different possible choices of rates in the market. Some fundamental products, whose evaluation depends only on the initially given curves and not on volatilities, such as bonds and interest-rate swaps, are introduced. A quick and informal account of fundamental derivatives depending on volatility such as caps and swaptions is also presented, mainly for motivating the following developments.

Chapter 2: No-Arbitrage Pricing and Numeraire Change. The chapter introduces the theoretical issues a model should deal with, namely the no-arbitrage condition and the change of numeraire technique. The change of numeraire is reviewed as a general and powerful theoretical tool that can be used in several situations, and, indeed, it will be often used in the book.

We remark how the standard Black models for either the cap or swaption markets, the two main markets of interest-rate derivatives, can be given a rigorous interpretation via suitable numeraires, as we will do later on in Chapter 6.

We finally hint at products involving more than one interest-rate curve at the same time, typically quanto-like products, and illustrate the no-arbitrage condition in this case.

Chapter 3: One-Factor Short-Rate Models. In this chapter, we begin to consider the dynamics of interest rates. The chapter is devoted to the short-rate world. In this context, one models the instantaneous spot interest rate via a possibly multi-dimensional driving diffusion process depending on some parameters. The whole yield-curve evolution is then characterized by the driving diffusion.

If the diffusion is one dimensional, with this approach one is directly modeling the short rate, and the model is said to be "one-factor". In this chapter, we focus on such models, leaving the development of the multi-dimensional (two-dimensional in particular) case to the next chapter.

As far as the dynamics of one-factor models is concerned, we observe the following. Since the short rate represents at each instant the initial point of the yield curve, one-factor short-rate models assume the evolution of the whole yield curve to be completely determined by the evolution of its initial point. This is clearly a dangerous assumption, especially when pricing products depending on the correlation between different rates of the yield curve at a certain time (this limitation is explicitly pointed out in the guided tour of the subsequent chapter).

We then illustrate the no-arbitrage condition for one-factor models and the fundamental notion of market price of risk connecting the objective world, where rates are observed, and the risk-neutral world, where expectations leading to prices occur. We also show how choosing particular forms for the market price of risk can lead to models to which one can apply both econometric techniques (in the objective world) and calibration to market prices (risk-neutral world). We briefly hint at this kind of approach and subsequently leave the econometric part, focusing on the market calibration.

A short-rate model is usually calibrated to some initial structures in the market, typically the initial yield curve, the caps volatility surface, the swaptions volatility surface, and possibly other products, thus determining the model parameters. We introduce the historical one-factor time-homogeneous models of Vasicek, Cox Ingersoll Ross (CIR), Dothan, and the Exponential Vasicek (EV) model. We hint at the fact that such models used to be calibrated only to the initial yield curve, without taking into account market volatility structures, and that the calibration can be very poor in many situations.

We then move to extensions of the above one-factor models to models including "time-varying coefficients", or described by inhomogeneous diffusions. In such a case, calibration to the initial yield curve can be made perfect, and the remaining model parameters can be used to calibrate the volatility structures. We examine classic one-factor extensions of this kind such as Hull and White's extended Vasicek (HW) model, classic extensions of the CIR model, Black and Karasinski's (BK) extended EV model and a few more.

We discuss the volatility structures that are relevant in the market and explain how they are related to short-rate models. We discuss the issue of a humped volatility structure for short-rate models and give the relevant definitions. We also present the Mercurio-Moraleda short-rate model, which allows for a parametric humped-volatility structure while exactly calibrating the initial yield curve, and briefly hint at the Moraleda-Vorst model.

We then present a method of ours for extending pre-existing time-homogeneous models to models that perfectly calibrate the initial yield curve while keeping free parameters for calibrating volatility structures. Our method preserves the possible analytical tractability of the basic model. Our extension is shown to be equivalent to HW for the Vasicek model, whereas it is original in case of the CIR model. We call CIR++ the CIR model being extended through our procedure. We also show how to extend the Dothan and EV models, as possible alternatives to the use of the popular BK model.

We explain how to price coupon-bearing bond options and swaptions with models that satisfy a specific tractability assumption, and give general comments and a few specific instructions on Monte Carlo pricing with short-rate models.

We finally analyze how the market volatility structures implied by some of the presented models change when varying the models parameters. We conclude with an example of calibration of different models to market data.

Chapter 4: Two-Factor Short-Rate Models. If the short rate is obtained as a function of all the driving diffusion components (typically a summation, leading to an additive multi-factor model), the model is said to be "multi-factor".

We start by explaining the importance of the multi-factor setting as far as more realistic correlation and volatility structures in the evolution of the interest-rate curve are concerned.

We then move to analyze two specific two-factor models.

First, we apply our above deterministic-shift method for extending pre-existing time-homogeneous models to the two-factor additive Gaussian case (G2). In doing so, we calibrate perfectly the initial yield curve while keeping five free parameters for calibrating volatility structures. As usual, our method preserves the analytical tractability of the basic model. Our extension G2++ is shown to be equivalent to the classic two-factor Hull and White model. We develop several formulas for the G2++ model and also explain how both a binomial and a trinomial tree for the two-dimensional dynamics can be obtained. We discuss the implications of the chosen dynamics as far as volatility and correlation structures are concerned, and finally present an example of calibration to market data.

The second two-factor model we consider is a deterministic-shift extension of the classic two-factor CIR (CIR2) model, which is essentially the same as extending the Longstaff and Schwartz (LS) models. Indeed, we show that CIR2 and LS are essentially the same model, as is well known. We call CIR2++ the CIR2/LS model being extended through our deterministic-shift procedure, and provide a few analytical formulas. We do not consider this model with the same level of detail devoted to the G2++ model, due to the fact that its volatility structures are less flexible than the G2++'s, at least in case one wishes to preserve analytical tractability.

Chapter 5: The Heath-Jarrow-Morton Framework. In this chapter we consider the Heath-Jarrow-Morton (HJM) framework. We introduce the general framework and point out how it can be considered the right theoretical framework for developing interest-rate theory and especially no-arbitrage. However, we also point out that the most significant models coming out concretely from such a framework are the same models we met in the short-rate approach.

We report conditions on volatilities leading to a Markovian process for the short rate. This is important for implementation of lattices, since one then obtains (linearly-growing) recombining trees, instead of exponentially-growing ones. We show that in the one-factor case, a general condition leading to Markovianity of the short rate yields the Hull-White model with all time-varying coefficients, thus confirming that, in practice, short-rate models already contained some of the most interesting and tractable cases.

We then introduce the Ritchken and Sankarasubramanian framework, which allows for Markovianity of an enlarged process, of which the short rate is a component. The related tree (Li, Ritchken and Sankarasubramanian) is presented. Finally, we present a different version of the Mercurio-Moraleda model obtained through a specification of the HJM volatility structure, pointing out its advantages for realistic volatility behavior and its analytical formula for bond options.

Chapter 6: The LIBOR and Swap Market Models (LFM and LSM).
This chapter presents one of the most popular families of interest-rate models: the market models. A paramount fact is that the lognormal forward-LIBOR model (LFM) prices caps with Black's cap formula, which is the standard formula employed in the cap market. Moreover, the lognormal forward-swap model (LSM) prices swaptions with Black's swaption formula, which is the standard formula employed in the swaption market. Now, the cap and swaption markets are the two main markets in the interest-rate-derivatives world, so that compatibility with the related market formulas is a very desirable property. However, even with rigorous separate compatibility with the caps and swaptions classic formulas, the LFM and LSM are not compatible with each other. Still, the separate compatibility above is so important that these models, and especially the LFM, are nowadays seen as the most promising area in interest-rate modeling.

We start the chapter with a guided tour presenting intuitively the main issues concerning the LFM and the LSM, and giving motivation for the developments to come.

We then introduce the LFM, the "natural" model for caps, modeling forward-LIBOR rates. We give several possible instantaneous-volatility structures for this model, and derive its dynamics under different measures. We explain how the model can be calibrated to the cap market, examining the impact of the different structures of instantaneous volatility on the calibration. We introduce rigorously the term structure of volatility, and again check the impact of the different parameterizations of instantaneous volatilities on its evolution in time. We point out the difference between instantaneous and terminal correlation, the latter depending also on instantaneous volatilities.

We then introduce the LSM, the "natural" model for swaptions, modeling forward-swap rates. We show that the LSM is distributionally incompatible with the LFM. We discuss possible parametric forms for instantaneous correlations in the LFM, their impact on swaptions prices, and how, in general, Monte Carlo simulation should be used to price swaptions with the LFM instead of the LSM. We derive several approximated analytical formulas for swaption prices in the LFM (Brace's, Rebonato's and Hull-White's).

We point out that terminal correlation depends on the particular measure chosen for the joint dynamics in the LFM. We derive two analytical formulas

XX   Preface

based on "freezing the drift" for terminal correlation. These formulas clarify the relationship between instantaneous correlations and volatilities on one side and terminal correlations on the other side. We develop a similar formula for transforming volatility data of semi-annual or quarterly forward rates in volatility data of annual forward rates, and test it against Monte Carlo simulation of the true quantities. This is useful for joint calibration to caps and swaptions, allowing one to consider only annual data.

We present two methods for obtaining forward LIBOR rates in the LFM over non-standard periods, i.e. over expiry/maturity pairs that are not in the family of rates modeled in the chosen LFM.

We conclude the first chapter devoted to the market models with smile modeling. We introduce the smile problem with a guided tour. We provide a little history and a few references on simile modeling, and then present three models for the caplets smile.

The first model is a shifted lognormal dynamics. Having only two parameters for each fixed maturity, the curves along the strike dimension that can be reproduced by such a model are rather poor.

Then we introduce the constant-elasticity-of-variance extension of the LFM as from Andersen and Andreasen. This is an improvement, since this too can be shifted and now there are three parameters for fitting a caplet smile along the strike dimension for a given maturity. However, three parameters are still too few to give a flexible enough model.

We finally introduce the (possibly shifted) lognormal-mixture dynamics of Brigo and Mercurio for the LFM model. This model is always tractable, resulting in prices (and Greeks) that are linear combinations of Black's prices (Greeks), and can include any number of parameters. Therefore, its fitting capabilities surpass those of the previous models.

Chapter 7: Cases of Calibration of the LIBOR Market Model. In this chapter, we start from a set of market data including zero-coupon curve, caps volatilities and swaptions volatilities, and calibrate the LFM by resorting to several parameterizations of instantaneous volatilities and by several constraints on instantaneous correlations. Swaptions are evaluated through the analytical approximations derived in the previous chapter. We examine the evolution of the term structure of volatilities and the ten-year terminal correlation coming out from each calibration session, in order to assess advantages and drawbacks of every parameterization.

We finally present a particular parameterization establishing a one-to-one correspondence between LFM parameters and swaption volatilities, such that the calibration is immediate by solving a cascade of algebraic second-order equations. No optimization is necessary in general and the calibration is instantaneous. However, if the initial swaptions data are misaligned because of illiquidity or other reasons, the calibration can lead to negative or imaginary volatilities. We show that smoothing the initial data leads again to positive real volatilities.

Chapter 8: Monte Carlo Tests for LFM Analytical Approximations.
In this chapter we test Rebonato's and Hull-White's analytical formulas for swaptions prices in the LFM, presented earlier in Chapter 6, by means of a Monte Carlo simulation of the true LFM dynamics. This is done under different parametric assumptions for instantaneous volatilities and under different instantaneous correlations. We conclude that the above formulas are accurate in non-pathological situations.

We also plot the real swap-rate distribution obtained by simulation against the lognormal distribution with variance obtained by the analytical approximation. The two distributions are close in most cases, showing that the previously remarked theoretical incompatibility between LFM and LSM (where swap rates are lognormal) does not transfer to practice in most cases.

We also test our approximated formulas for terminal correlations, and see that these too are accurate in non-pathological situations.

Chapter 9 Other Interest-Rate Models. We present a few interest-rate models that are particular in their assumptions or in the quantities they model, and that have not been treated elsewhere in the book. We do not give a detailed presentation of these models but point out their particular features, compared to the models examined earlier in the book.

Part II: PRICING DERIVATIVES IN PRACTICE

Chapter 10: Pricing Derivatives on a Single Interest-Rate Curve.
This chapter deals with pricing specific derivatives on a single interest-rate curve. Most of these are products that are found in the market and for which no standard pricing technique is available. The model choice is made on a case by case basis, since different products motivate different models. The differences are based on realistic behaviour, ease of implementation, analytical tractability and so on. For each product we present at least one model based on a compromise between the above features, and in some cases we present more models and compare their strong and weak points. We try and understand which model parameters affect prices with a large or small influence. The financial products we consider are: in-advance swaps, in-advance caps, autocaps, caps with deferred caplets, ratchets (one-way floaters), constant-maturity swaps (introducing also the convexity-adjustment technique), captions and floortions, zero-coupon swaptions, Eurodollar futures, accrual swaps, trigger swaps and Bermudan-style swaptions.

Chapter 11: Pricing Derivatives on Two Interest-Rate Curves. The chapter deals with pricing specific derivatives involving two interest-rate curves. Again, most of these are products that are found in the market and for which no standard pricing technique is available. As before, the model

XXII   Preface

choice is made on a case by case basis, since different products motivate different models. The used models reduce to the LFM and the G2++ shifted two-factor Gaussian short-rate model. Under the G2++ model, we are able to model correlation between the interest rate curves of the two currencies. The financial products we consider include differential swaps, quanto caps, quanto swaptions, quanto constant-maturity swaps. A market quanto adjustment and market formulas for basic quanto derivatives are also introduced.

Chapter 12: Pricing Equity Derivatives under Stochastic Interest Rates. The chapter treats equity-derivatives valuation under stochastic interest rates, presenting us with the challenging task of modeling stock prices and interest rates at the same time. Precisely, we consider a continuous-time economy where asset prices evolve according to a geometric Brownian motion and interest rates are either normally or lognormally distributed. Explicit formulas for European options on a given asset are provided when the instantaneous spot rate follows the Hull-White one-factor process. It is also shown how to build approximating trees for the pricing of more complex derivatives, under a more general short-rate process.

Part III: APPENDICES

Appendix A: a Crash Introduction to Stochastic Differential Equations. This appendix is devoted to a quick intuitive introduction on SDE's. We start from deterministic differential equation and gradually introduce randomness. We introduce intuitively Brownian motion and explain how it can be used to model the "random noise" in the differential equation. We observe that Brownian motion is not differentiable, and explain that SDE's must be understood in integral form. We quickly introduce the related Ito and Stratonovich integrals, and introduce the fundamental Ito formula.

We then introduce the Euler and Milstein schemes for the time-discretization of an SDE. These schemes are essential when in need of Monte Carlo simulating the trajectories of an Ito process whose transition density is not explicitly known.

We include two important theorems: the Feynman-Kac theorem and the Girsanov theorem. The former connects PDE's to SDE's, while the latter permits to change the drift coefficient in an SDE by changing the basic probability measure. The Girsanov theorem in particular is used in the book to derive the change of numeraire toolkit.

Appendix B: a Useful Calculation. This appendix reports the calculation of a particular integral against a standard normal density, which is useful when dealing with Gaussian models.

Appendix C: Approximating Diffusions with Trees. This appendix explains a general method to obtain a trinomial tree approximating the dynamics of a general diffusion process. This is then generalized to a two-dimensional diffusion process, which is approximated via a two-dimensional trinomial tree.

Appendix D: Talking to the Traders. This is the ideal conclusion of the book, consisting of an interview with a quantitative trader. Several issues are discussed, also to put the book in a larger perspective.

Abbreviations and Notation

- ATM = At the money;
- BK = Black-Karasinski model;
- CIR = Cox-Ingersoll-Ross model;
- CIR2++ = Shifted two-factor Cox-Ingersoll-Ross model;
- EEV = Shifted (extended) exponential-Vasicek model;
- EV = Exponential-Vasicek model;
- G2++ = Shifted two-factor Gaussian (Vasicek) model;
- HJM = Heath-Jarrow-Morton model;
- HW = Hull-White model;
- IRS = Interest Rate Swap (either payer or receiver);
- ITM = In the money;
- LFM = Lognormal forward-Libor model (Libor market model, BGM model);
- LS = Longstaff-Schwartz short-rate model;
- LSM = Lognormal forward-swap model (swap market model);
- MC = Monte Carlo;
- OTM = Out of the money;
- PDE = Partial differential equation;
- SDE = Stochastic differential equation;
- I_n: the $n \times n$ identity matrix;
- $B(t), B_t$: Money market account at time t, bank account at time t ;
- $D(t,T)$: Stochastic discount factor at time t for the maturity T;
- $P(t,T)$: Bond price at time t for the maturity T;
- $P^f(t,T)$: Foreign Bond price at time t for the maturity T;
- $r(t), r_t$: Instantaneous spot interest rate at time t;
- $B^d(t)$: Discretely rebalanced bank-account at time t;
- $R(t,T)$: Continuously compounded spot rate at time t for the maturity T;
- $L(t,T)$: Simply compounded (Libor) spot rate at time t for the maturity T;
- $f(t,T)$: Instantaneous forward rate at time t for the maturity T;
- $F(t;T,S)$: Simply compounded forward (Libor) rate at time t for the expiry–maturity pair T, S;
- $\mathrm{FP}(t;T,S)$: Forward zero-coupon-bond price at time t for maturity S as seen from expiry T, $\mathrm{FP}(t;T,S) = P(t,S)/P(t,T)$.
- $F^f(t;T,S)$: Foreign simply compounded forward (Libor) rate at time t for the expiry–maturity pair T, S;
- $f(t;T,S)$: Continuously compounded forward rate at time t for the expiry–maturity pair T, S;
- $T_1, T_2, \ldots, T_{i-1}, T_i, \ldots$: An increasing set of maturities;
- τ_i: The year fraction between T_{i-1} and T_i;
- $F_i(t)$: $F(t;T_{i-1},T_i)$;
- $S(t;T_i,T_j), S_{i,j}(t)$: Forward swap rate at time t for a swap with first reset date T_i and payment dates $T_{i+1}, \ldots, T_j$;

XXVI   Notation

- $C_{i,j}(t)$: Present value of a basis point associated to the forward–swap rate $S_{i,j}(t)$, i.e. $\sum_{k=i+1}^{j} \tau_k P(t, T_k)$;
- Q_0: Physical/Objective/Real–World measure;
- Q: Risk-neutral measure, equivalent martingale measure, risk-adjusted measure;
- Q^U: Measure associated with the numeraire U when U is an asset;
- Q^d: Spot LIBOR measure, measure associated with the discretely rebalanced bank-account numeraire;
- Q^T: T-forward adjusted measure, i.e. measure associated with the numeraire $P(\cdot, T)$;
- Q^i: T_i-forward adjusted measure;
- $Q^{i,j}$: Swap measure between T_i, T_j, associated with the numeraire $C_{i,j}$;
- $X_t, X(t)$: Foreign exchange rate between two currencies at time T;
- W_t, Z_t: Brownian motions under the Risk Neutral measure;
- W_t^U: A Brownian motion under the measure associated with the numeraire U when U is an asset;
- W_t^T: Brownian motions under the T forward adjusted measure;
- $W_t^i, Z_t^i, W^i(t), Z^i(t),$: Brownian motions under the T_i forward adjusted measure;
- $[x_1, \ldots, x_m]$: row vector with i-th component x_i;
- $[x_1, \ldots, x_m]'$: column vector with i-th component x_i;
- $'$: Transposition;
- $1_A, 1\{A\}$: Indicator function of the set A;
- $\#\{A\}$: Number of elements of the finite set A;
- E: Expectation under the risk-neutral measure;
- E^Q: Expectation under the probability measure Q;
- E^U: Expectation under the probability measure Q^U associated with the numeraire U; This may be denoted also by E^{Q^U};
- E^T: Expectation under the T-forward adjusted measure;
- E^i: Expectation under the T_i-forward adjusted measure;
- $E_t, E\{\cdot|\mathcal{F}_t\}, E[\cdot|\mathcal{F}_t], E(\cdot|\mathcal{F}_t)$: Expectation conditional on the $\mathcal{F}_t$ σ–field;
- $\text{Corr}^i(X, Y)$: correlation between X and Y under the T_i forward adjusted measure Q^i; i can be omitted if clear from the context or under the risk-neutral measure;
- $\text{Var}^i(X)$: Variance of X under the T_i forward adjusted measure Q^i; i can be omitted if clear from the context or under the risk-neutral measure;
- $\text{Cov}^i(X)$: covariance matrix of the random vector X under the T_i forward adjusted measure Q^i; i can be omitted if clear from the context or under the risk-neutral measure;
- $\text{Std}^i(X)$: covariance matrix of the random vector X under the T_i forward adjusted measure Q^i; i can be omitted if clear from the context or under the risk-neutral measure;
- $\sim$: distributed as;
- $\mathcal{N}(\mu, V)$: Multivariate normal distribution with mean vector μ and covariance matrix V;
- Φ: Cumulative distribution function of the standard Gaussian distribution;
- $\chi^2(\cdot; r, \rho)$: Cumulative distribution function of the noncentral chi-squared distribution with r degrees of freedom and noncentrality parameter ρ;
- $\text{Bl}(K, F, v)$: The core of Black's formula:

$$\text{Bl}(K, F, v, \omega) = F\omega\Phi(\omega d_1(K, F, v)) - K\omega\Phi(\omega d_2(K, F, v)),$$
$$d_1(K, F, v) = \frac{\ln(F/K) + v^2/2}{v},$$
$$d_2(K, F, v) = \frac{\ln(F/K) - v^2/2}{v},$$

Notation   XXVII

where ω is either -1 or 1 and is meant to be 1 when omitted. The arguments of d_1 and d_2 may be omitted if clear from the context.
- **CB**$(t, \mathcal{T}, \tau, N, c)$: Coupon bond price at time t for a bond paying coupons $c = [c_1, \ldots, c_n]$ at times $\mathcal{T} = [T_1, \ldots, T_n]$ with year fractions $\tau = [\tau_1, \ldots, \tau_n]$ and nominal amount N; When assuming a unit nominal amount N can be omitted; When year fractions are clear from the context, τ can be omitted;
- **ZBC**(t, T, S, τ_0, N, K): Price at time t of an European call option with maturity $T > t$ and strike–price K on a Zero–coupon bond with nominal amount N, maturing at time $S > T$; τ_0 is the year fraction between T and S and can be omitted; When assuming a unit nominal amount N can be omitted;
- **ZBP**(t, T, S, τ_0, N, K): Same as above but for a put option;
- **ZBO**$(t, T, S, \tau_0, N, K, \omega)$: Unified notation for the price at time t of an European option with maturity $T > t$ and strike–price K on a Zero–coupon bond with face value N maturing at time $S > T$; τ_0 is the year fraction between T and S and can be omitted; ω is $+1$ for a call option and -1 for a put option, and can be omitted;
- **CBC**$(t, T, \mathcal{T}, \tau_0, \tau, N, c, K)$: Price at time t of an European call option with maturity $T > t$ and strike–price K on the coupon bond **CB**$(t, \mathcal{T}, \tau, N, c)$; τ_0 is the year fraction between T and T_1 and can be omitted; When the nominal N is one it is omitted;
- **CBP**$(t, T, \mathcal{T}, \tau_0, \tau, N, c, K)$: Same as above but for a put option;
- **CBO**$(t, T, \mathcal{T}, \tau_0, \tau, N, c, K, \omega)$: Unified notation for the price at time t of an European option with maturity $T > t$ and strike–price K on the coupon bond **CB**$(t, \mathcal{T}, \tau, c)$; τ_0 is the year fraction between T and T_1 and can be omitted. ω is $+1$ for a call option and -1 for a put option, and can be omitted;
- **Cpl**(t, T, S, τ_0, N, y): Price at time t of a caplet resetting at time T and paying at time S at a fixed strike–rate y; As usual τ_0 is the year fraction between T and S and can be omitted, and N is the nominal amount and can be omitted;
- **Fll**(t, T, S, τ_0, N, y): Price at time t of a floorlet resetting at time T and paying at time S at a fixed rate y; As usual τ_0 is the year fraction between T and S and can be omitted, and N is the nominal amount and can be omitted;
- **Cap**$(t, \mathcal{T}, \tau, N, y)$: Price at time t of a cap first resetting at time T_1 and paying at times $T_2, \ldots, T_n$ at a fixed rate y; As usual τ_i is the year fraction between T_{i-1} and T_i and can be omitted, and N is the nominal amount and can be omitted;
- **Flr**$(t, \mathcal{T}, \tau, N, y)$: Price at time t of a floor first resetting at time T_1 and paying at times $T_2, \ldots, T_n$ at a fixed rate y; As usual τ_i is the year fraction between T_{i-1} and T_i and can be omitted, and N is the nominal amount and can be omitted;
- **FRA**(t, T, S, τ, N, R): Price at time t of a forward–rate agreement with reset date T and payment date S at the fixed rate R; As usual τ is the year fraction between T and S and can be omitted, and N is the nominal amount and can be omitted;
- **PFS**$(t, \mathcal{T}, \tau, N, R)$: Price at time t of a payer forward–start interest rate swap with first reset date T_1 and payment dates $T_2, \ldots, T_n$ at the fixed rate R; As usual τ_i is the year fraction between T_{i-1} and T_i and can be omitted, and N is the nominal amount and can be omitted;
- **RFS**$(t, \mathcal{T}, \tau, N, R)$: Same as above but for a receiver swap;
- **PS**$(t, T, \mathcal{T}, \tau, N, R)$: Price of a payer swaption maturing at time T, which gives its holder the right to enter at time T an interest rate swap with first reset date T_1 and payment dates $T_2, \ldots, T_n$ (with $T_1 \geq T$) at the fixed strike–rate R; As usual τ_i is the year fraction between T_{i-1} and T_i and can be omitted, and N is the nominal amount and can be omitted;
- **RS**$(t, T, \mathcal{T}, \tau, N, R)$: Same as above but for a receiver swaption;

XXVIII Notation

- **ES**$(t, T, \mathcal{T}, \tau, N, R, \omega)$: Same as above but for a general European swaption; ω is $+1$ for a payer and -1 for a receiver, and can be omitted.
- **QCpl**$(t, T, S, \tau_0, N, y, \text{curr1}, \text{curr2})$: Price at time t of a quanto caplet resetting at time T and paying at time S at a fixed strike–rate y; Rates are related to the foreign "curr2" currency, whereas the payoff is an amount in domestic "curr1" currency. As usual τ_0 is the year fraction between T and S and can be omitted, and N is the nominal amount and can be omitted; the currency names "curr1" and "curr2" can be omitted.
- **QFll**$(t, T, S, \tau_0, N, y, \text{curr1}, \text{curr2})$: As above but for a floorlet.
- Instantaneous (absolute) volatility of a process Y is $\eta(t)$ in

$$dY_t = (\ldots)dt + \eta(t)dW_t .$$

- Instantaneous level-proportional (or proportional or percentage or relative or return) volatility of a process Y is $\sigma(t)$ in

$$dY_t = (\ldots)dt + \sigma(t)Y_t dW_t .$$

- Level-proportional (or proportional or percentage or relative) drift (or drift rate) of a process Y is $\mu(t)$ in

$$dY_t = \mu(t)Y_t dt + (\ldots)dW_t .$$

Table of Contents

Preface VII
  Motivation VII
  Aims, Readership and Book Structure XII
  Final Word and Acknowledgments XIII
  Description of Contents by Chapter XVI

Abbreviations and Notation XXV

Part I. MODELS: THEORY AND IMPLEMENTATION

1. **Definitions and Notation** 1
   1.1 The Bank Account and the Short Rate 1
   1.2 Zero-Coupon Bonds and Spot Interest Rates 3
   1.3 Fundamental Interest-Rate Curves 8
   1.4 Forward Rates 10
   1.5 Interest-Rate Swaps and Forward Swap Rates 13
   1.6 Interest-Rate Caps/Floors and Swaptions 15

2. **No-Arbitrage Pricing and Numeraire Change** 23
   2.1 No-Arbitrage in Continuous Time 24
   2.2 The Change-of-Numeraire Technique 26
   2.3 A Change-of-Numeraire Toolkit 28
   2.4 The Choice of a Convenient Numeraire 32
   2.5 The Forward Measure 33
   2.6 The Fundamental Pricing Formulas 35
       2.6.1 The Pricing of Caps and Floors 36
   2.7 Pricing Claims with Deferred Payoffs 37
   2.8 Pricing Claims with Multiple Payoffs 38
   2.9 Foreign Markets and Numeraire Change 40

3. **One-factor short-rate models** 43
   3.1 Introduction and Guided Tour 43
   3.2 Classical Time-Homogeneous Short-Rate Models 48
       3.2.1 The Vasicek Model 50

- 3.2.2 The Dothan Model 54
- 3.2.3 The Cox, Ingersoll and Ross (CIR) Model 56
- 3.2.4 Affine Term-Structure Models...................... 60
- 3.2.5 The Exponential-Vasicek (EV) Model 61
- 3.3 The Hull-White Extended Vasicek Model 63
  - 3.3.1 The Short-Rate Dynamics 64
  - 3.3.2 Bond and Option Pricing........................... 66
  - 3.3.3 The Construction of a Trinomial Tree 69
- 3.4 Possible Extensions of the CIR Model 72
- 3.5 The Black-Karasinski Model 73
  - 3.5.1 The Short-Rate Dynamics 74
  - 3.5.2 The Construction of a Trinomial Tree 76
- 3.6 Volatility Structures in One-Factor Short-Rate Models 77
- 3.7 Humped-Volatility Short-Rate Models..................... 83
- 3.8 A General Deterministic-Shift Extension 86
  - 3.8.1 The Basic Assumptions 87
  - 3.8.2 Fitting the Initial Term Structure of Interest Rates ... 88
  - 3.8.3 Explicit Formulas for European Options............. 90
  - 3.8.4 The Vasicek Case 91
- 3.9 The CIR++ Model 93
  - 3.9.1 The Construction of a Trinomial Tree 96
  - 3.9.2 The Positivity of Rates and Fitting Quality 97
- 3.10 Deterministic-Shift Extension of Lognormal Models 100
- 3.11 Some Further Remarks on Derivatives Pricing.............. 102
  - 3.11.1 Pricing European Options on a Coupon-Bearing Bond 102
  - 3.11.2 The Monte Carlo Simulation 103
  - 3.11.3 Pricing Early-Exercise Derivatives with a Tree 106
  - 3.11.4 A Fundamental Case of Early Exercise: Bermudan-Style Swaptions. 111
- 3.12 Implied Cap Volatility Curves 114
  - 3.12.1 The Black and Karasinski Model 115
  - 3.12.2 The CIR++ Model 116
  - 3.12.3 The Extended Exponential-Vasicek Model 117
- 3.13 Implied Swaption Volatility Surfaces 119
  - 3.13.1 The Black and Karasinski Model 120
  - 3.13.2 The Extended Exponential-Vasicek Model 120
- 3.14 An Example of Calibration to Real-Market Data 121

4. **Two-Factor Short-Rate Models** 127
   - 4.1 Introduction and Motivation 127
   - 4.2 The Two-Additive-Factor Gaussian Model G2++........... 132
     - 4.2.1 The Short-Rate Dynamics 133
     - 4.2.2 The Pricing of a Zero-Coupon Bond 134
     - 4.2.3 Volatility and Correlation Structures in Two-Factor Models... 137

	4.2.4 The Pricing of a European Option on a Zero-Coupon Bond 143

        4.2.4 The Pricing of a European Option on a Zero-Coupon
               Bond 143
        4.2.5 The Analogy with the Hull-White Two-Factor Model . 149
        4.2.6 The Construction of an Approximating Binomial Tree. 152
        4.2.7 Examples of Calibration to Real-Market Data 156
  4.3 The Two-Additive-Factor Extended CIR/LS Model CIR2++ 165
        4.3.1 The Basic Two-Factor CIR2 Model 166
        4.3.2 Relationship with the Longstaff and Schwartz Model
               (LS) 167
        4.3.3 Forward-Measure Dynamics and Option Pricing for
               CIR2 168
        4.3.4 The CIR2++ Model and Option Pricing 168

5. **The Heath-Jarrow-Morton (HJM) Framework** 173
   5.1 The HJM Forward-Rate Dynamics......................... 175
   5.2 Markovianity of the Short-Rate Process 176
   5.3 The Ritchken and Sankarasubramanian Framework 177
   5.4 The Mercurio and Moraleda Model 181

6. **The LIBOR and Swap Market Models (LFM and LSM)** .. 183
   6.1 Introduction 183
   6.2 Market Models: a Guided Tour 184
   6.3 The Lognormal Forward-LIBOR Model (LFM) 192
        6.3.1 Some Specifications of the Instantaneous Volatility of
               Forward Rates 195
        6.3.2 Forward-Rate Dynamics under Different Numeraires .. 198
   6.4 Calibration of the LFM to Caps and Floors Prices 203
        6.4.1 Piecewise-Constant Instantaneous-Volatility Structures 206
        6.4.2 Parametric Volatility Structures 207
        6.4.3 Cap Quotes in the Market 208
   6.5 The Term Structure of Volatility 210
        6.5.1 Piecewise-Constant Instantaneous Volatility Structures 210
        6.5.2 Parametric Volatility Structures 215
   6.6 Instantaneous Correlation and Terminal Correlation 217
   6.7 Swaptions and the Lognormal Forward-Swap Model (LSM) .. 220
        6.7.1 Swaptions Hedging 224
        6.7.2 Cash-Settled Swaptions 226
   6.8 Incompatibility between the LFM and the LSM 227
   6.9 The Structure of Instantaneous Correlations 230
   6.10 Monte Carlo Pricing of Swaptions with the LFM 233
   6.11 Rank-One Analytical Swaption Prices 236
   6.12 Rank-r Analytical Swaption Prices 242
   6.13 A Simpler LFM Formula for Swaptions Volatilities.......... 246
   6.14 A Formula for Terminal Correlations of Forward Rates 249
   6.15 Calibration to Swaptions Prices 252

6.16 Connecting Caplet and $S \times 1$-Swaption Volatilities 254
6.17 Forward and Spot Rates over Non-Standard Periods 261
  6.17.1 Drift Interpolation 262
  6.17.2 The Bridging Technique 264
6.18 Including the Caplet Smile in the LFM 266
  6.18.1 A Mini-tour on the Smile Problem 266
  6.18.2 Modeling the Smile 270
  6.18.3 The Shifted-Lognormal Case 271
  6.18.4 The Constant Elasticity of Variance (CEV) Model 273
  6.18.5 A Mixture-of-Lognormals Model 276
  6.18.6 Shifting the Lognormal-Mixture Dynamics 280

7. **Cases of Calibration of the LIBOR Market Model** 283
  7.1 The Inputs 284
  7.2 Joint Calibration with Piecewise-Constant Volatilities as in
      TABLE 5 284
    7.2.1 Instantaneous Correlations: Narrowing the Angles 288
    7.2.2 Instantaneous Correlations: Fixing the Angles to Typical Values 290
    7.2.3 Instantaneous Correlations: Fixing the Angles to Atypical Values 292
    7.2.4 Instantaneous Correlations: Collapsing to One Factor . 293
  7.3 Joint Calibration with Parameterized Volatilities as in Formulation 7 295
    7.3.1 Formulation 7: Narrowing the Angles 297
    7.3.2 Formulation 7: Calibrating only to Swaptions 300
  7.4 Exact Swaptions Calibration with Volatilities as TABLE 1 ... 303
    7.4.1 Some Numerical Results 309
  7.5 Conclusions: Where Now? 314

8. **Monte Carlo Tests for LFM Analytical Approximations** ... 317
  8.1 The Specification of Rates 317
  8.2 The "Testing Plan" for Volatilities 318
  8.3 Test Results for Volatilities 321
    8.3.1 Case (1): Constant Instantaneous Volatilities 322
    8.3.2 Case (2): Volatilities as Functions of Time to Maturity 329
    8.3.3 Case (3): Humped and Maturity-Adjusted Instantaneous Volatilities Depending only on Time to Maturity, Typical Rank-Two Correlations 334
  8.4 The "Testing Plan" for Terminal Correlations 345
  8.5 Test Results for Terminal Correlations 353
    8.5.1 Case (i): Humped and Maturity-Adjusted Instantaneous Volatilities Depending only on Time to Maturity, Typical Rank-Two Correlations 353

        8.5.2   Case (ii): Constant Instantaneous Volatilities, Typical Rank-Two Correlations. 355
        8.5.3   Case (iii): Humped and Maturity-Adjusted Instantaneous Volatilities Depending only on Time to Maturity, Some Negative Rank-Two Correlations. 359
        8.5.4   Case (iv): Constant Instantaneous Volatilities, Some Negative Rank-Two Correlations.................... 363
        8.5.5   Case (v): Constant Instantaneous Volatilities, Perfect Correlations, Upwardly Shifted Φ's 365
  8.6   Test Results: Stylized Conclusions 367

9. Other Interest-Rate Models 369
  9.1   Brennan and Schwartz's Model 369
  9.2   Balduzzi, Das, Foresi and Sundaram's Model 370
  9.3   Flesaker and Hughston's Model 371
  9.4   Rogers's Potential Approach 373
  9.5   Markov Functional Models.............................. 373

Part II. PRICING DERIVATIVES IN PRACTICE

10. Pricing Derivatives on a Single Interest-Rate Curve 377
  10.1   In-Advance Swaps 378
  10.2   In-Advance Caps 379
       10.2.1 A First Analytical Formula (LFM) 380
       10.2.2 A Second Analytical Formula (G2++) 380
  10.3   Autocaps 381
  10.4   Caps with Deferred Caplets 382
       10.4.1 A First Analytical Formula (LFM) 382
       10.4.2 A Second Analytical Formula (G2++) 383
  10.5   Ratchets (One-Way Floaters) 384
  10.6   Constant-Maturity Swaps (CMS) 385
       10.6.1 CMS with the LFM 385
       10.6.2 CMS with the G2++ Model 386
  10.7   The Convexity Adjustment and Applications to CMS 386
       10.7.1 Natural and Unnatural Time Lags 386
       10.7.2 The Convexity-Adjustment Technique............... 387
       10.7.3 Deducing a Simple Lognormal Dynamics from the Adjustment 391
       10.7.4 Application to CMS 392
       10.7.5 Forward Rate Resetting Unnaturally and Average-Rate Swaps....................................... 393
  10.8   Captions and Floortions 395
  10.9   Zero-Coupon Swaptions 395
  10.10 Eurodollar Futures 399

        10.10.1 The Shifted Two-Factor Vasicek G2++ Model....... 400
        10.10.2 Eurodollar Futures with the LFM................. 402
    10.11 LFM Pricing with "In-Between" Spot Rates............... 402
        10.11.1 Accrual Swaps 403
        10.11.2 Trigger Swaps 406
    10.12 LFM Pricing with Early Exercise and Possible Path Dependence 408
    10.13 LFM: Pricing Bermudan Swaptions 412
        10.13.1 Longstaff and Schwartz's Approach 413
        10.13.2 Carr and Yang's Approach....................... 415
        10.13.3 Andersen's Approach............................ 416

11. **Pricing Derivatives on Two Interest-Rate Curves** 421
    11.1 The Attractive Features of G2++ for Multi-Curve Payoffs ... 421
        11.1.1 The Model 421
        11.1.2 Interaction Between Models of the Two Curves "1" and "2" 424
        11.1.3 The Two-Models Dynamics under a Unique Convenient Forward Measure............................ 425
    11.2 Quanto Constant-Maturity Swaps 427
        11.2.1 Quanto CMS: The Contract 427
        11.2.2 Quanto CMS: The G2++ Model 429
        11.2.3 Quanto CMS: Quanto Adjustment................. 435
    11.3 Differential Swaps 437
        11.3.1 The Contract 437
        11.3.2 Differential Swaps with the G2++ Model........... 438
        11.3.3 A Market-Like Formula 440
    11.4 Market Formulas for Basic Quanto Derivatives 440
        11.4.1 The Pricing of Quanto Caplets/Floorlets 440
        11.4.2 The Pricing of Quanto Caps/Floors................. 443
        11.4.3 The Pricing of Differential Swaps 444
        11.4.4 The Pricing of Quanto Swaptions.................. 444

12. **Pricing Equity Derivatives under Stochastic Rates** 453
    12.1 The Short Rate and Asset-Price Dynamics................. 453
        12.1.1 The Dynamics under the Forward Measure 456
    12.2 The Pricing of a European Option on the Given Asset 458
    12.3 A More General Model................................. 459
        12.3.1 The Construction of an Approximating Tree for r 460
        12.3.2 The Approximating Tree for S 462
        12.3.3 The Two-Dimensional Tree 463

Part III. APPENDICES

A. A Crash Introduction to Stochastic Differential Equations 469
  A.1 From Deterministic to Stochastic Differential Equations 469
  A.2 Ito's Formula 476
  A.3 Discretizing SDEs for Monte Carlo: Euler and Milstein Schemes 478
  A.4 Examples 480
  A.5 Two Important Theorems 482

B. A Useful Calculation 485

C. Approximating Diffusions with Trees 487

D. Talking to the Traders 493

References 501

Index 509

Part I

MODELS: THEORY AND IMPLEMENTATION

1. Definitions and Notation

In this first chapter we present the main definitions that will be used throughout the book. We will introduce the basic concepts in a rigorous way while providing at the same time intuition and motivation for their introduction. However, before starting with the definitions, a remark is in order.

Remark 1.0.1. **(Interbank vs government rates, LIBOR rates).** There are different types of interest rates, and a first distinction can be made between interbank rates and government rates. Government rates are usually deduced by bonds issued by governments. By "interbank rates" we denote instead rates at which deposits are exchanged between banks, and at which swap transactions (see below) between banks occur.

Zero-coupon rates (see below) can be "stripped" either from bonds in the government sector of the market or from products in the interbank sector of the market, resulting in two different zero-coupon curves. Once this has been done, mathematical modeling of the resulting rates is analogous in the two cases. We will focus on interbank rates here, although the mathematical apparatus described in this book can be usually applied to products involving government rates as well.

The most important interbank rate usually considered as a reference for contracts is the LIBOR (London InterBank Offered Rate) rate, fixing daily in London. However, there exist analogous interbank rates fixing in other markets (e.g. the EURIBOR rate, fixing in Brussels), and when we refer to "LIBOR" we actually intend any of these interbank rates.

1.1 The Bank Account and the Short Rate

> *Why then didn't you put my money on deposit,*
> *so that when I came back, I could have collected it with interest?*
> St. Luke XIX. 23

The concept of interest rate belongs to our every-day life and has entered our minds as something familiar we know how to deal with. When depositing a certain amount of money in a bank account, everybody expects that the amount grows (at some rate) as time goes by. The fact that lending money must be rewarded somehow, so that receiving a given amount of money tomorrow is not equivalent to receiving exactly the same amount today, is indeed common knowledge and wisdom. However, expressing such concepts in mathematical terms may be less immediate and many definitions have to be introduced to develop a consistent theoretical apparatus.

The first definition we consider is the definition of a bank account, or money-market account. A money-market account represents a (locally) riskless investments, where profit is accrued continuously at the risk-free rate prevailing in the market at every instant.

Definition 1.1.1. Bank account (Money-market account). *We define $B(t)$ to be the value of a bank account at time $t \geq 0$. We assume $B(0) = 1$ and that the bank account evolves according to the following differential equation:*

$$dB(t) = r_t B(t) dt, \quad B(0) = 1, \tag{1.1}$$

where r_t is a positive function of time. As a consequence,

$$B(t) = \exp\left(\int_0^t r_s ds\right). \tag{1.2}$$

The above definition tells us that investing a unit amount at time 0 yields at time t the value in (1.2), and r_t is the *instantaneous rate* at which the bank account accrues. This instantaneous rate is usually referred to as *instantaneous spot rate*, or briefly as *short rate*. In fact, a first order expansion in Δt gives

$$B(t + \Delta t) = B(t)(1 + r(t)\Delta t), \tag{1.3}$$

which amounts to say that, in any arbitrarily small time interval $[t, t + \Delta t)$,

$$\frac{B(t + \Delta t) - B(t)}{B(t)} = r(t)\Delta t.$$

It is then clear that the bank account grows at each time instant t at a rate $r(t)$.

The bank-account numeraire[1] B is important for relating amounts of currencies available at different times. To this end, consider the following fundamental question: What is the value at time t of one unit of currency available at time T?

Assume for simplicity that the interest rate process r, and hence B, are deterministic. We know that if we deposit A units of currency in the bank account at time 0, at time $t > 0$ we have $A \times B(t)$ units of currency. Similarly, at time $T > t$ we have $A \times B(T)$ units. If we wish to have exactly one unit of currency at time T, i.e., if we wish that

$$A B(T) = 1,$$

---

[1] We refer to the next chapter for a formal definition of a numeraire.

we have to initially invest the amount $A = 1/B(T)$, which is known since the process B is deterministic. Hence, the value at time t of the amount A invested at the initial time is

$$A\,B(t) = \frac{B(t)}{B(T)}.$$

We have thus seen that the value of one unit of currency payable at time T, as seen from time t, is $B(t)/B(T)$. Coming back to the initial assumption of a general (stochastic) interest-rate process, this leads to the following.

Definition 1.1.2. Stochastic discount factor. *The (stochastic) discount factor $D(t,T)$ between two time instants t and T is the amount at time t that is "equivalent" to one unit of currency payable at time T, and is given by*

$$D(t,T) = \frac{B(t)}{B(T)} = \exp\left(-\int_t^T r_s ds\right). \tag{1.4}$$

The probabilistic nature of r_t is important since it affects the nature of the basic asset of our discussion, the bank-account numeraire B. In many pricing applications, especially when applying the Black and Scholes formula in equity or foreign-exchange (FX) markets, r is assumed to be a deterministic function of time, so that both the bank account (1.2) and the discount factors (1.4) at any future time are deterministic functions of time. This is usually motivated by assuming that variability of interest rates contributes to the price of equity or FX options by a smaller order of magnitude with respect to the underlying's movements.[2]

However, when dealing with interest-rate products, the main variability that matters is clearly that of the interest rates themselves. It is therefore necessary to drop the deterministic setup and to start modeling the evolution of r in time through a stochastic process. As a consequence, the bank account (1.2) and the discount factors (1.4) will be stochastic processes, too. Some particular forms (i.e., stochastic differential equations) of possible evolutions for r will be discussed later on in the book.

We now turn to other basic definitions concerning the interest-rate world.

1.2 Zero-Coupon Bonds and Spot Interest Rates

Definition 1.2.1. Zero-coupon bond. *A T-maturity zero-coupon bond (pure discount bond) is a contract that guarantees its holder the payment of one unit of currency at time T, with no intermediate payments. The contract value at time $t < T$ is denoted by $P(t,T)$. Clearly, $P(T,T) = 1$ for all T.*

---

[2] We will consider the possibility of removing this assumption in Chapter 12.

If we are now at time t, a zero-coupon bond for the maturity T is a contract that establishes the present value of one unit of currency to be paid at time T (the maturity of the contract).

A natural question arising now is: What is the relationship between the discount factor $D(t,T)$ and the zero-coupon-bond price $P(t,T)$? The difference lies in the two objects being respectively an "equivalent amount of currency" and a "value of a contract".

If rates r are deterministic, then D is deterministic as well and necessarily $D(t,T) = P(t,T)$ for each pair (t,T). However, if rates are stochastic, $D(t,T)$ is a random quantity at time t depending on the future evolution of rates r between t and T. Instead, the zero-coupon-bond price $P(t,T)$, being the time t-value of a contract with payoff at time T, has to be known (deterministic) at time t. We will see later on in the book that the bond price $P(t,T)$ and the discount factor $D(t,T)$ are closely linked, in that $P(t,T)$ can be actually viewed as the *expectation* of the random variable $D(t,T)$ under a particular probability measure.

In the following, by slight abuse of notation, t and T will denote both times, as measured by a real number from an instant chosen as time origin 0, and dates expressed as days/months/years.

Definition 1.2.2. Time to maturity. *The* time to maturity $T - t$ *is the amount of time (in years) from the present time t to the maturity time $T > t$.*

The definition "$T-t$" makes sense as long as t and T are real numbers associated to two time instants. However, if t and T denote two dates expressed as day/month/year, say $D_1 = (d_1, m_1, y_1)$ and $D_2 = (d_2, m_2, y_2)$, we need to define the amount of time between the two dates in terms of the number of days between them. This choice, however, is not unique, and the market evaluates the time between t and T in different ways. Indeed, the number of days between D_1 and D_2 is calculated according to the relevant market convention, which tells you how to count these days, whether to include holidays in the counting, and so on.

Definitions 1.2.1. Year fraction, Day-count convention. *We denote by $\tau(t,T)$ the chosen time measure between t and T, which is usually referred to as* year fraction *between the dates t and T. When t and T are less than one-day distant (typically when dealing with limit quantities involving time to maturities tending to zero), $\tau(t,T)$ is to be interpreted as the time difference $T - t$ (in years). The particular choice that is made to measure the time between two dates reflects what is known as the* day-count convention.

A detailed discussion on day-count conventions is beyond the scope of the book. For a complete treatment of the subject, the interested reader is then referred to the book of Miron and Swannell (1991). However, to clarify things, we mention the following three examples of day-count conventions.

- Actual/365. With this convention a year is 365 days long and the year fraction between two dates is the actual number of days between them

1.2 Zero-Coupon Bonds and Spot Interest Rates

divided by 365. If we denote by $D_2 - D_1$ the actual number of days between the two dates, $D_1 = (d_1, m_1, y_1)$ included and $D_2 = (d_2, m_2, y_2)$ excluded, we have that the year fraction in this case is

$$\frac{D_2 - D_1}{365}.$$

For example, the year fraction between January 4, 2000 and July 4, 2000 (both Tuesdays) is 182/365=0.49863.

- Actual/360. A year is in this case assumed to be 360 days long. The corresponding year fraction is

$$\frac{D_2 - D_1}{360}.$$

Therefore, the year fraction between January 4, 2000 and July 4, 2000 is 182/360=0.50556.

- 30/360. With this convention, months are assumed 30 days long and years are assumed 360 days long. We have that the year fraction between D_1 and D_2 is in this case given by the following formula:

$$\frac{\max(30 - d_1, 0) + \min(d_2, 30) + 360 \times (y_2 - y_1) + 30 \times (m_2 - m_1 - 1)}{360}.$$

For example, the year fraction between January 4, 2000 and July 4, 2000 is now

$$\frac{(30 - 4) + 4 + 360 \times 0 + 30 \times 5}{360} = 0.5.$$

As already hinted at above, adjustments may be included in the conventions, in order to leave out holidays. If D_2 is a holiday date, it can be replaced with the first working date following it, and this changes the evaluation of the year fractions. Again we refer to Miron and Swannell (1991) for the details.

Having clarified the market practice of using different day-count conventions, we can now proceed and comment on the definition of a zero-coupon bond. It is clear that every time we need to know the present value of a future-time payment, the zero-coupon-bond price for that future time is the fundamental quantity to deal with. Zero-coupon-bond prices are the basic quantities in interest-rate theory, and all interest rates can be defined in terms of zero-coupon-bond prices, as we shall see now. Therefore, they are often used as basic auxiliary quantities from which all rates can be recovered, and in turn zero-coupon-bond prices can be defined in terms of any given family of interest rates. Notice, however, that interest rates are what is usually quoted in (interbank) financial markets, whereas zero-coupon bonds are theoretical instruments that, as such, are not directly observable in the market.

In moving from zero-coupon-bond prices to interest rates, and vice versa, we need to know two fundamental features of the rates themselves: The compounding type and the day-count convention to be applied in the rate definition. What we mean by "compounding type" will be clear from the definitions below, while the day-count convention has been discussed earlier.

1. Definitions and Notation

Definition 1.2.3. Continuously-compounded spot interest rate. *The continuously-compounded spot interest rate prevailing at time t for the maturity T is denoted by $R(t,T)$ and is the constant rate at which an investment of $P(t,T)$ units of currency at time t accrues continuously to yield a unit amount of currency at maturity T. In formulas:*

$$R(t,T) := -\frac{\ln P(t,T)}{\tau(t,T)} \qquad (1.5)$$

The continuously-compounded interest rate is therefore a constant rate that is consistent with the zero-coupon-bond prices in that

$$e^{R(t,T)\tau(t,T)} P(t,T) = 1, \qquad (1.6)$$

from which we can express the bond price in terms of the continuously-compounded rate R:

$$P(t,T) = e^{-R(t,T)\tau(t,T)}. \qquad (1.7)$$

The year fraction involved in continuous compounding is usually $\tau(t,T) = T - t$, the time difference expressed in years.

An alternative to continuous compounding is simple compounding, which applies when accruing occurs proportionally to the time of the investment. We indeed have the following.

Definition 1.2.4. Simply-compounded spot interest rate. *The simply-compounded spot interest rate prevailing at time t for the maturity T is denoted by $L(t,T)$ and is the constant rate at which an investment has to be made to produce an amount of one unit of currency at maturity, starting from $P(t,T)$ units of currency at time t, when accruing occurs proportionally to the investment time. In formulas:*

$$L(t,T) := \frac{1 - P(t,T)}{\tau(t,T) \, P(t,T)}. \qquad (1.8)$$

The market LIBOR rates are simply-compounded rates, which motivates why we denote by L such rates. LIBOR rates are typically linked to zero-coupon-bond prices by the Actual/360 day-count convention for computing $\tau(t,T)$.

Definition (1.8) immediately leads to the simple-compounding counterpart of (1.6), i.e.,

$$P(t,T)(1 + L(t,T)\tau(t,T)) = 1, \qquad (1.9)$$

so that a bond price can be expressed in terms of L as:

$$P(t,T) = \frac{1}{1 + L(t,T)\tau(t,T)}.$$

A further compounding method that is considered is annual compounding. Annual compounding is obtained as follows. If we invest today a unit of

1.2 Zero-Coupon Bonds and Spot Interest Rates 7

currency at the simply-compounded rate Y, in one year we will obtain the amount $A = 1(1+Y)$. Suppose that, after this year, we invest such an amount for one more year at the same rate Y, so that we will obtain $A(1+Y) = (1+Y)^2$ in two years. If we keep on reinvesting for n years, the final amount we obtain is $(1+Y)^n$. Based on this reasoning, we have the following.

Definition 1.2.5. Annually-compounded spot interest rate. *The annually-compounded spot interest rate prevailing at time t for the maturity T is denoted by $Y(t,T)$ and is the constant rate at which an investment has to be made to produce an amount of one unit of currency at maturity, starting from $P(t,T)$ units of currency at time t, when reinvesting the obtained amounts once a year. In formulas*

$$Y(t,T) := \frac{1}{[P(t,T)]^{1/\tau(t,T)}} - 1. \tag{1.10}$$

Analogously to (1.6) and (1.9), we then have

$$P(t,T)(1+Y(t,T))^{\tau(t,T)} = 1, \tag{1.11}$$

which implies that bond prices can be expressed in terms of annually-compounded rates as

$$P(t,T) = \frac{1}{(1+Y(t,T))^{\tau(t,T)}}. \tag{1.12}$$

A year fraction τ that can be associated to annual compounding is for example ACT/365.

A straightforward extension of the annual compounding case leads to the following definition, which is based on reinvesting k times per year.

Definition 1.2.6. k-times-per-year compounded spot interest rate. *The k-times-per-year compounded spot interest rate prevailing at time t for the maturity T is denoted by $Y^k(t,T)$ and is the constant rate (referred to a one-year period) at which an investment has to be made to produce an amount of one unit of currency at maturity, starting from $P(t,T)$ units of currency at time t, when reinvesting the obtained amounts k times a year. In formulas:*

$$Y^k(t,T) := \frac{k}{[P(t,T)]^{1/(k\tau(t,T))}} - k. \tag{1.13}$$

We then have

$$P(t,T)\left(1+\frac{Y^k(t,T)}{k}\right)^{k\tau(t,T)} = 1, \tag{1.14}$$

so that we can write

1. Definitions and Notation

$$P(t,T) = \frac{1}{\left(1 + \frac{Y^k(t,T)}{k}\right)^{k\tau(t,T)}}. \tag{1.15}$$

A fundamental property is that continuously-compounded rates can be obtained as the limit of k-times-per-year compounded rates for the number k of compounding times going to infinity. Indeed, we can easily show that

$$\lim_{k \to +\infty} \frac{k}{[P(t,T)]^{1/(k\tau(t,T))}} - k = -\frac{\ln(P(t,T))}{\tau(t,T)} = R(t,T),$$

which justifies the name (and definition) of the rate R. Notice also that, for each fixed Y:

$$\lim_{k \to +\infty} \left(1 + \frac{Y}{k}\right)^{k\tau(t,T)} = e^{Y\tau(t,T)}.$$

In fact, continuously-compounded rates are commonly defined through these limit relations. In this book, however, we preferred to follow a different approach.

We finally remark that all previous definitions of spot interest rates are equivalent in infinitesimal time intervals. Indeed, it can be easily proved that the short rate is obtainable as a limit of all the different rates defined above, that is, for each t,

$$\begin{aligned} r(t) &= \lim_{T \to t^+} R(t,T) \\ &= \lim_{T \to t^+} L(t,T) \\ &= \lim_{T \to t^+} Y(t,T) \\ &= \lim_{T \to t^+} Y^k(t,T) \text{ for each } k. \end{aligned}$$

1.3 Fundamental Interest-Rate Curves

> When dealing with curves, nothing ever goes straight
> Andrea Bugin, Product and Business Development, Banca IMI

A fundamental curve that can be obtained from the market data of interest rates is the zero-coupon curve at a given date t. This curve is the graph of the function mapping maturities into rates at times t. Precisely, we have the following.

Definition 1.3.1. Zero-coupon curve. *The zero-coupon curve (sometimes also referred to as "yield curve") at time t is the graph of the function*

$$T \mapsto \begin{cases} L(t,T) & t < T \leq t+1 \text{ (years)}, \\ Y(t,T) & T > t+1 \text{ (years)}. \end{cases} \tag{1.16}$$

Such a zero-coupon curve is also called the *term structure of interest rates* at time t. It is a plot at time t of simply-compounded interest rates for all maturities T up to one year and of annually compounded rates for maturities T larger than one year. An example of such a curve is shown in Figure 1.1. In this example, the curve is not monotonic, and its initially-inverted behaviour resurfaces periodically in the market. The Italian curve, for example, has been inverted for several years before becoming monotonic. The Euro curve has often shown a monotonic pattern, although in our figure we see an example of the opposite situation.

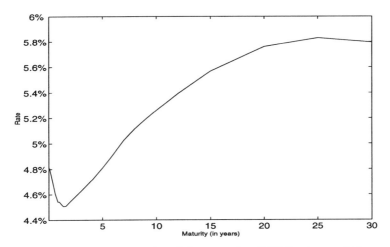

Fig. 1.1. Zero-coupon curve stripped from market EURO rates on February 13, 2001, at 5 p.m.

The term "yield curve" is often used to denote several different curves deduced from the interest-rate-market quotes, and is in fact slightly ambiguous. When used in the book, unless differently specified, it is intended to mean "zero-coupon curve".

Finally, at times, we may consider the same plot for rates with different compounding conventions, such as for example

$$T \mapsto R(t,T), \quad T > t.$$

The term "zero-coupon curve" will be used for all such curves, no matter the compounding convention being used. In the book we will often make use of the following.

Definition 1.3.2. Zero-bond curve. *The zero-bond curve at time t is the graph of the function*

$$T \mapsto P(t,T), \quad T > t,$$

10   1. Definitions and Notation

which, because of the positivity of interest rates, is a T-decreasing function starting from $P(t,t) = 1$. Such a curve is also referred to as term structure of discount factors.

An example of such a curve is shown in Figure 1.2. In a sense, at a quick

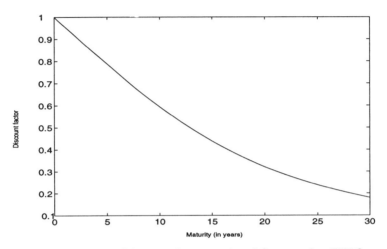

Fig. 1.2. Term structure of discount factors stripped from market EURO rates on February 13, 2001, at 5 p.m.

glance the zero-bond curve is less informative than the zero-coupon curve: The former is monotonic, while the latter can show several possible shapes. This is somehow clear from the definitions. For example, in the continuously-compounded case, the zero-coupon rates involve a logarithmic transformation of the "almost linear" zero-coupon bonds, which leads to a larger variability than that shown by the bonds themselves.

1.4 Forward Rates

We now move to the definition of forward rates. Forward rates are characterized by three time instants, namely the time t at which the rate is considered, its expiry T and its maturity S, with $t \leq T \leq S$. Forward rates are interest rates that can be locked in today for an investment in a future time period, and are set consistently with the current term structure of discount factors.

We can define a forward rate through a prototypical *forward-rate agreement* (FRA). A FRA is a contract involving three time instants: The current time t, the expiry time $T > t$, and the maturity time $S > T$. The contract gives its holder an interest-rate payment for the period between T and S. At the maturity S, a fixed payment based on a fixed rate K is exchanged against a floating payment based on the spot rate $L(T, S)$ resetting in T and with maturity S. Basically, this contract allows one to lock-in the interest rate between times T and S at a desired value K, with the rates in the contract that are simply compounded. Formally, at time S one receives $\tau(T, S)KN$ units of currency and pays the amount $\tau(T, S)L(T, S)N$, where N is the contract nominal value. The value of the contract in S is therefore

$$N\tau(T, S)(K - L(T, S)), \tag{1.17}$$

where we assume that both rates have the same day-count convention.[3] Clearly, if L is larger than K at time T, the contract value is negative, whereas in the other case it is positive. Recalling the expression (1.8) for L, we can rewrite this value as

$$N\left[\tau(T, S)K - \frac{1}{P(T, S)} + 1\right]. \tag{1.18}$$

Now consider the term $A = 1/P(T, S)$ as an amount of currency held at time S. Its value at time T is obtained by multiplying this amount A for the zero coupon price $P(T, S)$:

$$P(T, S)A = P(T, S)\frac{1}{P(T, S)} = 1,$$

so that this term is equivalent to holding one unit of currency at time T. In turn, one unit of currency at time T is worth $P(t, T)$ units of currency at time t. Therefore, the amount $1/P(T, S)$ in S is equivalent to an amount of $P(t, T)$ in t.

Then consider the other two terms in the contract value (1.18). The amount $B = \tau(T, S)K + 1$ at time S is worth

$$P(t, S)B = P(t, S)\tau(T, S)K + P(t, S)$$

at time t. The total value of the contract at time t is therefore

$$\mathbf{FRA}(t, T, S, \tau(T, S), N, K) = N\left[P(t, S)\tau(T, S)K - P(t, T) + P(t, S)\right]. \tag{1.19}$$

There is just one value of K that renders the contract fair at time t, i.e. such that the contract value is 0 in t. This value is of course obtained by equating to zero the FRA value. The resulting rate defines the (simply-compounded) forward rate.

---

[3] It is actually market practice to apply different day-count conventions to fixed and floating rates. Here, however, we assume the same convention for simplicity, without altering the treatment of the whole subject. The generalization to different day-count conventions for the fixed and floating rates is indeed straightforward.

1. Definitions and Notation

Definition 1.4.1. Simply-compounded forward interest rate. *The simply-compounded forward interest rate prevailing at time t for the expiry $T > t$ and maturity $S > T$ is denoted by $F(t; T, S)$ and is defined by*

$$F(t; T, S) := \frac{1}{\tau(T, S)} \left(\frac{P(t, T)}{P(t, S)} - 1 \right). \tag{1.20}$$

It is that value of the fixed rate in a prototypical FRA with expiry T and maturity S that renders the FRA a fair contract at time t.

Notice that we can rewrite the value of the above FRA (1.19) in terms of the just-defined simply compounded forward interest rate:

$$\mathbf{FRA}(t, T, S, \tau(T, S), N, K) = NP(t, S)\tau(T, S)(K - F(t; T, S)). \tag{1.21}$$

Therefore, to value a FRA, we just have to replace the LIBOR rate $L(T, S)$ in the payoff (1.17) with the corresponding forward rate $F(t; T, S)$, and then take the present value of the resulting (deterministic) quantity.

The forward rate $F(t; T, S)$ may thus be viewed as an estimate of the future spot rate $L(T, S)$, which is random at time t, based on market conditions at time t. In particular, we will see later on that $F(t; T, S)$ is the expectation of $L(T, S)$ at time t under a suitable probability measure.

When the maturity of the forward rate collapses towards its expiry, we have the notion of *instantaneous forward rate*. Indeed, let us consider the limit

$$\begin{aligned} \lim_{S \to T^+} F(t; T, S) &= -\lim_{S \to T^+} \frac{1}{P(t, S)} \frac{P(t, S) - P(t, T)}{S - T} \\ &= -\frac{1}{P(t, T)} \frac{\partial P(t, T)}{\partial T} \\ &= -\frac{\partial \ln P(t, T)}{\partial T}, \end{aligned} \tag{1.22}$$

where we use our convention that $\tau(T, S) = S - T$ when S is extremely close to T. This leads to the following.

Definition 1.4.2. Instantaneous forward interest rate. *The instantaneous forward interest rate prevailing at time t for the maturity $T > t$ is denoted by $f(t, T)$ and is defined as*

$$f(t, T) := \lim_{S \to T^+} F(t; T, S) = -\frac{\partial \ln P(t, T)}{\partial T}, \tag{1.23}$$

so that we also have

$$P(t, T) = \exp\left(-\int_t^T f(t, u) \, du \right).$$

Clearly, for this notion to make sense, we need to assume smoothness of the zero-coupon-price function $T \mapsto P(t,T)$ for all T's.

Intuitively, the instantaneous forward interest rate $f(t,T)$ is a forward interest rate at time t whose maturity is very close to its expiry T, say $f(t,T) \approx F(t;T,T+\Delta T)$ with ΔT small.

Instantaneous forward rates are fundamental quantities in the theory of interest rates. Indeed, it turns out that one of the most general ways to express "fairness" of an interest-rate model is to relate certain quantities in the expression for the evolution of f. The "fairness" we refer to is the absence of arbitrage opportunities, as we shall define it precisely later on in the book, and the theoretical framework expressing this absence of arbitrage is the celebrated Heath, Jarrow and Morton (1992) framework, to which Chapter 5 of the present book is devoted. This framework is based on focusing on the instantaneous forward rates f as fundamental quantities to be modeled.

1.5 Interest-Rate Swaps and Forward Swap Rates

Then, a few months later Dogan got nailed by the IRS.
John Grisham, The Chamber, 1994.

We have just considered a FRA, which is a particular contract whose "fairness" can be invoked to define forward rates. A generalization of the FRA is the Interest-Rate Swap (IRS). A prototypical Payer (Forward-start) Interest-Rate Swap (PFS) is a contract that exchanges payments between two differently indexed legs, starting from a future time instant. At every instant T_i in a prespecified set of dates $T_{\alpha+1}, \ldots, T_\beta$ the fixed leg pays out the amount

$$N\tau_i K,$$

corresponding to a fixed interest rate K, a nominal value N and a year fraction τ_i between T_{i-1} and T_i, whereas the floating leg pays the amount

$$N\tau_i L(T_{i-1}, T_i),$$

corresponding to the interest rate $L(T_{i-1}, T_i)$ resetting at the previous instant T_{i-1} for the maturity given by the current payment instant T_i, with T_α a given date. Clearly, the floating-leg rate resets at dates $T_\alpha, T_{\alpha+1}, \ldots, T_{\beta-1}$ and pays at dates $T_{\alpha+1}, \ldots, T_\beta$. We set $\mathcal{T} := \{T_\alpha, \ldots, T_\beta\}$ and $\tau := \{\tau_{\alpha+1}, \ldots, \tau_\beta\}$.

In this description, we are considering that fixed-rate payments and floating-rate payments occur at the same dates and with the same year fractions. Though the generalization to different payment dates and day-count conventions is straightforward, we prefer to present a simplified version to ease the notation.[4]

---

[4] Indeed, a typical IRS in the market has a fixed leg with annual payments and a floating leg with quarterly or semiannual payments.

When the fixed leg is paid and the floating leg is received the IRS is termed Payer IRS (PFS), whereas in the other case we have a Receiver IRS (RFS).

The *discounted* payoff at a time $t < T_\alpha$ of a PFS can be expressed as

$$\sum_{i=\alpha+1}^{\beta} D(t,T_i) N \tau_i (L(T_{i-1}, T_i) - K),$$

whereas the *discounted* payoff at a time $t < T_\alpha$ of a RFS can be expressed as

$$\sum_{i=\alpha+1}^{\beta} D(t,T_i) N \tau_i (K - L(T_{i-1}, T_i)).$$

If we view this last contract as a portfolio of FRAs, we can value each FRA through formulas (1.21) or (1.19) and then add up the resulting values. We thus obtain

$$\begin{aligned}
\mathbf{RFS}(t, \mathcal{T}, \tau, N, K) &= \sum_{i=\alpha+1}^{\beta} \mathbf{FRA}(t, T_{i-1}, T_i, \tau_i, N, K) \\
&= N \sum_{i=\alpha+1}^{\beta} \tau_i P(t, T_i) \left(K - F(t; T_{i-1}, T_i)\right) \\
&= -N P(t, T_\alpha) + N P(t, T_\beta) + N \sum_{i=\alpha+1}^{\beta} \tau_i K P(t, T_i).
\end{aligned}$$
(1.24)

The two legs of an IRS can be seen as two fundamental prototypical contracts. The fixed leg can be thought of as a coupon-bearing bond, and the floating leg can be thought of as a floating-rate note. An IRS can then be viewed as a contract for exchanging the coupon-bearing bond for the floating-rate note that are defined as follows.

Definition 1.5.1. Prototypical coupon-bearing bond. *A prototypical coupon-bearing bond is a contract that ensures the payment at future times $T_{\alpha+1}, \ldots, T_\beta$ of the deterministic amounts of currency (cash-flows) $c := \{c_{\alpha+1}, \ldots, c_\beta\}$. Typically, the cash flows are defined as $c_i = N\tau_i K$ for $i < \beta$ and $c_\beta = N\tau_\beta K + N$, where K is a fixed interest rate and N is bond nominal value. The last cash flow includes the reimbursement of the notional value of the bond.*

In case $K = 0$ the bond reduces to a zero-coupon bond with maturity T_β. Since each cash flow has to be discounted back to current time t from the payment times T, the current value of the bond is

$$\mathbf{CB}(t, \mathcal{T}, c) = \sum_{i=\alpha+1}^{\beta} c_i P(t, T_i).$$

Definition 1.5.2. Prototypical floating-rate note. *A prototypical floating-rate note is a contract ensuring the payment at future times $T_{\alpha+1}, \ldots, T_\beta$ of the LIBOR rates that reset at the previous instants $T_\alpha, \ldots, T_{\beta-1}$. Moreover, the note pays a last cash flow consisting of the reimbursement of the notional value of the note at final time T_β.*

The value of the note is obtained by changing sign to the above value of the RFS with $K = 0$ (no fixed leg) and by adding to it the present value $NP(t, T_\beta)$ of the cash flow N paid at time T_β. We thus obtain

$$-\mathbf{RFS}(t, \mathcal{T}, \tau, N, 0) + NP(t, T_\beta) = NP(t, T_\alpha),$$

meaning that a prototypical floating-rate note is always equivalent to N units of currency at its first reset date T_α. In particular, if $T = T_\alpha$, the value is N, so that the value of the floating-rate note at its first reset time is always equal to its nominal value. This holds as well for $t = T_i$, for all $i = \alpha+1, \ldots, \beta-1$, in that the value of the note at all these instants is N. This is sometimes expressed by saying that "a floating-rate note always trades at par".

We have seen before that requiring a FRA to be fair leads to the definition of forward rates. Analogously, we may require the above IRS to be fair at time t, and we look for the particular rate K such that the above contract value is zero. This defines a forward swap rate.

Definition 1.5.3. *The forward swap rate $S_{\alpha,\beta}(t)$ at time t for the sets of times $\mathcal{T}$ and year fractions τ is the rate in the fixed leg of the above IRS that makes the IRS a fair contract at the present time, i.e., it is the fixed rate K for which* $\mathbf{RFS}(t, \mathcal{T}, \tau, N, K) = 0$. We easily obtain

$$S_{\alpha,\beta}(t) = \frac{P(t, T_\alpha) - P(t, T_\beta)}{\sum_{i=\alpha+1}^{\beta} \tau_i P(t, T_i)}. \qquad (1.25)$$

Let us divide both the numerator and the denominator in (1.25) by $P(t, T_\alpha)$ and notice that the definition of F in terms of P's implies

$$\frac{P(t, T_k)}{P(t, T_\alpha)} = \prod_{j=\alpha+1}^{k} \frac{P(t, T_j)}{P(t, T_{j-1})} = \prod_{j=\alpha+1}^{k} \frac{1}{1 + \tau_j F_j(t)} \quad \text{for all } k > \alpha,$$

where we have set $F_j(t) = F(t; T_{j-1}, T_j)$. Formula (1.25) can then be written in terms of forward rates as

$$S_{\alpha,\beta}(t) = \frac{1 - \prod_{j=\alpha+1}^{\beta} \frac{1}{1+\tau_j F_j(t)}}{\sum_{i=\alpha+1}^{\beta} \tau_i \prod_{j=\alpha+1}^{i} \frac{1}{1+\tau_j F_j(t)}}.$$

1.6 Interest-Rate Caps/Floors and Swaptions

We conclude this chapter by introducing the two main derivative products of the interest-rates market, namely caps and swaptions. These two products

1. Definitions and Notation

will be described more extensively later on, especially in Chapter 6, devoted to market models.

Interest-Rate Caps/Floors

A *cap* is a contract that can be viewed as a payer IRS where each exchange payment is executed only if it has positive value. The cap discounted payoff is therefore given by

$$\sum_{i=\alpha+1}^{\beta} D(t,T_i) N \tau_i (L(T_{i-1}, T_i) - K)^+.$$

Analogously, a *floor* is equivalent to a receiver IRS where each exchange payment is executed only if it has positive value. The floor discounted payoff is therefore given by

$$\sum_{i=\alpha+1}^{\beta} D(t,T_i) N \tau_i (K - L(T_{i-1}, T_i))^+.$$

Where do the terms "cap" and "floor" originate from? Consider the cap case. Suppose a company is LIBOR indebted and has to pay at certain times $T_{\alpha+1}, \ldots, T_\beta$ the LIBOR rates resetting at times $T_\alpha, \ldots, T_{\beta-1}$, with associated year fractions $\tau = \{\tau_{\alpha+1}, \ldots, \tau_\beta\}$, and assume that the debt notional amount is one. Set $\mathcal{T} = \{T_\alpha, \ldots, T_\beta\}$. The company is afraid that LIBOR rates will increase in the future, and wishes to protect itself by locking the payment at a maximum "cap" rate K. In order to do this, the company enters a cap with the payoff described above, pays its debt in terms of the LIBOR rate L and receives $(L - K)^+$ from the cap contract. The difference gives what is paid when considering both contracts:

$$L - (L - K)^+ = \min(L, K).$$

This implies that the company pays at most K at each payment date, since its variable (L-indexed) payments have been *capped* to the fixed rate K, which is termed the strike of the cap contract or, more briefly, *cap rate*. The cap, therefore, can be seen as a contract that can be used to prevent losses from large movements of interest rates when indebted at a variable (LIBOR) rate.

A cap contract can be decomposed additively. Indeed, its discounted payoff is a sum of terms such as

$$D(t,T_i) N \tau_i (L(T_{i-1}, T_i) - K)^+.$$

Each such term defines a contract that is termed *caplet*. The *floorlet* contracts are defined in an analogous way.

It is market practice to price a cap with the following sum of Black's formulas (at time zero)

1.6 Interest-Rate Caps/Floors and Swaptions

$$\text{Cap}^{\text{Black}}(0, \mathcal{T}, \tau, N, K, \sigma_{\alpha,\beta}) = N \sum_{i=\alpha+1}^{\beta} P(0, T_i) \tau_i \text{Bl}(K, F(0, T_{i-1}, T_i), v_i, 1), \tag{1.26}$$

where, denoting by Φ the standard Gaussian cumulative distribution function,

$$\text{Bl}(K, F, v, \omega) = F\omega\Phi(\omega d_1(K, F, v)) - K\omega\Phi(\omega d_2(K, F, v)),$$

$$d_1(K, F, v) = \frac{\ln(F/K) + v^2/2}{v},$$

$$d_2(K, F, v) = \frac{\ln(F/K) - v^2/2}{v},$$

$$v_i = \sigma_{\alpha,\beta}\sqrt{T_{i-1}},$$

with the common volatility parameter $\sigma_{\alpha,\beta}$ that is retrieved from market quotes. Analogously, the corresponding floor is priced according to the formula

$$\text{Flr}^{\text{Black}}(0, \mathcal{T}, \tau, N, K, \sigma_{\alpha,\beta}) = N \sum_{i=\alpha+1}^{\beta} P(0, T_i) \tau_i \text{Bl}(K, F(0, T_{i-1}, T_i), v_i, -1). \tag{1.27}$$

The caps (floors) whose implied volatilities are quoted by the market are typically those with $\alpha = 0$, T_0 equal to three months, and all other T_i's equally three-month spaced, or those with T_0 equal to three months, the next T_i's up to one year equally three-month spaced, and all other T_i's equally six-month spaced.

An example of market cap volatility curve is shown in Figure 1.3, where we show a plot of the $\sigma_{\alpha,\beta}$'s against their corresponding T_β's for a fixed α.

The Black formulas (1.26) and (1.27) have both an historical and a formal justification, which will be extensively explained in the guided-tour Section 6.2 and in the rest of Chapter 6, devoted to the analysis of the (interest-rate) market models.

Definition 1.6.1. *Consider a cap (floor) with payment times $T_{\alpha+1}, \ldots, T_\beta$, associated year fractions $\tau_{\alpha+1}, \ldots, \tau_\beta$ and strike K. The cap (floor) is said to be at-the-money (ATM) if and only if*

$$K = K_{ATM} := S_{\alpha,\beta}(0) = \frac{P(0, T_\alpha) - P(0, T_\beta)}{\sum_{i=\alpha+1}^{\beta} \tau_i P(0, T_i)}.$$

The cap is instead said to be in-the-money (ITM) if $K < K_{ATM}$, and out-of-the-money (OTM) if $K > K_{ATM}$, with the converse holding for a floor.

Using the equality

$$(L - K)^+ - (K - L)^+ = L - K,$$

18   1. Definitions and Notation

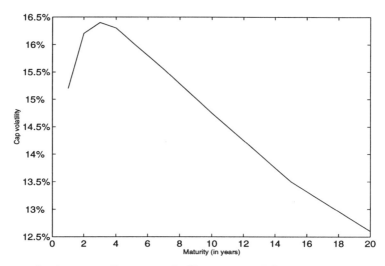

Fig. 1.3. At-the-money Euro cap volatility curve on February 13, 2001, at 5 p.m. The one-year cap has a period of three months, whereas the others (with maturities 2, 3, 4, 5, 7, 10, 15 and 20 years) of six months.

we can note that the difference between a cap and the corresponding floor is equivalent to a forward-start swap. It is, therefore, easy to prove that a cap (floor) is ATM if and only if its price equals that of the corresponding floor (cap).

Notice also that in case the cap has only one payment date ($\alpha + 1 = \beta$), the cap collapses to a single caplet. In such a case, the at-the-money caplet strike is $K_{\text{ATM}} = F(0, T_\alpha, T_{\alpha+1})$, and the caplet is ITM if $K < F(0, T_\alpha, T_{\alpha+1})$. The reason for these terms is particularly clear in the caplet case. In fact, if the contract terminal payoff is evaluated now with the current value of the underlying replacing the corresponding terminal value, we have a positive amount in the ITM case (so that we are "in the money"), whereas we have a non-positive amount in the other case (we are "out of the money"). Just replace the terminal underlying forward rate $F(T_\alpha, T_\alpha, T_{\alpha+1})$ with its current value $F(0, T_\alpha, T_{\alpha+1})$ in the caplet terminal payoff

$$(F(T_\alpha, T_\alpha, T_{\alpha+1}) - K)^+$$

in the ITM and OTM cases to check this.

Swaptions

We finally introduce the second class of basic derivatives on interest rates. These derivatives, termed swap options or more commonly swaptions, are options on an IRS. There are two main types of swaptions, a payer version and a receiver version.

1.6 Interest-Rate Caps/Floors and Swaptions

A European *payer swaption* is an option giving the right (and no obligation) to enter a payer IRS at a given future time, the swaption maturity. Usually the swaption maturity coincides with the first reset date of the underlying IRS. The underlying-IRS length ($T_\beta - T_\alpha$ in our notation) is called the *tenor* of the swaption. Sometimes the set of reset and payment dates is called the tenor structure.

We can write the discounted payoff of a payer swaption by considering the value of the underlying payer IRS at its first reset date T_α, which is also assumed to be the swaption maturity. Such a value is given by changing sign in formula (1.24), i.e.,

$$N \sum_{i=\alpha+1}^{\beta} P(T_\alpha, T_i)\tau_i(F(T_\alpha; T_{i-1}, T_i) - K).$$

The option will be exercised only if this value is positive, so that, to obtain the swaption payoff at time T_α, we have to apply the positive-part operator. The payer-swaption payoff, discounted from the maturity T_α to the current time, is thus equal to

$$ND(t, T_\alpha) \left(\sum_{i=\alpha+1}^{\beta} P(T_\alpha, T_i)\tau_i(F(T_\alpha; T_{i-1}, T_i) - K) \right)^+.$$

Contrary to the cap case, this payoff cannot be decomposed in more elementary products, and this is a fundamental difference between the two main interest-rate derivatives. Indeed, we have seen that caps can be decomposed into the sum of the underlying caplets, each depending on a single forward rate. One can deal with each caplet separately, deriving results that can be finally put together to obtain results on the cap. The same, however, does not hold for swaptions. From an algebraic point of view, this is essentially due to the fact that the summation is *inside* the positive part operator, $(\cdots)^+$, and not outside like in the cap case. Since the positive part operator is not distributive with respect to sums, but is a piece-wise linear and convex function, we have

$$\left(\sum_{i=\alpha+1}^{\beta} P(T_\alpha, T_i)\tau_i(F(T_\alpha; T_{i-1}, T_i) - K) \right)^+$$
$$\leq \sum_{i=\alpha+1}^{\beta} P(T_\alpha, T_i)\tau_i(F(T_\alpha; T_{i-1}, T_i) - K)^+,$$

with no equality in general, so that the additive decomposition is not feasible. As a consequence, in order to value and manage swaptions contracts, we will need to consider the *joint* action of the rates involved in the contract payoff. From a mathematical point of view, this implies that, contrary to the cap case, *terminal correlation* between different rates could be fundamental in handling swaptions. The adjective "terminal" is somehow redundant and is used to point out that we are considering the correlation between rates instead of

20   1. Definitions and Notation

correlations between *infinitesimal changes* in rates, the latter being typically a dynamical property. In other words, the term "terminal" is used to stress that we are not considering *instantaneous* correlations. We will discuss the relationship between the two types of correlations at large in Section 6.6 and in other places.

Notice also that the right-hand side of the above inequality can be thought of as an alternative expression for a cap terminal payoff. *This means that a payer swaption has a value that is always smaller than the value of the corresponding cap contract.*

It is market practice to value swaptions with a Black-like formula. Precisely, the price of the above payer swaption (at time zero) is

$$\mathbf{PS}^{\text{Black}}(0,\mathcal{T},\tau,N,K,\sigma_{\alpha,\beta}) = N\text{Bl}(K,S_{\alpha,\beta}(0),\sigma_{\alpha,\beta}\sqrt{T_\alpha},1) \sum_{i=\alpha+1}^{\beta} \tau_i P(0,T_i),$$
(1.28)

where $\sigma_{\alpha,\beta}$ is now a volatility parameter quoted in the market that is different from the corresponding $\sigma_{\alpha,\beta}$ in the caps/floors case. A similar formula is used for a *receiver swaption*, which gives the holder the right to enter at time T_α a receiver IRS, with payment dates in $\mathcal{T}$. Such a formula is

$$\mathbf{RS}^{\text{Black}}(0,\mathcal{T},\tau,N,K,\sigma_{\alpha,\beta}) = N\text{Bl}(K,S_{\alpha,\beta}(0),\sigma_{\alpha,\beta}\sqrt{T_\alpha},-1) \sum_{i=\alpha+1}^{\beta} \tau_i P(0,T_i).$$
(1.29)

An example of market swaption volatility surface is shown in Figure 1.4, where we plot volatilities $\sigma_{\alpha,\beta}$ against their corresponding maturities T_α and swap lengths (tenors) $T_\beta - T_\alpha$.

Similarly to the cap/floor case, also the Black formulas (1.28) and (1.29) have a historical and a formal justification. Also in this case, a more rigorous treatment and derivation can be found in the chapter devoted to market models, and precisely in Section 6.7.

Definition 1.6.2. *Consider a payer (respectively receiver) swaption with strike K giving the holder the right to enter at time T_α a payer (receiver) IRS with payment dates $T_{\alpha+1},\ldots,T_\beta$ and associated year fractions $\tau_{\alpha+1},\ldots,\tau_\beta$. The swaption (either payer or receiver) is then said to be at-the-money (ATM) if and only if*

$$K = K_{ATM} := S_{\alpha,\beta}(0) = \frac{P(0,T_\alpha) - P(0,T_\beta)}{\sum_{i=\alpha+1}^{\beta} \tau_i P(0,T_i)}.$$

The payer swaption is instead said to be in-the-money (ITM) if $K < K_{ATM}$, and out of the moncy (OTM) if $K > K_{ATM}$. The receiver swaption is ITM if $K > K_{ATM}$, and OTM if $K < K_{ATM}$.

As in the above cap case, we can note that the difference between a payer swaption and the corresponding receiver swaption is equivalent to a forward-

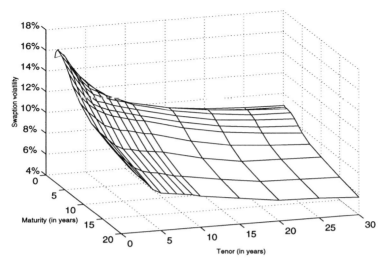

Fig. 1.4. At-the-money Euro swaption volatility surface on February 8, 2001, at 5 p.m.

start swap. Therefore, it is again easy to prove that a payer swaption is ATM if and only if its price equals that of the corresponding receiver swaption.

Finally, we point out that an alternative expression for the above discounted payer-swaption payoff, expressed in terms of the relevant forward swap rate, is at time $t = 0$

$$ND(0,T_\alpha)\left(S_{\alpha,\beta}(T_\alpha) - K\right)^+ \sum_{i=\alpha+1}^{\beta} \tau_i P(T_\alpha, T_i).$$

This alternative expression confirms the intuitive meaning of the ITM and OTM expressions. Indeed, if you evaluate this payoff by substituting the terminal swap rate $S_{\alpha,\beta}(T_\alpha)$ with its current forward value $S_{\alpha,\beta}(0)$ you obtain a positive multiple of

$$(S_{\alpha,\beta}(0) - K)^+.$$

This amount is strictly positive (so that you are "in the money") if $S_{\alpha,\beta}(0) > K$, whereas it is worthless (you are "out of the money") in the other case. The symmetric remark applies to receiver swaptions.

2. No-Arbitrage Pricing and Numeraire Change

"They have an expression on Earth that I believe applies to this situation."
"There ain't no such thing as a free lunch"
Adam Warlock and Moondragon, in
"Silver Surfer/Warlock: Resurrection" 1, 1993, Marvel Comics.

In the context of inflationary cosmology,
it is fair to say that the universe is the ultimate free lunch.
Guth, A.H., The Inflationary Universe, Addison-Wesley, 1997.

The fundamental economic assumption in the seminal paper by Black and Scholes (1973) is the absence of arbitrage opportunities in the considered financial market. Roughly speaking, absence of arbitrage is equivalent to the impossibility to invest zero today and receive tomorrow a nonnegative amount that is positive with positive probability. In other words, two portfolios having the same payoff at a given future date must have the same price today. By constructing a suitable portfolio having the same instantaneous return as that of a riskless investment, Black and Scholes could then conclude that the portfolio instantaneous return was indeed equal to the instantaneous risk-free rate, which immediately led to their celebrated partial differential equation and, through its solution, to their option-pricing formula.

The basic Black and Scholes argument was subsequently used by Vasicek (1977) to develop a model for the evolution of the term structure of interest rates and for the pricing of interest-rate derivatives (see also next Chapter 3). However, such argument deals with infinitesimal quantities in a way that only later works have justified. Indeed, the first rigorous and mathematically sound approach for the arbitrage-free pricing of general contingent claims was developed by Harrison and Kreps (1979) and Harrison and Pliska (1981, 1983). Their general theory has then inspired the work by Heath, Jarrow and Morton (1992) who developed a general framework for interest-rates dynamics.

In this chapter, we start by reviewing the main results of Harrison and Pliska (1981, 1983). We then describe the change-of-numeraire technique as developed by Geman et al. (1995) and provide a useful toolkit explaining how the various dynamics change when changing the numeraire. As an explicit example, the pricing of a cap and a swaption is considered. We conclude the

24   2. No-Arbitrage Pricing and Numeraire Change

chapter by providing some results on the change-of-numeraire technique in the presence of a foreign market.

The purpose of this chapter is highlighting some fundamental results concerning the arbitrage-free pricing of contingent claims. Our treatment is not meant to be extremely rigorous and essentially aims at providing the fundamental tools for the pricing purposes of the book. The reader interested in a more formal and detailed treatment of the no-arbitrage issue under stochastic interest rates is referred, for example, to Musiela and Rutkowski (1998) or Hunt and Kennedy (2000). Finally, the technical tools we will use are basically Ito's formula and Girsanov's theorem, and are briefly outlined together with equivalent measures, Radon-Nikodym derivatives and semimartingales in Appendix A at the end of the book.

2.1 No-Arbitrage in Continuous Time

I weave a delicate strategy which rash actions could rend. Patience, please.
Adam Warlock, "The Infinity Gauntlet", 1991, Marvel Comics.

We here briefly consider the case of the continuous-time economy analyzed by Harrison and Kreps (1979) and Harrison and Pliska (1981, 1983). We do this for historical reasons, but also for introducing a terminology that will be helpful later on.

We consider a time horizon $T > 0$, a probability space $(\Omega, \mathcal{F}, Q_0)$ and a right-continuous filtration $\mathbb{F} = \{\mathcal{F}_t : 0 \leq t \leq T\}$. In the given economy, $K+1$ non dividend paying securities are traded continuously from time 0 until time T. Their prices are modeled by a $K+1$ dimensional adapted semimartingale $S = \{S_t : 0 \leq t \leq T\}$, whose components $S^0, S^1, \ldots, S^K$ are positive. The asset indexed by 0 is a bank account. Its price then evolves according to

$$dS_t^0 = r_t S_t^0 dt,$$

with $S_0^0 = 1$ and where r_t is the instantaneous short-term rate at time t. In the book notation (see previous chapter), $S_t^0 = B(t)$ and $1/S_t^0 = D(0,t)$ for each t.

Definitions 2.1.1. *A* trading strategy *is a (K + 1 dimensional) process $\phi = \{\phi_t : 0 \leq t \leq T\}$, whose components $\phi^0, \phi^1, \ldots, \phi^K$ are locally bounded and predictable. The* value process *associated with a strategy ϕ is defined by*

$$V_t(\phi) = \phi_t S_t = \sum_{k=0}^{K} \phi_t^k S_t^k, \quad 0 \leq t \leq T,$$

and the gains process *associated with a strategy ϕ by*

$$G_t(\phi) = \int_0^t \phi_u dS_u = \sum_{k=0}^{K} \int_0^t \phi_u^k dS_u^k, \quad 0 \leq t \leq T.$$

2.1 No-Arbitrage in Continuous Time

The k-th component ϕ_t^k of the strategy ϕ_t at time t, for each t, is interpreted as the number of units of security k held by an investor at time t. The predictability condition on each ϕ^k means that the value ϕ_t^k is known immediately before time t. This is done to reduce the investor's freedom at any jump time. This issue, however, is not relevant if S has continuous paths (as happens for example in case S follows a diffusion process). Moreover, $V_t(\psi)$ and $G_t(\phi)$ are respectively interpreted as the market value of the portfolio ϕ_t and the cumulative gains realized by the investor until time t by adopting the strategy ϕ.

Definition 2.1.1. *A trading strategy ϕ is* self-financing *if $V(\phi) \geq 0$ and*

$$V_t(\phi) = V_0(\phi) + G_t(\phi), \quad 0 \leq t < T. \tag{2.1}$$

Intuitively, a strategy is self-financing if its value changes only due to changes in the asset prices. In other words, no additional cash inflows or outflows occur after the initial time.

A relation similar to (2.1) holds also when asset prices are all expressed in terms of the bank-account value. Indeed, Harrison and Pliska (1981) proved the following.

Proposition 2.1.1. *Let ϕ be a trading strategy. Then, ϕ is self-financing if and only if $D(0,t)V_t(\phi) = V_0(\phi) + \int_0^t \phi_u d(D(0,u)S_u)$.*

A key result in Harrison and Kreps (1979) and Harrison and Pliska (1981, 1983) is the established connection between the economic concept of absence of arbitrage and the mathematical property of existence of a probability measure, the equivalent martingale measure (or risk-neutral measure, or risk-adjusted measure), whose definition is given in the following.

Definition 2.1.2. *An* equivalent martingale measure Q *is a probability measure on the space $(\Omega, \mathcal{F})$ such that*

i) *Q_0 and Q are equivalent measures, that is $Q_0(A) = 0$ if and only if $Q(A) = 0$, for every $A \in \mathcal{F}$;*

ii) *the Radon-Nikodym derivative dQ/dQ_0 belongs to $L^2(\Omega, \mathcal{F}, Q_0)$ (i.e. it is square integrable with respect to Q_0).*

iii) *the "discounted asset price" process $D(0, \cdot)S$ is an $(\mathbb{F}, Q)$-martingale, that is $E(D(0,t)S_t^k | \mathcal{F}_u) = D(0,u)S_u^k$, for all $k = 0, 1, \ldots, K$ and all $0 \leq u \leq t \leq T$, with E denoting expectation under Q.*

An arbitrage opportunity is defined, in mathematical terms, as a self-financing strategy ϕ such that $V_0(\phi) = 0$ but $V_T(\phi) > 0$ Q_0-a.s. Harrison and Pliska (1981) then proved the fundamental result that the existence of an equivalent martingale measure implies the absence of arbitrage opportunities.

Definitions 2.1.2. *A* contingent claim *is a square-integrable and positive random variable on $(\Omega, \mathcal{F}, Q_0)$. A contingent claim H is* attainable *if there exists some ϕ such that $V_T(\phi) = H$. Such a ϕ is said to* generate H, *and $\pi_t = V_t(\phi)$ is the price at time t associated with H.*

The following proposition, proved by Harrison and Pliska (1981), provides the mathematical characterization of the unique no-arbitrage price associated with any attainable contingent claim.

Proposition 2.1.2. *Assume there exists an equivalent martingale measure Q and let H be an attainable contingent claim. Then, for each time t, $0 \leq t \leq T$, there exists a unique price π_t associated with H, i.e.,*

$$\pi_t = E(D(t,T)H|\mathcal{F}_t). \qquad (2.2)$$

When the set of all equivalent martingale measures is nonempty, it is then possible to derive a unique no-arbitrage price associated to any attainable contingent claim. Such a price is given by the expectation of the discounted claim payoff under the measure Q equivalent to Q_0. This result generalizes that of Black and Scholes (1973) to the pricing of any claim, which, in particular, may be path-dependent. Also the underlying-asset-price dynamics are quite general, which makes formula (2.2) applicable in quite different situations.

Definition 2.1.3. *A financial market is* complete *if and only if every contingent claim is attainable.*

Harrison and Pliska (1983) proved the following fundamental result. A financial market is (arbitrage free and) complete if and only if there exists a unique equivalent martingale measure. The existence of a unique equivalent martingale measure, therefore, not only makes the markets arbitrage free, but also allows the derivation of a unique price associated with any contingent claim.

We can summarize what stated above by the following well known stylized characterization of no-arbitrage theory via martingales.

- The market is free of arbitrage if (and only if) there exists a martingale measure;
- The market is complete if and only if the martingale measure is unique;
- In an arbitrage-free market, not necessarily complete, the price of any attainable claim is uniquely given, either by the value of the associated replicating strategy, or by the risk neutral expectation of the discounted claim payoff under any of the equivalent (risk-neutral) martingale measures.

2.2 The Change-of-Numeraire Technique

Formula (2.2) gives the unique no-arbitrage price of an attainable contingent claim H in terms of the expectation of the claim payoff under the selected martingale measure Q. Geman et al. (1995) noted, however, that an equivalent martingale measure Q, as in Definition 2.1.2, is not necessarily the most natural and convenient measure for pricing the claim H. Indeed, under

2.2 The Change-of-Numeraire Technique

stochastic interest rates, for example, the presence of the stochastic discount factor $D(t,T)$ complicates considerably the calculation of the expectation. In such cases, a change of measure can be quite helpful, and this is the approach followed, for instance, by Jamshidian (1989) in the calculation of a bond-option price under the Vasicek (1977) model.

Geman et al. (1995) introduced the following.

Definition 2.2.1. *A numeraire is any positive non-dividend-paying asset.*

In general, a numeraire Z is identifiable with a self-financing strategy ϕ in that $Z_t = V_t(\phi)$ for each t. Intuitively, a numeraire is a reference asset that is chosen so as to normalize all other asset prices with respect to it. Choosing a numeraire Z then implies that the relative prices S^k/Z, $k = 0, 1, \ldots, K$, are considered instead of the securities prices themselves. The value of the numeraire will be often used to denote the numeraire itself.

As proven by Geman et al. (1995), Proposition 2.1.1 can be extended to any numeraire, in that self-financing strategies remain self-financing after a numeraire change. Indeed, written in differential form, the self-financing condition

$$dV_t(\phi) = \sum_{k=0}^{K} \phi_t^k dS_t^k$$

implies that

$$d\left(\frac{V_t(\phi)}{Z_t}\right) = \sum_{k=0}^{K} \phi_t^k d\left(\frac{S_t^k}{Z_t}\right).$$

Therefore, an attainable claim is also attainable under any numeraire.

In the definition of equivalent martingale measure of the previous section, it has been implicitly assumed the bank account S^0 as numeraire. However, as already pointed out, this is just one of all possible choices and it turns out, in fact, that there can be more convenient numeraires as far as the calculation of claim prices is concerned. The following proposition by Geman et al. (1995) provides a fundamental tool for the pricing of derivatives and is the natural generalization of Proposition 2.1.2 to any numeraire.

Proposition 2.2.1. *Assume there exists a numeraire N and a probability measure Q^N, equivalent to the initial Q_0, such that the price of any traded asset X (without intermediate payments) relative to N is a martingale under Q^N, i.e.,*

$$\frac{X_t}{N_t} = E^N\left\{\frac{X_T}{N_T}|\mathcal{F}_t\right\} \quad 0 \leq t \leq T. \tag{2.3}$$

Let U be an arbitrary numeraire. Then there exists a probability measure Q^U, equivalent to the initial Q_0, such that the price of any attainable claim Y normalized by U is a martingale under Q^U, i.e.,

$$\frac{Y_t}{U_t} = E^U\left\{\frac{Y_T}{U_T}|\mathcal{F}_t\right\} \quad 0 \leq t \leq T. \tag{2.4}$$

28   2. No-Arbitrage Pricing and Numeraire Change

Moreover, the Radon-Nikodym derivative defining the measure Q^U is given by

$$\frac{dQ^U}{dQ^N} = \frac{U_T N_0}{U_0 N_T}. \qquad (2.5)$$

The derivation of (2.5) is outlined as follows. By definition of Q^N, we know that for any tradable asset price Z,

$$E^N\left[\frac{Z_T}{N_T}\right] = E^U\left[\frac{U_0}{N_0}\frac{Z_T}{U_T}\right] \qquad (2.6)$$

(both being equal to Z_0/N_0). By definition of Radon-Nikodym derivative, we know also that for all Z

$$E^N\left[\frac{Z_T}{N_T}\right] = E^U\left[\frac{Z_T}{N_T}\frac{dQ^N}{dQ^U}\right].$$

By comparing the right-hand sides of the last two equalities, from the arbitrariness of Z we obtain (2.5). The general formula (2.4) follows from immediate application of the Bayes rule for conditional expectations.

2.3 A Change-of-Numeraire Toolkit

In this section, we present some useful considerations and formulas on the change-of-numeraire technique developed in the previous section. The purpose of this section is to provide a useful toolkit for the derivation of the asset-price dynamics under different numeraires. The fundamental formula that will be derived here is presented in equation (2.12) below and reads, in a more stylized form,

$$\text{drift}^{\text{Num2}}_{\text{asset}} = \text{drift}^{\text{Num1}}_{\text{asset}} - \text{Vol}_{\text{asset}} \text{ Corr} \left(\frac{\text{Vol}_{\text{Num1}}}{\text{Num1}} - \frac{\text{Vol}_{\text{Num2}}}{\text{Num2}}\right)'.$$

This formula allows us to compute the drift in the dynamics of an asset price when moving from a first numeraire (Num1) to a second one (Num2), when we know the asset drift in the original numeraire, the asset volatility (that, as all instantaneous volatilities and correlations, does not depend on the numeraire), and the instantaneous correlation in the asset price dynamics, as well as the volatilities of the two numeraires.

More formally, we consider a numeraire S with its associated measure Q^S. We also consider an n-vector diffusion process X whose dynamics under Q^S is given by

$$dX_t = \mu_t^S(X_t)dt + \sigma_t(X_t)CdW_t^S, \quad Q^S,$$

where μ_t^S is a $n \times 1$ vector and σ_t is a $n \times n$ diagonal matrix, and where we explicitly point out the measure under which the dynamics is defined.

2.3 A Change-of-Numeraire Toolkit

Here W^S is a n-dimensional standard Brownian motion under Q^S, and the $n \times n$ matrix C is introduced to model correlation in the resulting noise (CdW is equivalent to an n-dimensional Brownian motion with instantaneous correlation matrix $\rho = CC'$).

We will drop the superscript in the Brownian motion when the measure is clear from the context.

Now suppose we are interested in expressing the dynamics of X under the measure associated with a new numeraire U. The new dynamics will then be

$$dX_t = \mu_t^U(X_t)dt + \sigma_t(X_t)CdW_t^U, \quad Q^U,$$

where W^U is a n-dimensional standard Brownian motion under Q^U.

We can employ Girsanov's theorem to deduce the Radon-Nikodym derivative between Q^S and Q^U from the dynamics of X under the two different measures:

$$\zeta_t := \frac{dQ^S}{dQ^U}\bigg|_{\mathcal{F}_t} = \exp\left(-\frac{1}{2}\int_0^t \left|(\sigma_s(X_s)C)^{-1}\left[\mu_s^S(X_s) - \mu_s^U(X_s)\right]\right|^2 ds \right.$$
$$\left. + \int_0^t \left\{(\sigma_s(X_s)C)^{-1}\left[\mu_s^S(X_s) - \mu_s^U(X_s)\right]\right\}' dW_s^U\right).$$

In doing so we have assumed for simplicity that C is an invertible (full-rank) matrix.

The Girsanov theorem is briefly reviewed in Appendix A in a simplified context.

The process ζ defines a measure Q^S under which X has the desired dynamics, given its dynamics under Q^U. We know that ζ is an exponential martingale, in that by setting

$$\alpha_t := \left[\mu_t^S(X_t) - \mu_t^U(X_t)\right]' \left((\sigma_t(X_t)C)^{-1}\right)'$$

we obtain the "exponential martingale" dynamics:

$$d\zeta_t = \alpha_t \zeta_t dW_t^U. \tag{2.7}$$

Now, by (2.5),

$$\zeta_T = \frac{dQ^S}{dQ^U}\bigg|_{\mathcal{F}_T} = \frac{U_0 S_T}{S_0 U_T}, \tag{2.8}$$

and, since ζ is a Q^U-martingale,

$$\zeta_t = E_t^{Q^U}(\zeta_T) = E_t^{Q^U}\left[\frac{U_0 S_T}{S_0 U_T}\right] = \frac{U_0 S_t}{S_0 U_t}. \tag{2.9}$$

It follows by differentiation that

30   2. No-Arbitrage Pricing and Numeraire Change

$$d\zeta_t = \frac{U_0}{S_0} d\left(\frac{S_t}{U_t}\right) = \frac{U_0}{S_0} \sigma_t^{S/U} C dW_t^U, \qquad (2.10)$$

where, since S/U is a martingale under Q^U, we have assumed the following martingale dynamics

$$d\left(\frac{S_t}{U_t}\right) = \sigma_t^{S/U} C dW_t^U, \quad Q^U,$$

where $\sigma_t^{S/U}$ is a $1 \times n$ vector. Comparing equations (2.7) and (2.10) to deduce that

$$\alpha_t \zeta_t = \frac{U_0}{S_0} \sigma_t^{S/U} C,$$

and by taking into account (2.9), we obtain the fundamental result

$$\frac{S_t}{U_t} \alpha_t = \sigma_t^{S/U} C,$$

or, by definition of α,

$$\mu_t^U(X_t) = \mu_t^S(X_t) - \frac{U_t}{S_t} \sigma_t(X_t) \rho (\sigma_t^{S/U})' \qquad (2.11)$$

with $\rho = CC'$, which gives the change in the drift of a stochastic process when changing numeraire from U to S.

A useful characterization of formula (2.11) is given in the following.

Proposition 2.3.1. *Let us assume that the two numeraires S and U evolve under Q^U according to*[1]

$$dS_t = (\ldots)dt + \sigma_t^S C dW_t^U, \quad Q^U$$
$$dU_t = (\ldots)dt + \sigma_t^U C dW_t^U, \quad Q^U,$$

where both σ_t^S and σ_t^U are $1 \times n$ vectors, W^U is the usual n-dimensional driftless (under Q^U) standard Brownian motion and $CC' = \rho$. Then, the drift of the process X under the numeraire U is

$$\boxed{\mu_t^U(X_t) = \mu_t^S(X_t) - \sigma_t(X_t) \rho \left(\frac{\sigma_t^S}{S_t} - \frac{\sigma_t^U}{U_t}\right)'.} \qquad (2.12)$$

Proof. Use the stochastic Leibnitz rule to compute $d(S_t/U_t)$ as

$$d\frac{S_t}{U_t} = \frac{1}{U_t} dS_t + S_t d\frac{1}{U_t} + dS_t \, d\frac{1}{U_t},$$

---
[1] Or under any other equivalent measure, since the volatility coefficients in the dynamics do not depend on the particular equivalent measure that has been chosen

in combination with Ito's formula
$$d\frac{1}{U_t} = -\frac{1}{U_t^2}dU_t + \frac{1}{U_t^3}dU_t\,dU_t$$
and substitute the above dynamics for S and U to arrive easily at
$$\sigma_t^{S/U} = \frac{\sigma_t^S}{U_t} - \frac{S_t}{U_t}\frac{\sigma_t^U}{U_t},$$
which combined with (2.11) gives (2.12). □

It is sometimes helpful to consider what happens in terms of "shocks". Indeed, through straightforward passages, we can write
$$C dW_t^S = C dW_t^U - \rho\left(\frac{\sigma_t^S}{S_t} - \frac{\sigma_t^U}{U_t}\right)'dt. \tag{2.13}$$

Another interesting issue is the derivation of the dynamics of a tradable asset under the measure featuring the asset itself as numeraire. Let the asset be S, with diffusion coefficient as in Proposition 2.3.1, and consider the risk-neutral probability measure Q_0 associated with the money-market account numeraire B. We know that under $Q_0 = Q^B$ the process S_t/B_t is a martingale. It follows that the dynamics of S under the risk-neutral measure is
$$dS_t = r_t S_t dt + \sigma_t^S C dW_t^0.$$
We now apply formula (2.13) and obtain that the desired dynamics of S under Q^S is
$$dS_t = \left[r_t S_t + \frac{\sigma_t^S \rho (\sigma_t^S)'}{S_t}\right]dt + \sigma_t^S C dW_t^S.$$
A fundamental particular case is treated in the following.

Proposition 2.3.2. *Let us assume a "level-proportional" functional form for volatilities (typical of the lognormal case), i.e.*
$$\sigma_t^S = v_t^S S_t,$$
$$\sigma_t^U = v_t^U U_t,$$
$$\sigma_t(X_t) = \mathrm{diag}(X_t)\,\mathrm{diag}(v_t^X),$$
where the v's are deterministic $1 \times n$-vector functions of time, and $\mathrm{diag}(X)$ denotes a diagonal matrix whose diagonal elements are the entries of vector X. We then obtain
$$\mu_t^U(X_t) = \mu_t^S(X_t) - \mathrm{diag}(X_t)\,\mathrm{diag}(v_t^X)\rho\left(v_t^S - v_t^U\right)'$$
$$= \mu_t^S(X_t) - \mathrm{diag}(X_t)\frac{d\langle \ln X, \ln(S/U)'\rangle_t}{dt}, \tag{2.14}$$
where the quadratic covariation and the logarithms, when applied to vectors, are meant to act componentwise.

32   2. No-Arbitrage Pricing and Numeraire Change

In the "fully lognormal" case, where also the drift of X under Q^S is deterministically level proportional, i.e.,

$$\mu_t^S(X_t) = \text{diag}(X_t)\, m_t^S,$$

with m^S a deterministic $n \times 1$ vector, it turns out that the drift under the new numeraire measure Q^U is of the same kind, i.e.,

$$\mu_t^U(X_t) = \text{diag}(X_t)\, m_t^U,$$

with

$$m_t^U = m_t^S - \text{diag}(v_t^X)\rho\left(v_t^S - v_t^U\right)' = m_t^S - \frac{d\langle \ln X, \ln(S/U)' \rangle_t}{dt},$$

which is often written as

$$m_t^U\, dt = m_t^S\, dt - (d\ln X_t)\,(d\ln(S_t/U_t)). \qquad (2.15)$$

It is easy to see that this last formula connecting the drift rates m^U and m^S holds also in the more general case where the dynamics of the numeraires S and U are not lognormal-like. In this case one can still start from a lognormal X with deterministic drift rate m^S under Q^S, but one obtains a stochastic drift rate m^U for X under Q^U given by the above formula (where this time the covariance term is not deterministic). We will usually consider the above formula in this wider sense.

Remark 2.3.1. The covariation term in (2.14) is usually called "Vaillant brackets" in Rebonato (1998), who defines

$$[X, Y]_t := \frac{d\langle \ln X, \ln Y \rangle_t}{dt}.$$

However, this square-brackets operator from Rebonato (1998) is not to be confused with the one from semimartingale theory.

2.4 The Choice of a Convenient Numeraire

As far as pricing derivatives is concerned, the change-of-numeraire technique is typically employed as follows. A payoff $h(X_T)$ is given, which depends on an underlying variable X (an interest rate, an exchange rate, a commodity price, etc.) at time T. Pricing such a payoff amounts to compute the risk-neutral expectation

$$E_0\{D(0,T)h(X_T)\}.$$

The risk-neutral numeraire is the money-market account

$$B(t) = D(0,t)^{-1} = \exp\left(\int_0^t r_s ds\right).$$

By using formula (2.6) for pricing under a new numeraire S, we obtain

$$E_0\{D(0,T)h(X_T)\} = S_0 E^{Q^S}\left\{\frac{h(X_T)}{S_T}\right\}. \tag{2.16}$$

Motivated by the above formula, we look for a numeraire S with the following two properties.

1. $X_t S_t$ is (the price of) a tradable asset (in such a case S is sometimes termed the *natural payoff* of X).
2. The quantity $h(X_T)/S_T$ is conveniently simple.

Why are these properties desirable?

The first condition ensures that $(X_t S_t)/S_t = X_t$ is a martingale under Q^S, so that one can assume for example a lognormal martingale dynamics for X:

$$dX_t = \sigma(t) X_t dW_t, \quad Q^S$$

with the consequent ease in computing expected values of functions of X. Indeed, in this case, the distribution of X under Q^S is known, and we have

$$\ln X_t \sim \mathcal{N}\left(\ln X_0 - \frac{1}{2}\int_0^t \sigma(s)^2 ds, \int_0^t \sigma(s)^2 ds\right).$$

The second condition ensures that the new numeraire renders the computation of the right hand side of (2.16) simpler, instead of complicating it.

Remark 2.4.1. Two standard applications of the above method are in the forward LIBOR and swap market models of Chapter 6, where the Black formulas for caps or swaptions are retrieved within a consistent no-arbitrage framework. The interested reader is therefore referred to such a chapter.

2.5 The Forward Measure

In many concrete situations, a useful numeraire is the zero-coupon bond whose maturity T coincides with that of the derivative to price. In such a case, in fact, $S_T = P(T,T) = 1$, so that pricing the derivative can be achieved by calculating an expectation of its payoff (divided by one). The measure associated with the bond maturing at time T is referred to as T-forward risk-adjust measure, or more briefly as T-forward measure, and will be denoted by Q^T. The related expectation is denoted by E^T.

Denoting by π_t the price of the derivative at time t, formula (2.4) applied to the above numeraire yields

$$\pi_t = P(t,T)E^T\{H_T|\mathcal{F}_t\},\qquad(2.17)$$

for $0 \leq t \leq T$, and where H_T is the claim payoff at time T.

The reason why the measure Q^T is called forward measure is justified by the following.

Proposition 2.5.1. *Any simply-compounded forward rate spanning a time interval ending in T is a martingale under the T-forward measure, i.e.,*

$$E^T\{F(t;S,T)|\mathcal{F}_u\} = F(u;S,T),$$

for each $0 \leq u \leq t \leq S < T$. In particular, the forward rate spanning the interval $[S,T]$ is the Q^T-expectation of the future simply-compounded spot rate at time S for the maturity T, i.e,

$$E^T\{L(S,T)|\mathcal{F}_t\} = F(t;S,T),\qquad(2.18)$$

for each $0 \leq t \leq S < T$.

Proof. From the definition of a simply-compounded forward rate

$$F(t;S,T) = \frac{1}{\tau(S,T)}\left[\frac{P(t,S)}{P(t,T)} - 1\right],$$

with $\tau(S,T)$ the year fraction from S to T, we have that

$$F(t;S,T)P(t,T) = \frac{1}{\tau(S,T)}[P(t,S) - P(t,T)]$$

is the price at time t of a traded asset, since it is a multiple of the difference of two bonds. Therefore, by definition of the T-forward measure

$$\frac{F(t;S,T)P(t,T)}{P(t,T)} = F(t;S,T)$$

is a martingale under such a measure. The relation (2.18) then immediately follows from the equality $F(S;S,T) = L(S,T)$. □

The equality (2.18) can be extended to instantaneous rates as well. This is done in the following.

Proposition 2.5.2. *The expected value of any future instantaneous spot interest rate, under the corresponding forward measure, is equal to related instantaneous forward rate, i.e.,*

$$E^T\{r_T|\mathcal{F}_t\} = f(t,T),$$

for each $0 \leq t \leq T$.

Proof. Let us apply (2.17) with $H_T = r_T$ and remember the risk-neutral valuation formula (2.2). We then have

$$\begin{aligned} E^T\{r_T|\mathcal{F}_t\} &= \frac{1}{P(t,T)} E\left\{r_T e^{-\int_t^T r_s ds} |\mathcal{F}_t\right\} \\ &= -\frac{1}{P(t,T)} E\left\{\frac{\partial}{\partial T} e^{-\int_t^T r_s ds} |\mathcal{F}_t\right\} \\ &= -\frac{1}{P(t,T)} \frac{\partial P(t,T)}{\partial T} \\ &= f(t,T). \end{aligned}$$

□

2.6 The Fundamental Pricing Formulas

The results of the previous sections are developed in case of a finite number of basic market securities. A (theoretical) bond market, however, has a continuum of basic assets (the zero-coupon bonds), one for each possible maturity until a given time horizon. The no-arbitrage theory in such a situation is more complicated even though quite similar in spirit. A thorough analysis of this theory is, however, beyond the scope of this book. The interested reader is again referred, for example, to Musiela and Rutkowski (1998) or Hunt and Kennedy (2000). What we will do is simply to assume the existence of a risk-neutral measure Q under which the price at time t of any attainable contingent claim with payoff H_T at time $T > t$ is given by

$$\boxed{\pi_t = E\left(e^{-\int_t^T r_s ds} H_T | \mathcal{F}_t\right)}, \tag{2.19}$$

consistently with formula (2.2), where a claim is now meant to be attainable if it can be replicated by a self-financing strategy involving a finite number of basic assets at every trading time.

Remark 2.6.1. **(Attainability Assumption)** For all contingent claims that will be priced in the book we will assume attainability to hold. Therefore, when a claim is priced, we are assuming the existence of a suitable self-financing strategy that replicates the claim. However, an example of derivation of an explicit replicating strategy will be considered in Section 6.7.1.

The particular case of a European call option with maturity T, strike X and written on a unit-principal zero-coupon bond with maturity $S > T$ leads to the pricing formula

$$\mathbf{ZBC}(t,T,S,X) = E\left(e^{-\int_t^T r_s ds}(P(T,S) - X)^+ | \mathcal{F}_t\right). \tag{2.20}$$

36   2. No-Arbitrage Pricing and Numeraire Change

Moreover, the change-of-numeraire technique can be applied exactly as before, since it is just based on changing the underlying probability measure. For example, the forward measure Q^T is defined by the Radon-Nikodym derivative

$$\frac{dQ^T}{dQ} = \frac{P(T,T)B(0)}{P(0,T)B(T)} = \frac{e^{-\int_0^T r_s ds}}{P(0,T)} = \frac{D(0,T)}{P(0,T)},$$

so that the price at time t of the above claim is also given by (2.17), that is

$$\boxed{\pi_t = P(t,T)E^T(H_T|\mathcal{F}_t)}. \tag{2.21}$$

In the case of the above call option, we then have

$$\mathbf{ZBC}(t,T,S,X) = P(t,T)E^T\left((P(T,S) - X)^+|\mathcal{F}_t\right). \tag{2.22}$$

Of course, formula (2.21), and hence formula (2.22), turn out to be useful when the quantities defining the claim payoff have known dynamics (or distributions) under the forward measure Q^T. Typically, if $P(T,S)$ has a lognormal distribution conditional on $\mathcal{F}_t$ under the T-forward measure, the above expectation reduces to a Black-like formula with a suitable volatility input.

2.6.1 The Pricing of Caps and Floors

We now show that a cap (floor) is actually equivalent to a portfolio of European zero-coupon put (call) options. This equivalence will be repeatedly used to derive explicit formulas for cap/floor prices under the analytically tractable short-rate models we will consider in the next chapters.

We denote by $D = \{d_1, d_2, \ldots, d_n\}$ the set of the cap/floor payment dates and by $\mathcal{T} = \{t_0, t_1, \ldots, t_n\}$ the set of the corresponding times, meaning that t_i is the difference in years between d_i and the settlement date t, and where t_0 is the first reset time. Moreover, we denote by τ_i the year fraction from d_{i-1} to d_i, $i = 1, \ldots, n$ and by N the cap/floor nominal value, and we set $\tau = \{\tau_1, \ldots, \tau_n\}$. The arbitrage-free price of the i-th caplet is then given by

$$\mathbf{Cpl}(t, t_{i-1}, t_i, \tau_i, N, X)$$
$$= E\left(e^{-\int_t^{t_i} r_s ds} N\tau_i(L(t_{i-1}, t_i) - X)^+ | \mathcal{F}_t\right)$$
$$= NE\left(e^{-\int_t^{t_{i-1}} r_s ds} P(t_{i-1}, t_i)\tau_i(L(t_{i-1}, t_i) - X)^+ | \mathcal{F}_t\right),$$

where the second equality comes from iterated conditioning, which will be explained extensively in Section 2.7.

Using the definition of the LIBOR rate $L(t_{i-1}, t_i)$, we obtain

$$\mathbf{Cpl}(t, t_{i-1}, t_i, \tau_i, N, X)$$
$$= NE\left(e^{-\int_t^{t_{i-1}} r_s ds} P(t_{i-1}, t_i) \left[\frac{1}{P(t_{i-1}, t_i)} - 1 - X\tau_i\right]^+ \Big| \mathcal{F}_t\right)$$
$$= NE\left(e^{-\int_t^{t_{i-1}} r_s ds} [1 - (1 + X\tau_i) P(t_{i-1}, t_i)]^+ \Big| \mathcal{F}_t\right),$$

so that
$$\mathbf{Cpl}(t, t_{i-1}, t_i, \tau_i, N, X) = N_i' \mathbf{ZBP}(t, t_{i-1}, t_i, X_i'), \tag{2.23}$$
where
$$X_i' = \frac{1}{1 + X\tau_i},$$
$$N_i' = N(1 + X\tau_i).$$

Therefore, the caplet price can be written as a multiple of the price of a European put with maturity t_{i-1}, strike X_i' and written on a zero-coupon bond with maturity t_i and unit nominal amount. Alternatively, we can also write
$$\mathbf{Cpl}(t, t_{i-1}, t_i, \tau_i, N, X) = \mathbf{ZBP}(t, t_{i-1}, t_i, N_i', N). \tag{2.24}$$

The analogous formulas for the corresponding floor are then given by
$$\mathbf{Fll}(t, t_{i-1}, t_i, \tau_i, N, X) = N_i' \mathbf{ZBC}(t, t_{i-1}, t_i, X_i'),$$
$$= \mathbf{ZBC}(t, t_{i-1}, t_i, N_i', N). \tag{2.25}$$

Finally, cap and floor prices are simply obtained by summing up the prices of the underlying caplets and floorlets, respectively. We thus obtain
$$\mathbf{Cap}(t, \mathcal{T}, \tau, N, X) = \sum_{i=1}^n N_i' \mathbf{ZBP}(t, t_{i-1}, t_i, X_i')$$
$$\mathbf{Flr}(t, \mathcal{T}, \tau, N, X), = \sum_{i=1}^n N_i' \mathbf{ZBC}(t, t_{i-1}, t_i, X_i'). \tag{2.26}$$

2.7 Pricing Claims with Deferred Payoffs

Many real-market interest-rate derivatives have (random) payoffs that depend on some interest rate whose value is set on some date prior to the derivative maturity. An easy example is given by a FRA, whose definition and pricing are described in Chapter 1. Standard interest-rate swaps themselves feature rates that usually reset a period earlier than the corresponding payment dates. In these cases, the associated time lag between reset and payment is named "natural", whereas when the time lag is different it is termed "unnatural", see Section 10.7.1 for more details.

38   2. No-Arbitrage Pricing and Numeraire Change

Pricing more general claims with deferred payoffs, however, can be less straightforward. For instance, when using a tree to approximate the basic interest-rate process, the typical pricing procedure is to calculate the claim payoff on the final nodes in the tree and proceed backwards until the unique node at the initial time. In such a case, we immediately see that the payoff dependence on previously reset rates contrasts with a backwards calculation, since it would require the knowledge at time T of values that are only known at time $t < T$.

We now show a very simple way to address this type of problem.

If $t < \tau < T$ are three times and H_τ is known at time τ, then the time t-value of the payoff H_τ at time T is, by (2.19) and the tower property of conditional expectations,

$$\begin{aligned}\pi_t &= E\left(e^{-\int_t^T r_s ds} H_\tau | \mathcal{F}_t\right) \\ &= E\left[E\left(e^{-\int_t^T r_s ds} H_\tau | \mathcal{F}_\tau\right) | \mathcal{F}_t\right] \\ &= E\left[e^{-\int_t^\tau r_s ds} H_\tau E\left(e^{-\int_\tau^T r_s ds} | \mathcal{F}_\tau\right) | \mathcal{F}_t\right] \\ &= E\left[e^{-\int_t^\tau r_s ds} H_\tau P(\tau, T) | \mathcal{F}_t\right].\end{aligned}$$

This implies that the time-t value of the payoff H_τ "payable" at time T (first expression) is equal to the time-t value of the payoff $H_\tau P(\tau, T)$ "payable" at time τ (last expression). The above problem, therefore, can be addressed by acting as if the payoff were anticipated and multiplied by the proper discount factor. Such a result is quite intuitive and actually confirms the basic economic principle according to which receiving on a future date an amount known today is equivalent to receiving today the present value of such an amount.

2.8 Pricing Claims with Multiple Payoffs

It can be interesting to consider also the case of those interest-rate derivatives that have (random) payoffs occurring at different dates. A typical example is a cap that gives its holder a stream of option-like cashflows on some pre-defined dates until the cap maturity. In such cases, assuming there are no early-exercise features, each single cashflow can be priced by taking expectation under the associated forward measure. However, when there are path-dependent features so that we have to resort to Monte Carlo pricing, dealing with different measures is usually quite burdensome and time consuming. In this situations, it is advisable, therefore, to change measure and act as if all cash flows occurred at the same time.

We here notice that, similarly to the previous section case, there is a very simple way of doing so by means of zero-coupon bonds. Indeed, from a

2.8 Pricing Claims with Multiple Payoffs

basic financial point of view, receiving a known amount today is equivalent to receiving the forward value of this amount at the future date corresponding to the forward maturity. Equivalently, the value at time t of a payoff x "payable" at time $T > t$, with x known at time T, is equal to the value at time t of the payoff $x/P(T,S)$ "payable" at time $S > T$. This is formally proven in the following.

Proposition 2.8.1. *If H is a T-measurable random variable, we have the identity:*

$$E[D(t,T)H|\mathcal{F}_t] = E\left[\frac{D(t,S)H}{P(T,S)}\Big|\mathcal{F}_t\right], \quad (2.27)$$

for all $t < T < S$.

Proof. From the tower property of conditional expectations, and remembering that $D(t,S) = D(t,T)D(T,S)$, we immediately have

$$E\left[\frac{D(t,S)H}{P(T,S)}\Big|\mathcal{F}_t\right] = E\left(E\left[\frac{D(t,T)D(T,S)H}{P(T,S)}\Big|\mathcal{F}_T\right]\Big|\mathcal{F}_t\right)$$

$$= E\left[\frac{D(t,T)H}{P(T,S)}E(D(T,S)|\mathcal{F}_T)\Big|\mathcal{F}_t\right]$$

$$= E\left[\frac{D(t,T)H}{P(T,S)}P(T,S)\Big|\mathcal{F}_t\right]$$

$$= E[D(t,T)H|\mathcal{F}_t].$$

□

As an example, we consider n times $T_1 < T_2 \ldots < T_n$ and an interest-rate derivative that at each time T_i pays out the quantity H_i, which is known at time T_i, with no early-exercise features. The derivative price at time $t < T_1$ is therefore

$$\pi_t = \sum_{i=1}^{n} E\{D(t,T_i)H_i|\mathcal{F}_t\}$$

$$= \sum_{i=1}^{n} P(t,T_i)E^{T_i}\{H_i|\mathcal{F}_t\}.$$

The previous proposition, however, enables us to consider just one forward measure, precisely that relative to the longest maturity T_n. Indeed, by (2.27),

$$E\{D(t,T_i)H_i|\mathcal{F}_t\} = E\left\{\frac{D(t,T_n)H_i}{P(T_i,T_n)}\Big|\mathcal{F}_t\right\},$$

so that we obtain

40   2. No-Arbitrage Pricing and Numeraire Change

$$\pi_t = \sum_{i=1}^{n} E\left\{\frac{D(t,T_n)H_i}{P(T_i,T_n)}|\mathcal{F}_t\right\}$$

$$= P(t,T_n)E^{T_n}\left\{\sum_{i=1}^{n}\frac{H_i}{P(T_i,T_n)}|\mathcal{F}_t\right\}.$$

Therefore, when resorting to Monte Carlo pricing, the last equation can be used to simulate the evolution of the underlying variables under a unique measure, namely the T_n-forward measure, often called the "terminal (forward) measure".

2.9 Foreign Markets and Numeraire Change

In this section we derive the Radon-Nikodym derivative that defines the change of measure between a foreign risk-neutral probability measure and the domestic risk-neutral probability measure. We then interpret this measure change as a change of numeraire. These results can be helpful when pricing multi-currency interest-rate derivatives, as we will show in Chapter 11.

Consider a foreign market where an asset with price X^f is traded. Denote by Q^f the corresponding (foreign) risk-adjusted martingale measure. Assume that the foreign money-market account evolves according to the process B^f. Also consider a domestic market and assume that the domestic money-market account evolves according to the process B and the exchange rate between the two corresponding currencies is modeled through the process $\mathcal{Q}$, in that 1 unit of the foreign currency is worth $\mathcal{Q}_t$ units of domestic currency at time t. Denote by $\mathbb{F} = \{\mathcal{F}_t : 0 \leq t \leq T\}$, the filtration generated by all the above processes, with $\mathcal{F}_0$ the trivial sigma-field. Assume that all expectations below are well defined.

Thinking of X^f as a derivative that pays out X_T^f at time T, from the previous sections we know that the arbitrage-free price of X^f at time t in the foreign market is (E^f denotes expectation under Q^f)

$$V_t^f = B_t^f E^f\left\{\frac{X_T^f}{B_T^f}|\mathcal{F}_t\right\},$$

which expressed in terms of the domestic currency becomes

$$V_t = \mathcal{Q}_t B_t^f E^f\left\{\frac{X_T^f}{B_T^f}|\mathcal{F}_t\right\}.$$

From the perspective of a domestic investor, the asset X^f is perfectly equivalent to a derivative that pays out $X_T^f \mathcal{Q}_T$ at time T. In fact, since X^f is denominated in foreign currency, the actual payoff at time T for a domestic investor that buys this asset is $X_T^f Q_T$. Therefore, to avoid arbitrage, the arbitrage-free price of the domestic asset $X^f Q$ at time t must be equal to V_t, the foreign-currency price of X^f (i.e., V_t^f) multiplied by the exchange rate at time t. In formulas, (as usual, E denotes expectation under Q)

$$Q_t B_t^f E^f \left\{ \frac{X_T^f}{B_T^f} | \mathcal{F}_t \right\} = B_t E \left\{ \frac{X_T^f Q_T}{B_T} | \mathcal{F}_t \right\}. \qquad (2.28)$$

Theorem 2.9.1. *The Radon-Nikodym derivative $\frac{dQ^f}{dQ}$ defining the change of measure from the foreign risk-neutral probability measure Q^f and the domestic risk-neutral probability measure Q is given by*

$$\frac{dQ^f}{dQ} = \frac{Q_T B_T^f}{Q_0 B_T}. \qquad (2.29)$$

Proof. Combining equation (2.28) at time 0

$$E^f \left\{ \frac{Q_0 X_T^f}{B_T^f} \right\} = E \left\{ \frac{X_T^f Q_T}{B_T} \right\}$$

with the following immediate property of the measure change

$$E^f \left\{ \frac{Q_0 X_T^f}{B_T^f} \right\} = E \left\{ \frac{dQ^f}{dQ} \frac{Q_0 X_T^f}{B_T^f} \right\},$$

we have that (2.29) gives the right candidate for defining $\frac{dQ^f}{dQ}$. We then prove that this candidate is a positive martingale with mean one. This immediately follows from the definition of Q. In fact, we know that under Q all domestic assets divided by the domestic money-market account are martingales:

$$E \left\{ \frac{Q_T B_T^f}{Q_0 B_T} | \mathcal{F}_t \right\} = \frac{1}{Q_0} E \left\{ \frac{Q_T B_T^f}{B_T} | \mathcal{F}_t \right\}$$
$$= \frac{1}{Q_0} \frac{Q_t B_t^f}{B_t},$$

since $B^f Q$ is a domestic asset, which proves the martingale property. As for the expectation being equal to one, just notice that the constant expected value equals the initial value of the martingale, which is trivially seen to be one. □

Corollary 2.9.1. *Changing the measure from Q^f to Q is equivalent to changing the numeraire from B^f to $\frac{B}{Q}$.*

Proof. Following the discussions seen in the change-of-numeraire-toolkit section, it is enough to remember that the Radon-Nikodym derivative when changing the numeraire from B^f to any U is

$$\frac{dQ^U}{dQ^f} = \frac{U_T}{U_0 B_T^f}$$

and that

$$\frac{dQ}{dQ^f} = \frac{Q_0 B_T}{Q_T B_T^f}.$$

□

This corollary tells us that moving from the foreign measure Q^f to the domestic measure Q amounts to changing the numeraire from the foreign bank account to the domestic bank account translated into foreign currency through the related exchange rate.

The above result also follows from the definition of a martingale measure associated with a numeraire. In fact, under the measure associated with the numeraire B/Q, for any (foreign) traded asset Y^f, the process

$$\left\{ \frac{Y_t^f Q_t}{B_t} : 0 \le t \le T \right\}$$

is a martingale. Analogously, under the measure associated with the numeraire B, for any (domestic) traded asset Y, the process

$$\left\{ \frac{Y_t}{B_t} = \frac{Y_t}{Q_t} \frac{Q_t}{B_t} : 0 \le t \le T \right\}$$

is a martingale. Then, we just have to notice that Y^f is a foreign traded asset if and only if $Y^f Q_t$ is a domestic traded asset and Y is a domestic traded asset if and only if Y/Q is a foreign traded asset.

3. One-factor short-rate models

*"It will be short, the interim is mine.
And a man's life is no more than to say 'one' "*
Hamlet, V.2

3.1 Introduction and Guided Tour

The theory of interest-rate modeling was originally based on the assumption of specific one-dimensional dynamics for the instantaneous spot rate process r. Modeling directly such dynamics is very convenient since all fundamental quantities (rates and bonds) are readily defined, by no-arbitrage arguments, as the expectation of a functional of the process r. Indeed, the existence of a risk-neutral measure implies that the arbitrage-free price at time t of a contingent claim with payoff H_T at time T is given by

$$H_t = E_t\left\{D(t,T)\,H_T\right\} = E_t\left\{e^{-\int_t^T r(s)ds}H_T\right\}, \qquad (3.1)$$

with E_t denoting the time t-conditional expectation under that measure. In particular, the zero-coupon-bond price at time t for the maturity T is characterized by a unit amount of currency available at time T, so that $H_T = 1$ and we obtain

$$P(t,T) = E_t\left\{e^{-\int_t^T r(s)ds}\right\}. \qquad (3.2)$$

From this last expression it is clear that whenever we can characterize the distribution of $e^{-\int_t^T r(s)ds}$ in terms of a chosen dynamics for r, conditional on the information available at time t, we are able to compute bond prices P. As we have seen earlier in Chapter 1, from bond prices all kind of rates are available, so that indeed the whole zero-coupon curve is characterized in terms of distributional properties of r.

The pioneering approach proposed by Vasicek (1977) was based on defining the instantaneous-spot-rate dynamics under the real-world measure. His derivation of an arbitrage-free price for any interest-rate derivative followed from using the basic Black and Scholes (1973) arguments, while taking into account the non-tradable feature of interest rates.

3. One-factor short-rate models

The construction of a suitable locally-riskless portfolio, as in Black and Scholes (1973), leads to the existence of a stochastic process that only depends on the current time and instantaneous spot rate and not on the maturities of the claims constituting the portfolio. Such process, which is commonly referred to as *market price of risk*, defines a Girsanov change of measure from the real-world measure to the risk-neutral one also in case of more general dynamics than Vasicek's. Precisely, let us assume that the instantaneous spot rate evolves under the real-world measure Q_0 according to

$$dr(t) = \mu(t, r(t))dt + \sigma(t, r(t))dW^0(t),$$

where μ and σ are well-behaved functions and W^0 is a Q_0-Brownian motion. It is possible to show[1] the existence of a stochastic process λ such that if

$$dP(t,T) = \mu^T(t, r(t))dt + \sigma^T(t, r(t))dW^0(t),$$

then

$$\frac{\mu^T(t, r(t)) - r(t)P(t,T)}{\sigma^T(t, r(t))} = \lambda(t)$$

for each maturity T, with λ that may depend on r but not on T. Moreover, there exists a measure Q that is equivalent to Q_0 and is defined by the Radon-Nikodym derivative

$$\left.\frac{dQ}{dQ_0}\right|_{\mathcal{F}_t} = \exp\left(-\frac{1}{2}\int_0^t \lambda^2(s)ds - \int_0^t \lambda(s)dW^0(s)\right),$$

where $\mathcal{F}_t$ is the σ-field generated by r up to time t. As a consequence, the process r evolves under Q according to

$$dr(t) = [\mu(t, r(t)) - \lambda(t)\sigma(t, r(t))]dt + \sigma(t, r(t))dW(t),$$

where $W(t) = W^0(t) + \int_0^t \lambda(s)ds$ is a Brownian motion under Q.[2]

Let us comment briefly on this setup. The above equation for dP actually expresses the bond-price dynamics in terms of the short rate r. It expresses how the bond price P evolves over time. Now recall that r is the instantaneous-return rate of a risk-free investment, so that the difference $\mu - r$ represents a difference in returns. It tells us how much better we are doing with respect to the risk-free case, i.e. with respect to putting our money in a riskless bank account. When we divide this quantity by σ^T, we are dividing by the amount of risk we are subject to, as measured by the bond-price volatility σ^T. This is why λ is referred to as "market price of risk". An alternative term could be "excess return with respect to a risk-free investment per unit

---
[1] See for instance Björk (1997).
[2] The Radon-Nikodym derivative and the Girsanov change of measure are briefly reviewed in Appendix A. For a formal treatment see Musiela and Rutkowski (1998).

of risk". The crucial observation is that in order to specify completely the model, we have to provide λ. In effect, the market price of risk λ connects the real-world measure to the risk-neutral measure as the main ingredient in the mathematical object dQ/dQ_0 expressing the connection between these two "worlds". The way of moving from one world to the other is characterized by our choice of λ. However, if we are just concerned with the pricing of (interest-rate) derivatives, we can directly model the rate dynamics under the measure Q, so that λ will be implicit in our dynamics. We put ourselves in the world Q and we do not bother about the way of moving to the world Q_0. Then we would be in troubles only if we needed to move under the objective measure, but for pricing derivatives, the objective measure is not necessary, so that we can safely ignore it. Indeed, the value of the model parameters under the risk-neutral measure Q is what really matters in the pricing procedure, given also that the zero-coupon bonds are themselves derivatives under the above framework. We have then decided to present all the models we consider in this chapter under the risk-neutral measure, even when their original formulation was under the measure Q_0. We will hint at the relationship between the two measures only occasionally, and will explore the interaction of the dynamics under the two different measures in the Vasicek case as an illustration.

We start the chapter by introducing, in chronological order, some classical short-rate models: the Vasicek (1977) model, the Dothan (1978) model and the Cox, Ingersoll and Ross (1985) model followed by what we refer to as the Exponential-Vasicek model. These are all endogenous term-structure models, meaning that the current term structure of rates is an output rather than an input of the model. Their introduction is justified both for historical reasons and for easing the exposition of the more general models that follow. For example, the Vasicek model will be defined, under the risk-neutral measure Q, by the dynamics

$$dr(t) = k[\theta - r(t)]dt + \sigma dW(t), \quad r(0) = r_0 .$$

This dynamics has some peculiarities that make the model attractive. The equation is linear and can be solved explicitly, the distribution of the short rate is Gaussian, and both the expressions and the distributions of several useful quantities related to the interest-rate world are easily obtainable. Besides, the endogenous nature of the model is now clear. Since the bond price $P(t,T) = E_t \left\{ e^{-\int_t^T r(s)ds} \right\}$ can be computed as a simple expression depending on k, θ, σ and $r(t)$, once the function $T \mapsto P(t,T; k, \theta, \sigma, r(t))$ is known, we know the whole interest-rate curve at time t. This means that, if $t = 0$ is the initial time, the initial interest rate curve is an output of the model, depending on the parameters k, θ, σ in the dynamics (and on the initial condition r_0). However, this model features also some drawbacks. For example, rates can assume negative values with positive probability. What we are actually trying to point out with this initial hint at the Vasicek model is that the choice of a particular dynamics has several important consequences, which must be kept

3. One-factor short-rate models

in mind when designing or choosing a particular short-rate model. A typical comparison is for example with the Cox Ingersoll Ross (CIR) model. Assume we take as dynamics of r the following square root process:

$$dr(t) = k[\theta - r(t)]dt + \sigma\sqrt{r(t)}dW(t), \quad r(0) = r_0 > 0.$$

For the parameters k, θ and σ ranging in a reasonable region, this model implies positive interest rates, and the instantaneous rate is characterized by a noncentral chi-squared distribution. Moreover, this model maintains a certain degree of analytical tractability. However, the model is less tractable than the Vasicek model, especially as far as the extension to the multifactor case with correlation is concerned (see the following chapter). Therefore, the CIR dynamics has both some advantages and disadvantages with respect to the Vasicek model. In particular, when choosing a model, one should pose the following questions:

- Does the dynamics imply positive rates, i.e., $r(t) > 0$ a.s. for each t?
- What distribution does the dynamics imply for the short rate r? Is it, for instance, a fat-tailed distribution?
- Are bond prices $P(t,T) = E_t\left\{e^{-\int_t^T r(s)ds}\right\}$ (and therefore spot rates, forward rates and swap rates) explicitly computable from the dynamics?
- Are bond-option (and cap, floor, swaption) prices explicitly computable form the dynamics?
- Is the model mean reverting, in the sense that the expected value of the short rate tends to a constant value as time grows towards infinity, while its variance does not explode?
- How do the volatility structures implied by the model look like?
- Does the model allow for explicit short-rate dynamics under the forward measures?
- How suited is the model for Monte Carlo simulation?
- How suited is the model for building recombining lattices?
- Does the chosen dynamics allow for historical estimation techniques to be used for parameter estimation purposes?

These points are essential for the understanding of the theoretical and practical implications of any interest rate model. In this chapter, therefore, we will try to give an answer to the questions above for each considered short-rate model. Of course, the richness of details will vary according to the importance and practical usefulness of the model.

A classic problem with the above models is their endogenous nature. If we have the initial zero-coupon bond curve $T \mapsto P^M(0,T)$ from the market, and we wish our model to incorporate this curve, we need forcing the model parameters to produce a model curve as close as possible to the market curve. For example, again in the Vasicek case, we need to run an optimization to find the values of k, θ and σ such that the model initial curve $T \mapsto P(0,T; k, \theta, \sigma, r(0))$ is as close as possible to the market curve

$T \mapsto P^M(0,T)$. Although the values $P^M(0,T)$ are actually observed only at a finite number of maturities $P^M(0,T_i)$, three parameters are not enough to reproduce satisfactorily a given term structure. Moreover, some shapes of the zero-coupon curve $T \mapsto L^M(0,T)$ (like an inverted shape) can never be obtained with the Vasicek model, no matter the values of the parameters in the dynamics that are chosen.

The point of this digression is making clear that these models are quite hopeless: they cannot reproduce satisfactorily the initial yield curve, and so speaking of volatility structures and realism in other respects becomes partly pointless.

To improve this situation, exogenous term structure models are usually considered. Such models are built by suitably modifying the above endogenous models. The basic strategy that is used to transform an endogenous model into an exogenous model is the inclusion of "time-varying" parameters. Typically, in the Vasicek case, one does the following:

$$dr(t) = k[\theta - r(t)]dt + \sigma dW(t) \longrightarrow dr(t) = k[\vartheta(t) - r(t)]dt + \sigma dW(t) \ .$$

Now the function of time $\vartheta(t)$ can be defined in terms of the market curve $T \mapsto L^M(0,T)$ in such a way that the model reproduces exactly the curve itself at time 0.

In the chapter we then consider the description of some major exogenous term-structure models, i.e., models where the current term structure of rates is exogenously given. We analyze: the Hull and White (1990) extended Vasicek model, possible extensions of the Cox, Ingersoll and Ross (1985) model, the Black and Karasinski (1991) model and some humped-volatility short-rate models. Finally, we show how to extend a general time-homogeneous model so as to exactly reproduce the initial term structure of rates, with a special focus on the extensions of the Cox, Ingersoll and Ross (1985) model, the Dothan (1978) model and the Exponential-Vasicek model.

We investigate the analytical features of each model, providing analytical formulas for zero-coupon bonds, options on zero-coupon bonds, and hence caps and floors, whenever they exist. When possible, we also explicitly write the short-rate dynamics under the forward-adjusted measure and, for few selected models, we illustrate how to build an approximating trinomial tree. When just the price of a European call is provided, the price of the corresponding put can be obtained through the put-call parity for bond options. Indeed, if the options have maturity T, strike K and are written on a zero coupon bond maturing at time τ, their prices at time t satisfy

$$\mathbf{ZBC}(t,T,\tau,K) + KP(t,T) = \mathbf{ZBP}(t,T,\tau,K) + P(t,\tau), \quad (3.3)$$

so that

$$\mathbf{ZBP}(t,T,\tau,K) = \mathbf{ZBC}(t,T,\tau,K) - P(t,\tau) + KP(t,T). \quad (3.4)$$

48   3. One-factor short-rate models

We then devote a section to the analytical pricing of coupon bearing bond options, and hence European swaptions, and we show how to price path-dependent derivatives through a Monte Carlo procedure.

We conclude the chapter by reporting some empirical results concerning the Black and Karasinski (1991) model and the above extensions of the Cox, Ingersoll and Ross (1985) model and the Exponential-Vasicek model. We first show and comment the cap volatility curves and swaption volatility surfaces that are implied by these models. We then consider a specific example of the models calibration to real market data and compare the resulting fitting qualities.

Throughout the chapter, we assume that the term structure of discount factors that is currently observed in the market is given by the sufficiently-smooth function $t \mapsto P^M(0,t)$. We then denote by $f^M(0,t)$ the market instantaneous forward rates at time 0 for a maturity t as associated with the bond prices $\{P^M(0,t) : t \geq 0\}$, i.e.,

$$f^M(0,t) = -\frac{\partial \ln P^M(0,t)}{\partial t}.$$

The relevant properties of the instantaneous short rate models we will analyze in this chapter are summarized in Table 3.1, where V, CIR, D, EV, HW, BK, MM, CIR++, EVV stand respectively for the Vasicek (1977) model, the Cox, Ingersoll and Ross (1985) model, the Dothan (1978) model, the Exponential Vasicek model, the Hull and White (1990) model, the Black and Karasinski (1991) model, the Mercurio and Moraleda (2000) model, the CIR++ model and the Extended Exponential Vasicek model; N and Y stand respectively for "No" and "Yes", whereas Y* means that rates are positive under suitable conditions for the deterministic function φ; $\mathcal{N}$, $L\mathcal{N}$, $NC\chi^2$, $SNC\chi^2$, $SL\mathcal{N}$ denote respectively normal, lognormal, noncentral χ^2, shifted noncentral χ^2 and shifted lognormal distributions; AB(O) stands for Analytical Bond (Option) price.

3.2 Classical Time-Homogeneous Short-Rate Models

Has fate become so imaginatively bankrupt that I am now doomed to naught but tedious recapitulation of the past?
Adam Warlock in "Rune" 1, 1995, Malibu Comics.

The first instantaneous short rate models being proposed in the financial literature were time-homogeneous, meaning that the assumed short rate dynamics depended only on constant coefficients. The success of models like that of Vasicek (1977) and that of Cox, Ingersoll and Ross (1985) was mainly due to their possibility of pricing analytically bonds and bond options. However, as observed in the introduction, these models produce an endogenous

3.2 Classical Time-Homogeneous Short-Rate Models

Model	Dynamics	$r>0$	$r \sim$	AB	AO
V	$dr_t = k[\theta - r_t]dt + \sigma dW_t$	N	$\mathcal{N}$	Y	Y
CIR	$dr_t = k[\theta - r_t]dt + \sigma\sqrt{r_t}dW_t$	Y	NCχ^2	Y	Y
D	$dr_t = ar_t dt + \sigma r_t dW_t$	Y	L$\mathcal{N}$	Y	N
EV	$dr_t = r_t[\eta - a\ln r_t]dt + \sigma r_t dW_t$	Y	L$\mathcal{N}$	N	N
HW	$dr_t = k[\theta_t - r_t]dt + \sigma dW_t$	N	$\mathcal{N}$	Y	Y
BK	$dr_t = r_t[\eta_t - a\ln r_t]dt + \sigma r_t dW_t$	Y	L$\mathcal{N}$	N	N
MM	$dr_t = r_t\left[\eta_t - \left(\lambda - \frac{\gamma}{1+\gamma t}\right)\ln r_t\right]dt + \sigma r_t dW_t$	Y	L$\mathcal{N}$	N	N
CIR++	$r_t = x_t + \varphi_t,\ dx_t = k[\theta - x_t]dt + \sigma\sqrt{x_t}dW_t$	Y*	SNCχ^2	Y	Y
EEV	$r_t = x_t + \varphi_t,\ dx_t = x_t[\eta - a\ln x_t]dt + \sigma x_t dW_t$	Y*	SL$\mathcal{N}$	N	N

Table 3.1. Summary of instantaneous short rate models.

term structure of interest rates,[3] in that the initial term structure of (e.g. continuously-compounded) rates $T \mapsto R(0,T) = -(\ln P(0,T))/T$ does not necessarily match that observed in the market, no matter how the model parameters are chosen.

Moreover, as we pointed out earlier, the small number of model parameters prevents a satisfactory calibration to market data and even the zero-coupon curve is quite likely to be badly reproduced, also because some typical shapes, like that of an inverted yield curve, may not be reproduced by the model.

In this section we present three classical time-homogeneous short-rate models, namely the Vasicek (1977), the Dothan (1978) and the Cox, Ingersoll and Ross (1985) models, and we finally introduce the Exponential-Vasicek model. As already mentioned, these models are described not only for their historical importance but also for letting us treat in a clearer way the extensions we shall illustrate in the sequel.

As to the analytical tractability of the first three models, we want to remark the following. The original derivation of the explicit formulas for

---
[3] This is the reason why they have been also referred to as "endogenous term structure models".

bond prices was based on solving the PDE that, by no-arbitrage arguments, must be satisfied by the bond price process. In the presence of a Gaussian distribution, however, we can price bonds also by directly computing the expectation (3.2). The derivation of explicit formulas for bond options in the first and third model relied instead on a suitable change of the underlying probability measure. Indeed, once the distribution of the instantaneous short rate is known under the desired forward measure, any payoff can be priced by calculating the expectation (2.21).

Finally, in the Vasicek case we will study a few facts concerning possible uses of the dynamics under the objective measure, in order to give a feeling for the kind of considerations involved in combining these pricing and hedging short-rate models with historical data. We develop this theme only for the Vasicek model because in the rest of the book we will not use historical estimation techniques.

3.2.1 The Vasicek Model

Vasicek (1977) assumed that the instantaneous spot rate under the real-world measure evolves as an Ornstein-Uhlenbeck process with constant coefficients. For a suitable choice of the market price of risk (more on this later, see equation (3.11)), this is equivalent to assume that r follows an Ornstein-Uhlenbeck process with constant coefficients under the risk-neutral measure as well, that is

$$dr(t) = k[\theta - r(t)]dt + \sigma dW(t), \quad r(0) = r_0, \tag{3.5}$$

where r_0, k, θ and σ are positive constants.

Integrating equation (3.5), we obtain, for each $s \leq t$,

$$r(t) = r(s)e^{-k(t-s)} + \theta\left(1 - e^{-k(t-s)}\right) + \sigma\int_s^t e^{-k(t-u)}dW(u), \tag{3.6}$$

so that $r(t)$ conditional on $\mathcal{F}_s$ is normally distributed with mean and variance given respectively by

$$\begin{aligned} E\{r(t)|\mathcal{F}_s\} &= r(s)e^{-k(t-s)} + \theta\left(1 - e^{-k(t-s)}\right) \\ \text{Var}\{r(t)|\mathcal{F}_s\} &= \frac{\sigma^2}{2k}\left[1 - e^{-2k(t-s)}\right]. \end{aligned} \tag{3.7}$$

This implies that, for each time t, the rate $r(t)$ can be negative with positive probability. The possibility of negative rates is indeed a major drawback of the Vasicek model. However, the analytical tractability that is implied by a Gaussian density is hardly achieved when assuming other distributions for the process r.

As a consequence of (3.7), the short rate r is mean reverting, since the expected rate tends, for t going to infinity, to the value θ. The fact that

3.2 Classical Time-Homogeneous Short-Rate Models

θ can be regarded as a long term average rate could be also inferred from the dynamics (3.5) itself. Notice, indeed, that the drift of the process r is positive whenever the short rate is below θ and negative otherwise, so that r is pushed, at every time, to be closer on average to the level θ.

The price of a pure-discount bond can be derived by computing the expectation (3.2). We obtain

$$P(t,T) = A(t,T)e^{-B(t,T)r(t)}, \qquad (3.8)$$

where

$$A(t,T) = \exp\left\{\left(\theta - \frac{\sigma^2}{2k^2}\right)[B(t,T) - T + t] - \frac{\sigma^2}{4k}B(t,T)^2\right\}$$

$$B(t,T) = \frac{1}{k}\left[1 - e^{-k(T-t)}\right].$$

If we fix a maturity T, the change-of-numeraire toolkit developed in Section (2.3), and formula (2.12) in particular (with $S_t = B(t)$, the bank-account numeraire, $U_t = P(t,T)$, the T-bond numeraire, and $X_t = r_t$) imply that under the T-forward measure Q^T

$$dr(t) = [k\theta - B(t,T)\sigma^2 - kr(t)]dt + \sigma dW^T(t), \qquad (3.9)$$

where the Q^T-Brownian motion W^T is defined by

$$dW^T(t) = dW(t) + \sigma B(t,T)dt,$$

so that, for $s \le t \le T$,

$$r(t) = r(s)e^{-k(t-s)} + M^T(s,t) + \sigma\int_s^t e^{-k(t-u)}dW^T(u),$$

with

$$M^T(s,t) = \left(\theta - \frac{\sigma^2}{k^2}\right)\left(1 - e^{-k(t-s)}\right) + \frac{\sigma^2}{2k^2}\left[e^{-k(T-t)} - e^{-k(T+t-2s)}\right].$$

Therefore, under Q^T, the transition distribution of $r(t)$ conditional on $\mathcal{F}_s$ is still normal with mean and variance given by

$$E^T\{r(t)|\mathcal{F}_s\} = r(s)e^{-k(t-s)} + M^T(s,t)$$

$$\text{Var}^T\{r(t)|\mathcal{F}_s\} = \frac{\sigma^2}{2k}\left[1 - e^{-2k(t-s)}\right].$$

The price at time t of a European option with strike X, maturity T and written on a pure discount bond maturing at time S has been derived by Jamshidian (1989). Using the known distribution of $r(T)$ under Q^T, the calculation of the expectation (3.1), where $H_T = (P(T,S)-X)^+$, yields, through the general formula (B.2),

52   3. One-factor short-rate models

$$\mathbf{ZBO}(t,T,S,X) = \omega \left[P(t,S)\Phi(\omega h) - XP(t,T)\Phi(\omega(h-\sigma_p)) \right], \quad (3.10)$$

where $\omega = 1$ for a call and $\omega = -1$ for a put, $\Phi(\cdot)$ denotes the standard normal cumulative distribution function, and

$$\sigma_p = \sigma \sqrt{\frac{1-e^{-2k(T-t)}}{2k}} B(T,S),$$

$$h = \frac{1}{\sigma_p} \ln \frac{P(t,S)}{P(t,T)X} + \frac{\sigma_p}{2}.$$

Objective measure dynamics and historical estimation We can consider the objective measure dynamics of the Vasicek model as a process of the form

$$dr(t) = [k\,\theta - (k+\lambda\,\sigma)r(t)]dt + \sigma dW^0(t), \quad r(0) = r_0\,, \quad (3.11)$$

where λ is a new parameter, contributing to the market price of risk. Compare this Q_0 dynamics to the Q-dynamics (3.5). Notice that for $\lambda = 0$ the two dynamics coincide, i.e. there is no difference between the risk neutral world and the objective world. More generally, the above Q_0-dynamics is expressed again as a linear Gaussian stochastic differential equation, although it depends on the new parameter λ. This is a tacit assumption on the form of the market price of risk process. Indeed, requiring that the dynamics be of the same nature under the two measures, imposes a Girsanov change of measure of the following kind to go from (3.5) to (3.11) :

$$\left.\frac{dQ}{dQ_0}\right|_{\mathcal{F}_t} = \exp\left(-\frac{1}{2}\int_0^t \lambda^2\,r(s)^2 ds + \int_0^t \lambda\,r(s)dW^0(s)\right)$$

(see Appendix A for an introduction to Girsanov's theorem).

In other terms, we are assuming that the market price of risk process $\lambda(t)$ has the functional form

$$\lambda(t) = \lambda\,r(t)$$

in the short rate. Of course, in general there is no reason why this should be the case. However, under this choice we obtain a short rate process that is tractable under both measures.[4]

It is clear why tractability under the risk-neutral measure is a desirable property: claims are priced under that measure, so that the possibility to compute expectations in a tractable way with the Q-dynamics (3.5) is important. Yet, why do we find it desirable to have a tractable dynamics under Q_0 too? In order to answer this question, suppose for a moment that we are provided with a series $r_0, r_1, r_2, \ldots, r_n$ of daily observations of a proxy of $r(t)$ (say a monthly rate, $r(t) \approx L(t, t+1m)$), and that we wish to incorporate

---

[4] Indeed, the market price of risk under the Vasicek model is usually chosen to be constant, i.e., $\lambda(t) = \lambda$. However, this is just another possible formulation.

information from this series in our model. We can estimate the model parameters on the basis of this daily series of data. However, data are collected in the real world, and their statistical properties characterize the distribution of our interest-rate process $r(t)$ under the objective measure Q_0. Therefore, what is to be estimated from historical observations is the Q_0 dynamics. The estimation technique can provide us with estimates for the objective parameters k, λ, θ and σ, or more precisely for combinations thereof.

On the other hand, prices are computed through expectations under the risk-neutral measure. When we observe prices, we observe expectations under the measure Q. Therefore, when we calibrate the model to derivative prices we need to use the Q dynamics (3.5), thus finding the parameters k, θ and σ involved in the Q-dynamics.

We could then combine the two approaches. For example, since the diffusion coefficient is the same under the two measures, we might estimate σ from historical data through a maximum-likelihood estimator, while finding k and θ through calibration to market prices. However, this procedure may be necessary when very few prices are available. Otherwise, it might be used to deduce historically a σ which can be used as initial guess when trying to find the three parameters that match the market prices of a given set of instruments.

We conclude the section by presenting the maximum-likelihood estimator for the Vasicek model. Rewrite the dynamics (3.11) as

$$dr(t) = [b - ar(t)]dt + \sigma dW^0(t), \tag{3.12}$$

with b and a suitable constants. As usual, by integration we obtain, between two any instants s and t,

$$r(t) = r(s)e^{-a(t-s)} + \frac{b}{a}(1 - e^{-a(t-s)}) + \sigma \int_s^t e^{-a(t-u)} dW^0(u). \tag{3.13}$$

As noticed earlier, conditional on $\mathcal{F}_s$, the variable $r(t)$ is normally distributed with mean $r(s)e^{-a(t-s)} + \frac{b}{a}(1 - e^{-a(t-s)})$ and variance $\frac{\sigma^2}{2a}(1 - e^{-2a(t-s)})$.

It is natural to estimate the following functions of the parameters: $\beta := b/a$, $\alpha := e^{-a\delta}$ and $V^2 = \frac{\sigma^2}{2a}(1 - e^{-2a\delta})$, where δ denotes the time-step of the observed proxies $r_0, r_1, \ldots, r_n$ of r (typically $\delta = 1$ day). The maximum likelihood estimators for α, β and V^2 are easily derived as

$$\widehat{\alpha} = \frac{n \sum_{i=1}^n r_i r_{i-1} - \sum_{i=1}^n r_i \sum_{i=1}^n r_{i-1}}{n \sum_{i=1}^n r_i^2 - \left(\sum_{i=1}^n r_{i-1}\right)^2}, \tag{3.14}$$

$$\widehat{\beta} = \frac{\sum_{i=1}^n [r_i - \widehat{\alpha} r_{i-1}]}{n(1 - \widehat{\alpha})}, \tag{3.15}$$

$$\widehat{V^2} = \frac{1}{n} \sum_{i=1}^n \left[r_i - \widehat{\alpha} r_{i-1} - \widehat{\beta}(1 - \widehat{\alpha})\right]^2. \tag{3.16}$$

3.2.2 The Dothan Model

In his original paper, Dothan started from a driftless geometric Brownian motion as short-rate process under the objective probability measure Q_0:

$$dr(t) = \sigma r(t) dW^0(t), \quad r(0) = r_0,$$

where r_0 and σ are positive constants.

Subsequently, Dothan introduced a constant market price of risk, which is equivalent to directly assuming a risk-neutral dynamics of type

$$dr(t) = ar(t)dt + \sigma r(t)dW(t), \tag{3.17}$$

where a is a real constant, thus yielding a continuous-time version of the Rendleman and Bartter (1980) model.

The dynamics (3.17) are easily integrated as follows

$$r(t) = r(s) \exp\left\{\left(a - \frac{1}{2}\sigma^2\right)(t-s) + \sigma(W(t) - W(s))\right\}, \tag{3.18}$$

for $s \leq t$. Hence, $r(t)$ conditional on $\mathcal{F}_s$ is lognormally distributed with mean and variance given by

$$\begin{aligned} E\{r(t)|\mathcal{F}_s\} &= r(s)e^{a(t-s)}, \\ \text{Var}\{r(t)|\mathcal{F}_s\} &= r^2(s)e^{2a(t-s)}\left(e^{\sigma^2(t-s)} - 1\right). \end{aligned} \tag{3.19}$$

The lognormal distribution implies that $r(t)$ is always positive for each t, so that a main drawback of the Vasicek (1977) model is here addressed. However, as we can easily infer from (3.19), the process (3.17) is mean reverting if and only if $a < 0$ with the mean-reversion level that must be necessarily equal to zero.[5] This is a restriction on mean reversion that will be addressed by the exponential Vasicek model.

The Dothan (1978) model is the only lognormal short rate model in the literature with analytical formulas for pure discount bonds. This is the key feature that led us to consider such model and the relative extension we shall propose in a later section.

The zero-coupon bond price derived by Dothan is given by

---

[5] We should stress that mean reversion under Q does not necessarily imply mean reversion under Q_0. However, we can assume that the change of measure does not affect the asymptotic behavior of the process r.

3.2 Classical Time-Homogeneous Short-Rate Models

$$P(t,T) = \frac{\bar{r}^p}{\pi^2} \int_0^\infty \sin(2\sqrt{\bar{r}}\sinh y) \int_0^\infty f(z)\sin(yz)dzdy + \frac{2}{\Gamma(2p)}\bar{r}^p K_{2p}(2\sqrt{\bar{r}}) \tag{3.20}$$

where

$$f(z) = \exp\left[\frac{-\sigma^2(4p^2+z^2)(T-t)}{8}\right] z \left|\Gamma\left(-p+i\frac{z}{2}\right)\right|^2 \cosh\frac{\pi z}{2},$$

$$\bar{r} = \frac{2r(t)}{\sigma^2},$$

$$p = \frac{1}{2} - a,$$

and K_q denotes the modified Bessel function of the second kind of order q.

Concerning the model analytical tractability we need, however, to remark the following. Though somehow explicit, formula (3.20) is rather complex since it depends on two integrals of functions involving hyperbolic sines and cosines. A double numerical integration is needed so that the advantage of having an "explicit" formula is dramatically reduced. In particular, as far as computational issues are concerned, implementing an approximating tree for the process r may be conceptually easier and not necessarily more time consuming.

The dynamics of the process r under any T-forward measure can be derived by applying formula (2.12).

No analytical formula for an option on a zero-coupon bond is available in this model.

Finally, we need to remark a problem concerning the Dothan model, and lognormal models in general.

Explosion of the bank account for lognormal short-rate models.
Assume we are at time 0 and we put one unit of currency in the bank account, for a small time Δt. We know that the expected value of our position at time Δt will be

$$E_0 B(\Delta t) = E_0\left\{e^{\int_0^{\Delta t} r(s)ds}\right\} \approx \ldots$$

Now if Δt is small, we can approximate the integral as follows:

$$\approx E_0\left\{e^{\Delta t\,[r(0)+r(\Delta t)]/2}\right\}.$$

Given that the short rate $r(\Delta t)$ is lognormally distributed, we face an expectation of the type

$$E_0\left\{\exp(\exp(Y))\right\}$$

where Y is normally distributed. It is easy to see that such an expectation is infinite, so that we conclude

$$E_0\left\{B(\Delta t)\right\} = E_0\left\{e^{\int_0^{\Delta t} r(s)ds}\right\} = \infty.$$

This means that in an arbitrarily small time we can make infinite money on average starting from one unit of currency. This drawback is common to all models where r is lognormally distributed. The Black Karasinski model to be introduced later on, the Dothan model, the EV model and their extensions that will be explored in the following, all share this problem. As a consequence, the price of a Eurodollar future is also infinite for all these models, too. However, this explosion problem is partially overcome when using an approximating tree, because one deals with a finite number of states, and hence with finite expectations. Since these models are always applied via trees in practice, this drawback's impact is usually less dramatic than one would expect in the first place. The problem of explosion in these models where the *continuously-compounded* instantaneous rate r is modeled through a lognormal process has been studied by Sandmann and Sondermann (1997), who observe that the problem of explosion can be avoided by modeling rates with a *strictly-positive compounding period* to be lognormal instead.

3.2.3 The Cox, Ingersoll and Ross (CIR) Model

The general equilibrium approach developed by Cox, Ingersoll and Ross (1985) led to the introduction of a "square-root" term in the diffusion coefficient of the instantaneous short-rate dynamics proposed by Vasicek (1977). The resulting model has been a benchmark for many years because of its analytical tractability and the fact that, contrary to the Vasicek (1977) model, the instantaneous short rate is always positive.

The model formulation under the risk-neutral measure Q is

$$dr(t) = k(\theta - r(t))dt + \sigma\sqrt{r(t)}\,dW(t), \quad r(0) = r_0, \qquad (3.21)$$

with r_0, k, θ, σ positive constants. The condition

$$2k\theta > \sigma^2$$

has to be imposed to ensure that the origin is inaccessible to the process (3.21), so that we can grant that r remains positive.

We now consider a little digression on a tractable form for the market price of risk in this model. If we need to model the objective measure dynamics Q_0 of the model, it is a good idea to adopt the following formulation:

$$dr(t) = [k\theta - (k + \lambda\sigma)r(t)]dt + \sigma\sqrt{r(t)}\,dW^0(t), \quad r(0) = r_0. \qquad (3.22)$$

Notice that in moving from Q to Q_0 the drift has been modified exactly as in the Vasicek case (3.11), and exactly for the same reason: preserving the same structure under the two measures. While in the Vasicek case the change of measure was designed so as to maintain a linear dynamics, here it has been designed so as to maintain a square-root-process structure. Since the diffusion

3.2 Classical Time-Homogeneous Short-Rate Models

coefficient is different, the change of measure is also different. In particular, we have

$$\left.\frac{dQ}{dQ_0}\right|_{\mathcal{F}_t} = \exp\left(-\frac{1}{2}\int_0^t \lambda^2 r(s)ds + \int_0^t \lambda\sqrt{r(s)}dW^0(s)\right).$$

In other terms, we are assuming the market price of risk process $\lambda(t)$ to be of the particular functional form

$$\lambda(t) = \lambda\sqrt{r(t)}$$

in the short rate. Of course, in general there is no reason why this should be the case. However, under this choice we obtain a short-rate process which is tractable under both measures. As for the Vasicek case, tractability under the objective measure can be helpful for historical-estimation purposes.

Let us now move back to the risk-neutral measure Q. The process r features a noncentral chi-squared distribution. Precisely, denoting by p_Y the density function of the random variable Y,

$$p_{r(t)}(x) = p_{\chi^2(v,\,\lambda_t)/c_t}(x) = c_t p_{\chi^2(v,\,\lambda_t)}(c_t x),$$

$$c_t = \frac{4k}{\sigma^2(1-\exp(-kt))},$$

$$v = 4k\theta/\sigma^2,$$

$$\lambda_t = c_t r_0 \exp(-kt),$$

where the noncentral chi-squared distribution function $\chi^2(\cdot, v, \lambda)$ with v degrees of freedom and non-centrality parameter λ has density

$$p_{\chi^2(v,\,\lambda)}(z) = \sum_{i=0}^{\infty} \frac{e^{-\lambda/2}(\lambda/2)^i}{i!} p_{\Gamma(i+v/2,\,1/2)}(z),$$

$$p_{\Gamma(i+v/2,\,1/2)}(z) = \frac{(1/2)^{i+v/2}}{\Gamma(i+v/2)} z^{i-1+v/2} e^{-z/2} = p_{\chi^2(v+2i)}(z),$$

with $p_{\chi^2(v+2i)}(z)$ denoting the density of a (central) chi-squared distribution function with $v+2i$ degrees of freedom.[6]

The mean and the variance of $r(t)$ conditional on $\mathcal{F}_s$ are given by

$$E\{r(t)|\mathcal{F}_s\} = r(s)e^{-k(t-s)} + \theta\left(1 - e^{-k(t-s)}\right),$$

$$\text{Var}\{r(t)|\mathcal{F}_s\} = r(s)\frac{\sigma^2}{k}\left(e^{-k(t-s)} - e^{-2k(t-s)}\right) + \theta\frac{\sigma^2}{2k}\left(1 - e^{-k(t-s)}\right)^2.$$
(3.23)

---

[6] A useful identity concerning densities of χ^2 distributions is

$$p_{\chi^2(v,\,\lambda)}(bz) = \exp\left(\tfrac{1}{2}(1-b)(z-\lambda)\right) b^{v/2-1} p_{\chi^2(v,\,b\lambda)}(z).$$

3. One-factor short-rate models

The price at time t of a zero-coupon bond with maturity T is

$$P(t,T) = A(t,T)e^{-B(t,T)r(t)}, \qquad (3.24)$$

where

$$A(t,T) = \left[\frac{2h\exp\{(k+h)(T-t)/2\}}{2h + (k+h)(\exp\{(T-t)h\} - 1)}\right]^{2k\theta/\sigma^2},$$

$$B(t,T) = \frac{2(\exp\{(T-t)h\} - 1)}{2h + (k+h)(\exp\{(T-t)h\} - 1)}, \qquad (3.25)$$

$$h = \sqrt{k^2 + 2\sigma^2}.$$

Under the risk-neutral measure Q, the bond price dynamics can be easily obtained via Ito's formula:

$$dP(t,T) = r(t)P(t,T)dt - B(t,T)P(t,T)\sigma\sqrt{r(t)}dW(t).$$

By inverting the bond-price formula, thus deriving r from P, we obtain

$$dP(t,T)$$
$$= \frac{1}{B(t,T)}\ln\left[\frac{A(t,T)}{P(t,T)}\right]P(t,T)dt - \sigma P(t,T)\sqrt{B(t,T)\ln\left[\frac{A(t,T)}{P(t,T)}\right]}dW(t)$$

which is better readable as

$$d\ln P(t,T) = \left(\frac{1}{B(t,T)} - \frac{1}{2}\sigma^2 B(t,T)\right)[\ln A(t,T) - \ln P(t,T)]dt$$
$$- \sigma\sqrt{B(t,T)}[\ln A(t,T) - \ln P(t,T)]dW(t).$$

We notice that the bond-price percentage volatility is not a deterministic function, but depends on the current level of the bond price.

The price at time t of a European call option with maturity $T > t$, strike price X, written on a zero-coupon bond maturing at $S > T$, and with the instantaneous rate at time t given by $r(t)$, is (see Cox, Ingersoll and Ross (1985))

$$\mathbf{ZBC}(t,T,S,X)$$
$$= P(t,S)\chi^2\left(2\bar{r}[\rho + \psi + B(T,S)]; \frac{4k\theta}{\sigma^2}, \frac{2\rho^2 r(t)\exp\{h(T-t)\}}{\rho + \psi + B(T,S)}\right)$$
$$- XP(t,T)\chi^2\left(2\bar{r}[\rho + \psi]; \frac{4k\theta}{\sigma^2}, \frac{2\rho^2 r(t)\exp\{h(T-t)\}}{\rho + \psi}\right)$$
$$\qquad (3.26)$$

where

3.2 Classical Time-Homogeneous Short-Rate Models

$$\rho = \rho(T-t) := \frac{2h}{\sigma^2(\exp[h(T-t)]-1)},$$

$$\psi = \frac{k+h}{\sigma^2},$$

$$\bar{r} = \bar{r}(S-T) := \frac{\ln(A(T,S)/X)}{B(T,S)}.$$

By applying formula (2.12) (with $S_t = B(t)$, the bank-account numeraire, $U_t = P(t,T)$, the T-bond numeraire, and $X_t = r_t$), we obtain that the short-rate dynamics under the T-forward measure Q^T is

$$dr(t) = [k\theta - (k + B(t,T)\sigma^2)r(t)]dt + \sigma\sqrt{r(t)}dW^T(t), \qquad (3.27)$$

where the Q^T-Brownian motion W^T is defined by

$$dW^T(t) = dW(t) + \sigma B(t,T)\sqrt{r(t)}dt.$$

It is also possible to show that, under Q^T, the distribution of the short rate $r(t)$ conditional on $r(s)$, $s \leq t \leq T$, is given by

$$\begin{aligned} p^T_{r(t)|r(s)}(x) &= p_{\chi^2(v,\delta(t,s))/q(t,s)}(x) = q(t,s)p_{\chi^2(v,\delta(t,s))}(q(t,s)x), \\ q(t,s) &= 2[\rho(t-s) + \psi + B(t,T)], \\ \delta(t,s) &= \frac{4\rho(t-s)^2 r(s)e^{h(t-s)}}{q(t,s)}. \end{aligned} \qquad (3.28)$$

This can be shown by differentiating the call-option price with respect to the strike price and by suitable decompositions.

We can also derive the forward-rate dynamics implied by the CIR short-rate dynamics. Indeed, consider the simply-compounded forward rate at time t with expiry T and maturity S, as defined by

$$F(t;T,S) = \frac{1}{\gamma(T,S)}\left[\frac{P(t,T)}{P(t,S)} - 1\right],$$

where $\gamma(T,S)$ is the year fraction between T and S. By Ito's formula it is easy to check that under the forward measure Q^S the (driftless) dynamics of $F(t;T,S)$ follows

$$\begin{aligned} &dF(t;T,S) \\ &= \sigma\frac{A(t,T)}{A(t,S)}(B(t,S) - B(t,T))\exp\{-(B(t,T) - B(t,S))r(t)\}\sqrt{r(t)}dW^S(t). \end{aligned}$$

This last equation can be easily rewritten as

$$\begin{aligned} &dF(t;T,S) \\ &= \sigma\left(F(t;T,S) + \frac{1}{\gamma(T,S)}\right) \\ &\quad \cdot\sqrt{(B(t,S) - B(t,T))\ln\left[(\gamma(T,S)F(t;T,S)+1)\frac{A(t,S)}{A(t,T)}\right]}dW^S(t). \end{aligned}$$

Notice that this is rather different from the lognormal dynamics assumed for F when pricing caps and floors with the LIBOR market model, where typically

$$dF(t;T,S) = \sigma(t)\, F(t;T,S)\, dW^S(t)$$

for a deterministic time function σ.

3.2.4 Affine Term-Structure Models

Affine term-structure models are interest-rate models where the continuously-compounded spot rate $R(t,T)$ is an affine function in the short rate $r(t)$, i.e.

$$R(t,T) = \alpha(t,T) + \beta(t,T)r(t),$$

where α and β are deterministic functions of time. If this happens, the model is said to possess an affine term structure. This relationship is always satisfied when the zero–coupon bond price can be written in the form

$$P(t,T) = A(t,T)e^{-B(t,T)r(t)},$$

since then clearly it suffices to set

$$\alpha(t,T) = -(\ln A(t,T))/(T-t), \quad \beta(t,T) = B(t,T)/(T-t).$$

Both the Vasicek and CIR models we have seen earlier are affine models, since the bond price has an expression of the above form in both cases. The Dothan model is not an affine model.

A computation that will be helpful in the following is the instantaneous absolute volatility of instantaneous forward rates in affine models. Since in general

$$f(t,T) = -\frac{\partial \ln P(t,T)}{\partial T},$$

for affine models we have

$$f(t,T) = -\frac{\partial \ln A(t,T)}{\partial T} + \frac{\partial B(t,T)}{\partial T}r(t),$$

so that clearly

$$df(t,T) = (\cdots)dt + \frac{\partial B(t,T)}{\partial T}\sigma(t,r(t))dW(t),$$

where $\sigma(t,r(t))$ is the diffusion coefficient in the short rate dynamics. It follows that the absolute volatility of the instantaneous forward rate $f(t,T)$ at time t in a short rate model with an affine term structure is

$$\sigma_f(t,T) = \frac{\partial B(t,T)}{\partial T}\sigma(t,r(t)). \tag{3.29}$$

In particular, for the Vasicek and CIR models we can obtain explicit expressions for this quantity by using the known expressions for B. We will see later on why this volatility function is important.

Given that by inspection one sees that the Vasicek and CIR models are affine models whereas Dothan is not, one may wonder whether there is a relationship between the coefficients in the risk-neutral dynamics of the short rate and affinity of the term structure in the above sense. Assume we have a general time-homogeneous risk-neutral dynamics for the short rate,

$$dr(t) = b(t, r(t))dt + \sigma(t, r(t))dW(t) \ .$$

We may wonder whether there exist conditions on b and σ such that the resulting model displays an affine term structure. The answer is simply that the coefficients b and σ^2 need be affine functions themselves (see for example Björk (1997) or Duffie (1996)). If the coefficients b and σ^2 are of the form

$$b(t, x) = \lambda(t)x + \eta(t), \quad \sigma^2(t, x) = \gamma(t)x + \delta(t)$$

for suitable deterministic time functions $\lambda, \eta, \gamma, \delta$, then the model has an affine term structure, with α and β (or A and B) above depending on the chosen functions $\lambda, \eta, \gamma, \delta$. The functions A and B can be obtained from the coefficients $\lambda, \eta, \gamma, \delta$ by solving the following differential equations:

$$\frac{\partial}{\partial t}B(t,T) + \lambda(t)B(t,T) - \tfrac{1}{2}\gamma(t)B(t,T)^2 + 1 = 0, \quad B(T,T) = 0,$$

$$\frac{\partial}{\partial t}[\ln A(t,T)] - \eta(t)B(t,T) + \tfrac{1}{2}\delta(t)B(t,T)^2 = 0, \quad A(T,T) = 1.$$

The first equation is a Riccati differential equation that, in general, needs to be solved numerically. However, in the particular cases of Vasicek ($\lambda(t) = -k, \eta(t) = k\theta, \gamma(t) = 0, \delta(t) = \sigma^2$) or CIR ($\lambda(t) = -k, \eta(t) = k\theta, \gamma(t) = \sigma^2, \delta(t) = 0$), we have that the equations are explicitly solvable, yielding the expressions for A and B we have written in the previous sections.

Therefore affinity in the coefficients translates into affinity of the term structure. The converse is also true, but in the time-homogeneous case. Precisely, it is possible to prove that if a model has an affine term structure and has time-homogeneous coefficients $b(t, x) = b(x)$ and $\sigma(t, x) = \sigma(x)$, then these coefficients are necessarily affine functions of x:

$$b(x) = \lambda x + \eta, \quad \sigma^2(x) = \gamma x + \delta \ ,$$

for suitable constants $\lambda, \eta, \gamma, \delta$. The relation between affine-term-structure models (ATS), affine-coefficients models (AC) and time-homogeneous models (TH) is visualized through the diagrams displayed in Figure 3.1.

3.2.5 The Exponential-Vasicek (EV) Model

A natural way to obtain a lognormal short-rate model that is alternative to that of Dothan (1977) is by assuming that the logarithm of r follows an

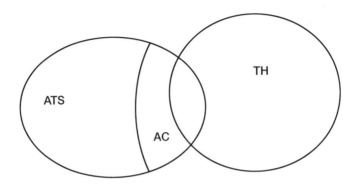

Fig. 3.1. Relation between ATS, AC and TH models.

Ornstein-Uhlenbeck process y under the risk-neutral measure Q. Precisely, let y be defined by

$$dy(t) = [\theta - ay(t)]dt + \sigma dW(t), \quad y(0) = y_0,$$

where θ, a and σ are positive constants and y_0 is a real number. Then if we set $r(t) = \exp(y(t))$, for each time t, we have the following dynamics for the short rate:

$$dr(t) = r(t)\left[\theta + \frac{\sigma^2}{2} - a \ln r(t)\right] dt + \sigma r(t) dW(t). \tag{3.30}$$

Since the instantaneous short rate is defined as the exponential of a process that is perfectly equivalent to that of Vasicek (1977), we shall refer to this model as to the Exponential-Vasicek model.

Remembering (3.6), the process r, for each $s \leq t$, is explicitly given by

$$r(t) = \exp\left\{\ln r(s)e^{-a(t-s)} + \frac{\theta}{a}\left(1 - e^{-a(t-s)}\right) + \sigma \int_s^t e^{-a(t-u)}dW(u)\right\},$$

so that $r(t)$ conditional on $\mathcal{F}_s$ is lognormally distributed with first and second moments given respectively by

$$E_s\{r(t)\} = \exp\left\{\ln r(s)e^{-a(t-s)} + \frac{\theta}{a}\left(1 - e^{-a(t-s)}\right) + \frac{\sigma^2}{4a}\left[1 - e^{-2a(t-s)}\right]\right\}$$

$$E_s\{r^2(t)\} = \exp\left\{2\ln r(s)e^{-a(t-s)} + 2\frac{\theta}{a}\left(1 - e^{-a(t-s)}\right) + \frac{\sigma^2}{a}\left[1 - e^{-2a(t-s)}\right]\right\}.$$

Therefore, contrary to the Dothan (1978) process (3.17), this r is always mean reverting since

$$\lim_{t\to\infty} E\{r(t)|\mathcal{F}_s\} = \exp\left(\frac{\theta}{a} + \frac{\sigma^2}{4a}\right).$$

Notice that also the variance converges to a finite value since

$$\lim_{t\to\infty} \text{Var}\{r(t)|\mathcal{F}_s\} = \exp\left(\frac{2\theta}{a} + \frac{\sigma^2}{2a}\right)\left[\exp\left(\frac{\sigma^2}{2a}\right) - 1\right]. \quad (3.31)$$

The Exponential-Vasicek model does not imply explicit formulas for either pure-discount bonds or options on them. However, when proposing in a later section an extension of this model that exactly fits the current term structure of rates, we will show how to implement fast numerical procedures that make the extension quite appealing in some practical market situations.

The EV model is not an affine term-structure model, as is clear from the criterion given in Section 3.2.4. Finally, since it implies a lognormal distribution for r, the EV model shares the explosion problem that was pointed out in the Dothan case.

3.3 The Hull-White Extended Vasicek Model

The poor fitting of the initial term structure of interest rates implied by the Vasicek model has been addressed by Hull and White in their 1990 and subsequent papers. Ho and Lee (1986) have been the first to propose an exogenous term-structure model as opposed to models that endogenously produce the current term structure of rates. However, their model was based on the assumption of a binomial tree governing the evolution of the entire term structure of rates, and even its continuous-time limit, as derived by Dybvig (1988) and Jamshidian (1988), cannot be regarded as a proper extension of the Vasicek model because of the lack of mean reversion in the short-rate dynamics.

The need for an exact fit to the currently-observed yield curve, led Hull and White to the introduction of a time-varying parameter in the Vasicek model. Notice indeed that matching the model and the market term structures of rates at the current time is equivalent to solving a system with an infinite number of equations, one for each possible maturity. Such a system can be solved in general only after introducing an infinite number of parameters, or equivalently a deterministic function of time.

By considering a further time-varying parameter, Hull and White (1990b) proposed an even more general model that is also able to fit a given term structure of volatilities. Such a model, however, may be somehow dangerous when applied to concrete market situations as we will hint at below. This is the main reason why in this section we stick to the extension where only one parameter, corresponding to the Vasicek θ, is chosen to be a deterministic function of time.

The model we analyze implies a normal distribution for the short-rate process at each time. Moreover, it is quite analytically tractable in that zero-coupon bonds and options on them can be explicitly priced. The Gaussian distribution of continuously-compounded rates then allows for the derivation of analytical formulas and the construction of efficient numerical procedures for pricing a large variety of derivative securities. On the other hand, the possibility of negative rates and the one-factor formulation make the model hardly applicable to concrete pricing problems.

However, the Hull and White extension of the Vasicek model is one of the historically most important interest-rate models, being still nowadays used for risk-management purposes. From a theoretical point of view, moreover, it allows the development of some general tools and procedures that can be easily borrowed by other short-rate models, as we will show in the following sections.

3.3.1 The Short-Rate Dynamics

Hull and White (1990) assumed that the instantaneous short-rate process evolves under the risk-neutral measure according to

$$dr(t) = [\vartheta(t) - a(t)r(t)]dt + \sigma(t)dW(t), \tag{3.32}$$

where ϑ, a and σ are deterministic functions of time.

Such a model can be fitted to the term structure of interest rates and the term structure of spot or forward-rate volatilities. However, if an exact calibration to the current yield curve is a desirable feature, the perfect fitting to a volatility term structure can be rather dangerous and must be carefully dealt with. The reason is two-fold. First, not all the volatilities that are quoted in the market are significant: some market sectors are less liquid, with the associated quotes that may be neither informative nor reliable. Second, the future volatility structures implied by (3.32) are likely to be unrealistic in that they do not conform to typical market shapes, as was remarked by Hull and White (1995b) themselves.

We therefore concentrate on the following extension of the Vasicek model being analyzed by Hull and White (1994a)

$$dr(t) = [\vartheta(t) - ar(t)]dt + \sigma dW(t), \tag{3.33}$$

where a and σ are now positive constants and ϑ is chosen so as to exactly fit the term structure of interest rates being currently observed in the market. It can be shown that, denoting by $f^M(0,T)$ the market instantaneous forward rate at time 0 for the maturity T, i.e.,

$$f^M(0,T) = -\frac{\partial \ln P^M(0,T)}{\partial T},$$

with $P^M(0,T)$ the market discount factor for the maturity T, we must have

3.3 The Hull-White Extended Vasicek Model

$$\vartheta(t) = \frac{\partial f^M(0,t)}{\partial T} + af^M(0,t) + \frac{\sigma^2}{2a}(1 - e^{-2at}). \quad (3.34)$$

Equation (3.33) can be easily integrated so as to yield

$$\begin{aligned}r(t) &= r(s)e^{-a(t-s)} + \int_s^t e^{-a(t-u)}\vartheta(u)du + \sigma \int_s^t e^{-a(t-u)}dW(u) \\ &= r(s)e^{-a(t-s)} + \alpha(t) - \alpha(s)e^{-a(t-s)} + \sigma \int_s^t e^{-a(t-u)}dW(u),\end{aligned} \quad (3.35)$$

where

$$\alpha(t) = f^M(0,t) + \frac{\sigma^2}{2a^2}(1 - e^{-at})^2. \quad (3.36)$$

Therefore, $r(t)$ conditional on $\mathcal{F}_s$ is normally distributed with mean and variance given respectively by

$$\begin{aligned}E\{r(t)|\mathcal{F}_s\} &= r(s)e^{-a(t-s)} + \alpha(t) - \alpha(s)e^{-a(t-s)} \\ \text{Var}\{r(t)|\mathcal{F}_s\} &= \frac{\sigma^2}{2a}\left[1 - e^{-2a(t-s)}\right].\end{aligned} \quad (3.37)$$

Notice that defining the process x by

$$dx(t) = -ax(t)dt + \sigma dW(t), \quad x(0) = 0, \quad (3.38)$$

we immediately have that, for each $s < t$,

$$x(t) = x(s)e^{-a(t-s)} + \sigma \int_s^t e^{-a(t-u)}dW(u),$$

so that we can write $r(t) = x(t) + \alpha(t)$ for each t.

As mentioned above, the theoretical possibility of r going below zero is a clear drawback of the model (3.32) in general, and of (3.33) in particular. Indeed, for model (3.33), the risk-neutral probability of negative rates at time t is explicitly given by

$$Q\{r(t) < 0\} = \Phi\left(-\frac{\alpha(t)}{\sqrt{\frac{\sigma^2}{2a}[1 - e^{-2at}]}}\right),$$

with Φ denoting the standard normal cumulative distribution function. However, such probability is almost negligible in practice. As an example, we show in Figure 3.2 the evolution over time of the two standard-deviation window, under Q, for the instantaneous short rate r, with parameters calibrated to market data as of 2 June 1999.[7]

---

[7] We thank our colleague Francesco Rapisarda for kindly providing us with such a figure.

66   3. One-factor short-rate models

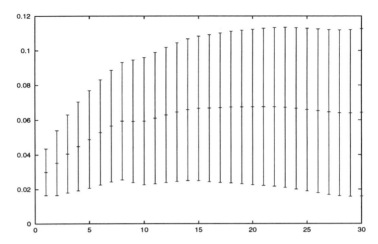

Fig. 3.2. The two standard-deviation window for the instantaneous short rate r as time goes by (market data as of 2 June 1999).

3.3.2 Bond and Option Pricing

The price at time t of a pure discount bond paying off 1 at time T is given by the expectation (3.2). Such expectation is relatively easy to compute under the dynamics (3.33). Notice indeed that, due to the Gaussian distribution of $r(T)$ conditional on $\mathcal{F}_t$, $t \leq T$, $\int_t^T r(u)du$ is itself normally distributed. Precisely we can show that

$$\int_t^T r(u)du | \mathcal{F}_t$$
$$\sim \mathcal{N}\left(B(t,T)[r(t) - \alpha(t)] + \ln\frac{P^M(0,t)}{P^M(0,T)} + \frac{1}{2}[V(0,T) - V(0,t)], V(t,T)\right),$$

where

$$B(t,T) = \frac{1}{a}\left[1 - e^{-a(T-t)}\right],$$
$$V(t,T) = \frac{\sigma^2}{a^2}\left[T - t + \frac{2}{a}e^{-a(T-t)} - \frac{1}{2a}e^{-2a(T-t)} - \frac{3}{2a}\right],$$

so that we obtain
$$P(t,T) = A(t,T)e^{-B(t,T)r(t)}, \qquad (3.39)$$

where

$$A(t,T) = \frac{P^M(0,T)}{P^M(0,t)}\exp\left\{B(t,T)f^M(0,t) - \frac{\sigma^2}{4a}(1 - e^{-2at})B(t,T)^2\right\}.$$

Similarly, the price **ZBC**(t,T,S,X) at time t of a European call option with strike X, maturity T and written on a pure discount bond maturing at time S

3.3 The Hull-White Extended Vasicek Model

is given by the expectation (2.20) or, equivalently, by (2.22). To compute the latter expectation, we need to know the distribution of the process r under the T-forward measure Q^T. Since the process x corresponds to the Vasicek's r with $\theta = 0$, we can use formula (3.9) to get

$$dx(t) = [-D(t,T)\sigma^2 - ax(t)]dt + \sigma dW^T(t),$$

where the Q^T-Brownian motion W^T is defined by $dW^T(t) = dW(t) + \sigma B(t,T)dt$, so that, for $s \leq t \leq T$,

$$x(t) = x(s)e^{-a(t-s)} - M^T(s,t) + \sigma \int_s^t e^{-a(t-u)} dW^T(u)$$

with

$$M^T(s,t) = \frac{\sigma^2}{a^2}\left[1 - e^{-a(t-s)}\right] - \frac{\sigma^2}{2a^2}\left[e^{-a(T-t)} - e^{-a(T+t-2s)}\right].$$

It is then easy to realize that the distribution of the short rate $r(t)$ conditional on $\mathcal{F}_s$ is, under the measure Q^T, still Gaussian with mean and variance given respectively by

$$E^T\{r(t)|\mathcal{F}_s\} = x(s)e^{-a(t-s)} - M^T(s,t) + \alpha(t),$$
$$\text{Var}^T\{r(t)|\mathcal{F}_s\} = \frac{\sigma^2}{2a}\left[1 - e^{-2a(t-s)}\right].$$

As a consequence, the European call-option price is

$$\mathbf{ZBC}(t,T,S,X) = P(t,S)\Phi(h) - XP(t,T)\Phi(h - \sigma_p), \quad (3.40)$$

where

$$\sigma_p = \sigma\sqrt{\frac{1 - e^{-2a(T-t)}}{2a}} B(T,S),$$
$$h = \frac{1}{\sigma_p} \ln \frac{P(t,S)}{P(t,T)X} + \frac{\sigma_p}{2}.$$

Analogously, the price $\mathbf{ZBP}(t,T,S,X)$ at time t of a European put option with strike X, maturity T and written on a pure discount bond maturing at time S is given by

$$\mathbf{ZBP}(t,T,S,X) = XP(t,T)\Phi(-h + \sigma_p) - P(t,S)\Phi(-h). \quad (3.41)$$

Through formulas (3.40) and (3.41), we can also price caps and floors since they can be viewed as portfolios of zero-bond options. To this end, we denote by $D = \{d_1, d_2, \ldots, d_n\}$ the set of the cap/floor payment dates and by $\mathcal{T} = \{t_0, t_1, \ldots, t_n\}$ the set of the corresponding times, meaning that t_i is the difference in years between d_i and the settlement date t, and where t_0 is the first reset time. Moreover, we denote by τ_i the year fraction from d_{i-1} to d_i, $i = 1, \ldots, n$. Applying formula (2.23), we then obtain that the price at time $t < t_0$ of the cap with cap rate (strike) X, nominal value N and set of times $\mathcal{T}$ is given by

$$\mathbf{Cap}(t, \mathcal{T}, N, X) = N \sum_{i=1}^{n} (1 + X\tau_i) \mathbf{ZBP}\left(t, t_{i-1}, t_i, \frac{1}{1 + X\tau_i}\right),$$

or, more explicitly,

$$\mathbf{Cap}(t, \mathcal{T}, N, X) = N \sum_{i=1}^{n} \left[P(t, t_{i-1}) \Phi(-h_i + \sigma_p^i) - (1 + X\tau_i) P(t, t_i) \Phi(-h_i) \right],$$
(3.42)

where

$$\sigma_p^i = \sigma \sqrt{\frac{1 - e^{-2a(t_{i-1}-t)}}{2a}} B(t_{i-1}, t_i),$$

$$h_i = \frac{1}{\sigma_p^i} \ln \frac{P(t, t_i)(1 + X\tau_i)}{P(t, t_{i-1})} + \frac{\sigma_p^i}{2}.$$

Analogously, the price of the corresponding floor is

$$\mathbf{Flr}(t, \mathcal{T}, N, X) = N \sum_{i=1}^{n} \left[(1 + X\tau_i) P(t, t_i) \Phi(h_i) - P(t, t_{i-1}) \Phi(h_i - \sigma_p^i) \right].$$
(3.43)

European options on coupon-bearing bonds can be explicitly priced by means of Jamshidian's (1989) decomposition. To this end, consider a European option with strike X and maturity T, written on a bond paying n coupons after the option maturity. Denote by T_i, $T_i > T$, and by c_i the payment time and value of the i-th cash flow after T. Let $\mathcal{T} := \{T_1, \ldots, T_n\}$ and $c := \{c_1, \ldots, c_n\}$. Denote by r^* the value of the spot rate at time T for which the coupon-bearing bond price equals the strike and by X_i the time-T value of a pure-discount bond maturing at T_i when the spot rate is r^*. Then the option price at time $t < T$ is

$$\mathbf{CBO}(t, T, \mathcal{T}, c, X) = \sum_{i=1}^{n} c_i \mathbf{ZBO}(t, T, T_i, X_i).$$
(3.44)

For a formal prove of this result we refer to Section 3.11.1.

Given the analytical formula (3.44), also European swaptions can be analytically priced, since a European swaption can be viewed as an option on a coupon-bearing bond. Indeed, consider a payer swaption with strike rate X, maturity T and nominal value N, which gives the holder the right to enter at time $t_0 = T$ an interest rate swap with payment times $\mathcal{T} = \{t_1, \ldots, t_n\}$,

$t_1 > T$, where he pays at the fixed rate X and receives LIBOR set "in arrears". We denote by τ_i the year fraction from t_{i-1} to t_i, $i = 1, \ldots, n$ and set $c_i := X\tau_i$ for $i = 1, \ldots, n-1$ and $c_n := 1 + X\tau_n$. Denoting by r^* the value of the spot rate at time T for which

$$\sum_{i=1}^{n} c_i A(T, t_i) e^{-B(T,t_i)r^*} = 1,$$

and setting $X_i := A(T, t_i)\exp(-B(T, t_i)r^*)$, the swaption price at time $t < T$ is then given by

$$\mathbf{PS}(t, T, \mathcal{T}, N, X) = N \sum_{i=1}^{n} c_i \mathbf{ZBP}(t, T, t_i, X_i). \tag{3.45}$$

Analogously, the price of the corresponding receiver swaption is

$$\mathbf{RS}(t, T, \mathcal{T}, N, X) = N \sum_{i=1}^{n} c_i \mathbf{ZBC}(t, T, t_i, X_i). \tag{3.46}$$

As a final remark, before moving to three construction techniques, we observe that the HW model is an affine term-structure model in the sense we have seen in Section 3.2.4.

3.3.3 The Construction of a Trinomial Tree

Even so, every good tree produces good fruit, but a bad tree produces bad fruit, a good tree cannot bear bad fruit, nor can a bad tree bear good fruit
St. Matthew VII.17-18

We now illustrate a procedure for the construction of a trinomial tree that approximates the evolution of the process r. This is a two-stage procedure that is basically based on those suggested by Hull and White (1993d, 1994a).

Let us fix a time horizon T and the times $0 = t_0 < t_1 < \cdots < t_N = T$, and set $\Delta t_i = t_{i+1} - t_i$, for each i. The time instants t_i need not be equally spaced. This is an essential feature when employing the tree for practical purposes.

The first stage consists in constructing a trinomial tree for the process x in (3.38) along the procedure illustrated in Appendix C. For further justifications and details we refer to such appendix.

We denote the tree nodes by (i, j) where the time index i ranges from 0 to N and the space index j ranges from some $\underline{j}_i < 0$ to some $\overline{j}_i > 0$. We denote by $x_{i,j}$ the process value on node (i, j).

Remembering formulas (3.37) and that $x(t) = r(t) - \alpha(t)$ for each t, we have

3. One-factor short-rate models

$$E\{x(t_{i+1})|x(t_i) = x_{i,j}\} = x_{i,j}e^{-a\Delta t_i} =: M_{i,j}$$

$$\text{Var}\{x(t_{i+1})|x(t_i) = x_{i,j}\} = \frac{\sigma^2}{2a}\left[1 - e^{-2a\Delta t_i}\right] =: V_i^2. \quad (3.47)$$

We then set $x_{i,j} = j\Delta x_i$, where

$$\Delta x_i = V_{i-1}\sqrt{3} = \sigma\sqrt{\frac{3}{2a}[1 - e^{-2a\Delta t_{i-1}}]}. \quad (3.48)$$

Assuming that at time t_i we are on node (i, j) with associated value $x_{i,j}$, the process can move to $x_{i+1,k+1}$, $x_{i+1,k}$ or $x_{i+1,k-1}$ at time t_{i+1} with probabilities p_u, p_m and p_d, respectively. The central node is therefore the k-th node at time t_{i+1}, where the level k is chosen so that $x_{i+1,k}$ is as close as possible to $M_{i,j}$, i.e.,

$$k = \text{round}\left(\frac{M_{i,j}}{\Delta x_{i+1}}\right), \quad (3.49)$$

where round(x) is the closest integer to the real number x. This definition fully determines the geometry of this initial tree for x. In particular, the minimum and the maximum levels $\underline{j}_i$ and $\overline{j}_i$ at each time step i are perfectly defined.

We now derive the probability p_u, p_m and p_d such that the conditional mean and variance in (3.47) match those in the tree. We obtain

$$\begin{cases} p_u = \frac{1}{6} + \frac{\eta_{j,k}^2}{6V_i^2} + \frac{\eta_{j,k}}{2\sqrt{3}V_i}, \\ p_m = \frac{2}{3} - \frac{\eta_{j,k}^2}{3V_i^2}, \\ p_d = \frac{1}{6} + \frac{\eta_{j,k}^2}{6V_i^2} - \frac{\eta_{j,k}}{2\sqrt{3}V_i}, \end{cases} \quad (3.50)$$

where $\eta_{j,k} = M_{i,j} - x_{i+1,k}$, with the dependence on i being omitted to lighten notation.

We can easily see that both p_u and p_d are positive for every value of $\eta_{j,k}$, whereas p_m is positive if and only if $|\eta_{j,k}| \leq V_i\sqrt{2}$. However, the definition of k implies that $|\eta_{j,k}| \leq V_i\sqrt{3}/2$, hence the middle probability p_m is positive, too. Therefore, (3.50) are actual probabilities such that the discrete process described by the tree has conditional mean and variance that match those of the process x. An example of such a tree geometry, with varying time step, is shown in Figure 3.3.[8]

The second stage of our construction procedure consists in displacing the tree nodes to obtain the corresponding tree for r. An easy way to do so is by means of the explicit formula (3.36). This has been suggested by Pelsser (1996) and Kijima and Nagayama (1994). However, combining this exact formula with the approximate nature of the tree prevents us from retrieving the correct market discount factors at time 0. For example, the analytical

---

[8] We thank our colleague Gianvittorio Mauri (aka " The Master") for kindly providing us with such a graphic "masterpiece"!

3.3 The Hull-White Extended Vasicek Model

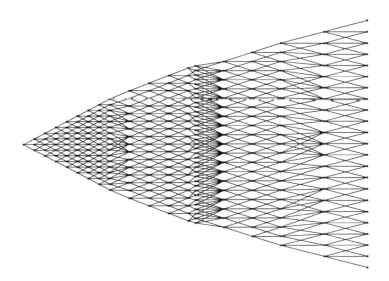

Fig. 3.3. A possible geometry for the tree approximating x.

displacement at time 0 is $\alpha(0) = r(0)$, so that the price of the zero-coupon bond with maturity t_1, as calculated in the tree, would be $\exp(-r(0)t_1)$, which is different in general from $P^M(0,t_1) = \exp(-R(0,t_1)t_1)$, with $R(0,t_1)$ the continuously-compounded rate at time 0 for the maturity t_1. This is the main reason why the original method proposed by Hull and White (1994a) relies on applying the displacements that perfectly reproduce the market zero-coupon curve at time 0. Notice indeed that even small errors in the pricing of discount bonds can lead to non-negligible errors in bond-option prices.

We denote by α_i the displacement at time t_i, which is common to all nodes $(i, \cdot)$. The quantity α_i is numerically calculated as follows. We denote by $Q_{i,j}$ the present value of an instrument paying 1 if node (i,j) is reached and zero otherwise (somehow discrete analogous to "Arrow-Debreu prices"). The values of α_i and $Q_{i,j}$ are calculated recursively from α_0 that is set so as to retrieve the correct discount factor for the maturity t_1, i.e., $\alpha_0 = -\ln(P^M(0,t_1))/t_1$. As soon as the value of α_i is known, the values $Q_{i+1,j}$, $j = \underline{j}_{i+1}, \ldots, \overline{j}_{i+1}$, are calculated through

$$Q_{i+1,j} = \sum_h Q_{i,h} q(h,j) \exp(-(\alpha_i + h\Delta x_i)\Delta t_i),$$

where $q(h,j)$ is the probability of moving from node (i,h) to node $(i+1,j)$ and the sum is over all values of h for which such probability is non-zero. After deriving the value of $Q_{i,j}$, for each $j = \underline{j}_i, \ldots, \overline{j}_i$, the value of α_i is calculated by solving

72   3. One-factor short-rate models

$$P(0, t_{i+1}) = \sum_{j=\underline{j}_i}^{\overline{j}_i} Q_{i,j} \exp(-(\alpha_i + j\Delta x_i)\Delta t_i),$$

that leads to

$$\alpha_i = \frac{1}{\Delta t_i} \ln \frac{\sum_{j=\underline{j}_i}^{\overline{j}_i} Q_{i,j} \exp(-j\Delta x_i \Delta t_i)}{P(0, t_{i+1})}.$$

We finally end up with a tree where each node (i,j) has associated value $r_{i,j} = x_{i,j} + \alpha_i$. This tree geometry is displayed in Figure 3.4.

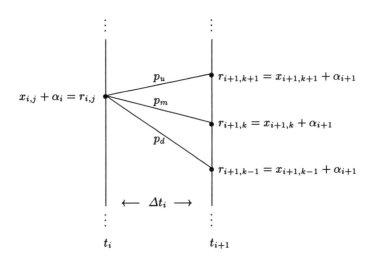

Fig. 3.4. Evolution of the process r starting from $r_{i,j}$ at time t_i and moving to $r_{i+1,k+1}$, $r_{i+1,k}$ or $r_{i+1,k-1}$ at time t_{i+1} with probabilities p_u, p_m and p_d, respectively.

3.4 Possible Extensions of the CIR Model

Besides their extension of the Vasicek (1977) model, Hull and White (1990b) proposed an extension of the Cox, Ingersoll and Ross (1985) model based on the same idea of considering time dependent coefficients. The short rate dynamics are then given by

$$dr(t) = [\vartheta(t) - a(t)r(t)]dt + \sigma(t)\sqrt{r(t)}dW(t), \tag{3.51}$$

where a, ϑ and σ are deterministic functions of time. Such extension however is not analytically tractable. Indeed, one can show that, for $t < T$, the pure-discount-bond price can be written as

$$P(t,T) = A(t,T)e^{-B(t,T)r(t)},$$

where B solves a Riccati equation and A solves a linear differential equation subject to some boundary conditions. However, the Riccati equation can be explicitly solved only for constant coefficients, so that, under the general dynamics (3.51), one has to resort to numerical procedures.[9]

Of course the same drawback holds for the simplified dynamics with constant volatility parameters

$$dr(t) = [\vartheta(t) - ar(t)]dt + \sigma\sqrt{r(t)}dW(t), \qquad (3.52)$$

where a and σ are now positive constants, and only the function ϑ is assumed to be time dependent so as to exactly fit the initial term structure of interest rates. To our knowledge, no analytical expression for $\vartheta(t)$ in terms of the observed yield curve is available in the literature. Furthermore, there is no guarantee that a numerical approximation of $\vartheta(t)$ would keep the rate r positive, hence that the diffusion coefficient would always be well defined. These are the major reasons why this extension has been less successful than its Gaussian counterpart.

A simple version of (3.51) that turns out to be analytically tractable has been proposed by Jamshidian (1995). He assumed that, for each t, the ratio $\vartheta(t)/\sigma^2(t)$ is equal to a positive constant δ, which must be greater than $1/2$ to ensure that the origin is inaccessible.

Maghsoodi (1996) and Rogers (1995) have shown that this simple version can be obtained from the classical constant coefficient CIR model by applying a deterministic time change and then multiplying the resulting process by a positive deterministic function of time. However the acquired analytical tractability is paid by having European bond-option prices that explicitly depend on the instantaneous short-term rate. This is a characteristics of all square-root models. Gaussian models feature option prices depending on the instantaneous rate only implicitly through $P(t,T)$, since $P(t,T)$ is a function of $r(t)$. Instead, in the CIR model, and in all its extensions, $r(t)$ appears explicitly, i.e, outside expressions of long-term rates such as $L(t,T)$ or $R(t,T)$ (or, equivalently, $P(t,T)$). While it is natural that long-dated option prices at time t depend on long-term rates, it may be undesirable that they depend on the instantaneous rate explicitly. This can be undesirable when pricing and hedging long-dated options.

3.5 The Black-Karasinski Model

The drawback of negative rates has been also addressed by Black and Karasinski (1991) in their celebrated lognormal short rate model. Black and Karasin-

---

[9] Maghsoodi (1996) derived different formulas for the prices of bonds and bond options, but still relying on numerical integration.

ski assumed that the instantaneous short rate process evolves as the exponential of an Ornstein-Uhlenbeck process with time dependent coefficients.[10] Since the market formulas for caps and swaptions are based on the assumption of lognormal rates, it seemed reasonable to choose the same distribution for the instantaneous short-rate process.[11] Moreover, the rather good fitting quality of the model to market data, and especially to the swaption volatility surface, has made the model quite popular among practitioners and financial engineers. However, analogously to the Exponential-Vasicek model, the Black-Karasinski (1991) model is not analytically tractable. This renders the model calibration to market data more burdensome than in the Hull and White (1990) Gaussian model, since no analytical formulas for bonds are available. Indeed, when using a tree to price an option on a zero-coupon bond, one has to construct the tree until the bond maturity, which may actually be much longer than that of the option. The same inconvenience occurs when we need to simulate rates which are not instantaneous. If we need for example to simulate the four-year rate in one year, we may simulate the short rate up to one year but then we are not done. With Hull and White's model for example, when we have simulated $r(1y)$, we can easily compute $P(1y, 5y)$ (and therefore $L(1y, 5y)$) algebraically from (3.39). The same holds for the Vasicek model, the CIR model and their deterministic shift extensions we will explore in the following. However, this desirable feature is not shared by the Black-Karasinski model, the EV model or its extension. When we have simulated $r(1y)$, we need to compute $P(1y, 5y)$ numerically for each simulated realization of r. In practice, we need to use a tree for each simulated r, and this renders the Monte Carlo approach much heavier than in the above-mentioned tractable models.

A further, and more fundamental, drawback of the model is that the expected value of the money-market account is infinite no matter which maturity is considered, as a consequence of the lognormal distribution of r. This was already remarked in the section devoted to the Dothan model and is a problem of lognormal models in general. As a consequence, the price of a Eurodollar future is infinite, too. However, this problem is partially overcome when using an approximating tree, because one deals with a finite number of states, and hence with finite expectations.

3.5.1 The Short-Rate Dynamics

Black and Karasinski (1991) assumed that the logarithm $\ln(r(t))$ of the instantaneous spot rate evolves under the risk neutral measure Q according to

---

[10] The Black and Karasinski (1991) model is actually a generalization of the continuous-time formulation of the Black, Derman and Toy (1990) model.

[11] Notice, however, that a lognormal instantaneous short-rate process does not lead to lognormal simple forward rates or lognormal swap rates.

3.5 The Black-Karasinski Model

$$d\ln(r(t)) = [\theta(t) - a(t)\ln(r(t))]dt + \sigma(t)dW(t), \quad r(0) = r_0, \quad (3.53)$$

where r_0 is a positive constant, $\theta(t)$, $a(t)$ and $\sigma(t)$ are deterministic functions of time that can be chosen so as to exactly fit the initial term structure of interest rates and some market volatility curves.

As for the Hull and White (1990b) model, one can choose to set $a(t) = a$ and $\sigma(t) = \sigma$, with a and σ positive constants, leading to

$$d\ln(r(t)) = [\theta(t) - a\ln(r(t))]dt + \sigma dW(t), \quad r(0) = r_0. \quad (3.54)$$

This choice can be motivated by arguments similar to those reported in Section 3.3. By letting θ be the only time dependent function, we decide to exactly fit the current term structure of rates and to keep the other two parameters at our disposal for the calibration to option data.

As in previous models, the coefficients a and σ can be interpreted as follows: a gives a measure of the "speed" at which the logarithm of $r(t)$ tends to its long-term value; σ is the standard-deviation rate of $dr(t)/r(t)$, namely the standard deviation per time unit of the instantaneous return of $r(t)$. Note also that σ, denoting the volatility of the instantaneous spot rate, must not be confused either with the volatility of the forward rate or with the volatility of the forward swap rate that must be plugged into the Black formulas for caps/floors and swaptions, respectively.

From (3.54), by Ito's lemma, we obtain

$$dr(t) = r(t)\left[\theta(t) + \frac{\sigma^2}{2} - a\ln r(t)\right]dt + \sigma r(t)dW(t).$$

whose explicit solution satisfies, for each $s \leq t$,

$$r(t) = \exp\left\{\ln r(s)e^{-a(t-s)} + \int_s^t e^{-a(t-u)}\theta(u)du + \sigma\int_s^t e^{-a(t-u)}dW(u)\right\}.$$

Therefore, $r(t)$ conditional on $\mathcal{F}_s$ is lognormally distributed with first and second moments given respectively by

$$E_s\{r(t)\} = \exp\left\{\ln r(s)e^{-a(t-s)} + \int_s^t e^{-a(t-u)}\theta(u)du + \frac{\sigma^2}{4a}\left[1 - e^{-2a(t-s)}\right]\right\}$$

$$E_s\{r^2(t)\} = \exp\left\{2\ln r(s)e^{-a(t-s)} + 2\int_s^t e^{-a(t-u)}\theta(u)du + \frac{\sigma^2}{a}\left[1 - e^{-2a(t-s)}\right]\right\}.$$

Moreover, setting

$$\alpha(t) = \ln(r_0)e^{-at} + \int_0^t e^{-a(t-u)}\theta(u)du, \quad (3.55)$$

we have that

$$\lim_{t\to\infty} E(r(t)) = \exp\left(\lim_{t\to\infty}\alpha(t) + \frac{\sigma^2}{4a}\right).$$

The limit on the left hand side cannot be computed analytically. However, the numerical procedure below allows for the extrapolation of an asymptotic value of $\alpha(t)$.

3.5.2 The Construction of a Trinomial Tree

From the withered tree, a flower blooms
Zen saying

As we have already pointed out, the Black and Karasinski model does not yield analytical formulas either for discount bonds or for options on bonds. The pricing of these (and other more general) instruments, therefore, must be performed through numerical procedures. An efficient numerical procedure has been suggested by Hull and White (1994a) and is based on a straightforward transformation of the trinomial tree we have illustrated in Section 3.3.3. Notice, indeed, that we can write

$$r(t) = e^{\alpha(t)+x(t)}, \qquad (3.56)$$

where α and x are defined as in (3.55) and (3.38), respectively. As for the Hull and White (1994a) model, we first construct a trinomial tree for x and then use (3.56) to displace the tree nodes so as to exactly retrieve the initial zero-coupon curve. For a better understanding of the construction procedure we refer to Section 3.3.3 and to Appendix C in particular.

Let us fix a time horizon T and the times $0 = t_0 < t_1 < \cdots < t_N = T$, and set $\Delta t_i = t_{i+1} - t_i$, for each i. As before, the time instants t_i need not be equally spaced. Again, we denote the tree nodes by (i,j) where the time index i ranges from 0 to N and the space index j ranges from some $\underline{j}_i < 0$ to some $\overline{j}_i > 0$.

We denote by $x_{i,j}$ the process value on node (i,j) and set $x_{i,j} = j\Delta x_i$, where Δx_i is defined as in (3.48).

Assuming that at time t_i we are on node (i,j) with associated value $x_{i,j}$, the process can move to $x_{i+1,k+1}$, $x_{i+1,k}$ or $x_{i+1,k-1}$ at time t_{i+1} with probabilities p_u, p_m and p_d, respectively. The central node is therefore the k-th node at time t_{i+1}, where k is defined as in (3.49). The probabilities p_u, p_m and p_d are defined as in (3.50). These definitions completely specify the initial tree geometry, and in particular the minimum and the maximum levels $\underline{j}_i$ and $\overline{j}_i$ at each time step i.

The final stage in our construction procedure consists in suitably shifting the tree nodes in order to obtain the proper tree for r through formula (3.56). Contrary to the Hull and White (1994a) case, the function α cannot be evaluated analytically. However, as already explained in Section 3.3.3, a numerical procedure is anyway required to exactly reproduce the initial term structure of discount factors.

We again denote by α_i the displacement at time t_i, which is common to all nodes $(i,\cdot)$. The quantity α_i is numerically calculated as in Section 3.3.3 with the only difference that now (3.56) holds. We again denote by $Q_{i,j}$ the present value of an instrument paying 1 if node (i,j) is reached and zero otherwise. The values of α_i and $Q_{i,j}$ are calculated recursively from α_0 that is set so as to retrieve the correct discount factor for the maturity t_1, i.e.,

$\alpha_0 = \ln(-\ln(P^M(0,t_1))/t_1)$. As soon as the value of α_i is known, the values $Q_{i+1,j}$, $j = \underline{j}_{i+1}, \ldots, \overline{j}_{i+1}$, are calculated through

$$Q_{i+1,j} = \sum_h Q_{i,h} q(h,j) \exp(-\exp(\alpha_i + h\Delta x_i)\Delta t_i),$$

where $q(h,j)$ is the probability of moving from node (i,h) to node $(i+1,j)$ and the sum is over all values of h for which such probability is non-zero. After deriving the value of $Q_{i,j}$, for each $j = \underline{j}_i, \ldots, \overline{j}_i$, the value of α_i is calculated by numerically solving

$$\psi(\alpha_i) := P(0, t_{i+1}) - \sum_{j=\underline{j}_i}^{\overline{j}_i} Q_{i,j} \exp(-\exp(\alpha_i + j\Delta x_i)\Delta t_i) = 0.$$

Using, for instance, the Newton-Raphson procedure, it is helpful to employ the analytical formula for the first derivative of the function ψ, that is

$$\psi'(\alpha_i) = \sum_{j=\underline{j}_i}^{\overline{j}_i} Q_{i,j} \exp(-\exp(\alpha_i + j\Delta x_i)\Delta t_i) \exp(\alpha_i + j\Delta x_i)\Delta t_i.$$

Finally, we must apply the exponential function to each node value to end up with a tree where each node (i,j) has associated value $r_{i,j} = \exp(x_{i,j} + \alpha_i)$. This tree geometry is displayed in Figure 3.5.

3.6 Volatility Structures in One-Factor Short-Rate Models

'I know what's real and what's false. In fact... I define it'
Matthew Ryder of the Linear Men, "The Kingdom" 1, 1999, DC Comics.

The aim of the present section is to clarify which volatility structures are relevant as far as the short-rate model performances are concerned. We will also point out the different volatility structures that are usually considered in the market. We will come back to this problem again in the chapters devoted to multi-factor models and to the LIBOR market model.

When approaching the interest-rate option market from a practical point of view, one immediately realizes that the *volatility* is the fundamental quantity one has to deal with. Such a quantity is so important that it is not just a sheer parameter, as theoretical researchers are tempted to view it, but it becomes an actual asset that can be bought or sold in the market. However, despite its practical importance, it may be hard to retrieve a clear definition of volatility in quantitative terms, and some confusion is likely to arise. A typical situation is when one hears traders pronounce sentences like: "A

78   3. One-factor short-rate models

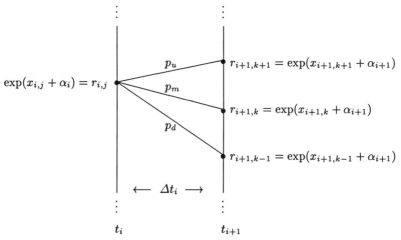

Fig. 3.5. Evolution of the process r starting from $r_{i,j}$ at time t_i and moving to $r_{i+1,k+1}$, $r_{i+1,k}$ or $r_{i+1,k-1}$ at time t_{i+1} with probabilities p_u, p_m and p_d, respectively.

satisfactory interest-rate model has to allow for a humped shape in the term structure of volatilities." In this section, we will then try to clarify statements of this type.

We now describe the path we adopt in this section in introducing volatility concepts. This path does not follow a fully logical development, because we are again adopting the "market/heuristics then rigor" approach. We prefer to give the reader motivation before we introduce new concepts, even if this creates a kind of time warp in the exposition. We adopt the following plan.

- Explain how the market considers (and defines) *caplet volatilities*.
- Explain how the market builds from these the *term structure of volatilities*.
- Explain what goes wrong if we simply transliterate the caplet-volatilities definition above, consistent with the market, to short-rate models language (model-intrinsic caplet volatility).
- Explain how we can modify the definition of caplet volatility for short rate models so that things work again (model-implied caplet volatility).
- Define, consequently, the term structure of volatility for short-rate models.

Caplet Volatilities in the Market

We have already seen in Chapter 1 that it is market practice to price caps and floors by means of the Black formulas (1.26) and (1.27) and to quote, instead of the price, the volatility parameter σ that enters such formulas. The market cap volatility is then simply defined as the parameter σ that must be plugged

3.6 Volatility Structures in One-Factor Short-Rate Models   79

into the Black cap formula to obtain the right market cap price. Similarly, a caplet volatility can be defined as the parameter σ to plug into the Black caplet formula to obtain the right caplet price. A fundamental difference is that the cap volatility assumes that all caplets concurring to a given cap share the same volatility, which is then set to a value matching the cap market price. On the contrary, caplet volatilities are allowed to be different also for caplets concurring to the same cap. In practice, caplet volatilities are stripped from the market cap volatilities along the procedure that will be explained in Sections 6.4 and 6.4.3 in particular, to which we refer for a more complete treatment.

Therefore, in this context, both cap and caplet volatilities are parameters obtained by inverting market prices through market established formulas. While for cap volatilities one usually stops here, the (market) caplet volatilities can instead be defined in an alternative way, which sheds further light on their actual meaning. Indeed, let us consider a caplet resetting at time T and paying at time $T+\tau$ the LIBOR rate $\tau L(T, T+\tau) = \tau F(T; T, T+\tau)$, where τ (measured in years) is typically three or six months. Denote the T-expiry caplet (percentage) volatility by

$$v_{T-\text{caplet}}^2 := \frac{1}{T} \int_0^T (d \ln F(t; T, T+\tau))(d \ln F(t; T, T+\tau))$$
$$= \frac{1}{T} \int_0^T \sigma(t; T, T+\tau)^2 \, dt.$$

The quantity $\sigma(t; T, T+\tau)$ is the (percentage) instantaneous volatility at time t of the simply-compounded forward rate $F(t; T, T+\tau)$ that underlies the T-expiry caplet. The instantaneous percentage volatilities $\sigma(t; T, T+\tau)$ are modeled deterministically in Black's market model for the cap market, so that the caplet volatility $v_{T-\text{caplet}}$, obtained by integrating deterministic functions, is also deterministic.

The Term Structure of (Caplet) Volatilities

The term structure of volatilities at time 0 is then to be intended as the graph of the map

$$T \mapsto v_{T-\text{caplet}},$$

and *this* is the graph that is observed to be humped most of times in the market.

An important question arises now: In what sense are caplet volatilities to be understood for short-rate models? There are two possible ways to define caplet (and cap) volatilities for short-rate models. We may call them the "model-intrinsic caplet volatility" and the "model-implied caplet volatility".

Why not Simply Transliterating the Definition to Short-Rate Models?

Let us begin with the "model-intrinsic caplet volatility". Before giving the formal definition, let us illustrate this idea by resorting to a specific model. Consider for example the CIR model. In a model like CIR the integrals in the definition of $v_{T-\text{caplet}}$ above are stochastic, since the corresponding function $\sigma(t; T, T+\tau)$ is not deterministic. One can indeed convince oneself of this by recalling the dynamics of F in the CIR model as from Section 3.2.3,

$$dF(t;T,S) = \sigma \left(F(t;T,S) + \frac{1}{\gamma(T,S)} \right)$$

$$\cdot \sqrt{(B(t,S) - B(t,T)) \ln \left[(\gamma(T,S)F(t;T,S) + 1) \frac{A(t,S)}{A(t,T)} \right]} \, dW^S(t).$$

Here we have a stochastic percentage instantaneous volatility that is given by

$$\sigma(t;T,S) = \sigma \left(1 + \frac{1}{\gamma(T,S)F(t;T,S)} \right)$$

$$\cdot \sqrt{(B(t,S) - B(t,T)) \ln \left[(\gamma(T,S)F(t;T,S) + 1) \frac{A(t,S)}{A(t,T)} \right]}.$$

Since the quantities $v_{T-\text{caplet}}$'s are (square roots of) integrals of the instantaneous variances $\sigma(t;T,T+\tau)^2$, in the CIR model the $v_{T-\text{caplet}}$'s are stochastic. More generally, we may define, for a short rate model,

Definition 3.6.1. Model intrinsic T-caplet volatility. *The* model intrinsic T-caplet volatility *at time 0 is defined as the random variable*

$$\sqrt{\frac{1}{T} \int_0^T \sigma(t;T,T+\tau)^2 \, dt} \ .$$

As the CIR example pointed out, however, this definition presents us with a problem: if we define caplet volatilities like this, by simply mimicking the definition we have seen above for the market, we obtain random caplet volatilities. But caplet volatilities are not random in the market. How can we modify the definition of caplet volatility for short-rate models so as to go back to a deterministic setup?

Modify the Definition so as to have Deterministic Caplet Volatilities

The necessity of dealing with deterministic quantities leads to the following procedure. Compute the model price of the at-the-money T-caplet at time

3.6 Volatility Structures in One-Factor Short-Rate Models   81

0, thus obtaining a function $\mathbf{Cpl}(0, T, T + \tau, F(0; T, T + \tau))$ of the model parameters, and then invert the T-expiry Black market formula for caplets to find the percentage Black volatility $v_{T-\text{caplet}}$ that once plugged into such a formula gives the model price. More precisely, one solves the following equation for $v_{T-\text{caplet}}^{\text{MODEL}}$:

$$P(0, T+\tau) \ \tau F(0; T, T+\tau) \left(2\Phi\left(\frac{v_{T-\text{caplet}}^{\text{MODEL}} \sqrt{T}}{2} \right) - 1 \right) \quad (3.57)$$
$$= \mathbf{Cpl}(0, T, T+\tau, F(0; T, T+\tau)).$$

The left-hand side is the market Black's formula for a T-expiry $T+\tau$-maturity at-the-money caplet, whereas the right-hand side is the corresponding model formula. We are now ready to introduce the following.

Definition 3.6.2. Model-implied T-caplet volatility. *The* model-implied T-caplet volatility *at time 0 is the (deterministic) solution* $v_{T-\text{caplet}}^{\text{MODEL}}$ *of the above equation (3.57).*

Implied cap volatilities can be defined in an analogous way. Precisely, let us consider a set of reset times $\{T_\alpha, \ldots, T_{\beta-1}\}$ with the final payment time T_β, and the set of the associated year fractions $\{\tau_\alpha, \ldots, \tau_\beta\}$. Then, setting $\mathcal{T}_i = \{T_\alpha, \ldots, T_i\}$ and $\bar{\tau}_i = \{\tau_{\alpha+1}, \ldots, \tau_i\}$, we have the following.

Definition 3.6.3. Model-implied $\mathcal{T}_i$-cap volatility. *The* model-implied $\mathcal{T}_i$-cap volatility *at time 0 is the (deterministic) solution* $v_{\mathcal{T}_i-\text{cap}}^{\text{MODEL}}$ *of the equation:*

$$\sum_{j=\alpha+1}^{i} P(0, T_j) \tau_j Bl(S_{\alpha,\beta}(0), F(0, T_{j-1}, T_j), v_{\mathcal{T}_i-\text{cap}}^{\text{MODEL}} \sqrt{T_{j-1}})$$
$$= \mathbf{Cap}(0, \mathcal{T}_i, \bar{\tau}_i, S_{\alpha,\beta}(0)).$$

where the forward swap rate $S_{\alpha,\beta}(0)$ is defined as in (1.25).

Term Structure of Volatilities for a Short-Rate Model

We can now define easily two types of term structure of volatilities associated with a short-rate model.

Definition 3.6.4. Term structure of caplet volatilities implied by a short-rate model. *The* term structure of caplet volatilities implied by a short-rate model *is the graph of the model-implied T-caplet volatility against T, i.e. the graph of the function $T \mapsto v_{T-\text{caplet}}^{\text{MODEL}}$.*

Definition 3.6.5. Term structure of cap volatilities implied by a short-rate model. *The* term structure of cap volatilities implied by a short-rate model *is the graph of the model-implied $\mathcal{T}_i$-cap volatility against T_i, i.e. the graph of the function $T_i \mapsto v_{\mathcal{T}_i-\text{cap}}^{\text{MODEL}}$.*

82   3. One-factor short-rate models

In the remainder of the book, when dealing with short-rate models, $v_{T-\text{caplet}}^{\text{MODEL}}$ will always denote the model-*implied* T-caplet volatility, unless differently specified, and we will often omit the "MODEL" superscript. Notice that an analogous graph for the model-intrinsic T-caplet volatility would consist of a bunch of curves, one for each trajectory ω of $F(\cdot; T, T+\tau)(\omega)$.

A similar argument applies to the model-*implied* T-cap volatility. Notice also that for a cap with a single payment in T_i (coinciding with a T_{i-1}-resetting caplet), we have $\mathcal{T}_i = \{T_i\}$, i.e. $v_{T_{i-1}-\text{caplet}} = v_{\{T_i\}-\text{cap}}$. A little attention can avoid being disoriented by this "i" versus "$i-1$" notation.

Now that we have motivated and introduced the relevant definitions, we continue the discussion. Going back to the "humped shape observed in the market", what is actually observed to have a humped shape is the curve of the cap volatilities for different maturities at time 0. From cap volatilities one can strip caplet volatilities, and also this second curve shows a humped shape when observed in the market.[12]

In practice, it is commonly seen that this term structure is able to feature large humps if the related absolute instantaneous volatilities of instantaneous forward rates,

$$T \mapsto \sqrt{\frac{\text{Var}(df(t,T))}{dt}} =: \sigma_f(t,T),$$

allow for a hump themselves. In other terms, there is a link between potentially large humps of the term structure of volatilities (as observed in the market) and possible humps in the volatility of instantaneous forward rates. We used the term "large humps" because *small* humps in the model caplet and cap curves $T \mapsto v_{T-\text{caplet}}$ and $T_i \mapsto v_{\mathcal{T}_i-\text{cap}}$ are possible even when the model instantaneous-forward volatility curve $T \mapsto \sigma_f(t,T)$ is monotonically decreasing. In short, it usually happens that:

1. No humps in $T \mapsto \sigma_f(t,T)$ $\Rightarrow$ only small humps for $T \mapsto v_{T-\text{caplet}}^{\text{MODEL}}$ are possible;
2. Humps in $T \mapsto \sigma_f(t,T)$ $\Rightarrow$ large humps for $T \mapsto v_{T-\text{caplet}}^{\text{MODEL}}$ are possible.

A typical example is the CIR++ model (3.76) we will introduce later on. After computing the absolute volatility of instantaneous forward rates $\sigma_f(t,T)$ through formula (3.29), a little analysis shows that under this model $T \mapsto \sigma_f(t,T)$ is indeed monotonically decreasing, while the model calibration to caps data usually leads to implied cap volatilities displaying a slightly humped shape (so that also the initial caplet volatilities display a small hump). See also the related section in this chapter. Summarizing, *small* humps in the caplet curve are possible even with monotonically decreasing instantaneous-forward volatilities. However, for these models *large* humps in caplets volatilities are often ruled out by monotonicity of $T \mapsto \sigma_f(t,T)$.

---

[12] What we are saying here is a little redundant with the material appearing in other chapters. This is done to maintain the single chapters self-contained to a certain degree.

These considerations on the amplitude of the cap-volatility hump mainly apply when the zero-coupon curve is increasing or slightly inverted, like that in Figure 1.1. However, in case of decreasing yield curves, it turns out that large humps can be produced even by those models for which the map $T \mapsto \sigma_f(t,T)$ is monotone. For instance, in case of the Hull and White model (3.33), we can prove that

$$v^{\text{HW}}_{T-\text{caplet}} \approx \sigma \frac{1 - e^{-a\tau}}{a} \left[1 + \frac{1}{\tau F(0, T, T+\tau)}\right], \quad (3.58)$$

for a and T positive and small enough. Therefore, in case of a (initially) decreasing forward rate curve $T \mapsto F(0, T, T+\tau)$, the implied caplet curve is initially upward sloping, and then necessarily humped since $v^{\text{HW}}_{T-\text{caplet}}$ goes to zero for T going to infinity.

We noticed above that confusion often arises when speaking of "allowing a humped shape for forward-rates volatilities". In other words, it is not clear what kind of volatilities one is considering: instantaneous-forward absolute volatilities, caplet volatilities, cap volatilities, etc. Some further specification is then needed, in general. As we have just seen, in fact, some models allow for humps in all of these structures. Some others, instead, only allow for small humps in the caplets or caps volatilities.

The framework just introduced can be used for calibration purposes. For instance, when the CIR++ model is calibrated to the cap market, the parameters k, θ, σ in the CIR dynamics are set to values such that the market cap volatilities $v^{\text{MKT}}_{T_i-\text{cap}}$ are as close as possible to the model volatilities $v^{\text{CIR}}_{T_i-\text{cap}}(k, \theta, \sigma)$ for all relevant i's. Alternatively, if one has already derived caplet volatilities from cap quotes along the lines of the method suggested in Section 6.4.3, the calibration can be made on caplet volatilities directly. The parameters k, θ, σ in the CIR dynamics are then set to values such that the market caplet volatilities $v^{\text{MKT}}_{T-\text{caplet}}$ are as close as possible to the model volatilities $v^{\text{CIR}}_{T-\text{caplet}}(k, \theta, \sigma)$ for all (relevant) T's.

We finally stress that the considered short-rate model need not be analytically tractable. Indeed, we just have to notice that, in non analytically tractable models, we need to resort to numerical procedures in order to compute the caplet prices $\mathbf{Cpl}(0, T, T+\tau, F(0; T, T+\tau))$ and retrieve $v^{\text{MODEL}}_{T-\text{caplet}}$ through inversion of the Black caplet formula.

3.7 Humped-Volatility Short-Rate Models

As we have seen in Section 3.6, allowing for a humped shape in the volatility structure of instantaneous forward rates results in possibly large humps in the caplet or cap volatility curves, as is often observed in the market.

Motivated by such considerations, as well as by a number of empirical works, Mercurio and Moraleda (2000a) proposed an extension of the Hull and White model (3.33) that in general is outside the general Hull and White class (3.32) and that allows for humped shapes in the volatility structure of instantaneous forward rates. They selected a suitable volatility function in the Heath, Jarrow and Morton (1992) framework and derived closed form formulas for European bond option prices. Precisely, they assumed the following form for the absolute volatility of the instantaneous forward rate process $f(t,T)$

$$\sqrt{\frac{\text{Var}(df(t,T))}{dt}} = \sigma[1 + \gamma(T-t)]e^{-\lambda(T-t)}, \tag{3.59}$$

where σ, λ and γ are positive constants. Their model, which will be better described in Chapter 5, has however two major drawbacks. First, the implied instantaneous spot rate process is not Markov and no explicit formula for zero-coupon bond prices can be derived. Second, the process distribution is Gaussian so that negative rates are actually possible.

The first drawback has been addressed by Moraleda and Vorst (1997) who considered the following dynamics for the instantaneous spot rate

$$\begin{cases} dr(t) = [\theta(t) - \beta(t)r(t)]\,dt + \sigma dW(t) \\ \beta(t) = \lambda - \frac{\gamma}{1+\gamma t} \end{cases} \tag{3.60}$$

which is a particular case of the full Hull and White model (3.32) which coincides with (3.33) for $\gamma = 0$ (and setting $\lambda = a$), and where $\theta(t)$ is a deterministic function of time. Equivalently, for each $s \leq t$,

$$r(t) = r(s)e^{-\int_s^t \beta(u)du} + \int_s^t e^{-\int_u^t \beta(v)dv}\vartheta(u)du + \sigma \int_s^t e^{-\int_u^t \beta(v)dv}dW(u)$$

$$= r(s)\mathcal{B}(s,t) + \int_s^t \mathcal{B}(u,t)\vartheta(u)du + \sigma \int_s^t \mathcal{B}(u,t)dW(u)$$

where

$$\mathcal{B}(s,t) := e^{-\int_s^t \beta(u)du} = \frac{1+\gamma t}{1+\gamma s}e^{-\lambda(t-s)}. \tag{3.61}$$

The absolute volatility of the instantaneous-forward-rate process $f(t,T)$ implied by (3.60) is

$$\sqrt{\frac{\text{Var}(df(t,T))}{dt}} = \sigma \frac{1+\gamma T}{1+\gamma t}e^{-\lambda(T-t)} \tag{3.62}$$

and coincides, at first order in $T-t$, with (3.59). The model (3.60) is analytically tractable and leads to analytical formulas for zero coupon bonds as well. Indeed, Moraleda and Vorst proved that the price at time t of a pure discount bond with maturity T is given by

$$P(t,T) = \frac{P(0,T)}{P(0,t)}\exp\left\{-\frac{1}{2}\Lambda^2(t,T)\phi(t) + \Lambda(t,T)[f(0,t) - r(t)]\right\},$$

where

$$\Lambda(t,T) = \frac{1}{\lambda^2(1+\gamma t)}\left[\gamma\lambda t + \gamma + \lambda - (\gamma\lambda T + \gamma + \lambda)e^{-\lambda(T-t)}\right],$$

$$\phi(t) = \frac{\sigma^2(1+\gamma t)}{\gamma^2}\left[2\lambda(1+\gamma t)\text{Ei}\left(\frac{2\lambda(1+\gamma t)}{\gamma}\right)e^{-\frac{2\lambda(1+\gamma t)}{\gamma}} - \gamma\right],$$

$$-\frac{\sigma^2(1+\gamma t)^2}{\gamma^2}\left[2\lambda\text{Ei}\left(\frac{2\lambda}{\gamma}\right)e^{-\frac{2\lambda(1+\gamma t)}{\gamma}} - \gamma e^{-2\lambda t}\right],$$

with Ei denoting the exponential integral function

$$\text{Ei}(x) = \int_{-\infty}^{x}\frac{e^t}{t}dt.$$

It is clear that this model is affine in the sense explained in Section 3.2.4, as is immediate also by the drift and diffusion coefficients in its short-rate dynamics.

Moreover, the time-t price of a European call option with maturity T, strike price X and written on a pure discount bond with maturity S is given by

$$\textbf{ZBC}(t,T,S,X) = P(t,S)\Phi(h) - XP(t,T)\Phi(h - \sigma_p),$$

where

$$h = \frac{\ln\frac{P(t,S)}{XP(t,T)} + \frac{1}{2}\sigma_p^2}{\sigma_p},$$

$$\sigma_p^2 = \frac{\sigma^2 A^2}{\lambda^4\gamma^2}\left[B(T) - B(t)\right],$$

$$A = (\gamma\lambda S + \gamma + \lambda)e^{-\lambda S} - (\gamma\lambda T + \gamma + \lambda)e^{-\lambda T},$$

$$B(\tau) = 2\lambda e^{-\frac{2\lambda}{\gamma}}\text{Ei}\left(\frac{2\lambda(1+\gamma\tau)}{\gamma}\right) - \frac{\gamma e^{2\lambda\tau}}{1+\gamma\tau}.$$

However, the short-rate process is still normally distributed, and since (3.62) is not just a function of $T-t$, the humped shape of the forward-rate volatility is lost as soon as t increases.

The second drawback of the Mercurio and Moraleda (2000a) model has been addressed by Mercurio and Moraleda (2000b) by assuming the following dynamics for the instantaneous spot rate

$$\begin{cases} dx(t) = [\theta(t) - \beta(t)x(t)]\,dt + \sigma dW(t) \\ \beta(t) = \lambda - \frac{\gamma}{1+\gamma t} \\ r(t) = e^{x(t)} \end{cases} \quad (3.63)$$

where x is an underlying Gaussian process. This model can be viewed as a characterization of (3.53), where the function σ is set to be a positive constant

86   3. One-factor short-rate models

and $a(t) = \beta(t)$ for each t.[13] Indeed, for each $s \leq t$,

$$r(t) = \exp\left\{\ln r(s)\mathcal{B}(s,t) + \int_s^t \mathcal{B}(u,t)\theta(u)du + \sigma\int_s^t \mathcal{B}(u,t)dW(u)\right\},$$

with $\mathcal{B}(s,t)$ defined as in (3.61). Notice that, for $\gamma = 0$, we retrieve the model (3.54).

Though sharing the same drawbacks of the model (3.54), the model (3.63) implies, however, a much better fitting of the cap-volatility curve, which indeed justifies the introduction of the extra parameter γ for practical purposes. For related empirical results we refer to the original paper by Mercurio and Moraleda (2000b) and to a later section in this chapter. Such results are based on the implementation of a trinomial tree, which is constructed in a similar fashion to that of Section 3.5 by taking into account that the value of a changes at each time step.

3.8 A General Deterministic-Shift Extension

In this section we illustrate a simple method to extend any time-homogeneous short-rate model, so as to exactly reproduce any observed term structure of interest rates while preserving the possible analytical tractability of the original model. This method has been extensively described by Brigo and Mercurio (1998, 2001a). A similar approach has inspired the independent works by Scott (1995), Dybvig (1997), and Avellaneda and Newman (1998).

In the case of the Vasicek (1977) model, the extension is perfectly equivalent to that of Hull and White (1990b), see Section 3.3. In the case of the Cox-Ingersoll-Ross (1985) model, instead, the extension is more analytically tractable than those of Section 3.4 and avoids problems concerning the use of numerical solutions. In fact, we can exactly fit any observed term structure of interest rates and derive analytical formulas both for pure discount bonds and for European bond options. The unique drawback is that in principle we can guarantee the positivity of rates only through restrictions on the parameters, which might worsen the quality of the calibration to caps/floors or swaption prices.

The CIR model is the most relevant case to which this procedure can be applied. Indeed, the extension yields the unique short-rate model featuring the following three properties: i) Exact fit of any observed term structure; ii) Analytical formulas for bond prices, bond-option prices, swaptions and caps prices; iii) The distribution of the instantaneous spot rate has tails that are fatter than in the Gaussian case and, through restriction on the parameters, it is always possible to guarantee positive rates without worsening the

---

[13] The specification of the function a is motivated by its interpretation in the above Gaussian case.

3.8 A General Deterministic-Shift Extension

volatility calibration in most situations. Moreover, one further property of the extended model is that the term structure is affine in the short rate. The above uniqueness is the reason why we devote more space to the CIR case.

The extension procedure is also applied to the Dothan (1978) model, thus yielding a shifted lognormal short-rate model that fits any given yield curve and for which there exist analytical formulas for zero-coupon bonds.

Though conceived for analytically tractable models, the method we describe in this section can be employed to extend more general time-homogeneous models. As a clarification, we consider the example of an original short-rate process that evolves as an Exponential-Vasicek process. The only requirement that is needed in general is a numerical procedure for pricing interest-rate derivatives under the original model.

3.8.1 The Basic Assumptions

We consider a time-homogeneous stochastic process x^α, whose dynamics under a given measure Q^x is expressed by

$$dx_t^\alpha = \mu(x_t^\alpha; \alpha)dt + \sigma(x_t^\alpha; \alpha)dW_t^x , \qquad (3.64)$$

where W^x is a standard Brownian motion, x_0^α is a given real number, $\alpha = \{\alpha_1, \ldots, \alpha_n\} \in \mathbb{R}^n$, $n \geq 1$, is a vector of parameters, and μ and σ are sufficiently well behaved real functions. We set $\mathcal{F}_t^x$ to be the sigma-field generated by x^α up to time t.

We assume that the process x^α describes the evolution of the instantaneous spot interest rate under the risk-adjusted martingale measure, and refer to this model as to the "reference model". We denote by $P^x(t,T)$ the price at time t of a zero-coupon bond maturing at T and with unit face value, so that

$$P^x(t,T) = E^x \left\{ \exp\left[-\int_t^T x_s^\alpha ds \right] | \mathcal{F}_t^x \right\},$$

where E^x denotes the expectation under the risk-adjusted measure Q^x.

We also assume there exists an explicit real function Π^x, defined on a suitable subset of $\mathbb{R}^{n+3}$, such that

$$P^x(t,T) = \Pi^x(t,T,x_t^\alpha;\alpha). \qquad (3.65)$$

The best known examples of spot-rate models satisfying the assumptions above are the Vasicek (1977) model (3.5), the Dothan (1978) model (3.17) and the Cox-Ingersoll-Ross (1985) model (3.21).

We now illustrate a simple approach for extending the time-homogeneous spot-rate model (3.64), in such a way that the extended version preserves the analytical tractability of the initial model while exactly fitting the observed term structure of interest rates. Precisely, we define the instantaneous short rate under the risk neutral measure Q by

88   3. One-factor short-rate models

$$r_t = x_t + \varphi(t;\alpha), \quad t \geq 0, \tag{3.66}$$

where x is a stochastic process that has under Q the same dynamics as x^α under Q^x, and φ is a deterministic function, depending on the parameter vector (α, x_0), that is integrable on closed intervals. Notice that x_0 is one more parameter at our disposal. We are free to select its value as long as

$$\varphi(0;\alpha) = r_0 - x_0 .$$

The function φ can be chosen so as to fit exactly the initial term structure of interest rates. We set $\mathcal{F}_t$ to be the sigma-field generated by x up to time t.

We notice that the process r depends on the parameters $\alpha_1, \ldots, \alpha_n, x_0$ both through the process x and through the function φ. As a common practice, we can determine $\alpha_1, \ldots, \alpha_n, x_0$ by calibrating the model to the current term structure of volatilities, fitting for example cap and floor prices or a few swaption prices.

Notice that, if φ is differentiable, the stochastic differential equation for the short-rate process (3.66) is,

$$dr_t = \left[\frac{d\varphi(t;\alpha)}{dt} + \mu(r_t - \varphi(t;\alpha);\alpha)\right] dt + \sigma(r_t - \varphi(t;\alpha);\alpha) dW_t .$$

As we have seen in Section 3.2.4, in case of time-homogeneous coefficients an affine term structure in the short-rate is equivalent to affine drift and squared diffusion coefficients. It follows immediately that if the reference model has an affine term structure so does the extended model. We may then anticipate that the shifted Vasicek (equivalent to HW) and CIR (CIR++) models will be affine term-structure models.

3.8.2 Fitting the Initial Term Structure of Interest Rates

Definition (3.66) immediately leads to the following.

Theorem 3.8.1. *The price at time t of a zero-coupon bond maturing at T is*

$$P(t,T) = \exp\left[-\int_t^T \varphi(s;\alpha)ds\right] \Pi^x(t,T,r_t - \varphi(t;\alpha);\alpha). \tag{3.67}$$

Proof. Denoting by E the expectation under the measure Q, we simply have to notice that

$$P(t,T) = E\left\{\exp\left[-\int_t^T (x_s + \varphi(s;\alpha))ds\right] \Big| \mathcal{F}_t\right\}$$

$$= \exp\left[-\int_t^T \varphi(s;\alpha)ds\right] E\left\{\exp\left[-\int_t^T x_s ds\right] \Big| \mathcal{F}_t\right\}$$

$$= \exp\left[-\int_t^T \varphi(s;\alpha)ds\right] \Pi^x(t,T,x_t;\alpha),$$

3.8 A General Deterministic-Shift Extension

where in the last step we use the equivalence of the dynamics of x under Q and x^α under Q^x. □

If we denote by $f^x(0, t; \alpha)$ the instantaneous forward rate at time 0 for a maturity t as associated with the bond price $\{P^x(0,t) : t \geq 0\}$, i.e.,

$$f^x(0, t; \alpha) = -\frac{\partial \ln P^x(0,t)}{\partial t} = -\frac{\partial \ln \Pi^x(0, t, x_0; \alpha)}{\partial t},$$

we then have the following.

Corollary 3.8.1. *The model (3.66) fits the currently observed term structure of discount factors if and only if*

$$\varphi(t; \alpha) = \varphi^*(t; \alpha) := f^M(0,t) - f^x(0,t;\alpha), \tag{3.68}$$

i.e., if and only if

$$\exp\left[-\int_t^T \varphi(s;\alpha)ds\right] = \Phi^*(t,T,x_0;\alpha) := \frac{P^M(0,T)}{\Pi^x(0,T,x_0;\alpha)}\frac{\Pi^x(0,t,x_0;\alpha)}{P^M(0,t)}. \tag{3.69}$$

Moreover, the corresponding zero-coupon-bond prices at time t are given by $P(t,T) = \Pi(t,T,r_t;\alpha)$, where

$$\Pi(t,T,r_t;\alpha) = \Phi^*(t,T,x_0;\alpha)\Pi^x(t,T,r_t - \varphi^*(t;\alpha);\alpha) \tag{3.70}$$

Proof. From the equality

$$P^M(0,t) = \exp\left[-\int_0^t \varphi(s;\alpha)ds\right]\Pi^x(0,t,x_0;\alpha),$$

we obtain (3.68) by taking the natural logarithm of both members and then differentiating. From the same equality, we also obtain (3.69) by noting that

$$\exp\left[-\int_t^T \varphi(s;\alpha)ds\right] = \exp\left[-\int_0^T \varphi(s;\alpha)ds\right]\exp\left[\int_0^t \varphi(s;\alpha)ds\right]$$

$$= \frac{P^M(0,T)}{\Pi^x(0,T,x_0;\alpha)}\frac{\Pi^x(0,t,x_0;\alpha)}{P^M(0,t)},$$

which, combined with (3.67), gives (3.70). □

Notice that by choosing $\varphi(t;\alpha)$ as in (3.68), the model (3.66) exactly fits the observed term structure of interest rates, no matter which values of α and x_0 are chosen.

3.8.3 Explicit Formulas for European Options

The extension (3.66) is even more interesting when the reference model (3.64) allows for analytical formulas for zero-coupon-bond options as well. It is easily seen that the extended model preserves the analytical tractability for option prices by means of analytical correction factors that are defined in terms of φ.

Noting that under the model (3.64), the price at time t of a European call option with maturity T, strike K and written on a zero-coupon bond maturing at time τ is

$$V^x(t,T,\tau,K) = E^x\left\{\exp\left[-\int_t^T x_s^\alpha ds\right](P^x(T,\tau)-K)^+ \Big| \mathcal{F}_t^x\right\},$$

we assume there exists an explicit real function Ψ^x defined on a suitable subset of $\mathbb{R}^{n+5}$, such that

$$V^x(t,T,\tau,K) = \Psi^x(t,T,\tau,K,x_t^\alpha;\alpha). \tag{3.71}$$

The best known examples of models (3.64) for which this holds are again the Vasicek (1977) model (3.5) and the Cox-Ingersoll-Ross (1985) model (3.21).

Straightforward algebra leads to the following.

Theorem 3.8.2. *Under the model (3.66), the price at time t of a European call option with maturity T, strike K and written on a zero-coupon bond maturing at time τ is*

$$\mathbf{ZBC}(t,T,\tau,K) = \exp\left[-\int_t^T \varphi(s;\alpha)ds\right]$$
$$\cdot \Psi^x\left(t,T,\tau,K\exp\left[\int_T^\tau \varphi(s;\alpha)ds\right], r_t - \varphi(t;\alpha);\alpha\right). \tag{3.72}$$

Proof. We simply have to notice that

$$\mathbf{ZBC}(t,T,\tau,K) = E\left\{\exp\left[-\int_t^T (x_s+\varphi(s;\alpha))ds\right](P(T,\tau)-K)^+ \Big| \mathcal{F}_t\right\}$$

$$= \exp\left[-\int_t^T \varphi(s;\alpha)ds\right] E\left\{\exp\left[-\int_t^T x_s ds\right]\right.$$

$$\left. \cdot \left(\exp\left[-\int_T^\tau \varphi(s;\alpha)ds\right] \Pi^x(T,\tau,x_T;\alpha) - K\right)^+ \Big| \mathcal{F}_t\right\}$$

3.8 A General Deterministic-Shift Extension

$$= \exp\left[-\int_t^\tau \varphi(s;\alpha)ds\right] E\left\{\exp\left[-\int_t^T x_s^\alpha ds\right]\right.$$

$$\left.\cdot\left(\Pi^x(T,\tau,x_T;\alpha) - K\exp\left[\int_T^\tau \varphi(s;\alpha)ds\right]\right)^+ \Big| \mathcal{F}_t^x\right\}$$

$$= \exp\left[-\int_t^\tau \varphi(s;\alpha)ds\right] \Psi^x\left(t,T,\tau,K\exp\left[\int_T^\tau \varphi(s;\alpha)ds\right], x_t;\alpha\right),$$

where in the last step we use the equivalence of the dynamics of x under Q and x^α under Q^x. □

The price of a European put option can be obtained through the put-call parity for bond options, and formula (3.4) in particular. As immediate consequence, caps and floors can be priced analytically as well.

The previous formula for a European option holds for any specification of the function φ. In particular, when exactly fitting the initial term structure of interest rates, the equality (3.68) must be used to yield $\mathbf{ZBC}(t,T,\tau,K) = \Psi(t,T,\tau,K,r_t;\alpha)$, where

$$\Psi(t,T,\tau,K,r_t;\alpha) = \Phi^*(t,\tau,x_0;\alpha)\Psi^x(t,T,\tau,K\Phi^*(\tau,T,x_0;\alpha), r_t - \varphi^*(t;\alpha);\alpha). \tag{3.73}$$

To this end, we notice that, if prices are to be calculated at time 0, we need not explicitly compute $\varphi^*(t;\alpha)$ since the relevant quantities are the discount factors at time zero.

Moreover, if Jamshidian (1989)'s decomposition for valuing coupon-bearing bond options, and hence swaptions, can be applied to the model (3.64), the same decomposition is still feasible under (3.66) through straightforward modifications, so that also in the extended model we can price analytically coupon-bearing bond options and swaptions. See also the later section in this chapter being devoted to such issue.

3.8.4 The Vasicek Case

The first application we consider is based on the Vasicek (1977) model. In this case, the basic time-homogeneous model x^α evolves according to (3.5), where the parameter vector is $\alpha = (k,\theta,\sigma)$. Extending this model through (3.66) amounts to have the following short-rate dynamics:

$$dr_t = \left[k\theta + k\varphi(t;\alpha) + \frac{d\varphi(t;\alpha)}{dt} - kr_t\right]dt + \sigma dW_t. \tag{3.74}$$

Moreover, under the specification (3.68), $\varphi(t;\alpha) = \varphi^{VAS}(t;\alpha)$, where

$$\varphi^{VAS}(t;\alpha) = f^M(0,t) + (e^{-kt} - 1)\frac{k^2\theta - \sigma^2/2}{k^2} - \frac{\sigma^2}{2k^2}e^{-kt}(1-e^{-kt}) - x_0 e^{-kt}.$$

The price at time t of a zero-coupon bond maturing at time T is

$$P(t,T)$$
$$= \frac{P^M(0,T)A(0,t)\exp\{-B(0,t)x_0\}}{P^M(0,t)A(0,T)\exp\{-B(0,T)x_0\}} A(t,T)\exp\{-B(t,T)[r_t - \varphi^{VAS}(t;\alpha)]\},$$

and the price at time t of a European call option with strike K, maturity T and written on a zero-coupon bond maturing at time τ is

$$\mathbf{ZBC}(t,T,\tau,K) = \frac{P^M(0,\tau)A(0,t)\exp\{-B(0,t)x_0\}}{P^M(0,t)A(0,\tau)\exp\{-B(0,\tau)x_0\}}$$
$$\cdot \Psi^{VAS}\left(t,T,\tau, K\frac{P^M(0,T)A(0,\tau)\exp\{-B(0,\tau)x_0\}}{P^M(0,\tau)A(0,T)\exp\{-B(0,T)x_0\}}, r_t - \varphi^{VAS}(t;\alpha); \alpha\right),$$

where

$$\Psi^{VAS}(t,T,\tau,X,x;\alpha)$$
$$= A(t,\tau)\exp\{-B(t,\tau)x\}\Phi(h) - XA(t,T)\exp\{-B(t,T)x\}\Phi(h-\bar{\sigma})$$

and

$$h = \frac{1}{\bar{\sigma}} \ln \frac{A(t,\tau)\exp\{-B(t,\tau)x\}}{XA(t,T)\exp\{-B(t,T)x\}} + \frac{\bar{\sigma}}{2},$$

$$\bar{\sigma} = \sigma B(T,\tau)\sqrt{\frac{1-e^{-2k(T-t)}}{2k}}$$

$$A(t,T) = \exp\left[\frac{(B(t,T)-T+t)(k^2\theta-\sigma^2/2)}{k^2} - \frac{\sigma^2 B(t,T)^2}{4k}\right],$$

$$B(t,T) = \frac{1-e^{-k(T-t)}}{k}.$$

We now notice that by defining

$$\vartheta(t) = \theta + \varphi(t;\alpha) + \frac{1}{k}\frac{d\varphi(t;\alpha)}{dt},$$

model (3.74) can be written as

$$dr_t = k(\vartheta(t) - r_t)dt + \sigma\, dW_t \tag{3.75}$$

which coincides with the Hull and White (1994a) extended Vasicek model (3.33).

Vice versa, from (3.75), one can obtain the extension (3.74) by setting

$$\varphi(t;\alpha) = e^{-kt}\varphi(0;\alpha) + k\int_0^t e^{-k(t-s)}\vartheta(s)ds - \theta(1-e^{-kt}).$$

Therefore, this extension of the Vasicek (1977) model is perfectly equivalent to that of Hull and White (1994a), as we can verify also by developing more explicitly our bond and option-pricing formulas. This equivalence is basically due to the linearity of the reference-model equation (3.5). In fact, the extra parameter θ turns out to be redundant since it is absorbed by the time-dependent function ϑ that is completely determined through the fitting of the current term structure of interest rates.

We notice that the function φ^{VAS}, for $\theta = 0$, is related to the function α in (3.36). The only difference is that, in (3.36), r_0 is completely absorbed by the function α, whereas here r_0 is partly absorbed by the reference-model initial condition x_0 and partly by $\varphi^{VAS}(0;\alpha)$.

We finally notice that in the Vasicek case keeping a general x_0 adds no further flexibility to the extended model, because of linearity of the short-rate equation. Therefore, here we can safely set $x_0 = r_0$ and $\varphi^{VAS}(0;\alpha) = 0$ without affecting the model fitting quality.

In the following sections we will develop further examples of models that can be obtained via the deterministic shift extension (3.66).

3.9 The CIR++ Model

The most relevant application of the results of the previous section is the extension of the Cox-Ingersoll-Ross (1985) model, referred to as CIR++. In this case, the process x^α is defined as in (3.21), where the parameter vector is $\alpha = (k, \theta, \sigma)$. The short-rate dynamics is then given by

$$dx(t) = k(\theta - x(t))dt + \sigma\sqrt{x(t)}dW(t), \quad x(0) = x_0,$$
$$r(t) = x(t) + \varphi(t), \qquad (3.76)$$

where x_0, k, θ and σ are positive constants such that $2k\theta > \sigma^2$, thus ensuring that the origin is inaccessible to x, and hence that the process x remains positive.

We calculate the analytical formulas implied by such extension, by simply retrieving the explicit expressions for Π^x and Ψ^x as given in (3.24) and (3.26). Then, assuming exact fitting of the initial term structure of discount factors, we have that $\varphi(t) = \varphi^{CIR}(t;\alpha)$ where

$$\varphi^{CIR}(t;\alpha) = f^M(0,t) - f^{CIR}(0,t;\alpha),$$
$$f^{CIR}(0,t;\alpha) = \frac{k\theta(\exp\{th\} - 1)}{2h + (k+h)(\exp\{th\} - 1)} + x_0 \frac{4h^2 \exp\{th\}}{[2h + (k+h)(\exp\{th\} - 1)]^2} \qquad (3.77)$$

with $h = \sqrt{k^2 + 2\sigma^2}$. Moreover, the price at time t of a zero-coupon bond maturing at time T is

94   3. One-factor short-rate models

$$P(t,T) = \bar{A}(t,T)e^{-B(t,T)r(t)},$$

where

$$\bar{A}(t,T) = \frac{P^M(0,T)A(0,t)\exp\{-B(0,t)x_0\}}{P^M(0,t)A(0,T)\exp\{-B(0,T)x_0\}} A(t,T) e^{B(t,T)\varphi^{CIR}(t;\alpha)},$$

and $A(t,T)$ and $B(t,T)$ are defined as in (3.25).

The spot interest rate at time t for the maturity T is therefore

$$R(t,T) = \frac{\ln \dfrac{P^M(0,t)A(0,T)\exp\{-B(0,T)x_0\}}{A(t,T)P^M(0,T)A(0,t)\exp\{-B(0,t)x_0\}}}{T-t}$$

$$- \frac{B(t,T)\varphi^{CIR}(t;\alpha) - B(t,T)r(t)}{T-t}$$

which is still affine in $r(t)$.

The price at time t of a European call option with maturity $T > t$ and strike price K on a zero-coupon bond maturing at $\tau > T$ is

$$\mathbf{ZBC}(t,T,\tau,K) = \frac{P^M(0,\tau)A(0,t)\exp\{-B(0,t)x_0\}}{P^M(0,t)A(0,\tau)\exp\{-B(0,\tau)x_0\}}$$

$$\cdot \Psi^{CIR}\left(t,T,\tau,K\frac{P^M(0,T)A(0,\tau)\exp\{-B(0,\tau)x_0\}}{P^M(0,\tau)A(0,T)\exp\{-B(0,T)x_0\}}, r(t) - \varphi^{CIR}(t;\alpha); \alpha\right),$$

where $\Psi^{CIR}(t,T,\tau,X,x;\alpha)$ is the CIR option price as defined in (3.26) with $r(t) = x$. By further simplifying this formula, we obtain

$$\mathbf{ZBC}(t,T,\tau,K) =$$

$$P(t,\tau)\chi^2\left(2\hat{r}[\rho + \psi + B(T,\tau)]; \frac{4k\theta}{\sigma^2}, \frac{2\rho^2[r(t) - \varphi^{CIR}(t;\alpha)]\exp\{h(T-t)\}}{\rho + \psi + B(T,\tau)}\right)$$

$$- KP(t,T)\chi^2\left(2\hat{r}[\rho + \psi]; \frac{4k\theta}{\sigma^2}, \frac{2\rho^2[r(t) - \varphi^{CIR}(t;\alpha)]\exp\{h(T-t)\}}{\rho + \psi}\right),$$

(3.78)

with

$$\hat{r} = \frac{1}{B(T,\tau)}\left[\ln \frac{A(T,\tau)}{K} - \ln \frac{P^M(0,T)A(0,\tau)\exp\{-B(0,\tau)x_0\}}{P^M(0,\tau)A(0,T)\exp\{-B(0,T)x_0\}}\right].$$

The analogous put-option price is obtained through the put-call parity (3.4).

Through formula (3.78), we can also price caps and floors since they can be viewed as portfolios of zero bond options.

Let us start by a single caplet value at time t. The caplet resets at time T, pays at time $T + \tau$ and has strike X and nominal amount N.

$$\mathbf{Cpl}(t,T,T+\tau,N,X) = N(1+X\tau)\mathbf{ZBP}\left(t,T,T+\tau,\frac{1}{1+X\tau}\right). \quad (3.79)$$

More generally, as far as a cap is concerned, we denote by $\mathcal{T} = \{t_0, t_1, \ldots, t_n\}$ the set of the cap payment times augmented with the first reset date t_0, and by τ_i the year fraction from t_{i-1} to t_i, $i = 1, \ldots, n$. Applying formulas (2.26), we then obtain that the price at time $t < t_0$ of the cap with cap rate (strike) X, nominal value N and set of times $\mathcal{T}$ is given by

$$\mathbf{Cap}(t,\mathcal{T},N,X) = N\sum_{i=1}^{n}(1+X\tau_i)\mathbf{ZBP}\left(t,t_{i-1},t_i,\frac{1}{1+X\tau_i}\right), \quad (3.80)$$

whereas the price of the corresponding floor is

$$\mathbf{Flr}(t,\mathcal{T},N,X) = N\sum_{i=1}^{n}(1+X\tau_i)\mathbf{ZBC}\left(t,t_{i-1},t_i,\frac{1}{1+X\tau_i}\right). \quad (3.81)$$

European swaptions can be explicitly priced by means of Jamshidian's (1989) decomposition. Indeed, consider a payer swaption with strike rate X, maturity T and nominal value N, which gives the holder the right to enter at time $t_0 = T$ an interest-rate swap with payment times $\mathcal{T} = \{t_1, \ldots, t_n\}$, $t_1 > T$, where he pays at the fixed rate X and receives LIBOR set "in arrears". We denote by τ_i the year fraction from t_{i-1} to t_i, $i = 1, \ldots, n$ and set $c_i := X\tau_i$ for $i = 1, \ldots, n-1$ and $c_n := 1 + X\tau_i$. Denoting by r^* the value of the spot rate at time T for which

$$\sum_{i=1}^{n} c_i \bar{A}(T,t_i) e^{-B(T,t_i)r^*} = 1,$$

and setting $X_i := \bar{A}(T,t_i)\exp(-B(T,t_i)r^*)$, the swaption price at time $t < T$ is then given by

$$\mathbf{PS}(t,T,\mathcal{T},N,X) = N\sum_{i=1}^{n} c_i \mathbf{ZBP}(t,T,t_i,X_i). \quad (3.82)$$

Analogously, the price of the corresponding receiver swaption is

$$\mathbf{RS}(t,T,\mathcal{T},N,X) = N\sum_{i=1}^{n} c_i \mathbf{ZBC}(t,T,t_i,X_i). \quad (3.83)$$

Remark 3.9.1. The T-forward-measure dynamics and distribution of the short rate are easily obtained through (3.66) from (3.27) and (3.28) where r is replaced with x.

3.9.1 The Construction of a Trinomial Tree

Legolas Greenleaf long under tree
In joy thou hast lived. Beware of the Sea!
J.R.R. Tolkien, "The Lord of The Rings"

A binomial tree for the CIR model has been suggested by Nelson and Ramaswamy (1990). Here, instead, we propose an alternative trinomial tree that is constructed along the procedure illustrated in Appendix C.

For convergence purposes, we define the process y as

$$y(t) = \sqrt{x(t)}, \qquad (3.84)$$

where the process x is defined in (3.76). By Ito's lemma,

$$dy(t) = \left[\left(\frac{k\theta}{2} - \frac{1}{8}\sigma^2\right)\frac{1}{y(t)} - \frac{k}{2}y(t)\right]dt + \frac{\sigma}{2}dW(t) \qquad (3.85)$$

We first construct a trinomial tree for y, we then use (3.84) and we finally displace the tree nodes so as to exactly retrieve the initial zero-coupon curve.

Let us fix a time horizon T and the times $0 = t_0 < t_1 < \cdots < t_N = T$, and set $\Delta t_i = t_{i+1} - t_i$, for each i. As usual, the time instants t_i need not be equally spaced. Again, we denote the tree nodes by (i,j) where the time index i ranges from 0 to N and the space index j ranges from some $\underline{j}_i$ to some $\overline{j}_i$.

We denote by $y_{i,j}$ the process value on node (i,j) and set $y_{i,j} = j\Delta y_i$, where $\Delta y_i := V_{i-1}\sqrt{3}$ and $V_i := \sigma\sqrt{\Delta t_i}/2$. We set

$$M_{i,j} := y_{i,j} + \left[\left(\frac{k\theta}{2} - \frac{1}{8}\sigma^2\right)\frac{1}{y_{i,j}} - \frac{k}{2}y_{i,j}\right]\Delta t_i.$$

Assuming that at time t_i we are on node (i,j), with associated value $y_{i,j}$, the process can move to $y_{i+1,k+1}$, $y_{i+1,k}$ or $y_{i+1,k-1}$ at time t_{i+1} with probabilities p_u, p_m and p_d, respectively. The central node is therefore the k-th node at time t_{i+1}, where k is defined by

$$k = \text{round}\left(\frac{M_{i,j}}{\Delta y_{i+1}}\right).$$

Setting $\eta_{j,k} = M_{i,j} - y_{i+1,k}$,[14] we finally obtain

$$\begin{cases} p_u = \frac{1}{6} + \frac{\eta_{j,k}^2}{6V_i^2} + \frac{\eta_{j,k}}{2\sqrt{3}V_i}, \\ p_m = \frac{2}{3} - \frac{\eta_{j,k}^2}{3V_i^2}, \\ p_d = \frac{1}{6} + \frac{\eta_{j,k}^2}{6V_i^2} - \frac{\eta_{j,k}}{2\sqrt{3}V_i}. \end{cases}$$

---
[14] We again omit to express the dependence on the index i to lighten notation.

However, the tree thus defined has the drawback that some nodes may lie below the zero level. Since the tree must approximate a positive process, we truncate the tree below some predefined level $\epsilon > 0$, which can be chosen arbitrarily close to zero, and then suitably define the tree geometry and probabilities around this level.

All the definitions above completely specify the initial-tree geometry, and in particular the minimum and the maximum levels $\underline{j}_i$ and $\overline{j}_i$ at each time step i. The tree for the process x is then built by remembering that, by (3.84), $x(t) = y^2(t)$.

The final stage in our construction procedure consists in suitably shifting the tree nodes in order to obtain the proper tree for r. To this end, we can either apply formula (3.76) with the analytical formula for φ given in (3.77), or employ a similar procedure to that illustrated in Section 3.3 to retrieve the correct discount factors at the initial time.

3.9.2 The Positivity of Rates and Fitting Quality

The use of CIR++ as a pricing model concerns mostly non-standard interest-rate derivatives. The analytical formulas given in this section are used to determine the model parameters α such that the model prices are as close as possible to the selected subset of market prices. An important issue for the CIR++ model is whether calibration to market prices is feasible while imposing positive rates. We know that the CIR++ rates are always positive if

$$\varphi^{CIR}(t;\alpha) > 0 \quad \text{for all } t \geq 0.$$

In turn, this condition is satisfied if

$$f^{CIR}(0,t;\alpha) < f^M(0,t) \quad \text{for all } t \geq 0. \tag{3.86}$$

Studying the behaviour of the function $t \mapsto f^{CIR}(0,t;\alpha)$ is helpful in addressing the positivity issue. In particular, we are interested in its supremum

$$f^*(\alpha) := \sup_{t \geq 0} f^{CIR}(0,t;\alpha) .$$

There are three possible cases of interest:

i) $x_0 \leq \theta h/k$: In this case $t \mapsto f^{CIR}(0,t;\alpha)$ is monotonically increasing and the supremum of all its values is

$$f_1^*(\alpha) = \lim_{t \to \infty} f^{CIR}(0,t;\alpha) = \frac{2k\theta}{k+h};$$

ii) $\theta h/k < x_0 < \theta$: In this case $t \mapsto f^{CIR}(0,t;\alpha)$ takes its maximum value in

$$t^* = \frac{1}{h} \ln \frac{(x_0 h + k\theta)(h-k)}{(x_0 h - k\theta)(h+k)} > 0$$

and such a value is given by

$$f_2^*(\alpha) = x_0 + \frac{(x_0 - \theta)^2 k^2}{2\sigma^2 x_0};$$

iii) $x_0 \geq \theta$: In this case $t \mapsto f^{CIR}(0,t;\alpha)$ is monotonically decreasing for $t > 0$ and the supremum of all its values is

$$f_3^*(\alpha) = f^{CIR}(0,0;\alpha) = x_0.$$

We can try and enforce positivity of φ^{CIR} analytically in a number of ways. To ensure (3.86) one can for example impose that the market curve $t \mapsto f^M(0,t)$ remains above the corresponding CIR curve by requiring

$$f^*(\alpha) \leq \inf_{t \geq 0} f^M(0,t).$$

This condition ensures (3.86), but appears to be too restrictive. To fix ideas, assume that, calibrating the model parameters to market cap and floor prices, we constrain the parameters to satisfy

$$x_0 > \theta.$$

Then we are in case iii) above. All we need for ensuring the positivity of φ^{CIR} in this case is

$$x_0 < \inf_{t \geq 0} f^M(0,t).$$

This amounts to say that the initial condition of the time-homogeneous part of the model has to be placed below the whole market forward curve. On the other hand, since $\theta < x_0$ and θ is the mean-reversion level of the time-homogeneous part, this means that the time-homogeneous part will tend to decrease, and indeed we have seen that in case iii) $t \mapsto f^{CIR}(0,t;\alpha)$ is monotonically decreasing. If, on the contrary, the market forward curve is increasing (as is happening in the most liquid markets nowadays) the "reconciling role" of φ^{CIR} between the time-homogeneous part and the market forward curve will be stronger. Part of the flexibility of the time-homogeneous part of the model is then lost in this strong "reconciliation", thus subtracting freedom which the model can otherwise use to improve calibration to caps and floors prices.

In general, it turns out in real applications that the above requirements are too strong. Constraining the parameters to satisfy

$$\theta < x_0 < \inf_{t \geq 0} f^M(0,t)$$

leads to a caps/floors calibration whose quality is much lower than we have with less restrictive constraints. For practical purposes it is in fact enough to impose weaker restrictions. Such restrictions, though not guaranteeing

3.9 The CIR++ Model

positivity of φ^{CIR} analytically, work well in all the market situation we tested. First consider the case of a monotonically increasing market curve $t \mapsto f^M(0,t)$. We can choose a similarly increasing $t \mapsto f^{CIR}(0,t;\alpha)$ (case i)) starting from below the market curve, $x_0 < r_0$, and impose that its asymptotic limit

$$\frac{2k\theta}{k+h} \approx \theta$$

be below the corresponding market-curve limit. If we do this, calibration results are satisfactory and we obtain usually positive rates. Of course we have no analytical certainty. The f^{CIR} curve might increase quicker than the market one, f^M, for small t's, cross it, and then increase more slowly so as to return below f^M for large t's. In such a case φ^{CIR} would be negative in the in-between interval. However, this situation is very unlikely and never occurred in the real-market situations we tested.

Next, consider a case with a decreasing market curve $t \mapsto f^M(0,t)$. In this case we take again $x_0 < r_0$ and we impose the same condition as before on the terminal limit for $t \to \infty$.

Similar considerations apply in the case of an upwardly humped market curve $t \mapsto f^M(0,t)$, which can be reproduced qualitatively by the time-homogeneous-model curve $t \mapsto f^{CIR}(0,t;\alpha)$. In this case one makes sure that the initial point, the analytical maximum and the asymptotic value of the time-homogeneous-CIR curve remain below the corresponding points of the market curve.

Finally, the only critical situation is the case of an inverted yield curve, as was observed for example in the Italian market in the past years. The forward curve $t \mapsto f^{CIR}(0,t;\alpha)$ of the CIR model cannot mimic such a shape. Therefore, either we constrain it to stay below the inverted market curve by choosing a decreasing CIR curve (case iii)) starting below the market curve, i.e. $x_0 < r_0$, or we make the CIR curve start from a very small x_0 and increase, though not too steeply. The discrepancy between f^{CIR} and f^M becomes very large for large t's in the first case and for small t's in the second one. This feature lowers the quality of the caps/floors fitting in the case of an inverted yield curve if one wishes to maintain positive rates. However, highly inverted curves are not so common in liquid markets, so that this problem can be generally avoided.

As far as the quality of fitting for the caps volatility curve is concerned, we notice the following. If the initial point in the market caps volatility curve is much smaller than the second one, so as to produce a large hump, the CIR++ model has difficulties in fitting the volatility structure. We observed through numerical simulations that, ceteris paribus, lowering the initial point of the caps volatility structure implied by the model roughly amounts to lowering the parameter x_0. This is consistent with the following formula

$$\sigma \sqrt{x_0} \; \frac{2h \, \exp(Th)}{[2h + (k+h)(\exp\{Th\} - 1)]^2}$$

for the volatility of instantaneous forward rates $f^{CIR}(t,T;\alpha)$ at the initial time $t = 0$ for the maturity T. Therefore, a desirable fitting of the caps volatility curve can require low values of x_0, in agreement with one of the conditions needed to preserve positive rates.

3.10 Deterministic-Shift Extension of Lognormal Models

The third example we consider is the extension of the Dothan (1978) model. This extension yields a "quasi" lognormal short-rate model that fits any given yield curve and for which there exist analytical formulas for zero-coupon bonds, i.e.,

$$r(t) = x(t) + \varphi(t),$$
$$x(t) = x_0 \exp\left\{\left(a - \frac{1}{2}\sigma^2\right)t + \sigma W(t)\right\}, \qquad (3.87)$$

where x_0, a and σ are real constants and φ is a deterministic function of time.

The introduction of a deterministic shift in the Dothan (1978) short-rate dynamics (3.17) implies that the expected long-term rate is not just zero or infinity, depending on the sign of the parameter a, but can have any value, given the asymptotic behaviour of the function φ.

The analytical formulas for bond prices in the extended Dothan model are simply obtained by combining (3.70) with (3.20). Equivalently to (3.20), however, the resulting bond price formula is rather involved, requiring double numerical integrations, so that implementing an approximating tree for the process x may be easier and not necessarily more time consuming.

The construction of a binomial tree for pricing interest-rate derivatives under this extended Dothan model is rather straightforward. In fact, we just have to build a tree for x, the time-homogeneous part of the process r, and then shift the tree nodes at each time period by the corresponding value of φ.

Since (3.17) is a geometric Brownian motion, the well consolidated Cox-Ross-Rubinstein (1979) procedure can be used here, thus rendering the tree construction extremely simple while simultaneously ensuring well known results of convergence in law. To this end, we denote by N the number of time-steps in the tree and with T a fixed maturity. We then define the following coefficients

$$u = e^{\sigma\sqrt{\frac{T}{N}}}, \qquad (3.88)$$
$$d = e^{-\sigma\sqrt{\frac{T}{N}}}, \qquad (3.89)$$
$$p = \frac{e^{a\frac{T}{N}} - d}{u - d}, \qquad (3.90)$$

3.10 Deterministic-Shift Extension of Lognormal Models

and build inductively the tree for x starting from x_0. Denoting by $\hat{x}$ the value of x at a certain node of the tree, the value of x in the subsequent period can either go up to $\hat{x}u$ with probability p or go down to $\hat{x}d$ with probability $1-p$. Notice that the probabilities p and $1-p$ are always well defined for a sufficiently large N, both tending to $\frac{1}{2}$ for N going to infinity.

The tree for the short-rate process r is finally constructed by displacing the previous nodes through the function φ. When exactly fitting the current term-structure of interest rates, the displacement of the tree nodes can be done, for instance, through a procedure that is similar to that illustrated in Section 3.3.3.

The last example we consider is the extension of the Exponential-Vasicek model (3.30). To this end, we remark that the extension procedure of the previous section, though particularly meaningful when the original model is time-homogeneous and analytically tractable, is quite general and can be in principle applied to any endogenous term-structure model.

The short-rate process under such extension is explicitly given by

$$r(t) = x(t) + \varphi(t)$$
$$x(t) = \exp\left\{\ln x_0 e^{-at} + \frac{\theta}{a}\left(1 - e^{-at}\right) + \sigma \int_0^t e^{-a(t-u)} dW(u)\right\}, \quad (3.91)$$

where x_0, θ, a and σ are real constants and φ is a deterministic function of time.

The Exponential-Vasicek model does not imply explicit formulas for pure discount bonds. However, we can easily construct an approximating tree for the reference process x, whose dynamics are defined by (3.30). In fact, defining the process z as

$$dz(t) = -az(t)dt + \sigma dW(t), \quad z(0) = 0,$$

by Ito's lemma, we obtain

$$x(t) = \exp\left(z(t) + \left(\ln x_0 - \frac{\theta}{a}\right)e^{-at} + \frac{\theta}{a}\right). \quad (3.92)$$

We can then use the procedure illustrated in Section 3.3.3 to construct a trinomial model for z, apply (3.92) and finally displace the tree node so as to exactly retrieve the initial term structure of discount factors.

This Extended Exponential-Vasicek (EEV) model leads to a fairly good calibration to cap prices in that the resulting model prices lie within the band formed by the market bid and ask prices. Indeed, this model clearly outperforms the classical lognormal short-rate model (3.54). An example of the quality of the model fitting to at-the-money cap prices is shown later in this chapter.

A final word has to be spent on a further advantage of the above extensions of classical lognormal model. In fact, the assumption of a shifted lognormal

102   3. One-factor short-rate models

distribution can be quite helpful when fitting volatility smiles. Notice, indeed, that the Black formula for cap prices is based on a lognormal distribution for the forward rates, and that a practical way to obtain implied-volatility smiles is by shifting the support of this distribution by a quantity to be suitably determined. However, the major drawback of these extensions is that the positivity of the short-rate process cannot be guaranteed any more since its distribution is obtained by shifting a lognormal distribution by a possibly negative quantity.

3.11 Some Further Remarks on Derivatives Pricing

In this section we first show how to price analytically a European option on a coupon-bearing bond under an analytically-tractable short-rate model. The method we describe is a natural generalization of the Jamshidian (1989) decomposition, originally developed for the Vasicek (1977) model. We then show how to price path-dependent derivatives through a Monte Carlo simulation approach. We finally explain how to price early-exercise products by means of a (trinomial) tree.

3.11.1 Pricing European Options on a Coupon-Bearing Bond

We assume that the short-rate model is analytically tractable and denote the analytical price at time t of the zero coupon bond maturing at time T by $\Pi(t, T, r(t))$, where the dependence on the short rate at time t is explicitly written.

We consider a coupon-bearing bond paying the cash flows $\mathcal{C} = [c_1, \ldots, c_n]$ at maturities $\mathcal{T} = [T_1, \ldots, T_n]$. Let $T \leq T_1$. The price of our coupon-bearing bond in T is given by

$$\mathbf{CB}(T, \mathcal{T}, \mathcal{C}) = \sum_{i=1}^{n} c_i P(T, T_i) = \sum_{i=1}^{n} c_i \Pi(T, T_i, r(T)).$$

We want to price at time t a European put option on the coupon-bearing bond with strike price K and maturity T. The option payoff is

$$[K - \mathbf{CB}(T, \mathcal{T}, \mathcal{C})]^+ = \left[K - \sum_{i=1}^{n} c_i \Pi(T, T_i, r(T))\right]^+.$$

Jamshidian (1989) devised a simple method to convert this positive part of a sum into the sum of positive parts. His trick was based on finding the solution r^* of the following equation

$$\sum_{i=1}^{n} c_i \Pi(T, T_i, r^*) = K,$$

3.11 Some Further Remarks on Derivatives Pricing   103

and rewriting the payoff as

$$\left[\sum_{i=1}^{n} c_i\bigl(\Pi(T,T_i,r^*) - \Pi(T,T_i,r(T))\bigr)\right]^+.$$

To achieve the desired decomposition we need a condition that was automatically verified in the case of the Vasicek (1977) model treated by Jamshidian. We therefore assume that the short-rate model satisfies the following assumption:

$$\frac{\partial \Pi(t,s,r)}{\partial r} < 0 \quad \text{for all } 0 < t < s.$$

It is easy to see, for instance, that both the Hull and White (3.33) and the CIR++ (3.76) models satisfy this assumption. Under such condition, the payoff can be rewritten as

$$\sum_{i=1}^{n} c_i \bigl[\Pi(T,T_i,r^*) - \Pi(T,T_i,r(T))\bigr]^+,$$

so that pricing our coupon-bond option becomes equivalent to value a portfolio of put options on zero-coupon bonds. If we take the risk-neutral expectation of the discounted payoff, we obtain the price at time t of the coupon-bearing-bond option with maturity T and strike K:

$$\mathbf{CBP}(t,T,\mathcal{T},\mathcal{C},K) = \sum_{i=1}^{n} c_i \mathbf{ZBP}(t,T,T_i,\Pi(T,T_i,r^*)).$$

An analogous relationship holds for a European call so that we can write

$$\mathbf{CBO}(t,T,\mathcal{T},\mathcal{C},K,\omega) = \sum_{i=1}^{n} c_i \mathbf{ZBO}(t,T,T_i,\Pi(T,T_i,r^*),\omega),$$

where $\omega = 1$ for a call and $\omega = -1$ for a put.

3.11.2 The Monte Carlo Simulation

This subsection comments on the use of a short-rate model for Monte Carlo pricing of path-dependent interest-rate derivatives. We may start with a few remarks on the Monte Carlo method. We will reconsider this type of remarks in Section 10.12.

The category of products that is usually considered for Monte Carlo pricing is the family of "path-dependent" payoffs. These products are to be exercised only at a final time, and their final payoffs involve the history of the underlying variable up to the final time, and not only the final value of the

104   3. One-factor short-rate models

underlying variable. The Monte Carlo method works through forward propagation in time of the key variables, by simulating their transition density between dates where the key-variables history matters to the final payoff. Monte Carlo is thus ideally suited to "travel forward in time". Instead, the Monte Carlo method has problems with early exercise. Since with Monte Carlo we propagate trajectories forward in time, we have no means to know whether at a certain point in time it is optimal to continue or to exercise. Therefore, standard Monte Carlo cannot be used for products involving early exercise, although in Section 10.12 we present a recent method that proposes a remedy to this situation.

Now assume we want to price, at the current time $t = 0$, a path-dependent payoff with European exercise features. The payoff is a function of the values $r(t_1), \ldots, r(t_m)$ of the instantaneous interest rate at preassigned time instants $0 < t_1 < t_2 < \ldots < t_m = T$, where T is the final maturity. Let us denote the given discounted payoff (payments of different additive components occur at different times and are discounted accordingly) by

$$\sum_{j=1}^{m} \exp\left[-\int_0^{t_j} r(s)\,ds\right] H(r(t_1), \ldots, r(t_j)). \tag{3.93}$$

Usually the payoff depends on the instantaneous rates $r(t_i)$ through the simple spot rates $L(t_i, t_i + \tau, r(t_i)) := [1/P(t_i, t_i + \tau) - 1]/\tau$ (typically six-month rates, i.e. $\tau = 0.5$). The availability of analytical formulas for such L's simplifies considerably the Monte Carlo pricing. Indeed, without analytical formulas one should compute the L's from bond prices obtained through trees or through numerical approximations of the solution of the bond price PDE. This should be done for every path, so as to increase dramatically the computational burden.

Pricing the generic additive term in the payoff (3.93) by a Monte Carlo method involves simulation of p paths of the short-rate process r and computing the arithmetic mean of the p values assumed by the discounted payoff along each path.[15]

These paths are simulated after defining $q + 1$ sampling times $0 = s_0 < s_1 < s_2 < \ldots < s_q = T$, which have of course to include the times t_j's. We set $\Delta s_i := s_{i+1} - s_i$, $i = 0, \ldots, q - 1$.

When pricing under the risk-neutral measure one typically approximates a discounting term like

$$\exp\left[-\int_{s_k}^{s_j} r(s)ds\right]$$

with

---
[15] If no variance reduction technique is employed, p typically ranges from $p = 100,000$ to $p = 1,000,000$.

3.11 Some Further Remarks on Derivatives Pricing

$$\exp\left(-\sum_{i=k}^{j-1} r(s_i)\Delta s_i\right).$$

In order for the approximation to be accurate, the Δs_i's have to be small, which can be quite burdensome from the computational point of view. If the short-rate model is analytically tractable, it is therefore advisable to work under the T-forward adjusted measure, since (see also Chapter 2)

$$E\left\{\exp\left[-\int_0^{t_j} r(s)ds\right] H(r(t_1),\ldots,r(t_j))\right\} = P(0,T)E^T\left\{\frac{H(r(t_1),\ldots,r(t_j))}{P(t_j,T)}\right\}. \tag{3.94}$$

Notice that the term $P(t_j,T)$ is determined analytically from the simulated $r(t_j)$, so that no further simulation is required to compute it.

In order to obtain the simulated paths, two approaches are possible:

1. Sample at each time step the exact transition density from $r(s_i)$ to $r(s_{i+1})$, $i = 0,\ldots,q-1$, under the T-forward-adjusted measure;
2. Discretize the SDE for r, under the T-forward-adjusted measure, via the Euler or the Milstein scheme (see Klöden and Platen (1995)).

In the latter case, one has simply to sample at each time s_i the distribution of $W^T_{s_i+\Delta s_i} - W^T_{s_i}$, which is normal with mean 0 and variance Δs_i. Since the increments of the Brownian motion are independent, these normal samples are drawn from variables which are independent in different time intervals, so that one can generate a priori $p \times q$ independent realizations of such Gaussian random variables.

Considering as an example the CIR++ model, we can then either sample the exact noncentral-chi-squared transition density from s_i to s_{i+1}, as given by combining (3.66) with (3.28) and where $s = s_i$ and $t = s_{i+1}$, or employ the following Milstein scheme for the reference process x

$$x(s_i + \Delta s_i) = x(s_i) + [k\theta - (k + B(s_i,T)\sigma^2)x(s_i)]\,\Delta s_i$$
$$- \frac{\sigma^2}{4}\left((W^T_{s_i+\Delta s_i} - W^T_{s_i})^2 - \Delta s_i\right) + \sigma\sqrt{x(s_i)}\left(W^T_{s_i+\Delta s_i} - W^T_{s_i}\right).$$

The price of our derivative is finally obtained by calculating the payoff value

$$h_i := \sum_{j=1}^{m} \frac{H(r(t_1),\ldots,r(t_j))}{P(t_j,T)}$$

associated with each simulated path, averaging all payoff values and then discounting to the current time, thus yielding

$$P(0,T)\frac{\sum_{i=1}^{p} h_i}{p}.$$

3.11.3 Pricing Early-Exercise Derivatives with a Tree

Soldati
Bosco di Courton luglio 1918
Si sta come
d'autunno
sugli alberi
le foglie

[Soldiers

We remain
like leaves
on the trees
in autumn]

Giuseppe Ungaretti, translation by William Fense Weaver.

All the short-rate models we have described in this chapter can be approximated with lattices. For some of the models we introduced, we have detailed the construction of a trinomial tree based on the discretization of the rate dynamics under the risk-neutral measure. In this subsection, we show how to price a general (non-path-dependent) derivative with such a numerical procedure. This is particularly useful when pricing early-exercise products, as we will also see by analyzing the particular case of a Bermudan-style swaption. For a general and informal introduction to the use of trees as opposed to Monte Carlo and other methods see also Section 10.12. We now present a few general remarks on pricing with trees, and then move to the details.

The pricing of early-exercise products can be carried out through binomial/trinomial trees, when the fundamental underlying variable is low-dimensional (say one or two-dimensional). This is the case of the short-rate models we are considering here and in the next chapter. In these cases, the tree is the ideal instrument, given its "backward-in-time" nature. The value of the payoff in each final node is known, and we can move backward in time, thus updating the value of continuation through discounting. At each node of the tree we compare the "backwardly-cumulated" value of continuation with the payoff evaluated at that node ("immediate-exercise value"), thus deciding whether exercise is to be considered or not at that point. Once this exercise decision has been taken, the backward induction restarts and we continue to propagate backwards the updated value. Upon reaching the initial node of the tree (at time 0) we have the approximated price of our early-exercise product. It is thus clear that trees are ideally suited to "travel backward in time".

Trees may have problems with path-dependent products, for which we have seen Monte Carlo to be ideally suited. Indeed, in this case when we try and propagate backwards the contract value from the final nodes we are immediately in trouble, since to value the payoff at any final node we need to know the past history of the underlying variable. But this past history is not determined yet, since we move backward in time. Although there are ad-hoc procedures to render trees able to price particular path-dependent products in the Black and Scholes setup (barrier and lookback options), in general there is no efficient and universally accepted method for using a tree with path-dependent payoffs.

Now assume we have selected a short-rate model and a related tree. In order to fix ideas, assume we have selected a trinomial tree, constructed

3.11 Some Further Remarks on Derivatives Pricing

according to our general procedure of Appendix C. We thus have a finite set of times $0 = t_0 < t_1 < \cdots < t_n = T$ and, at each time t_i, a finite number of states. The time-horizon T is the longest maturity that is relevant for pricing a given derivative. The j-th node at time t_i is denoted by (i,j), with associated short-rate value $r_{i,j}$, and $i = 0, \ldots, n$ with j ranging, for each fixed i, from some $\underline{j}_i$ to some $\overline{j}_i$.

The tree branching is shown in Figure 3.6: assuming that, at time t_{i-1}, we are on node $(i-1,j)$, we can move to the three nodes $(i, k+1)$, (i,k) or $(i, k-1)$ at time t_i, with probabilities p_u, p_m and p_d, respectively, where the indices i, j and k in the probabilities are omitted for brevity.

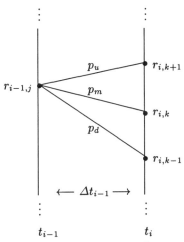

Fig. 3.6. Evolution of the process r starting from $r_{i-1,j}$ at time t_{i-1} and moving to $r_{i,k+1}$, $r_{i,k}$ or $r_{i,k-1}$ at time t_i with probabilities p_u, p_m and p_d, respectively.

A first simple payoff. We start by considering the simple case of a derivative whose payoff at time T is given by a function $H(T, r_T)$. Denoting by $h(t, r_t)$ the arbitrage-free price of the claim at time t, when the short rate is r_t, we have clearly $h(T, r_T) = H(T, r_T)$ and

$$\begin{aligned} h(0, r_0) &= E\left\{ e^{-\int_0^T r_t \, dt} h(T, r_T) \right\} \\ &= E\left\{ e^{-\int_0^{t_{n-1}} r_t \, dt} E\left[e^{-\int_{t_{n-1}}^T r_t \, dt} h(T, r_T) | \mathcal{F}_{t_{n-1}} \right] \right\} \\ &= E\left\{ e^{-\int_0^{t_{n-1}} r_t \, dt} h(t_{n-1}, r_{t_{n-1}}) \right\} \\ &= E\left\{ e^{-\int_0^{t_1} r_t \, dt} h(t_1, r_{t_1}) \right\}, \end{aligned}$$

where we have applied the tower property of conditional expectations. The generic value $h(t_i, r_{t_i})$ is calculated by iteratively taking expectations of the discounted values at later times, i.e.

$$h(t_{i-1}, r_{t_{i-1}}) = E\left[e^{-\int_{t_{i-1}}^{t_i} r_t dt} h(t_i, r_{t_i}) | \mathcal{F}_{t_{i-1}}\right], \tag{3.95}$$

so that the derivative value at any time can be calculated iteratively, starting from its time-T payoff. This is the property that is employed for the price calculation with the tree. Precisely, the following procedure based on backward induction is used.

We denote by $h_{i,j}$ the derivative value on node (i,j) and set at the final nodes

$$h_{n,j} := h(T, r_{n,j}) = H(T, r_{n,j}). \tag{3.96}$$

At the final time $t = t_n = T$, the derivative values on the tree nodes are known through this payoff condition (3.96). We now mode backwards to the time-t_{n-1} nodes and apply the general rule (3.95) with $i = n$.

Starting from the lowest level $j = \underline{j}_{n-1}$ up to $j = \overline{j}_{n-1}$, we use the approximation

$$h(t_{n-1}, r_{t_{n-1}}) \approx e^{-r_{t_{n-1}}(T - t_{n-1})} E\left[h(T, r_T) | \mathcal{F}_{t_{n-1}}\right]$$

to calculate the derivative value on the generic node $(n-1, j)$ as the discounted expectation of the corresponding values on nodes $(n, k+1)$, (n, k) and $(n, k-1)$:

$$h_{n-1,j} = e^{-r_{n-1,j}(T - t_{n-1})}[p_u h_{n,k+1} + p_m h_{n,k} + p_d h_{n,k-1}].$$

We then move to time-t_{n-2} nodes, apply again (3.95) with $i = n - 1$, and calculate discounted expectations as in the previous step.

Proceeding like this, the generic step backwards between time t_{i+1} and t_i is described by

$$h_{i,j} = e^{-r_{i,j}(t_{i+1} - t_i)}[p_u h_{i+1,k+1} + p_m h_{i+1,k} + p_d h_{i+1,k-1}],$$

and is illustrated in Figure 3.7. We remind that the indices i, j and k in the probabilities have been omitted for brevity.

We keep on moving backwards along the tree until we reach the initial node $(0,0)$, whose associated value $h_{0,0}$ gives the desired approximation for the derivative price $h(0, r_0)$.

A first simple payoff with early exercise. Now take the same payoff as before but assume that the holder of the contract has the right to exercise it at any instant t before final time T, receiving upon exercise at time t the amount $H(t, r_t)$ (typical examples are American options). Since we plan to use a tree with time instants t_i's, with no loss of generality we assume that early exercise can occur "only" at the instants t_i's.

3.11 Some Further Remarks on Derivatives Pricing

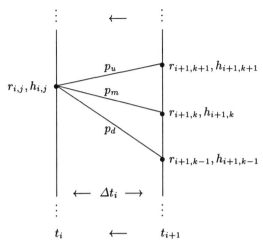

Fig. 3.7. Backward-induction step for the payoff evaluation on the tree moving back from time t_{i+1} to time t_i.

We need to modify the generic backward-induction step in the tree as follows: the generic step backwards between time t_{i+1} and t_i is now described by

$$h_{i,j} = \max\left(e^{-r_{i,j}(t_{i+1}-t_i)}[p_u h_{i+1,k+1} + p_m h_{i+1,k} + p_d h_{i+1,k-1}], H(t_i, r_{i,j})\right).$$

Indeed, what we are doing is the following. We roll back the "backwardly-cumulated" value $h_{i+1,\cdot}$ at time t_{i+1} down to time t_i, and then compare this backwardly-cumulated value

$$e^{-r_{i,j}(t_{i+1}-t_i)}[p_u h_{i+1,k+1} + p_m h_{i+1,k} + p_d h_{i+1,k-1}]$$

with the "immediate-exercise" value we would obtain by exercising the contract, i.e. with

$$H(t_i, r_{i,j}).$$

We then take the best of the two choices, i.e. the one that maximizes the value. Some contracts such as Bermudan-style swaptions only allow for exercise at a selected set of times that are typically one-year spaced, thus resulting in a much reduced subset of exercise times than the tree's t_i's. Then, when moving backwards along the tree, the comparison between the "backwardly-cumulated" value and the "immediate-exercise" value occurs only at times where exercise is allowed. In the remaining instants, the backward-induction step is the same as in the previous "no-early-exercise" case.

110   3. One-factor short-rate models

A second payoff. A second slightly more complicated example of payoff at time T is the following:

$$g(T, P(T,T_1), P(T,T_2), \ldots, P(T,T_m)), \qquad (3.97)$$

where $P(T,T_1), P(T,T_2), \ldots, P(T,T_m)$ are the bond prices at time T for increasing maturities $T_1, T_2, \ldots, T_m$, with $T_1 > T$. This is the typical case one encounters for payoffs depending, for instance, on LIBOR or swap rates, since they can be written in terms of zero-coupon bond prices.

If we initially selected an analytically-tractable short-rate model, for example the Hull-White (1994a) model (3.33) or the CIR++ model (3.76), explicit formulas for bond prices are available as functions of time and short-rate value: $P(t,S) = \Pi(t,S;r_t)$ with Π an explicit function. Therefore, in such a case, the payoff (3.97) can be replaced with some $\hat{g}(T, r_T)$ and thus be priced exactly as the first payoff above, with the tree being constructed until time T.

Instead, if we are dealing with a model like Black-Karasinski's (3.54), for instance, no analytical formula for bonds is available. In such a case, we are then compelled to construct the tree until the last relevant maturity, i.e. T_m.

Each bond value at time T is obtained, through backward induction, by assigning value 1 to all the tree nodes at the corresponding maturity. Notice that we need to propagate the whole vector of bond prices backwards through time, together with the short-rate and the backwardly-cumulated value. Therefore, in each tree node at a given time t_i we need to store all bond prices that have already come to life at that time, i.e. all $P(t_i, T_l)$ with $T_l > t_i$, and keep on propagating all of them backwards. Moreover, each time we reach a new maturity $t_i = T_l$, we need to add a component to the vector, and set to one the value of that new component. This component represents the bond $P(\cdot, T_l)$, that has come to life now and whose (terminal) current value is clearly 1. We thus see that, as we move backwards, the dimension of the bond vector to be stored at each node may increase. For an example see the scheme for Bermudan swaptions below.

Since in each node we have the whole zero-bond curve at the relevant instants, now from time T down to time 0 the calculation procedure is exactly equivalent to the previous ones. And again, in case early exercise is introduced, in the tree nodes at each time where early exercise is allowed we have to take the maximum between the backwardly-cumulated value and the immediate-exercise value.

Remark 3.11.1. (**Pricing early-exercise derivatives with a tree in multi-factor short-rate models**) The backward-induction paradigm illustrated here can be adapted to the two-factor models we will discuss in the next chapter, provided the tree for r is replaced with the two-dimensional tree for the two factors x and y and that the backward induction is modified accordingly.

3.11.4 A Fundamental Case of Early Exercise: Bermudan-Style Swaptions.

An important case is given by Bermudan swaptions, which we discuss now. Bermudan swaptions will be defined in larger detail in Section 10.13. We briefly introduce them here in order to illustrate how the tree works in an important case. Consider an interest-rate swap first resetting in T_α and paying at $T_{\alpha+1}, \ldots, T_\beta$, with fixed rate K, and year fractions $\tau_{\alpha+1}, \ldots, \tau_\beta$. Assume one has the right to enter the IRS at any of the following reset times:

$$T_h, T_{h+1}, \ldots, T_k, \quad \text{with} \quad T_\alpha \leq T_h < T_k < T_\beta.$$

The contract thus described is a Bermudan swaption of payer (receiver) type if the IRS is a payer (receiver) IRS.

Here we have a clear case of early exercise. Suppose we hold the contract. We can either wait for the final maturity T_k and then exercise our right to enter the "final" IRS with first reset date T_k and last payment in T_β, or exercise at any earlier time T_l (with $T_h \leq T_l < T_k$) and then enter the IRS with first reset date T_l and last payment in T_β.

In order to value such a contract, we have to discretize time.

For each l, select a set of times t^l forming a partition of $[T_l, T_{l+1}]$ as follows:

$$T_l = t_1^l < t_2^l < \cdots < t_{d(l)-1}^l < t_{d(l)}^l = T_{l+1}.$$

We then price the Bermudan swaption through the following method.

Build the usual trinomial tree for the short rate r with time instants

$$t_1^1, \ldots, t_{d(1)}^1 = t_1^2, \ldots, t_{d(2)}^2 = t_1^3, \ldots, t_{d(\beta)}^\beta.$$

Denote the short-rate value on the generic j-node at time t_i^l by $r_{i,j}^l$. Denote by $P_{i,j}^l(T_s)$ the bond price $P(t_i^l, T_s)$ for the maturity $T_s > t_i^l$ in the generic tree node hosting $r_{i,j}^l$. Lest one be lost in indices, we recall that, in $P_{i,j}^l(T_s)$:

- l specifies inside which $[T_l, T_{l+1}]$ discretization is occurring;
- i specifies at which time instant t_i inside $[T_l, T_{l+1}]$ the current discretization step is occurring;
- j specifies in which (vertical) "spatial" node in the "time-t_i column" of the tree we are acting;
- T_s specifies which bond we are propagating.

The pricing scheme runs as follows.

1. Set $l + 1 = \beta$.
2. (Positioning in a new $[T_l, T_{l+1}]$ and adding a new zero-coupon bond). Set $P_{d(l),j}^l(T_{l+1}) = 1$ for all j. Set $i + 1 = d(l)$.

112   3. One-factor short-rate models

3. (Backward induction inside $[T_l, T_{l+1}]$).
   While going backwards from time t^l_{i+1} to t^l_i in the tree, propagate backwards the vector of bond prices as follows:

   $$P^l_{i,j}(T_s) = e^{-r^l_{i,j}(t^l_{i+1}-t^l_i)}[p_u P^l_{i+1,k+1}(T_s) + p_m P^l_{i+1,k}(T_s) \\ + p_d P^l_{i+1,k-1}(T_s)],$$

   for all $s = l+1, \ldots, \beta$. Store in the j-node of the current time t^l_i the values of the above bond prices. This step is completely analogous to rolling back a general payoff h, as we have seen in earlier sections and as is illustrated in Figure 3.7. The notation is slightly more complicated due to the presence of the l partition index.

4. if $i > 1$ then decrease i by one and go back to the preceding point.

5. (Initial time T_l of the current partition reached).
   Since now $i = 1$, we have reached $t^l_1 = T_l$. If $l > k$, decrease l by one and go back to point 2, otherwise move on to point 6.

6. (Last exercise-time reached).
   Since now $l = k$, by going backwards we have reached the last point in time (first in our backward direction) where the swaption can be exercised.

7. (Checking the exercise opportunity in each node of the current time).
   Compute, for each level j in the current "column" of the tree, the time-T_l value of the underlying IRS with first reset date T_l and last payment in T_β, based on the backwardly-propagated bond prices

   $$P^l_{1,j}(T_{l+1}), P^l_{1,j}(T_{l+2}), \ldots, P^l_{1,j}(T_\beta),$$

   i.e.

   $$\text{IRS}^l_j = 1 - P^l_{1,j}(T_\beta) - \sum_{s=l+1}^{\beta} \tau_s K P^l_{1,j}(T_s).$$

   If $l = k$ then define the backwardly-Cumulated value from Continuation (CC) of the Bermudan swaption as this IRS value in each node j of the current time level in the tree,

   $$\text{CC}^{k-1}_{d(k-1),j} := \text{IRS}^k_j.$$

   Else, if $l < k$, check the exercise opportunity as follows: if the underlying IRS is larger than the backwardly-cumulated value from continuation, then set the CC value equal to the IRS value. In our notation, for each node j in the current column of the tree:
   If $\text{IRS}^l_j > \text{CC}^l_{1,j}$ then $\text{CC}^l_{1,j} \leftarrow \text{IRS}^l_j$.
   Store such CC values in the corresponding nodes of the tree. Decrease l by one and move to the next step.

3.11 Some Further Remarks on Derivatives Pricing

8. (Positioning in a new $[T_l, T_{l+1}]$).
   Set $P^l_{d(l),j}(T_{l+1}) = 1$ for all nodes j in the current time level. Set $i+1 = d(l)$.
9. (Backward induction inside $[T_l, T_{l+1}]$).
   Calculate backwards from time t^l_{i+1} to t^l_i: i) the vector of bond prices, and ii) the backwardly-cumulated value from continuation (CC) of the Bermudan swaption.
   The P's propagation is as in point 3 above, and the CC update is analogous:

$$CC^l_{i,j} = e^{-r^l_{i,j}(t^l_{i+1}-t^l_i)}[p_u CC^l_{i+1,k+1} + p_m CC^l_{i+1,k} + p_d CC^l_{i+1,k-1}]. \quad (3.98)$$

   Store in each node j of the current time t^l_i the values of the above bond prices and the value from continuation $CC^l_{i,j}$. Again, this step is completely analogous to rolling back a general payoff h, as we have seen in earlier sections and as is illustrated in Figure 3.7.
10. if $i > 1$ then decrease i by one and go back to the preceding point 9, otherwise move on to the next point.
11. (New exercise time T_l reached).
    Since now $i = 1$, we have reached $t^l_1 = T_l$. If $l > h$, we are still in the exercise region: move back to point 7 to check the related exercise opportunity and go on from there. Otherwise, if $l = h$, check the last (in a backward sense) exercise opportunity (indeed the first one specified by the contract) as follows. In each node j of the current column of the tree, if the value of the underlying IRS, namely

$$IRS^h_j = 1 - P^h_{1,j}(T_\beta) - \sum_{s=h+1}^{\beta} \tau_s K P^h_{1,j}(T_s),$$

    is larger than the backwardly-cumulated value from continuation of the Bermudan swaption, set the CC value equal to the IRS value. In our notation:
    If $IRS^h_j > CC^h_{1,j}$ then $CC^h_{1,j} \leftarrow IRS^h_j$.
    Then move on to the next point.
12. (No more exercise times left).
    We have reached the first allowed exercise time T_h, and now there are no exercise opportunities left as we keep on moving backward in time. We can now start rolling backwards the current backwardly cumulated value from continuation from current time $t^h_1 = T_h$ until time 0 along the tree as in (3.98) above. Now it is no longer necessary to propagate backwards the P's, but only CC.
13. When we reach time 0, the value of CC at the corresponding initial node of the tree is the Bermudan-option price.

We conclude this example with a few remarks. The method we just sketched allows for different discretizations in different partitions $[T_l, T_{l+1}]$. As l

changes, we are free to consider either a more refined or a coarser discretization in the tree construction. This is to say that we may decide to refine the discretization in zones of the tree that are more relevant to the payoff evaluation. Typically, one takes a finer discretization in the regions where early exercise is to be checked, and a coarser (but not too coarse) discretization after the last possible exercise date (Mauri (2001)).

An example of tree with different discretization steps is shown in Figure 3.3. Here it is clearly seen that, in different regions, the discretization is different. Notice in particular that in the initial part of the tree the discretization is particularly refined. Notice also that at the end of the tree the discretization step is larger.

Finally, as far as the above scheme is concerned, points 1 to 6 can be avoided if the short-rate model being used allows for analytical formulas for zero-coupon-bond prices. Indeed, in such a case, knowledge of the short rate amounts to knowledge of the bond price itself through the related analytical formulas, and therefore we do not need backward propagation to obtain the zero-coupon bonds at the last exercise date, since we can directly start from the short rate at that date. Also, in principle, there would be no need to store the vector of bond prices at each node, since they can be recovered analytically in terms of the short-rate value at each node. Nonetheless, there are times where it can be convenient to keep on propagating the bond-price vectors rather than resorting to the analytical formulas, for a series of reasons involving, for instance, the tradeoff between speed of convergence and computational complexity (Mauri (2001)).

3.12 Implied Cap Volatility Curves

When using an interest-rate model for pricing related derivatives, it can be useful to understand how the implied volatility structures vary due to changes in the model parameters values. To this end, we study the cap volatility curves that are implied by some of the short-rate models we have previously described. Precisely, we concentrate our analysis on the Black-Karasinski (3.54), the CIR++ (3.76) and the Extended Exponential-Vasicek (3.91) models.

The implied cap volatility curves are obtained by first pricing market caps with the chosen model and then inverting the Black formula (1.26) to retrieve the implied volatility associated with each cap maturity.

The model prices are calculated with the trinomial trees we have proposed, for each model, in the related sections. The correct market conventions and payment dates are taken into account, with the variable time steps being roughly equal to 0.02 years. The initial values for each model parameters are displayed in Table 3.2.[16] We carry out our analysis by moving the model

---

[16] These parameters values derive from the model calibration to the actual Euro ATM caps volatility curve on January 17, 2000.

3.12 Implied Cap Volatility Curves   115

BK	CIR++	EEV
$a = 0.0771118$	$x_0 = 0.0059379$	$x_0 = 1.009423$
$\sigma = 0.2286$	$k = 0.394529$	$a = 0.4512079$
	$\theta = 0.271373$	$\theta = -1.47198$
	$\sigma = 0.0545128$	$\sigma = 0.6926841$

Table 3.2. The initial set of parameters for the BK, CIR++ and EEV models

parameters around their starting values. We then look for a qualitative relationship between shifts in the parameters' values and changes in shape of the implied cap volatility curve. Our conclusions, however, should not be regarded as irrefutable statements, but simply as helpful remarks for a better understanding of the above models behaviour.

In our examples, we consider caps with integer maturities ranging from 1 to 15 years. The 2 to 15 year caps have semiannual frequency, whereas the frequency of the 1 year cap is quarterly.

3.12.1 The Black and Karasinski Model

The cap volatility curves that are implied by the Black and Karasinski model (3.54) are displayed in Figure 3.8. The graphs on the left are plotted by keeping the σ parameter fixed and varying a, whereas those on the right are plotted by keeping the a parameter fixed and varying σ. These graphs suggest

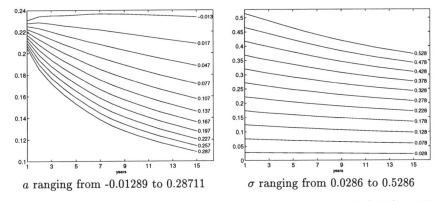

a ranging from -0.01289 to 0.28711 σ ranging from 0.0286 to 0.5286

Fig. 3.8. Cap volatility curves implied by the Black and Karasinski (1991) model.

that cap volatilities increase as σ increases and decrease as a increases. In addition, a has a bigger influence on longer maturities. Such features are indeed consistent with the theoretical meaning of the model parameters.

If we plot, for each given maturity, cap volatilities against σ, we almost obtain straight lines passing by the origin, as is shown in Figure 3.9. This

116   3. One-factor short-rate models

suggests that the functional dependence of the implied cap volatilities on the model parameters is roughly of type $f(a,\sigma) = \sigma g(a)$.

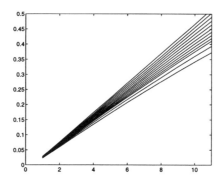

Fig. 3.9. Volatility-vs-σ plots for each maturity in the Black Karasinski (1991) model.

3.12.2 The CIR++ Model

Examples of the cap volatility curves that are implied by the CIR++ model (3.76) are displayed in Figure 3.10, where we set $\bar{\theta} := k\theta$.[17] The volatility curves are often humped and tend to be decreasing from the five to six year maturities on. More precisely, we can state the following.

Increasing x_0 makes the volatilities increase. The effect is much more pronounced for short-maturity volatilities. Indeed, x_0 does not affect the asymptotic variance of the spot rate r since, for a positive k,

$$\lim_{t\to\infty} \text{Var}\{r(t)\} = \theta \frac{\sigma^2}{2k}.$$

It is reasonable, therefore, that long-maturity volatilities are less sensitive to changes in x_0 and that increasing x_0 has the effect to decrease the initial slope of the volatility curve.

Increasing the mean reverting parameter k makes the volatility level decrease. The effect is less pronounced for short maturity caps. Accordingly, the volatility curve is initially increasing when k is small and decreasing when k is large.

Increasing the parameter θ makes the volatility level increase. The absolute change is fairly similar for all maturities so that the curve shape hardly changes with θ. Moreover, plotting, for each fixed maturity, implied volatilities against θ gives almost linear graphs. We may then guess that, for each

---
[17] Notice that, when studying the implied cap volatility curve, varying θ is equivalent to varying $\bar{\theta}$ as soon as k is fixed and both θ and k are positive.

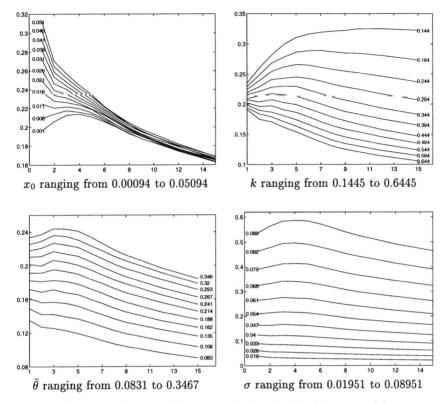

Fig. 3.10. Cap volatility curves implied by the CIR++ model.

fixed maturity, the implied volatility has roughly a functional dependence on the model parameters of the following type:

$$f(x_0, k, \theta, \sigma) = g(x_0, k, \sigma) + \theta h(x_0, k, \sigma).$$

The parameter σ affects the implied volatility curve as expected. Volatilities increase as σ grows and the volatility curve moves with almost parallel shifts. In our examples, the four and five year volatilities grow a bit faster, so that, as σ increases, a humped shape appears. Plotting, for each fixed maturity, implied volatilities against σ we obtain the increasing convex curves that are shown in Figure 3.11. Indeed, there is an empirical confirmation that the functional dependence of the implied volatility on the parameter σ is of type $f(\sigma) = k_1 \sigma^2 + k_2$.

3.12.3 The Extended Exponential-Vasicek Model

Examples of the cap volatility curves that are implied by the Extended Exponential-Vasicek model (3.91) are plotted in Figure 3.12, where we set

118   3. One-factor short-rate models

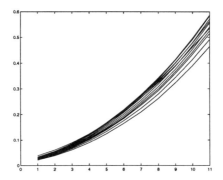

Fig. 3.11. Volatility-vs-σ plot for each maturity in the CIR++ model

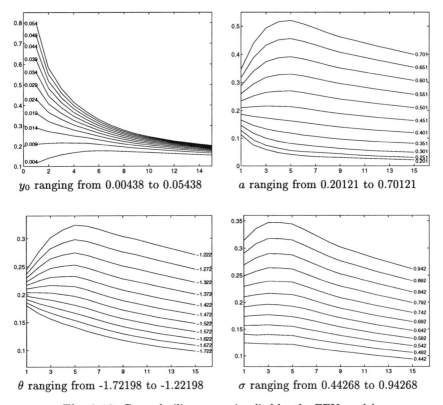

Fig. 3.12. Cap volatility curves implied by the EEV model.

$y_0 := \ln x_0$. Similarly to the CIR++ case, also in the examples for the EEV model we often find implied volatility curves that are humped and tend to be decreasing from the six year maturity on. More precisely, we have the following.

Increasing x_0 makes the volatilities increase and the absolute change is more relevant for short-maturity volatilities. Indeed, as in the CIR++ case, x_0 does not affect the asymptotic variance of the spot rate since, for a positive a (see also (3.31)),

$$\lim_{t\to\infty} \text{Var}\{r(t)\} = \exp\left(\frac{2\theta}{a} + \frac{\sigma^2}{2a}\right)\left[\exp\left(\frac{\sigma^2}{2a}\right) - 1\right].$$

Again, it is reasonable that long-maturity volatilities are less sensitive to changes in x_0 and that increasing x_0 has the effect to decrease the initial slope of the volatility curve.

The drift parameters a and θ have a similar impact on the implied volatility curve, making mostly the five-year volatility change, with the one-year volatility that is the least affected. This naturally leads to a humped shape as soon as these parameters grow.

Finally σ behaves as expected. The implied volatilities increase as σ increases, and changes in σ lead to almost parallel shifts in the volatility curve. In the given examples, the five-year volatility is subject to larger moves, which leads to a hump as soon as σ increases. Plotting, for each fixed maturity, implied volatilities against σ produces almost linear graphs. However, the hypothesis that implied volatilities depend linearly on the parameter σ does not have a fully satisfactory empirical confirmation.

3.13 Implied Swaption Volatility Surfaces

Similarly to the implied cap volatility curve, it can be useful to understand how the implied swaption volatility structures vary due to changes in the model-parameters values. To this end, we study the swaptions volatility curves and surfaces that are implied by the Black and Karasinski (3.54) and the Extended Exponential-Vasicek (3.91) models.[18]

The implied swaption volatility curves are obtained by first pricing market swaptions with the model and then inverting the Black formula (1.28) to retrieve the implied volatility associated with the selected pair of maturity and tenor.

As before, the model prices are calculated with the corresponding trinomial trees. The correct market conventions and payment dates are taken into account, with the variable time steps being again roughly equal to 0.02 years.

We consider ATM European swaptions whose maturities go from one to twenty years and whose tenors (i.e., the durations of the underlying swaps) go from one to twenty years as well. For our examples, we first choose the

---

[18] We just concentrate on these two models for practical purposes. Indeed, among the short rate models developed in this chapter, the BK and EEV models are likely to imply the best fitting to the swaption volatility surfaces in many concrete market situations.

initial set of parameters shown in Table 3.3 and shock each parameter ceteris paribus.[19] We then plot the implied volatility curves that are obtained by fixing respectively the one, five and ten-year tenors. We also show the implied volatility surfaces that correspond to the highest and lowest values being chosen for each parameter.

As for the cap-volatilities case, the purpose of this section is to comment on how the model parameters affect the shape of the implied volatility structures. Again, our conclusions are far from being "universal statements" and simply aim to provide some intuitions on the practical effects of the parameters changes.

BK	EEV
a= 0.1318301	x_0= 1.014098
σ=0.2342142	a = 0.1073888
	θ =-0.457712
	σ = 0.692684

Table 3.3. The initial set of parameters for the BK and EEV models

3.13.1 The Black and Karasinski Model

Figure 3.13 shows the swaption volatility curves that are obtained by fixing the tenor and by separately varying the two parameters a and σ. As to these examples, we can infer the following.

Implied volatilities decrease as a increases. We can see that, for small tenors, a has a small effect on short maturity volatilities, whereas for large tenors, changing a seems to affect short and long maturities with the same extent.

Volatilities also grow as σ grows. Similarly to the cap-volatilities case, if we fix tenor and maturity and plot volatilities against σ, we obtain an almost linear graph. However, if the maturity is large enough, this linearity is lost, revealing a more complex relation between volatilities and σ. Examples of the swaption volatility surfaces that are implied by the Black and Karasinski (1991) model are shown in Figure 3.14 for the displayed parameter values.

3.13.2 The Extended Exponential-Vasicek Model

Examples of the swaption volatility curves that, for fixed tenors, are implied by the extended exponential-Vasicek model (3.91) are plotted in Figure 3.15, where again $y_0 := \ln x_0$. The swaption volatility surfaces in this case are

---

[19] These initial parameters values derive from the models calibration to the actual Euro ATM swaption volatility surface on January 17, 2000.

3.14 An Example of Calibration to Real-Market Data

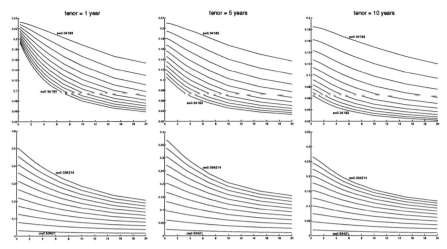

Fig. 3.13. Fixed-tenor volatility curves implied by the BK model. In the first row a varies ceteris paribus from 0.04183 to 0.34183 with step 0.03, whereas in the second row σ varies ceteris paribus from 0.034214 to 0.534214 with step 0.05.

more complex than those implied by the Black and Karasinski (1991) model. Indeed, our examples reveal that humped shapes are possible along the maturity dimension. More specifically, we have the following.

The dependence of the volatilities on the y_0 parameter is similar to that in the caps-volatilities case. Only short-maturity volatilities seem to be affected, increasing as y_0 increases. Moreover, the tenor does not seem to affect the dependence on y_0, and the volatilities decrease as the tenor increases.

When a increases, the volatility level decreases. Moreover, increasing a leads to a humped shape, and this holds for each tenor. The influence of the parameter a on the swaption volatilities is smaller when both maturity and tenor are small.

The influence of the parameter θ on the implied volatility curves is similar to that of a. The volatilities increase as θ increases. For small tenors, the long-maturity volatilities change more than the short-maturity ones. Finally, a hump appears as soon as θ increases.

As expected, increasing σ causes volatilities to increase. Indeed, for each fixed tenor, changing σ produces an almost parallel shift in the implied volatility curve.

Examples of the swaption volatility surfaces implied by the EEV model are displayed in Figure 3.17.

3.14 An Example of Calibration to Real-Market Data

We conclude the chapter with an example of calibration to real-market volatility data, which illustrates the fitting capability of the main one-factor

122   3. One-factor short-rate models

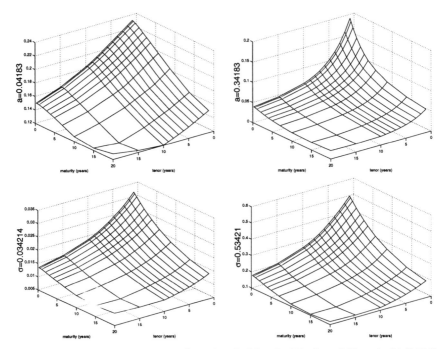

Fig. 3.14. Swaption volatility surfaces implied by the Black and Karasinski (1991) model.

(exogenous-term-structure) short-rate models we have reviewed in the previous sections. To this end, we use the at-the-money Euro cap-volatility quotes on February 13, 2001, at 5 p.m., which are reported in Table 3.4.. The zero-

Maturity	Volatility
1	0.152
2	0.162
3	0.164
4	0.163
5	0.1605
7	0.1555
10	0.1475
15	0.135
20	0.126

Table 3.4. At-the-money Euro cap-volatility quotes on February 13, 2001, at 5 p.m.

coupon curve, on the same day and time, is that shown in Figure 1.1 of Chapter 1.

We focus on the Hull and White model (3.33), the Black and Karasinski model (3.54), the Mercurio and Moraleda model (3.63), the CIR++ model

3.14 An Example of Calibration to Real-Market Data   123

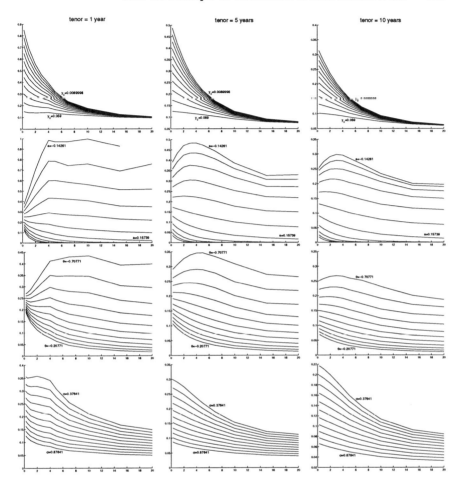

Fig. 3.15. Fixed-tenor swaption volatility curves implied by the EEV model. In the first row, y_0 varies ceteris paribus from 0.0089998 to 0.0599998 with step 0.005; In the second row, a varies ceteris paribus from -0.14261 to 0.15739 with step 0.03; In the third row, θ varies ceteris paribus from -0.70771 to -0.20771 with step 0.05; In the fourth row σ varies ceteris paribus from 0.37641 to 0.87641 with step 0.05.

(3.76) and the Extended exponential-Vasicek model (3.91). Minimizing the sum of the square percentage differences between model and market cap prices,[20] we obtain the results that are shown in Figure 3.16, where the models implied cap volatility curves are compared to the market one. Recall that a model cap implied volatility is the volatility parameter to be plugged into Black's formula (1.26) to match the observed model price.

---
[20] We use analytical formulas for the HW model, whereas, for the other models, we build a trinomial tree with a variable (small) time step in the first year, and an average of 50 time steps per year afterwards.

124   3. One-factor short-rate models

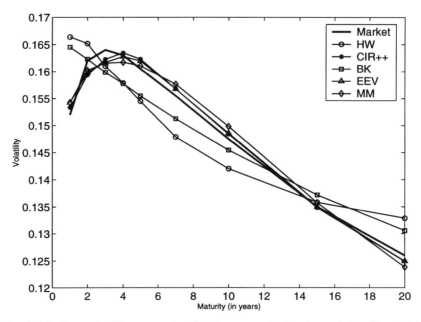

Fig. 3.16. Cap volatility curves implied by some short-rate models calibrated to the at-the-money Euro cap volatility curve on February 13, 2001, at 5 p.m.

Now examine carefully Figure 3.16, and notice that both the HW model and the BK model imply decreasing cap volatility curves as best fitting to the humped market curve. In addition, the calibrated value of the mean-reversion parameter a in the HW model is negative, actually leading to a mean-diverging model where the volatility of instantaneous forward rates diverges too. This is a common situation when calibrating the HW model to cap volatilities, and has often been interpreted as the result of a possible predominance of the increasing part of the market volatility curve.

We also see that the introduction of an extra parameter (in a suitable time-dependent function), as in the MM model, helps recover the typical shape of the market cap-volatility curve. We can then see that the "award for the best fitting quality" is shared by the CIR++ model and the EEV model, whose implied curves are almost overlapped. This could be expected since both models have the highest number of parameters, amounting to four.

Remark 3.14.1. A model fitting quality can be deeply affected by the specific market conditions one is trying to reproduce. We must indeed remember that in some particular conditions even the HW and BK models can lead to an implied cap volatility curve that follows the market hump. This happens, for instance, in case of a decreasing forward-rate curve. However, the "usual" situation illustrated in Figure 3.16 represents what we have been commonly witnessing in the Euro market in the last years.

3.14 An Example of Calibration to Real-Market Data 125

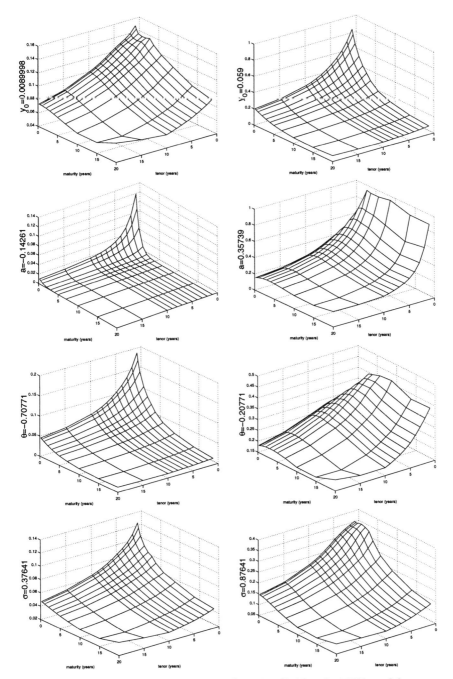

Fig. 3.17. Swaption volatility surfaces implied by the EEV model.

4. Two-Factor Short-Rate Models

"Entering secondary dimension". Adam Warlock, in:
"Silver Surfer/Warlock: Resurrection" 1, 1993, Marvel Comics.

4.1 Introduction and Motivation

However, motivation never harmed anyone (least of all, pure mathematicians), [...]
L.C.G. Rogers and D. Williams, in Chapter III.4 of "Diffusions, Markov Processes and Martingales", Vol. 1, 1994, Wiley and Sons

In the present chapter we introduce two major two-factor short-rate models. Before starting with the actual models, we would like to motivate two-factor models by pointing out the weaknesses of the one-factor models of the previous chapter. This is the purpose of this introductory section.

We have seen previously that the short rate r_t may constitute the fundamental coordinate with which the whole yield econometric curve can be characterized. Indeed, knowledge of the short rate and of its distributional properties leads to knowledge of bond prices as from the usual relationship

$$P(t,T) = E_t \left[\exp\left(-\int_t^T r_s \, ds \right) \right] .$$

From all bond prices $P(t,T)$ at a given time t one can reconstruct the whole zero-coupon interest-rate curve at the same time t, so that indeed the evolution of the whole curve is characterized by the evolution of the single quantity r. However, choosing a poor model for the evolution of r will result in a poor model for the evolution of the yield curve. In order to clarify this rather important point, let us consider for a moment the Vasicek model 3.5.

Recall from the previous chapter that the Vasicek model assumes the evolution of the short-rate process r to be given by the linear-Gaussian SDE

$$dr_t = k(\theta - r_t)dt + \sigma dW_t .$$

Recall also the bond price formula $P(t,T) = A(t,T) \exp(-B(t,T)r_t)$, from which all rates can be computed in terms of r. In particular, continuously-compounded spot rates are given by the following affine transformation of the fundamental quantity r

$$R(t,T) = -\frac{\ln A(t,T)}{T-t} + \frac{B(t,T)}{T-t}r_t =: a(t,T) + b(t,T)r_t .$$

Consider now a payoff depending on the joint distribution of two such rates at time t. For example, we may set $T_1 = t+1$ years and $T_2 = t+10$ years. The payoff would then depend on the joint distribution of the one-year and ten-year continuously-compounded spot interest rates at "terminal" time t. In particular, since the joint distribution is involved, the correlation between the two rates plays a crucial role. With the Vasicek model such terminal correlation is easily computed as

$$\text{Corr}(R(t,T_1), R(t,T_2)) = \text{Corr}(a(t,T_1) + b(t,T_1)r_t, a(t,T_2) + b(t,T_2)r_t) = 1,$$

so that at every time instant all the maturities of the curve are perfectly correlated. For example, the thirty year interest rate at a given instant is perfectly correlated with the three-month rate at the same instant. This means that a shock to the interest rate curve at time t is transmitted equally through all maturities, and the curve, when its initial point (the short rate r_t) is shocked, moves almost rigidly in the same direction. Clearly, it is hard to accept this perfect-correlation feature of the model. Truly, interest rates are known to exhibit some decorrelation (i.e. non-perfect correlation), so that a more satisfactory model of curve evolution has to be found.

One-factor models such as HW, BK, CIR++, EEV may still prove useful when the product to be priced does not depend on the correlations of different rates but depends at every instant on a single rate of the whole interest-rate curve (say for example the six-month rate). Otherwise, the approximation can still be acceptable, especially for "risk-management-like" purposes, when the rates that jointly influence the payoff at every instant are close (say for example the six-month and one-year rates). Indeed, the real correlation between such near rates is likely to be rather high anyway, so that the perfect correlation induced by the one-factor model will not be unacceptable in principle.

But in general, whenever the correlation plays a more relevant role, or when a higher precision is needed anyway, we need to move to a model allowing for more realistic correlation patterns. This can be achieved with multi-factor models, and in particular with two-factor models. Indeed, suppose for a moment that we replace the Gaussian Vasicek model with its hypothetical two-factor version (G2):

$$\begin{aligned} r_t &= x_t + y_t, \\ dx_t &= k_x(\theta_x - x_t)dt + \sigma_x dW_1(t), \\ dy_t &= k_y(\theta_y - y_t)dt + \sigma_y dW_2(t), \end{aligned} \tag{4.1}$$

with instantaneously-correlated sources of randomness, $dW_1 dW_2 = \rho\, dt$. Again, we will see later on in the chapter that also for this kind of models the bond price is an affine function, this time of the two factors x and y,

$$P(t,T) = A(t,T)\exp(-B^x(t,T)x_t - B^y(t,T)y_t),$$

where quantities with the superscripts "x" or "y" denote the analogous quantities for the one-factor model where the short rate is given by x or y, respectively. Taking this for granted at the moment, we can see easily that now

$$\text{Corr}(R(t,T_1), R(t,T_2))$$
$$= \text{Corr}(b^x(t,T_1)x_t + b^y(t,T_1)y_t, b^x(t,T_2)x_t + b^y(t,T_2)y_t),$$

and this quantity is not identically equal to one, but depends crucially on the correlation between the two factors x and y, which in turn depends, among other quantities, on the instantaneous correlation ρ in their joint dynamics. How much flexibility is gained in the correlation structure and whether this is sufficient for practical purposes will be debated in the chapter. It is however clear that the choice of a multi-factor model is a step forth in that correlation between different rates of the curve at a given instant is not necessarily equal to one.

Another question that arises naturally is: How many factors should one use for practical purposes? Indeed, what we have suggested with two factors can be extended to three or more factors. The choice of the number of factors then involves a compromise between numerically-efficient implementation and capability of the model to represent realistic correlation patterns (and covariance structures in general) and to fit satisfactorily enough market data in most concrete situations.

Usually, historical analysis of the whole yield curve, based on principal component analysis or factor analysis, suggests that under the objective measure two components can explain 85% to 90% of variations in the yield curve, as illustrated for example by Jamshidian and Zhu (1997) (in their Table 1), who consider JPY, USD and DEM data. They show that one principal component explains from 68% to 76% of the total variation, whereas three principal components can explain from 93% to 94%. A related analysis is carried out in Chapter 3 of Rebonato (1998) (in his Table 3.2) for the UK market, where results seem to be more optimistic: One component explains 92% of the total variance, whereas two components already explain 99.1% of the total variance. In some works an interpretation is given to the components in terms of average level, slope and curvature of the zero-coupon curve, see for example again Jamshidian and Zhu (1997).

What we learn from these analyses is that, in the objective world, a two- or three-dimensional process is needed to provide a realistic evolution of the whole zero-coupon curve. Since the instantaneous-covariance structure of the same process when moving from the objective probability measure to the risk-neutral probability measure does not change, we may guess that also in the risk-neutral world a two- or three-dimensional process may be needed in order

130   4. Two-Factor Short-Rate Models

to obtain satisfactory results. This is a further motivation for introducing a two- or three-factor model for the short rate. In this book, we have decided to focus on two-factor models for their better tractability and implementability, especially as far as recombining lattices are concerned. In particular, we will consider additive models of the form

$$r_t = x_t + y_t + \varphi(t), \qquad (4.2)$$

where φ is a deterministic shift which is added in order to fit exactly the initial zero-coupon curve, as in the one-factor case of Section 3.8. This formulation encompasses the classical Hull and White two-factor model as a deterministically-shifted two-factor Vasicek (G2++), and an extension of the Longstaff and Schwartz (LS) model that is capable of fitting the initial term structure of rates (CIR2++), where the basic LS model is obtained as a two-factor additive CIR model. These are the two-factor models we will consider, and we will focus especially on the two-factor additive Gaussian model G2++. The main advantage of the G2++ model (4.2) with x and y as in (4.1) over the shifted Longstaff and Schwartz CIR2++ given by (4.2) with x and y as in

$$\begin{aligned} dx_t &= k_x(\theta_x - x_t)dt + \sigma_x\sqrt{x_t}dW_1(t), \\ dy_t &= k_y(\theta_y - y_t)dt + \sigma_y\sqrt{y_t}dW_2(t), \end{aligned} \qquad (4.3)$$

is that in the latter we are forced to take $dW_1 dW_2 = 0\,dt$ in order to maintain analytical tractability, whereas in the former we do not need to do so. The reason why we are forced to take $\rho = 0$ in the CIR2++ case lies in the fact that square-root non-central chi-square processes do not work as well as linear-Gaussian processes when adding nonzero instantaneous correlations. Requiring $dW_1 dW_2 = \rho\,dt$ with $\rho \neq 0$ in the above CIR2++ model would indeed destroy analytical tractability: It would no longer be possible to compute analytically bond prices and rates starting from the short-rate factors. Moreover, the distribution of r would become more involved than that implied by a simple sum of independent non-central chi-square random variables. Why is the possibility that the parameter ρ be different than zero so important as to render G2++ preferable to CIR2++? As we said before, the presence of the parameter ρ renders the correlation structure of the two-factor model more flexible. Moreover, $\rho < 0$ allows for a humped volatility curve of the instantaneous forward rates. Indeed, if we consider at a given time instant t the graph of the T function

$$T \mapsto \sqrt{\mathrm{Var}[d\,f(t,T)]/dt}$$

where the instantaneous forward rate $f(t,T)$ comes from the G2++ model, it can be seen that for $\rho = 0$ this function is decreasing and upwardly concave. This function can assume a humped shape for suitable values of k_x and k_y only when $\rho < 0$. Since such a humped shape is a desirable feature of the model which is in agreement with market behaviour, it is important to allow for nonzero instantaneous correlation in the G2++ model. The situation is somewhat analogous in the CIR2++ case: Choosing $\rho = 0$ does not allow for humped shapes in the curve

$$T \mapsto \sqrt{\text{Var}[d\ f(t,T)]/dt},$$

which consequently results monotonically decreasing and upwardly concave, exactly as in the G2++ case with $\rho = 0$, as we will see later on in the chapter. In turn, the advantage of CIR2++ over G2++ is that, as in the one-factor case where HW is compared to CIR++, it can maintain positive rates through reasonable restrictions on the parameters. Moreover, the distribution of the short rate is the distribution of the sum of two independent noncentral chi-square variables, and as such it has fatter tails than the Gaussian distribution in G2++. This is considered a desirable property, especially because in such a way (continuously-compounded) spot rates for any maturity are affine transformations of such non-central chi-squared variables and are closer to the lognormal distribution than the Gaussian distribution for the same rates implied by the G2++ model. Therefore, both from a point of view of positivity and distribution of rates, the CIR2++ model would be preferable to the G2++ model. However, the humped shape for the instantaneous forward rates volatility curve is very important for the model to be able to fit market data in a satisfactory way. Furthermore, the G2++ model is more analytically tractable and easier to implement. These overall considerations then imply that the G2++ model is more suitable for practical applications, even though we should not neglect the advantages that a model like CIR2++ may have.

In general, when analyzing an interest rate model from a practical point of view, one should try to answer questions like the following. Is a two-factor model like G2++ flexible enough to be calibrated to a large set of swaptions, or even to caps and swaptions at the same time? How many swaptions can be calibrated in a sufficiently satisfactory way? What is the evolution of the term structure of volatilities as implied by the calibrated model? Is this realistic? How can one implement trees for models such as G2++? Is Monte Carlo simulation feasible? Can the model be profitably used for quanto-like products and for products depending on more than an interest rate curve when taking into account correlations between different interest-rate curves and also with exchange rates?

In this chapter, we will focus mainly on the G2++ model and we will try to deal with some of the above questions. Other questions will be then addressed in the second part of the book. We will then deal with the CIR2++ model, but for this model we will not consider the same level of detail we devoted to the G2++ model. We will try and keep the two sections on the two different models as self-contained as possible, in order for a reader interested only in one of the models to be able to concentrate on the interested section

132   4. Two-Factor Short-Rate Models

while skipping the other one. This will cause of course a little overlap in the presentation of the two models.

4.2 The Two-Additive-Factor Gaussian Model G2++

In this section we consider an interest-rate model where the instantaneous-short-rate process is given by the sum of two correlated Gaussian factors plus a deterministic function that is properly chosen so as to exactly fit the current term structure of discount factors. The model is quite analytically tractable in that explicit formulas for discount bonds, European options on pure discount bonds, hence caps and floors, can be readily derived.

Gaussian models like this G2++ model are very useful in practice, despite their unpleasant feature of the theoretical possibility of negative rates. Indeed, their analytical tractability considerably ease the task of pricing exotic products. The Gaussian distribution allows the derivation of explicit formulas for a number of non-plain-vanilla instruments and, combined with the analytical expression for zero-coupon bonds, leads to efficient and fairly fast numerical procedures for pricing any possible payoff. Also, finite spot and forward rates at a given time for any maturity and accrual conventions can be given an explicit analytical expression in terms of the short-rate factors at the relevant instant. This allows for easy propagation of the whole zero-coupon curve in terms of the two factors. Another consequence of the presence of two factors is that the actual variability of market rates is described in a better way: Among other improvements, a non-perfect correlation between rates of different maturities is introduced. This results in a more precise calibration to correlation-based products like European swaptions. These major advantages are the main reason why we devote so much attention to a two-factor Gaussian model. Such a model can also be helpful when pricing out-of-the-money exotic instruments after calibration to at-the-money plain-vanilla products. In fact, the smile effect that is present in the market can be better captured by a distribution with a significant mass around zero. This is the case, for instance, of a receiver out-of-the-money Bermudan swaption with the model parameters being calibrated to the market prices of the corresponding European at-the-money swaptions.

The Gaussian model of this section is naturally related to the Hull-White (1994c) two-factor model in that we can actually prove the equivalence between these two approaches. However, the formulation with two additive factors leads to less complicated formulas and is easier to implement in practice, even though we may lose some insight and intuition on the nature and the interpretation of the two factors.

This section is structured as follows. In the first subsection, we introduce the short-rate dynamics and explain the resulting distributional features. In the second subsection, we derive an analytical formula for the price of a zero-coupon bond. In the third subsection we derive the dynamics of forward rates and analyze their volatility and correlation structures. In the fourth subsection, we derive the short-rate dynamics and distribution under a general forward measure. As a consequence, we price European options on zero-coupon bonds, caps and floors and finally swaptions. The knowledge of the short-rate distribution under any forward measure will be quite helpful when pricing most of the specific interest-rate derivatives we analyze in the second part of the book. In the fifth subsection we study the analogy of this model with the two-factor model being proposed by Hull and White (1994c). In the sixth subsection we show how to construct a two-dimensional binomial tree that approximates the short-rate dynamics. The longest proofs are written in separate appendices.

4.2.1 The Short-Rate Dynamics

We assume that the dynamics of the instantaneous-short-rate process under the risk-adjusted measure Q is given by

$$r(t) = x(t) + y(t) + \varphi(t), \quad r(0) = r_0, \tag{4.4}$$

where the processes $\{x(t) : t \geq 0\}$ and $\{y(t) : t \geq 0\}$ satisfy

$$\begin{aligned} dx(t) &= -ax(t)dt + \sigma dW_1(t), \quad x(0) = 0, \\ dy(t) &= -by(t)dt + \eta dW_2(t), \quad y(0) = 0, \end{aligned} \tag{4.5}$$

where (W_1, W_2) is a two-dimensional Brownian motion with instantaneous correlation ρ as from

$$dW_1(t)dW_2(t) = \rho dt,$$

where r_0, a, b, σ, η are positive constants, and where $-1 \leq \rho \leq 1$. The function φ is deterministic and well defined in the time interval $[0, T^*]$, with T^* a given time horizon, typically 10, 30 or 50 (years). In particular, $\varphi(0) = r_0$. We denote by $\mathcal{F}_t$ the sigma-field generated by the pair (x, y) up to time t.

Simple integration of equations (4.5) implies that for each $s < t$

$$r(t) = x(s)e^{-a(t-s)} + y(s)e^{-b(t-s)}$$
$$+ \sigma \int_s^t e^{-a(t-u)} dW_1(u) + \eta \int_s^t e^{-b(t-u)} dW_2(u) + \varphi(t),$$

meaning that $r(t)$ conditional on $\mathcal{F}_s$ is normally distributed with mean and variance given respectively by

$$E\{r(t)|\mathcal{F}_s\} = x(s)e^{-a(t-s)} + y(s)e^{-b(t-s)} + \varphi(t),$$
$$\text{Var}\{r(t)|\mathcal{F}_s\} = \frac{\sigma^2}{2a}\left[1 - e^{-2a(t-s)}\right] + \frac{\eta^2}{2b}\left[1 - e^{-2b(t-s)}\right] \tag{4.6}$$
$$+ 2\rho\frac{\sigma\eta}{a+b}\left[1 - e^{-(a+b)(t-s)}\right].$$

134   4. Two-Factor Short-Rate Models

In particular

$$r(t) = \sigma \int_0^t e^{-a(t-u)} dW_1(u) + \eta \int_0^t e^{-b(t-u)} dW_2(u) + \varphi(t). \qquad (4.7)$$

The dynamics of the processes x and y can be also expressed in terms of two independent Brownian motions $\widetilde{W}_1$ and $\widetilde{W}_2$ as follows:[1]

$$\begin{aligned} dx(t) &= -ax(t)dt + \sigma d\widetilde{W}_1(t), \\ dy(t) &= -by(t)dt + \eta\rho d\widetilde{W}_1(t) + \eta\sqrt{1-\rho^2} d\widetilde{W}_2(t), \end{aligned} \qquad (4.8)$$

where

$$\begin{aligned} dW_1(t) &= d\widetilde{W}_1, \\ dW_2(t) &= \rho d\widetilde{W}_1(t) + \sqrt{1-\rho^2} d\widetilde{W}_2(t), \end{aligned}$$

so that we can also write

$$r(t) = x(s)e^{-a(t-s)} + y(s)e^{-b(t-s)} + \sigma \int_s^t e^{-a(t-u)} d\widetilde{W}_1(u)$$
$$+ \eta\rho \int_s^t e^{-b(t-u)} d\widetilde{W}_1(u) + \eta\sqrt{1-\rho^2} \int_s^t e^{-b(t-u)} d\widetilde{W}_2(u) + \varphi(t).$$

4.2.2 The Pricing of a Zero-Coupon Bond

We denote by $P(t,T)$ the price at time t of a zero-coupon bond maturing at T and with unit face value, so that

$$P(t,T) = E\left\{ e^{-\int_t^T r_s ds} | \mathcal{F}_t \right\},$$

where E denotes the expectation under the risk-adjusted measure Q. In order to explicitly compute this expectation, we need the following.

Lemma 4.2.1. *For each t, T the random variable*

$$I(t,T) := \int_t^T [x(u) + y(u)] du$$

conditional to the sigma-field $\mathcal{F}_t$ is normally distributed with mean $M(t,T)$ and variance $V(t,T)$, respectively given by

$$M(t,T) = \frac{1-e^{-a(T-t)}}{a} x(t) + \frac{1-e^{-b(T-t)}}{b} y(t) \qquad (4.9)$$

---
[1] This is equivalent to performing a Cholesky decomposition on the variance-covariance matrix of the pair $(W_1(t), W_2(t))$.

and

$$V(t,T) = \frac{\sigma^2}{a^2}\left[T-t+\frac{2}{a}e^{-a(T-t)} - \frac{1}{2a}e^{-2a(T-t)} - \frac{3}{2a}\right]$$
$$+ \frac{\eta^2}{b^2}\left[T-t+\frac{2}{b}e^{-b(T-t)} - \frac{1}{2b}e^{-2b(T-t)} - \frac{3}{2b}\right]$$
$$+ 2\rho\frac{\sigma\eta}{ab}\left[T-t+\frac{e^{-a(T-t)}-1}{a}+\frac{e^{-b(T-t)}-1}{b}-\frac{e^{-(a+b)(T-t)}-1}{a+b}\right]. \tag{4.10}$$

Proof. See Appendix A in this chapter. □

Theorem 4.2.1. *The price at time t of a zero-coupon bond maturing at time T and with unit face value is*

$$P(t,T) = \exp\left\{-\int_t^T \varphi(u)du - \frac{1-e^{-a(T-t)}}{a}x(t)\right.$$
$$\left. -\frac{1-e^{-b(T-t)}}{b}y(t) + \frac{1}{2}V(t,T)\right\}. \tag{4.11}$$

Proof. Being φ a deterministic function, the theorem follows from straightforward application of Lemma 4.2.1 and the fact that if Z is a normal random variable with mean m_Z and variance σ_Z^2, then $E\{\exp(Z)\} = \exp(m_Z + \frac{1}{2}\sigma_Z^2)$. □

Let us now assume that the term structure of discount factors that is currently observed in the market is given by the sufficiently smooth function $T \mapsto P^M(0,T)$.

If we denote by $f^M(0,T)$ the instantaneous forward rate at time 0 for a maturity T implied by the term structure $T \mapsto P^M(0,T)$, i.e.,

$$f^M(0,T) = -\frac{\partial \ln P^M(0,T)}{\partial T},$$

we then have the following.

Corollary 4.2.1. *The model (4.4) fits the currently-observed term structure of discount factors if and only if, for each T,*

$$\varphi(T) = f^M(0,T) + \frac{\sigma^2}{2a^2}\left(1-e^{-aT}\right)^2$$
$$+ \frac{\eta^2}{2b^2}\left(1-e^{-bT}\right)^2 + \rho\frac{\sigma\eta}{ab}\left(1-e^{-aT}\right)\left(1-e^{-bT}\right), \tag{4.12}$$

i.e., if and only if

136   4. Two-Factor Short-Rate Models

$$\exp\left\{-\int_t^T \varphi(u)du\right\} = \frac{P^M(0,T)}{P^M(0,t)} \exp\left\{-\frac{1}{2}[V(0,T) - V(0,t)]\right\}, \quad (4.13)$$

so that the corresponding zero-coupon-bond prices at time t are given by

$$P(t,T) = \frac{P^M(0,T)}{P^M(0,t)} \exp\{\mathcal{A}(t,T)\}$$

$$\mathcal{A}(t,T) := \frac{1}{2}[V(t,T) - V(0,T) + V(0,t)] \quad (4.14)$$

$$- \frac{1-e^{-a(T-t)}}{a} x(t) - \frac{1-e^{-b(T-t)}}{b} y(t).$$

Proof. The model (4.4) fits the currently-observed term structure of discount factors if and only if for each maturity $T \leq T^*$ the discount factor $P(0,T)$ produced by the model (4.4) coincides with the one observed in the market, i.e., if and only if

$$P^M(0,T) = \exp\left\{-\int_0^T \varphi(u)du + \frac{1}{2}V(0,T)\right\}.$$

Now let us take logs of both sides and differentiate with respect to T, so as to obtain (4.12) by noting that (see also Appendix A in this chapter)

$$V(t,T) = \frac{\sigma^2}{a^2} \int_t^T \left[1 - e^{-a(T-u)}\right]^2 du + \frac{\eta^2}{b^2} \int_t^T \left[1 - e^{-b(T-u)}\right]^2 du$$

$$+ 2\rho \frac{\sigma\eta}{ab} \int_t^T \left[1 - e^{-a(T-u)}\right]\left[1 - e^{-b(T-u)}\right] du.$$

Equality (4.13) follows from noting that, under the specification (4.12),

$$\exp\left\{-\int_t^T \varphi(u)du\right\} = \exp\left\{-\int_0^T \varphi(u)du\right\} \exp\left\{\int_0^t \varphi(u)du\right\}$$

$$= \frac{P^M(0,T)\exp\{-\frac{1}{2}V(0,T)\}}{P^M(0,t)\exp\{-\frac{1}{2}V(0,t)\}}.$$

Equality (4.14) immediately follows from (4.11) and (4.13).   □

Remark 4.2.1. **(Is it really necessary to derive the market instantaneous forward curve?)** Notice that, at a first sight, one may have the impression that in order to implement the G2++ model we need to derive the whole φ curve, and therefore the market instantaneous forward curve $T \mapsto f^M(0,T)$. Now, this curve involves differentiating the market discount curve $T \mapsto P^M(0,T)$, which is usually obtained from a finite set of maturities via interpolation. Interpolation and differentiation may induce a certain

4.2 The Two-Additive-Factor Gaussian Model G2++

degree of approximation, since the particular interpolation technique being used has a certain impact on (first) derivatives.

However, it turns out that one does not really need the whole φ curve. Indeed, what matters is the integral of φ between two given instants. This integral has been computed in (4.13). From this expression, we see that the only curve needed is the market discount curve, which need not be differentiated, and only at times corresponding to the maturities of the bond prices and rates desired, thus limiting also the need for interpolation.

Remark 4.2.2. **(Short-rate distribution and probability of negative rates).** By fitting the currently-observed term structure of discount factors, i.e. by applying (4.12), we obtain that the expected instantaneous short rate at time t, $\mu_r(t)$, is

$$\mu_r(t) := E\{r(t)\} = f^M(0,t) + \frac{\sigma^2}{2a^2}\left(1 - e^{-at}\right)^2 + \frac{\eta^2}{2b^2}\left(1 - e^{-bt}\right)^2$$
$$+ \rho\frac{\sigma\eta}{ab}\left(1 - e^{-at}\right)\left(1 - e^{-bt}\right),$$

while the variance $\sigma_r^2(t)$ of the instantaneous short rate at time t, see (4.6), is

$$\sigma_r^2(t) = \text{Var}\{r(t)\} = \frac{\sigma^2}{2a}\left(1 - e^{-2at}\right) + \frac{\eta^2}{2b}\left(1 - e^{-2bt}\right) + 2\frac{\rho\sigma\eta}{a+b}\left(1 - e^{-(a+b)t}\right).$$

This implies that the risk-neutral probability of negative rates at time t is

$$Q\{r(t) < 0\} = \Phi\left(-\frac{\mu_r(t)}{\sigma_r(t)}\right),$$

which is often negligible in many concrete situations, with Φ denoting the standard normal cumulative distribution function.

Furthermore, we have that the limit distribution of the process r is Gaussian with mean $\mu_r(\infty)$ and variance $\sigma_r^2(\infty)$ given by

$$\mu_r(\infty) := \lim_{t\to\infty} E\{r(t)\} = f^M(0,\infty) + \frac{\sigma^2}{2a^2} + \frac{\eta^2}{2b^2} + \rho\frac{\sigma\eta}{ab},$$
$$\sigma_r^2(\infty) := \lim_{t\to\infty} \text{Var}\{r(t)\} = \frac{\sigma^2}{2a} + \frac{\eta^2}{2b} + 2\rho\frac{\sigma\eta}{ab},$$

where

$$f^M(0,\infty) = \lim_{t\to\infty} f^M(0,t).$$

4.2.3 Volatility and Correlation Structures in Two-Factor Models

We now derive the dynamics of forward rates under the risk-neutral measure to obtain an equivalent formulation of the two-additive-factor Gaussian model in the Heath-Jarrow-Morton (1992) framework. In particular, we explicitly derive the volatility structure of forward rates. This also allows us to understand which market-volatility structures can be fitted by the model.

Let us define $A(t,T)$ and $B(z,t,T)$ by

$$A(t,T) = \frac{P^M(0,T)}{P^M(0,t)} \exp\left\{\frac{1}{2}[V(t,T) - V(0,T) + V(0,t)]\right\},$$

$$B(z,t,T) = \frac{1 - e^{-z(T-t)}}{z},$$

so that we can write

$$P(t,T) = A(t,T) \exp\left\{-B(a,t,T)x(t) - B(b,t,T)y(t)\right\}. \quad (4.15)$$

The (continuously-compounded) instantaneous forward rate at time t for the maturity T is then given by

$$f(t,T) = -\frac{\partial}{\partial T} \ln P(t,T)$$

$$= -\frac{\partial}{\partial T} \ln A(t,T) + \frac{\partial B}{\partial T}(a,t,T)x(t) + \frac{\partial B}{\partial T}(b,t,T)y(t),$$

whose differential form can be written as

$$df(t,T) = \ldots dt + \frac{\partial B}{\partial T}(a,t,T)\sigma dW_1(t) + \frac{\partial B}{\partial T}(b,t,T)\eta dW_2(t).$$

Therefore

$$\frac{\text{Var}(df(t,T))}{dt}$$

$$= \left(\frac{\partial B}{\partial T}(a,t,T)\sigma\right)^2 + \left(\frac{\partial B}{\partial T}(b,t,T)\eta\right)^2 + 2\rho\sigma\eta \frac{\partial B}{\partial T}(a,t,T)\frac{\partial B}{\partial T}(b,t,T)$$

$$= \sigma^2 e^{-2a(T-t)} + \eta^2 e^{-2b(T-t)} + 2\rho\sigma\eta e^{-(a+b)(T-t)},$$

which implies that the absolute volatility of the instantaneous forward rate $f(t,T)$ is

$$\sigma_f(t,T) = \sqrt{\sigma^2 e^{-2a(T-t)} + \eta^2 e^{-2b(T-t)} + 2\rho\sigma\eta e^{-(a+b)(T-t)}}. \quad (4.16)$$

From (4.16), we immediately see that the desirable feature, as far as calibration to the market is concerned, of a humped volatility structure similar to what is commonly observed in the market for the caplets volatility, may be only reproduced for negative values of ρ. Notice indeed that if ρ is positive, the terms $\sigma^2 e^{-2a(T-t)}$, $\eta^2 e^{-2b(T-t)}$ and $2\rho\sigma\eta e^{-(a+b)(T-t)}$ are all decreasing functions of the time to maturity $T - t$ and no hump is possible. This does not mean, in turn, that every combination of the parameter values with a

4.2 The Two-Additive-Factor Gaussian Model G2++

negative ρ leads to a volatility hump. A simple study of $\sigma_f(t,T)$ as a function of $T-t$, however, shows that there exist suitable choices of the parameter values that produce the desired shape.

Notice that the instantaneous-forwards humped-volatility shape has been considered, also in relation with market calibration, in Mercurio and Moraleda (2000a, 2000b) and in references given therein. See also Sections 3.6 and 3.7 of the present book.

Lest confusion may arise in the reader, we detail the humped-shape issue further, as we did in Section 3.6 for the one-factor case. We will allow for a little redundancy in the treatment, so as to preserve partially self-contained chapters.

We will go through the following points.

- Recall how cap and caplet volatilities are defined in the market model for caps.
- Recall how the term structure of volatility in the market is obtained by these, and recall its usually humped shape.
- Transliterate the market definition of caplet volatility above to short-rate models (model-intrinsic caplet volatility) and show what goes wrong.
- Modify the definition of caplet volatility for short-rate models so that things work again (model-implied caplet volatility and model-implied cap volatility), and then explain how the term structure of volatilities is consequently defined for a short-rate model.

Recall the Market-Like Definition of Caplet Volatilities

As we have seen earlier, cap or single caplet volatilities can be retrieved from market prices by inverting the related market formulas (see also Sections 6.4 and 6.4.3 in particular). Alternatively, caplet volatilities can be defined as suitable integrals (averages) as

$$v^2_{T-\text{caplet}} := \frac{1}{T} \int_0^T (d\ln F(t;T,T+\tau))(d\ln F(t;T,T+\tau))$$
$$= \frac{1}{T} \int_0^T \sigma(t;T,T+\tau)^2 \, dt,$$

where τ is typically six months, and where $\sigma(t;T,T+\tau)$ is the (percentage) instantaneous volatility at time t of the simply-compounded forward rate $F(t;T,T+\tau)$ underlying the T-expiry $(T+\tau)$-maturity caplet. The instantaneous percentage volatilities $\sigma(t;T,T+\tau)$ are deterministic in Black's model for the cap market, so that the caplet volatility $v_{T-\text{caplet}}$ is also deterministic.

What is observed to have a humped shape in the market is the curve of the caplet volatilities for different maturities at time 0, i.e., $T \mapsto v_{T-\text{caplet}}$.

Recall the Term Structure of Volatilities in the Market

The above-mentioned caplet curve $T \mapsto v_{T-\text{caplet}}$ is called the term structure of (caplet) volatilities at time 0.

Transliterating the Market Definition to Short-Rate Models: What Goes Wrong

Recall also that, in a model like G2++, the integrals in the above definition of $v_{T-\text{caplet}}$ would be stochastic since $\sigma(t; T, T+\tau)$ is not deterministic. This is an important point. One can convince oneself of this by deriving the expression for

$$d \ln F(t; T, T + \tau) = d \ln \left(\frac{P(t, T)}{P(t, T + \tau)} - 1 \right)$$

in the G2++ model through Ito's formula when expressing $P(t, T)$ and $P(t, T + \tau)$ according to (4.15).[2] It is easy to see that in both cases the (log) volatility (diffusion coefficient) will not be deterministic, but rather a stochastic quantity depending on $x(t)$ and $y(t)$. As usual, the logarithm is considered because the diffusion coefficient for the logarithm of a certain process amounts to the process percentage instantaneous volatility.

Since the quantity $v_{T-\text{caplet}}$ is an integral of the instantaneous variance $\sigma(t; T, T+\tau)^2$, in the G2++ model the $v_{T-\text{caplet}}$ would be stochastic if computed as

$$\sqrt{\frac{1}{T} \int_0^T \sigma(t; T, T+\tau)^2 \, dt}.$$

This is what we defined as the model-intrinsic T-caplet volatility in Section 3.6. We also pointed out that this is not the way caplet volatilities can be defined for models such as G2++.

Modify the Definition of Caplet Volatility for Short-Rate Models so that Things Work Again

Indeed, for models different from Black's market model for caps, such as the G2++ model considered here, the $v_{T-\text{caplet}}$'s are defined so as to be again deterministic and are usually understood as implied volatilities. One prices an at-the-money T-expiry caplet $\mathbf{Cpl}(0, T, T + \tau, F(0; T, T + \tau))$ with the model, and then inverts the T-expiry Black's market formula and finds the percentage Black volatility $v_{T-\text{caplet}}$ that, plugged into such a formula, yields the model price. More precisely, one solves the following equation for $v_{T-\text{caplet}}^{\text{G2++}}$:

---

[2] Below we consider an analogous calculation for the corresponding continuously-compounded forward rate $f(t; T, T + \tau)$.

4.2 The Two-Additive-Factor Gaussian Model G2++

$$P(0, T+\tau)F(0; T, T+\tau)\left(2\Phi\left(\frac{v_{T-\text{caplet}}^{\text{G2++}}\sqrt{T}}{2}\right) - 1\right)$$
$$= \mathbf{Cpl}(0, T, T+\tau, F(0; T, T+\tau)).$$

The left-hand side is the market Black's formula for a T-expiry $T+\tau$-maturity at-the-money caplet, whereas the right-hand side is the corresponding G2++ model formula (4.27) given later on. When this is done for all expiries T one can plot the term structure of (caplet) volatilities implied by the G2++ model,

$$T \mapsto v_{T-\text{caplet}}^{\text{G2++}}.$$

Implied cap volatilities can be defined similarly by considering a set of reset times $\{T_\alpha, \ldots, T_{\beta-1}\}$ with the final payment time T_β, and the set of the associated year fractions $\{\tau_\alpha, \ldots, \tau_\beta\}$. Then, setting $\mathcal{T}_i = \{T_\alpha, \ldots, T_i\}$ and $\bar{\tau}_i = \{\tau_{\alpha+1}, \ldots, \tau_i\}$, the *model implied $\mathcal{T}_i$-cap volatility* at time 0 is the (deterministic) solution $v_{\mathcal{T}_i-\text{cap}}^{\text{G2++}}$ of the equation:

$$\sum_{j=\alpha+1}^{i} P(0, T_j)\tau_j \text{Bl}(S_{\alpha,\beta}(0), F(0, T_{j-1}, T_j), v_{\mathcal{T}_i-\text{cap}}^{\text{G2++}}\sqrt{T_{j-1}})$$
$$= \mathbf{Cap}(0, \mathcal{T}_i, \bar{\tau}_i, S_{\alpha,\beta}(0)).$$

where the forward swap rate $S_{\alpha,\beta}(0)$ is defined in (1.25).

When this is done for all i's, one can plot the term structure of cap volatilities implied by the G2++ model,

$$T_i \mapsto v_{\mathcal{T}_i-\text{cap}}^{\text{G2++}}.$$

When the G2++ model is calibrated to the cap market, the parameters a, σ, b, η, ρ in the G2++ dynamics are set to values such that the model volatilities $v_{\mathcal{T}_i-\text{cap}}^{\text{G2++}}(a, \sigma, b, \eta, \rho)$ are as close as possible to the market cap volatilities $v_{\mathcal{T}_i-\text{cap}}^{\text{MKT}}$.

Alternatively, if one has already obtained caplet volatilities by stripping them from cap volatilities along the lines of Section 6.4.3, one can calibrate the model directly to caplet volatilities. The parameters a, σ, b, η, ρ in the G2++ dynamics are then set to values such that the model volatilities $v_{T-\text{caplet}}^{\text{G2++}}(a, \sigma, b, \eta, \rho)$ are as close as possible to the market caplet volatilities $v_{T-\text{caplet}}^{\text{MKT}}$.

We have also observed in Section 3.6 that, for this term structure to be able to feature large humps, the model (absolute) instantaneous volatilities of instantaneous forward rates,

$$T \mapsto \sigma_f(t, T),$$

usually need to allow for a hump themselves. We have also seen that in order to obtain *small* humps in such a term structure, it is not necessary to allow for a corresponding hump in the model absolute volatility of instantaneous forward rates. In short, one usually observes the following.

142   4. Two-Factor Short-Rate Models

1. No humps in $T \mapsto \sigma_f(t,T)$ $\Rightarrow$ only small humps for $T \mapsto v_{T-\text{caplet}}^{\text{MODEL}}$ are possible;
2. Humps in $T \mapsto \sigma_f(t,T)$ $\Rightarrow$ large humps for $T \mapsto v_{T-\text{caplet}}^{\text{MODEL}}$ are possible.

We have remarked that, in the typical example of the CIR++ model, a little analysis of the related analytical formulas shows how $T \mapsto \sigma_f(t,T)$ is monotonically decreasing, thus usually implying only small humps in the model caplet (and cap) volatilities.

Let us repeat once again that confusion often arises when speaking of "allowing a humped shape for forward-rates volatilities". Indeed, one has to specify what kind of volatilities are considered: instantaneous-forward absolute volatilities, caplet volatilities, cap volatilities. As we have seen, some models allow for humps in all of these structures; some other models only allow for small humps in the caplets or caps volatilities.

After the instantaneous volatility of forward rates, we can consider an analogous calculation for the instantaneous covariance per unit time between the two forward rates $f(t,T_1)$ and $f(t,T_2)$, obtaining

$$\frac{\text{Cov}(df(t,T_1), df(t,T_2))}{dt}$$
$$= \sigma^2 \frac{\partial B}{\partial T}(a,t,T_1) \frac{\partial B}{\partial T}(a,t,T_2) + \eta^2 \frac{\partial B}{\partial T}(b,t,T_1) \frac{\partial B}{\partial T}(b,t,T_2)$$
$$+ \rho \sigma \eta \left[\frac{\partial B}{\partial T}(a,t,T_1) \frac{\partial B}{\partial T}(b,t,T_2) + \frac{\partial B}{\partial T}(a,t,T_2) \frac{\partial B}{\partial T}(b,t,T_1) \right]$$
$$= \sigma^2 e^{-a(T_1+T_2-2t)} + \eta^2 e^{-b(T_1+T_2-2t)}$$
$$+ \rho \sigma \eta \left[e^{-aT_1-bT_2+(a+b)t} + e^{-aT_2-bT_1+(a+b)t} \right],$$

so that the instantaneous correlation between the two forward rates $f(t,T_1)$ and $f(t,T_2)$ is

$$\text{Corr}(df(t,T_1), df(t,T_2)) = \frac{\sigma^2 e^{-a(T_1+T_2-2t)} + \eta^2 e^{-b(T_1+T_2-2t)}}{\sigma_f(t,T_1)\sigma_f(t,T_2)}$$

$$+ \frac{\rho \sigma \eta \left[e^{-aT_1-bT_2+(a+b)t} + e^{-aT_2-bT_1+(a+b)t} \right]}{\sigma_f(t,T_1)\sigma_f(t,T_2)}.$$

As expected, the absolute value of such a correlation is smaller than one for general parameter values for which the model is non-degenerate.[3] The previous analyses and remarks also apply to forward rates spanning finite time intervals. We have the following results.

The (continuously-compounded) forward rate at time t between times T_1 and T_2 is

---

[3] The model is said to be degenerate if the two underlying factors are driven by the same noise.

$$f(t, T_1, T_2) = \frac{\ln P(t, T_1) - \ln P(t, T_2)}{T_2 - T_1},$$

whose differential form can be written as

$$df(t, T_1, T_2) = \ldots dt + \frac{B(a, t, T_2) - B(a, t, T_1)}{T_2 - T_1} \sigma dW_1(t)$$
$$+ \frac{B(b, t, T_2) - B(b, t, T_1)}{T_2 - T_1} \eta dW_2(t).$$

Therefore the absolute volatility of the forward rate $f(t, T_1, T_2)$ is

$$\sigma_f(t, T_1, T_2)$$
$$= \sqrt{\sigma^2 \beta(a, t, T_1, T_2)^2 + \eta^2 \beta(b, t, T_1, T_2)^2 + 2\rho\sigma\eta\beta(a, t, T_1, T_2)\beta(b, t, T_1, T_2)},$$

where

$$\beta(z, t, T_1, T_2) = \frac{B(z, t, T_2) - B(z, t, T_1)}{T_2 - T_1}.$$

Analogously, the instantaneous covariance per unit time between the two forward rates $f(t, T_1, T_2)$ and $f(t, T_3, T_4)$ is

$$\frac{\text{Cov}(df(t, T_1, T_2), df(t, T_3, T_4))}{dt}$$
$$= \sigma^2 \frac{B(a, t, T_2) - B(a, t, T_1)}{T_2 - T_1} \frac{B(a, t, T_4) - B(a, t, T_3)}{T_4 - T_3}$$
$$+ \eta^2 \frac{B(b, t, T_2) - B(b, t, T_1)}{T_2 - T_1} \frac{B(b, t, T_4) - B(b, t, T_3)}{T_4 - T_3}$$
$$+ \rho\sigma\eta \left[\frac{B(a, t, T_2) - B(a, t, T_1)}{T_2 - T_1} \frac{B(b, t, T_4) - B(b, t, T_3)}{T_4 - T_3} \right.$$
$$+ \left. \frac{B(a, t, T_4) - B(a, t, T_3)}{T_4 - T_3} \frac{B(b, t, T_2) - B(b, t, T_1)}{T_2 - T_1} \right].$$

4.2.4 The Pricing of a European Option on a Zero-Coupon Bond

The price at time t of a European call option with maturity T and strike K, written on a zero-coupon bond with unit face value and maturity τ is

$$\mathbf{ZBC}(t, T, S, K) = E\left\{ e^{-\int_t^T r(s)ds}(P(T, S) - K)^+ \Big| \mathcal{F}_t \right\}.$$

In order to explicitly compute this expectation we need to change probability measure as indicated by Jamshidian (1989) and more generally by Geman et al. (1995). Precisely, for any fixed maturity T, we denote by Q^T the probability measure defined by the Radon-Nikodym derivative (see Appendix A at the end of the book on the Radon-Nikodym derivative)

144   4. Two-Factor Short-Rate Models

$$\frac{dQ^T}{dQ} = \frac{B(0)P(T,T)}{B(T)P(0,T)} = \frac{\exp\left\{-\int_0^T r(u)du\right\}}{P(0,T)}$$

$$= \frac{\exp\left\{-\int_0^T \varphi(u)du - \int_0^T [x(u)+y(u)]du\right\}}{P(0,T)} \quad (4.17)$$

$$= \exp\left\{-\frac{1}{2}V(0,T) - \int_0^T [x(u)+y(u)]du\right\},$$

where B here is the bank-account numeraire. The measure Q^T is the well known T-forward (risk-adjusted) measure. The following lemma yields the dynamics of the processes x and y under Q^T.

Lemma 4.2.2. *The processes x and y under the forward measure Q^T evolve according to*

$$dx(t) = \left[-ax(t) - \frac{\sigma^2}{a}(1 - e^{-a(T-t)}) - \rho\frac{\sigma\eta}{b}(1 - e^{-b(T-t)})\right]dt + \sigma dW_1^T(t),$$

$$dy(t) = \left[-by(t) - \frac{\eta^2}{b}(1 - e^{-b(T-t)}) - \rho\frac{\sigma\eta}{a}(1 - e^{-a(T-t)})\right]dt + \eta dW_2^T(t),$$

(4.18)

where W_1^T and W_2^T are two correlated Brownian motions under Q^T with $dW_1^T(t)dW_2^T(t) = \rho\, dt$.

Moreover, the explicit solutions of equations (4.18) are, for $s \leq t \leq T$,

$$x(t) = x(s)e^{-a(t-s)} - M_x^T(s,t) + \sigma\int_s^t e^{-a(t-u)}dW_1^T(u)$$

$$y(t) = y(s)e^{-b(t-s)} - M_y^T(s,t) + \eta\int_s^t e^{-b(t-u)}dW_2^T(u),$$

(4.19)

where

$$M_x^T(s,t) = \left(\frac{\sigma^2}{a^2} + \rho\frac{\sigma\eta}{ab}\right)\left[1 - e^{-a(t-s)}\right] - \frac{\sigma^2}{2a^2}\left[e^{-a(T-t)} - e^{-a(T+t-2s)}\right]$$

$$- \frac{\rho\sigma\eta}{b(a+b)}\left[e^{-b(T-t)} - e^{-bT-at+(a+b)s}\right],$$

$$M_y^T(s,t) = \left(\frac{\eta^2}{b^2} + \rho\frac{\sigma\eta}{ab}\right)\left[1 - e^{-b(t-s)}\right] - \frac{\eta^2}{2b^2}\left[e^{-b(T-t)} - e^{-b(T+t-2s)}\right]$$

$$- \frac{\rho\sigma\eta}{a(a+b)}\left[e^{-a(T-t)} - e^{-aT-bt+(a+b)s}\right],$$

so that, under Q^T, the distribution of $r(t)$ conditional on $\mathcal{F}_s$ is normal with mean and variance given respectively by

4.2 The Two-Additive-Factor Gaussian Model G2++

$$E^{Q^T}\{r(t)|\mathcal{F}_s\} = x(s)e^{-a(t-s)} - M_x^T(s,t) + y(s)e^{-b(t-s)} - M_y^T(s,t) + \varphi(t),$$

$$\mathrm{Var}^{Q^T}\{r(t)|\mathcal{F}_s\} = \frac{\sigma^2}{2a}\left[1 - e^{-2a(t-s)}\right] + \frac{\eta^2}{2b}\left[1 - e^{-2b(t-s)}\right]$$
$$+ 2\rho\frac{\sigma\eta}{a+b}\left[1 - e^{-(a+b)(t-s)}\right].$$
(4.20)

Proof. See Appendix B in this chapter for a detailed proof, or else apply directly formula (2.12) with $U = P(\cdot, T)$, $S = B$, and $X = [x \ y]'$. □

Formulas (4.20) are very useful when pricing path-dependent derivatives through Monte Carlo generation of scenarios. To this end, we refer to the second part of this book.

We can now state the following.

Theorem 4.2.2. *The price at time t of a European call option with maturity T and strike K, written on a zero-coupon bond with unit face value and maturity S is given by*

$$\mathbf{ZBC}(t,T,S,K) = P(t,S)\Phi\left(\frac{\ln\frac{P(t,S)}{KP(t,T)}}{\Sigma(t,T,S)} + \frac{1}{2}\Sigma(t,T,S)\right)$$
$$- P(t,T)K\Phi\left(\frac{\ln\frac{P(t,S)}{KP(t,T)}}{\Sigma(t,T,S)} - \frac{1}{2}\Sigma(t,T,S)\right),$$
(4.21)

where

$$\Sigma(t,T,S)^2 = \frac{\sigma^2}{2a^3}\left[1 - e^{-a(S-T)}\right]^2\left[1 - e^{-2a(T-t)}\right]$$
$$+ \frac{\eta^2}{2b^3}\left[1 - e^{-b(S-T)}\right]^2\left[1 - e^{-2b(T-t)}\right]$$
$$+ 2\rho\frac{\sigma\eta}{ab(a+b)}\left[1 - e^{-a(S-T)}\right]\left[1 - e^{-b(S-T)}\right]\left[1 - e^{-(a+b)(T-t)}\right].$$

Analogously, the price at time t of a European put option with maturity T and strike K, written on a zero-coupon bond with unit face value and maturity S is given by

$$\mathbf{ZBP}(t,T,S,K) = -P(t,S)\Phi\left(\frac{\ln\frac{KP(t,T)}{P(t,S)}}{\Sigma(t,T,S)} - \frac{1}{2}\Sigma(t,T,S)\right)$$
$$+ P(t,T)K\Phi\left(\frac{\ln\frac{KP(t,T)}{P(t,S)}}{\Sigma(t,T,S)} + \frac{1}{2}\Sigma(t,T,S)\right).$$
(4.22)

Proof. See Appendix C in this chapter. □

We then have the following obvious generalization to the case where the underlying bond has an arbitrary face value.

Corollary 4.2.2. *The price at time t of a European call option with maturity T and strike K, written on a zero-coupon bond with face value N and maturity S is given by*

$$\mathbf{ZBC}(t,T,S,N,K) = NP(t,S)\Phi\left(\frac{\ln\frac{NP(t,S)}{KP(t,T)}}{\Sigma(t,T,S)} + \frac{1}{2}\Sigma(t,T,S)\right)$$
$$- P(t,T)K\Phi\left(\frac{\ln\frac{NP(t,S)}{KP(t,T)}}{\Sigma(t,T,S)} - \frac{1}{2}\Sigma(t,T,S)\right). \quad (4.23)$$

Analogously, the price at time t of the corresponding put option is

$$\mathbf{ZBP}(t,T,S,N,K) = -NP(t,S)\Phi\left(\frac{\ln\frac{KP(t,T)}{NP(t,S)}}{\Sigma(t,T,S)} - \frac{1}{2}\Sigma(t,T,S)\right)$$
$$+ P(t,T)K\Phi\left(\frac{\ln\frac{KP(t,T)}{NP(t,S)}}{\Sigma(t,T,S)} + \frac{1}{2}\Sigma(t,T,S)\right). \quad (4.24)$$

The pricing of caplets and floorlets. Given the current time t and the future times T_1 and T_2, an "in-arrears" caplet pays off at time T_2

$$[L(T_1,T_2) - X]^+ \alpha(T_1,T_2)N, \quad (4.25)$$

where N is the nominal value, X is the caplet rate (strike), $\alpha(T_1,T_2)$ is the year fraction between times T_1 and T_2 and $L(T_1,T_2)$ is the LIBOR rate at time T_1 for the maturity T_2, i.e.,

$$L(T_1,T_2) = \frac{1}{\alpha(T_1,T_2)}\left[\frac{1}{P(T_1,T_2)} - 1\right]. \quad (4.26)$$

The no-arbitrage value at time t of the payoff (4.25) is, by (2.23) and (2.24),

$$\mathbf{Cpl}(t,T_1,T_2,N,X) = N'\mathbf{ZBP}(t,T_1,T_2,X')$$
$$= \mathbf{ZBP}(t,T_1,T_2,N',N), \quad (4.27)$$

where

$$X' = \frac{1}{1+X\alpha(T_1,T_2)},$$
$$N' = N(1+X\alpha(T_1,T_2)).$$

Explicitly,

4.2 The Two-Additive-Factor Gaussian Model G2++

$$\mathbf{Cpl}(t, T_1, T_2, N, X) = - N'P(t,T_2)\Phi\left(\frac{\ln \frac{NP(t,T_1)}{N'P(t,T_2)}}{\Sigma(t,T_1,T_2)} - \frac{1}{2}\Sigma(t,T_1,T_2)\right)$$

$$+ P(t,T_1)N\Phi\left(\frac{\ln \frac{NP(t,T_1)}{N'P(t,T_2)}}{\Sigma(t,T_1,T_2)} + \frac{1}{2}\Sigma(t,T_1,T_2)\right). \tag{4.28}$$

Analogously, the no-arbitrage value at time t of the floorlet that pays off

$$[X - L(T_1, T_2)]^+ \, \alpha(T_1, T_2) N$$

at time T_2 is

$$\mathbf{Fll}(t, T_1, T_2, N, X) = N'\mathbf{ZBC}(t, T_1, T_2, X') \\ = \mathbf{ZBC}(t, T_1, T_2, N', N).$$

Explicitly,

$$\mathbf{Fll}(t, T_1, T_2, N, X) = N'P(t,T_2)\Phi\left(\frac{\ln \frac{N'P(t,T_2)}{NP(t,T_1)}}{\Sigma(t,T_1,T_2)} + \frac{1}{2}\Sigma(t,T_1,T_2)\right)$$

$$- P(t,T_1)N\Phi\left(\frac{\ln \frac{N'P(t,T_2)}{NP(t,T_1)}}{\Sigma(t,T_1,T_2)} - \frac{1}{2}\Sigma(t,T_1,T_2)\right).$$

The pricing of caps and floors. We denote by $\mathcal{T} = \{T_0, T_1, T_2, \ldots, T_n\}$ the set of the cap/floor payment dates, augmented with the first reset date T_0, and by $\tau = \{\tau_1, \ldots, \tau_n\}$ the set of the corresponding year fractions, meaning that τ_i is the year fraction between T_{i-1} and T_i.

Since the price of a cap (floor) is the sum of the prices of the underlying caplets (floorlets), the price at time t of a cap with cap rate (strike) X, nominal value N, set of times $\mathcal{T}$ and year fractions τ is then given by

$$\mathbf{Cap}(t, \mathcal{T}, \tau, N, X)$$

$$= \sum_{i=1}^{n} \left[-N(1 + X\tau_i)P(t,T_i)\Phi\left(\frac{\ln \frac{P(t,T_{i-1})}{(1+X\tau_i)P(t,T_i)}}{\Sigma(t,T_{i-1},T_i)} - \frac{1}{2}\Sigma(t,T_{i-1},T_i)\right) \right.$$

$$\left. + P(t,T_{i-1})N\Phi\left(\frac{\ln \frac{P(t,T_{i-1})}{(1+X\tau_i)P(t,T_i)}}{\Sigma(t,T_{i-1},T_i)} + \frac{1}{2}\Sigma(t,T_{i-1},T_i)\right) \right], \tag{4.29}$$

and the price of the corresponding floor is

Flr$(t, \mathcal{T}, N, X)$

$$= \sum_{i=1}^{n} \left[N(1 + X\tau_i)P(t, T_i)\Phi\left(\frac{\ln \frac{(1+X\tau_i)P(t,T_i)}{P(t,T_{i-1})}}{\Sigma(t, T_{i-1}, T_i)} + \frac{1}{2}\Sigma(t, T_{i-1}, T_i)\right) \right.$$

$$\left. - P(t, T_{i-1})N\Phi\left(\frac{\ln \frac{(1+X\tau_i)P(t,T_i)}{P(t,T_{i-1})}}{\Sigma(t, T_{i-1}, T_i)} - \frac{1}{2}\Sigma(t, T_{i-1}, T_i)\right) \right].$$

(4.30)

The pricing of European swaptions.

> *God does not care about our mathematical difficulties.*
> *He integrates empirically.*
> Albert Einstein (1879-1955)

Consider a European swaption with strike rate X, maturity T and nominal value N, which gives the holder the right to enter at time $t_0 = T$ an interest-rate swap with payment times $\mathcal{T} = \{t_1, \ldots, t_n\}$, $t_1 > T$, where he pays (receives) at the fixed rate X and receives (pays) LIBOR set "in arrears". We denote by τ_i the year fraction from t_{i-1} to t_i, $i = 1, \ldots, n$ and set $c_i := X\tau_i$ for $i = 1, \ldots, n-1$ and $c_n := 1 + X\tau_n$. We then have the following theorem.

Theorem 4.2.3. *The arbitrage-free price at time $t = 0$ of the above European swaption is given by numerically computing the following one-dimensional integral:*

$$\mathbf{ES}(0, T, \mathcal{T}, N, X, \omega) =$$

$$N\omega P(0, T) \int_{-\infty}^{+\infty} \frac{e^{-\frac{1}{2}\left(\frac{x-\mu_x}{\sigma_x}\right)^2}}{\sigma_x \sqrt{2\pi}} \left[\Phi(-\omega h_1(x)) - \sum_{i=1}^{n} \lambda_i(x) e^{\kappa_i(x)} \Phi(-\omega h_2(x)) \right] dx,$$

(4.31)

where $\omega = 1$ ($\omega = -1$) for a payer (receiver) swaption,

$$h_1(x) := \frac{\bar{y} - \mu_y}{\sigma_y \sqrt{1 - \rho_{xy}^2}} - \frac{\rho_{xy}(x - \mu_x)}{\sigma_x \sqrt{1 - \rho_{xy}^2}}$$

$$h_2(x) := h_1(x) + B(b, T, t_i)\sigma_y \sqrt{1 - \rho_{xy}^2}$$

$$\lambda_i(x) := c_i A(T, t_i) e^{-B(a, T, t_i)x}$$

$$\kappa_i(x) := -B(b, T, t_i)\left[\mu_y - \frac{1}{2}(1 - \rho_{xy}^2)\sigma_y^2 B(b, T, t_i) + \rho_{xy}\sigma_y \frac{x - \mu_x}{\sigma_x}\right],$$

$\bar{y} = \bar{y}(x)$ is the unique solution of the following equation

$$\sum_{i=1}^{n} c_i A(T, t_i) e^{-B(a, T, t_i)x - B(b, T, t_i)\bar{y}} = 1,$$

and

$$\mu_x := -M_x^T(0,T),$$
$$\mu_y := -M_y^T(0,T),$$
$$\sigma_x := \sigma\sqrt{\frac{1-e^{-2aT}}{2a}},$$
$$\sigma_y := \eta\sqrt{\frac{1-e^{-2bT}}{2b}},$$
$$\rho_{xy} := \frac{\rho\sigma\eta}{(a+b)\sigma_x\sigma_y}\left[1 - e^{-(a+b)T}\right].$$

Proof. See Appendix D in this chapter. □

4.2.5 The Analogy with the Hull-White Two-Factor Model

The Hull-White (1994c) two-factor model assumes that the instantaneous short rate evolves in the risk-adjusted measure according to

$$dr(t) = [\theta(t) + u(t) - \bar{a}r(t)]dt + \sigma_1 dZ_1(t), \quad r(0) = r_0, \quad (4.32)$$

where the stochastic mean-reversion level satisfies

$$du(t) = -\bar{b}u(t)dt + \sigma_2 dZ_2(t), \quad u(0) = 0,$$

with (Z_1, Z_2) a two-dimensional Brownian motion with $dZ_1(t)dZ_2(t) = \bar{\rho}dt$, r_0, $\bar{a}$, $\bar{b}$, σ_1 and σ_2 positive constants, and $-1 \leq \bar{\rho} \leq 1$. The function θ is deterministic and properly chosen so as to exactly fit the current term structure of interest rates.

Simple integration leads to

$$r(t) = r(s)e^{-\bar{a}(t-s)} + \int_s^t \theta(v)e^{-\bar{a}(t-v)}dv + \int_s^t u(v)e^{-\bar{a}(t-v)}dv$$
$$+ \sigma_1 \int_s^t e^{-\bar{a}(t-v)}dZ_1(v),$$
$$u(t) = u(s)e^{-\bar{b}(t-s)} + \sigma_2 \int_s^t e^{-\bar{b}(t-v)}dZ_2(v).$$

Assuming $\bar{a} \neq \bar{b}$, we have

$$\int_s^t u(v)e^{-\bar{a}(t-v)}dv$$
$$= \int_s^t u(s)e^{-\bar{b}(v-s)-\bar{a}(t-v)}dv + \sigma_2 \int_s^t e^{-\bar{a}(t-v)}\int_s^v e^{-\bar{b}(v-x)}dZ_2(x)dv$$
$$= u(s)\frac{e^{-\bar{b}(t-s)} - e^{-\bar{a}(t-s)}}{\bar{a}-\bar{b}} + \sigma_2 e^{-\bar{a}t}\int_s^t e^{(\bar{a}-\bar{b})v}\int_s^v e^{\bar{b}x}dZ_2(x)dv.$$

By integration by parts we then have

$$\int_s^t e^{(\bar{a}-\bar{b})v} \int_s^v e^{\bar{b}x} dZ_2(x) dv$$

$$= \frac{1}{\bar{a}-\bar{b}} \int_s^t \left(\int_s^v e^{\bar{b}x} dZ_2(x) \right) d_v\left(e^{(\bar{a}-\bar{b})v}\right)$$

$$= \frac{1}{\bar{a}-\bar{b}} \left[e^{(\bar{a}-\bar{b})t} \int_s^t e^{\bar{b}x} dZ_2(x) - \int_s^t e^{(\bar{a}-\bar{b})v} d_v\left(\int_s^v e^{\bar{b}x} dZ_2(x) \right) \right]$$

$$= \frac{1}{\bar{a}-\bar{b}} \int_s^t \left[e^{(\bar{a}-\bar{b})t} - e^{(\bar{a}-\bar{b})v} \right] d_v\left(\int_s^v e^{\bar{b}x} dZ_2(x) \right)$$

$$= \frac{1}{\bar{a}-\bar{b}} \int_s^t \left[e^{\bar{a}t - \bar{b}(t-v)} - e^{\bar{a}v} \right] dZ_2(v),$$

so that we finally obtain

$$r(t) = r(s)e^{-\bar{a}(t-s)} + \int_s^t \theta(v) e^{-\bar{a}(t-v)} dv + \sigma_1 \int_s^t e^{-\bar{a}(t-v)} dZ_1(v)$$

$$+ u(s) \frac{e^{-\bar{b}(t-s)} - e^{-\bar{a}(t-s)}}{\bar{a}-\bar{b}} + \frac{\sigma_2}{\bar{a}-\bar{b}} \int_s^t \left[e^{-\bar{b}(t-v)} - e^{-\bar{a}(t-v)} \right] dZ_2(v),$$

and in particular,

$$r(t) = r_0 e^{-\bar{a}t} + \int_0^t \theta(v) e^{-\bar{a}(t-v)} dv + \sigma_1 \int_0^t e^{-\bar{a}(t-v)} dZ_1(v)$$

$$+ \frac{\sigma_2}{\bar{a}-\bar{b}} \int_0^t \left[e^{-\bar{b}(t-v)} - e^{-\bar{a}(t-v)} \right] dZ_2(v).$$

Now if we assume $\bar{a} > \bar{b}$ (the case "$\bar{a} < \bar{b}$" is analogous) and define

$$\sigma_3 = \sqrt{\sigma_1^2 + \frac{\sigma_2^2}{(\bar{a}-\bar{b})^2} + 2\bar{\rho}\frac{\sigma_1\sigma_2}{\bar{b}-\bar{a}}}$$

$$dZ_3(t) = \frac{\sigma_1 dZ_1(t) - \frac{\sigma_2}{\bar{a}-\bar{b}} dZ_2(t)}{\sigma_3}$$

$$\sigma_4 = \frac{\sigma_2}{\bar{a}-\bar{b}},$$

we can write

4.2 The Two-Additive-Factor Gaussian Model G2++

$$r(t) = r_0 e^{-\bar{a}t} + \int_0^t \theta(v) e^{-\bar{a}(t-v)} dv + \int_0^t e^{-\bar{a}(t-v)} \left[\sigma_1 dZ_1(v) + \frac{\sigma_2}{\bar{b} - \bar{a}} dZ_2(v) \right]$$

$$+ \frac{\sigma_2}{\bar{a} - \bar{b}} \int_0^t e^{-\bar{b}(t-v)} dZ_2(v)$$

$$= r_0 e^{-\bar{a}t} + \int_0^t \theta(v) e^{-\bar{a}(t-v)} dv$$

$$+ \sigma_3 \int_0^t e^{-\bar{a}(t-v)} dZ_3(v) + \sigma_4 \int_0^t e^{-\bar{b}(t-v)} dZ_2(v).$$

At this stage, the analogy with the G2++ model (4.4) becomes clear. Precisely, by setting

$$a = \bar{a}$$
$$b = \bar{b}$$
$$\sigma = \sigma_3$$
$$\eta = \sigma_4$$
$$\rho = \frac{\sigma_1 \bar{\rho} - \sigma_4}{\sigma_3}$$

$$\varphi(t) = r_0 e^{-\bar{a}t} + \int_0^t \theta(v) e^{-\bar{a}(t-v)} dv,$$

we exactly recover the expression (4.7) for the short rate in the G2++ model. Conversely, given the G2++ model (4.4), we can recover the classical two-factor Hull-White model (4.32), by setting

$$\bar{a} = a$$
$$\bar{b} = b$$
$$\sigma_1 = \sqrt{\sigma^2 + \eta^2 + 2\rho\sigma\eta}$$
$$\sigma_2 = \eta(a - b)$$
$$\bar{\rho} = \frac{\sigma\rho + \eta}{\sqrt{\sigma^2 + \eta^2 + 2\rho\sigma\eta}}$$
$$\theta(t) = \frac{d\varphi(t)}{dt} + a\varphi(t).$$

A different way to prove this analogy is by defining the new stochastic process

$$\chi(t) = r(t) + \delta u(t),$$

where $\delta = 1/(\bar{b} - \bar{a})$. In fact,

$$d\chi(t) = [\theta(t) + u(t) - \bar{a}r(t)]dt + \sigma_1 dZ_1(t) - \delta \bar{b} u(t)dt + \delta \sigma_2 dZ_2(t)$$
$$= [\theta(t) + u(t) - \bar{a}\chi(t) + \bar{a}\delta u(t) - \bar{b}\delta u(t)]dt + \sigma_1 dZ_1(t) + \delta \sigma_2 dZ_2(t)$$
$$= [\theta(t) - \bar{a}\chi(t)]dt + \sigma_3 dZ_3(t),$$

with σ_3 and dZ_3 defined previously.

Moreover, if we define
$$\psi(t) = \frac{u(t)}{\bar{a}-\bar{b}} = -\delta u(t),$$

then
$$d\psi(t) = -\frac{\bar{b}}{\bar{a}-\bar{b}}u(t)dt + \frac{\sigma_2}{\bar{a}-\bar{b}}dZ_2(t)$$
$$= -\bar{b}\psi(t)dt + \sigma_4 dZ_2(t),$$

with σ_4 defined previously. Therefore, we again obtain that $r(t)$ can be written as
$$r(t) = \tilde{\chi}(t) + \psi(t) + \varphi(t),$$

where
$$d\tilde{\chi}(t) = -\bar{a}\tilde{\chi}(t)dt + \sigma_3 dZ_3(t),$$
$$d\psi(t) = -\bar{b}\psi(t)dt + \sigma_4 dZ_2(t),$$
$$\varphi(t) = r_0 e^{-\bar{a}t} + \int_0^t \theta(v) e^{-\bar{a}(t-v)} dv.$$

4.2.6 The Construction of an Approximating Binomial Tree

The purpose of this section is the construction of an approximating tree for the G2++ process (4.4). Such a tree is a fundamental tool when pricing exotic interest rate derivatives.

A two-dimensional tree, trinomial in both dimensions, can be constructed according to the procedure suggested by Hull-White (1994c). We just have to follow the general method illustrated in Appendix C at the end of the book and apply it to the dynamics underlying (4.4).

Alternatively, we can build a simpler tree, which is binomial in both dimensions. The construction of such a tree is, for sake of completeness, outlined in this section.

We start by constructing two binomial trees approximating respectively the dynamics of the processes x and y given in (4.5).

We first remember that, for any t and $\Delta t > 0$, we have
$$\begin{cases} E\{x(t+\Delta t)|\mathcal{F}_t\} = x(t)e^{-a\Delta t}, \\ \text{Var}\{x(t+\Delta t)|\mathcal{F}_t\} = \frac{\sigma^2}{2a}(1-e^{-2a\Delta t}), \end{cases}$$

$$\begin{cases} E\{y(t+\Delta t)|\mathcal{F}_t\} = y(t)e^{-b\Delta t}, \\ \text{Var}\{y(t+\Delta t)|\mathcal{F}_t\} = \frac{\eta^2}{2b}(1-e^{-2b\Delta t}), \end{cases}$$

and

4.2 The Two-Additive-Factor Gaussian Model G2++

$$\text{Cov}\{x(t+\Delta t), y(t+\Delta t)|\mathcal{F}_t\}$$
$$= E\left\{[x(t+\Delta t) - E\{x(t+\Delta t)|\mathcal{F}_t\}][y(t+\Delta t) - E\{y(t+\Delta t)|\mathcal{F}_t\}]|\mathcal{F}_t\right\}$$
$$= \sigma\eta E\left\{\int_t^{t+\Delta t} e^{-a(t+\Delta t - u)}dW_1(u) \int_t^{t+\Delta t} e^{-b(t+\Delta t - u)}dW_2(u)|\mathcal{F}_t\right\}$$
$$= \sigma\eta\rho \int_t^{t+\Delta t} e^{-(a+b)(t+\Delta t - u)}du$$
$$= \frac{\sigma\eta\rho}{a+b}\left[1 - e^{-(a+b)\Delta t}\right].$$

By expanding up to first order in Δt, we have

$$\begin{cases} E\{x(t+\Delta t)|\mathcal{F}_t\} = x(t)(1 - a\Delta t), \\ \text{Var}\{x(t+\Delta t)|\mathcal{F}_t\} = \sigma^2 \Delta t, \end{cases} \quad \begin{cases} E\{y(t+\Delta t)|\mathcal{F}_t\} = y(t)(1 - b\Delta t), \\ \text{Var}\{y(t+\Delta t)|\mathcal{F}_t\} = \eta^2 \Delta t, \end{cases}$$

and
$$\text{Cov}\{x(t+\Delta t), y(t+\Delta t)|\mathcal{F}_t\} = \sigma\eta\rho\Delta t.$$

The Binomial Trees for x and y. The binomial trees approximating the processes x and y are reproduced in Figure 4.1. Precisely, we assume that if at time t we have a value $x(t)$ (resp. $y(t)$), then at time $t + \Delta t$, the process x (resp. y) can either move up to $x(t) + \Delta x$ (resp. $y(t) + \Delta y$) with probability p (resp. q) or down to $x(t) - \Delta x$ (resp. $y(t) - \Delta y$) with probability $1 - p$ (resp. $1 - q$). The quantities Δx, Δy, p and q are to be properly chosen in order to match (at first order in Δt) the conditional mean and variance of the (continuous-time) processes x and y. Precisely, we have to solve

$$\begin{cases} p(x(t) + \Delta x) + (1-p)(x(t) - \Delta x) = x(t)(1 - a\Delta t), \\ p(x(t) + \Delta x)^2 + (1-p)(x(t) - \Delta x)^2 - [x(t)(1 - a\Delta t)]^2 = \sigma^2 \Delta t, \end{cases}$$

and, equivalently,

$$\begin{cases} q(y(t) + \Delta y) + (1-q)(y(t) - \Delta y) = y(t)(1 - b\Delta t), \\ q(y(t) + \Delta y)^2 + (1-q)(y(t) - \Delta y)^2 - [y(t)(1 - b\Delta t)]^2 = \eta^2 \Delta t. \end{cases}$$

Neglecting all terms with higher order than $\sqrt{\Delta t}$, we obtain

$$\begin{cases} \Delta x = \sigma\sqrt{\Delta t}, \\ p = \frac{1}{2} - \frac{x(t)a\Delta t}{2\Delta x} = \frac{1}{2} - \frac{x(t)a}{2\sigma}\sqrt{\Delta t}, \end{cases}$$

$$\begin{cases} \Delta y = \eta\sqrt{\Delta t}, \\ q = \frac{1}{2} - \frac{y(t)b\Delta t}{2\Delta y} = \frac{1}{2} - \frac{y(t)b}{2\eta}\sqrt{\Delta t}, \end{cases}$$

so that both p and q only depend on the values of $x(t)$ and $y(t)$ respectively and not explicitly on time t. It is also easy to see that

4. Two-Factor Short-Rate Models

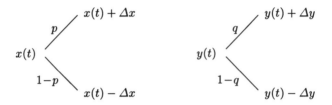

Fig. 4.1. Evolution of the processes x and y starting respectively from $x(t)$ and $y(t)$ at time t and moving upwards to $x(t) + \Delta x$ and to $y(t) + \Delta y$ at time $t + \Delta t$ with probabilities p and q, downwards to $x(t) - \Delta x$ and to $y(t) - \Delta y$ at time $t + \Delta t$ with probabilities $1 - p$ and $1 - q$.

$$0 \leq p \leq 1 \text{ if and only if } |x(t)| \leq \frac{\sigma}{a\sqrt{\Delta t}},$$

$$0 \leq q \leq 1 \text{ if and only if } |y(t)| \leq \frac{\eta}{b\sqrt{\Delta t}}.$$

The Approximating Tree for r. If the two factors x and y are both approximated through the previous binomial lattices, the process r can be approximated through a quadrinomial tree as represented in Figure 4.2. Precisely, we assume that if at time t we start from a pair $(x(t), y(t))$, then at time $t + \Delta t$, the pair (x, y) can move to

- $(x(t) + \Delta x, y(t) + \Delta y)$ with probability π_1;
- $(x(t) + \Delta x, y(t) - \Delta y)$ with probability π_2;
- $(x(t) - \Delta x, y(t) + \Delta y)$ with probability π_3;
- $(x(t) - \Delta x, y(t) - \Delta y)$ with probability π_4;

where $0 \leq \pi_1, \pi_2, \pi_3, \pi_4 \leq 1$ and $\pi_1 + \pi_2 + \pi_3 + \pi_4 = 1$. The probabilities π_1, π_2, π_3 and π_4 are to be chosen in order to match the marginal distributions of the binomial trees for x and y and the conditional covariance (at first order in Δt) between the (continuous-time) processes x and y. Matching the marginal distributions and imposing that the probabilities sum up to one, we get

4.2 The Two-Additive-Factor Gaussian Model G2++

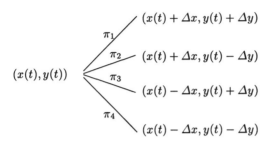

Fig. 4.2. Evolution of the pair process (x, y) starting from $(x(t), y(t))$ and moving to i) $(x(t) + \Delta x, y(t) + \Delta y)$ with probability π_1; ii) $(x(t) + \Delta x, y(t) - \Delta y)$ with probability π_2; iii) $(x(t) - \Delta x, y(t) + \Delta y)$ with probability π_3; iv) $(x(t) - \Delta x, y(t) - \Delta y)$ with probability π_4.

$$\begin{cases} \pi_1 + \pi_2 + \pi_3 + \pi_4 = 1 \\ \pi_1 + \pi_2 = p \\ \pi_3 + \pi_4 = 1 - p \\ \pi_1 + \pi_3 = q \\ \pi_2 + \pi_4 = 1 - q \end{cases}$$

which implies that

$$\begin{cases} \pi_1 = \pi_4 - 1 + q + p \\ \pi_2 = -\pi_4 + 1 - q \\ \pi_3 = -\pi_4 + 1 - p \\ \pi_4 = \pi_4 \end{cases} \quad (4.33)$$

Matching the conditional covariance, we have the additional constraint

$$(\Delta x + ax(t)\Delta t)(\Delta y + by(t)\Delta t)\pi_1 + (\Delta x + ax(t)\Delta t)(-\Delta y + by(t)\Delta t)\pi_2$$
$$+ (ax(t)\Delta t - \Delta x)(\Delta y + by(t)\Delta t)\pi_3 + (ax(t)\Delta t - \Delta x)(-\Delta y + by(t)\Delta t)\pi_4$$
$$= \rho\sigma\eta\Delta t \quad (4.34)$$

which leads to

156   4. Two-Factor Short-Rate Models

$$\begin{cases} \pi_1 = \dfrac{1+\rho}{4} - \dfrac{b\sigma y(t) + a\eta x(t)}{4\sigma\eta}\sqrt{\Delta t} \\ \pi_2 = \dfrac{1-\rho}{4} + \dfrac{b\sigma y(t) - a\eta x(t)}{4\sigma\eta}\sqrt{\Delta t} \\ \pi_3 = \dfrac{1-\rho}{4} - \dfrac{b\sigma y(t) - a\eta x(t)}{4\sigma\eta}\sqrt{\Delta t} \\ \pi_4 = \dfrac{1+\rho}{4} + \dfrac{b\sigma y(t) + a\eta x(t)}{4\sigma\eta}\sqrt{\Delta t} \end{cases}$$

It is clear, however, that, for fixed a, σ, b, η, ρ and Δt, the conditions

$$\begin{cases} 0 \leq \dfrac{1+\rho}{4} - \dfrac{b\sigma y(t) + a\eta x(t)}{4\sigma\eta}\sqrt{\Delta t} \leq 1 \\ 0 \leq \dfrac{1-\rho}{4} + \dfrac{b\sigma y(t) - a\eta x(t)}{4\sigma\eta}\sqrt{\Delta t} \leq 1 \\ 0 \leq \dfrac{1-\rho}{4} - \dfrac{b\sigma y(t) - a\eta x(t)}{4\sigma\eta}\sqrt{\Delta t} \leq 1 \\ 0 \leq \dfrac{1+\rho}{4} + \dfrac{b\sigma y(t) + a\eta x(t)}{4\sigma\eta}\sqrt{\Delta t} \leq 1 \end{cases} \qquad (4.35)$$

are not necessarily satisfied for every choice of $(x(t), y(t))$. This is exactly the same problem that Hull and White (1994c) encountered in the construction of their two-dimensional trinomial tree. As also suggested by them, a possible way out is to start from (4.33) and solve (4.34) by suitably changing the value of ρ in such a way that the (4.35)'s are fulfilled. This should not affect in a relevant way the pricing of a general claim if we choose Δt to be sufficiently small. Notice, in fact, that the limit values of the probabilities for Δt going to zero always fall in-between zero and one (since $|\rho| \leq 1$), so that, for sufficiently small Δt, the nodes in the quadrinomial tree where we impose a different (and wrong) correlation give actually a negligible contribution.

4.2.7 Examples of Calibration to Real-Market Data

We now propose two examples of calibration to real-market volatility data, which illustrate the fitting capability of our two-factor Gaussian model. We first consider a calibration to cap volatilities and then a calibration to swaption volatilities.

In our first example, we use the same at-the-money Euro cap-volatility data as in Section 3.14, see Table 3.4 or Table 4.1 below. The zero-coupon curve, on the same day and time, is again that shown in Figure 1.1 of Chapter 1.

Given its high number of parameters (five) and the possibility of humped structures in the volatility of instantaneous forward rates, one can expect the fitting quality of the model to be high in general. Indeed, according to our experience, the G2++ model can reproduce market cap-volatility data very accurately. This is confirmed by our calibration result, which is shown

4.2 The Two-Additive-Factor Gaussian Model G2++

in Table 4.1. An important point is that this accurate calibration did not require the introduction of some "all-fitting" time-varying parameters for the volatility part. This guarantees a more regular evolution in time of the market volatility structures associated with the calibrated dynamics. Indeed, analogously to what we have observed when commenting the one-factor dynamics (3.32) in Section 3.3.1, too many time-varying parameters can lead to over-fitting. Moreover, here too the future volatility structures implied by a G2++ model with more time-varying parameters are likely to be unrealistic. This is not to say that the basic G2++ model always produces realistic future structures, but simply that such models are easier to control in terms of the model parameters.

The calibration is performed by minimizing the sum of the squares of the percentage differences between model and market cap prices, and leads to the following parameters: $a = 0.543009105$, $b = 0.075716774$, $\sigma = 0.005837408$, $\eta = 0.011657837$ and $\rho = -0.991401219$.[4]

Contrary to its one-factor analogue (3.33), we here obtain positive best-fitting values for the mean-reversion parameters a and b. Indeed, the increasing part of the market hump can be retrieved with the help of a highly negative ρ.

However, as in this case, it often happens that the ρ value is quite close to minus one, which implies that the G2++ model tends to degenerate into a one-factor (non-Markov) short-rate process. This is also intuitive. In fact, as already explained in Chapter 1, caps prices do not depend on the correlation of forward rates, so that even a one-factor model that implies perfectly-correlated rates can fit caps data well in many situations. Notice, moreover, that the degenerate process for the short rate is still non-Markovian (if $a \neq b$), which explains what really makes the G2++ model outperform its one-factor version (3.33).

Maturity	Market volatility	G2++ implied volatility
1	0.1520	0.1520
2	0.1620	0.1622
3	0.1640	0.1631
4	0.1630	0.1631
5	0.1605	0.1614
7	0.1555	0.1554
10	0.1475	0.1472
15	0.1350	0.1349
20	0.1260	0.1261

Table 4.1. At-the-money Euro cap-volatility quotes on February 13, 2001, at 5 p.m., and corresponding volatilities implied by the G2++ model.

---

[4] The minimization is performed with a simulated-annealing method followed by a local-search algorithm to refine the last solution found.

158   4. Two-Factor Short-Rate Models

We now move to our second example of calibration. We again consider data as of February 13, 2001, at 5 p.m., with the related swaption-volatility quotes being shown in Table 4.2 below. Swaption maturities are one, two, three, four, five, seven and ten years, and the tenors of the underlying swaps go from one to ten years. For a detailed explanation of how ATM-swaptions data are organized in such a table, see Section 6.15 later on. Indeed, it is with market models that swaptions enter the picture at full power, and this is the reason why we present this explanation in the related Chapter 6.

	1y	2y	3y	4y	5y	6y	7y	8y	9y	10y
1y	0.1640	0.1550	0.1430	0.1310	0.1240	0.1190	0.1160	0.1120	0.1100	0.1070
2y	0.1600	0.1500	0.1390	0.1290	0.1220	0.1190	0.1160	0.1130	0.1100	0.1080
3y	0.1570	0.1450	0.1340	0.1240	0.1190	0.1150	0.1130	0.1100	0.1080	0.1060
4y	0.1480	0.1360	0.1260	0.1190	0.1140	0.1120	0.1090	0.1070	0.1050	0.1030
5y	0.1400	0.1280	0.1210	0.1140	0.1100	0.1070	0.1050	0.1030	0.1020	0.1000
7y	0.1300	0.1190	0.1130	0.1050	0.1010	0.0990	0.0970	0.0960	0.0950	0.0930
10y	0.1160	0.1070	0.1000	0.0930	0.0900	0.0890	0.0870	0.0860	0.0850	0.0840

Table 4.2. At-the-money Euro swaption-volatility quotes on February 13, 2001, at 5 p.m.

Minimization of the sum of the squares of the percentage differences between model and market swaption prices produces the following calibrated parameters: $a = 0.773511777$, $b = 0.082013014$, $\sigma = 0.022284644$, $\eta = 0.010382461$ and $\rho = -0.701985206$.

Notice that the value of ρ is now far from minus one. This could be expected since swaption prices contain information on the correlation between forward rates, and indeed a possible way to incorporate such information into the model consists in assigning a non-trivial value to the ρ coefficient.

The calibration results are summarized in Tables 4.3 and 4.4. In the first, we show the fitted swaption volatilities as implied by the G2++ model, whereas in the second we report the percentage differences

$$\frac{\text{G2++ implied volatility - market volatility}}{\text{market volatility}}.$$

We can see that the calibration result is rather satisfactory. Indeed, apart from few exceptions in the first two columns, the percentage errors are rather low given that we have tried to fit seventy prices with only five parameters.

A few concluding remarks are in order.

It may be a good idea to calibrate the G2++ model only to the most significant swaptions data, leaving out the illiquid entries (see again Section 6.15 for more details). Or, when in need to price a particular product that is influenced only by a certain set of swap rates, it may be reasonable to calibrate the model only to the relevant swaptions. For instance, to price a Bermudan swaption, one typically fits the model around the (diagonal) volatilities of the underlying European swaptions. It is however necessary to check how the

4.2 The Two-Additive-Factor Gaussian Model G2++

	1y	2y	3y	4y	5y	6y	7y	8y	9y	10y
1y	0.1870	0.1529	0.1395	0.1327	0.1276	0.1231	0.1190	0.1154	0.1120	0.1085
2y	0.1603	0.1427	0.1348	0.1295	0.1251	0.1210	0.1174	0.1139	0.1103	0.1068
3y	0.1509	0.1376	0.1307	0.1258	0.1216	0.1180	0.1146	0.1109	0.1073	0.1041
4y	0.1422	0.1311	0.1252	0.1210	0.1175	0.1142	0.1106	0.1069	0.1037	0.1005
5y	0.1335	0.1247	0.1199	0.1165	0.1134	0.1098	0.1062	0.1030	0.0998	0.0968
7y	0.1218	0.1161	0.1124	0.1087	0.1050	0.1018	0.0986	0.0954	0.0928	0.0902
10y	0.1090	0.1029	0.0997	0.0965	0.0934	0.0909	0.0884	0.0858	0.0833	0.0809

Table 4.3. G2++ calibrated swaptions volatilities.

	1y	2y	3y	4y	5y	6y	7y	8y	9y	10y
1y	14.01%	-1.35%	-2.43%	1.27%	2.86%	3.44%	2.59%	3.03%	1.81%	1.40%
2y	0.17%	-4.88%	-3.03%	0.39%	2.51%	1.67%	1.19%	0.84%	0.31%	-1.12%
3y	-3.91%	-5.10%	-2.49%	1.44%	2.21%	2.64%	1.42%	0.86%	-0.64%	-1.83%
4y	-3.92%	-3.59%	-0.61%	1.65%	3.08%	1.97%	1.44%	-0.07%	-1.26%	-2.41%
5y	-4.61%	-2.54%	-0.91%	2.23%	3.09%	2.65%	1.16%	-0.01%	-2.13%	-3.25%
7y	-6.29%	-2.47%	-0.50%	3.51%	3.94%	2.82%	1.65%	-0.58%	-2.33%	-3.01%
10y	-6.08%	-3.81%	-0.34%	3.80%	3.79%	2.17%	1.58%	-0.21%	-2.00%	-3.69%

Table 4.4. Swaptions calibration results: percentage differences.

resulting calibrated model prices swaptions that had not been included in the calibration set. This is important in order to verify that the behaviour of the model out of the calibration range is not wild or too weird.

We also point out that both for the cap and swaptions calibrations we have resorted to the related analytical formulas of the G2++ model. In particular, for the swaptions we have used formula (4.31), resorting to a numerical integration against a Gaussian distribution, whose support can be reduced to a convenient number of standard deviations. The related optimization requires a reasonable time when running on a PC. A global optimization involving the seventy swaptions requires a few minutes, whereas a local optimization takes about one minute. Clearly, such times reduce when considering only a few swaption instead of the whole table.

Finally, one can try a joint calibration to caps and swaptions data. Results are usually not completely satisfactory. This may be due both to misalignments between the two markets and to the low number of parameters in the model. In order to allow for a full-power joint calibration we will need to resort to the LIBOR market model. The related cases will be presented in Chapter 7.

Appendix A: Proof of Lemma 4.2.1

Stochastic integration by parts implies that

$$\int_t^T x(u)du = Tx(T) - tx(t) - \int_t^T u\,dx(u) = \int_t^T (T-u)dx(u) + (T-t)x(t). \tag{4.36}$$

By definition of x, the integral in the right-hand side can be written as

4. Two-Factor Short-Rate Models

$$\int_t^T (T-u)dx(u) = -a\int_t^T (T-u)x(u)du + \sigma\int_t^T (T-u)dW_1(u)$$

by substituting the expression for $dx(u)$, and

$$\int_t^T (T-u)x(u)du = x(t)\int_t^T (T-u)e^{-a(u-t)}du$$
$$+ \sigma\int_t^T (T-u)\int_t^u e^{-a(u-s)}dW_1(s)du.$$

by substituting the expression for $x(u)$. Calculating separately the last two integrals (multiplied by $-a$), we have

$$-ax(t)\int_t^T (T-u)e^{-a(u-t)}du = -x(t)(T-t) - \frac{e^{-a(T-t)}-1}{a}x(t)$$

and, again by integration by parts,

$$-a\sigma\int_t^T (T-u)\int_t^u e^{-a(u-s)}dW_1(s)du$$
$$= -a\sigma\int_t^T \left(\int_t^u e^{as}dW_1(s)\right)d_u\left(\int_t^u (T-v)e^{-av}dv\right)$$
$$= -a\sigma\left[\left(\int_t^T e^{au}dW_1(u)\right)\left(\int_t^T (T-v)e^{-av}dv\right)\right.$$
$$\left. - \int_t^T \left(\int_t^u (T-v)e^{-av}dv\right)e^{au}dW_1(u)\right]$$
$$= -a\sigma\int_t^T \left(\int_u^T (T-v)e^{-av}dv\right)e^{au}dW_1(u)$$
$$= -\sigma\int_t^T \left[(T-u) + \frac{e^{-a(T-u)}-1}{a}\right]dW_1(u),$$

where in the last step we have used the fact that

$$\int_u^T (T-v)e^{-av}dv = \frac{(T-u)e^{-au}}{a} + \frac{e^{-aT} - e^{-au}}{a^2}.$$

Adding up the previous terms, we obtain

$$\int_t^T x(u)du = \frac{1-e^{-a(T-t)}}{a}x(t) + \frac{\sigma}{a}\int_t^T \left[1 - e^{-u(T-u)}\right]dW_1(u).$$

The analogous expression for y is obtained by replacing a, σ and W_1 respectively with b, η and W_2, i.e.,

4.2 The Two-Additive-Factor Gaussian Model G2++

$$\int_t^T y(u)du = \frac{1-e^{-b(T-t)}}{b}y(t) + \frac{\eta}{b}\int_t^T \left[1 - e^{-b(T-u)}\right]dW_2(u),$$

so that (4.9) is immediately verified. As to the calculation of the conditional variance, we have

$$\begin{aligned}\operatorname{Var}\{I(t,T)|\mathcal{F}_t\} &= \operatorname{Var}\bigg\{\frac{\sigma}{a}\int_t^T\left[1-e^{-a(T-u)}\right]dW_1(u) \\ &\quad + \frac{\eta}{b}\int_t^T\left[1-e^{-b(T-u)}\right]dW_2(u)\bigg|\mathcal{F}_t\bigg\} \\ &= \frac{\sigma^2}{a^2}\int_t^T\left[1-e^{-a(T-u)}\right]^2 du + \frac{\eta^2}{b^2}\int_t^T\left[1-e^{-b(T-u)}\right]^2 du \\ &\quad + 2\rho\frac{\sigma\eta}{ab}\int_t^T\left[1-e^{-a(T-u)}\right]\left[1-e^{-b(T-u)}\right]du.\end{aligned}$$

Simple integration then leads to (4.10).

Appendix B: proof of Lemma 4.2.2

The Radon-Nikodym derivative (4.17) can be written in terms of the two independent Brownian motions $\widetilde{W}_1$ and $\widetilde{W}_2$ as follows

$$\begin{aligned}\frac{dQ^T}{dQ} = \exp\bigg\{&-\frac{1}{2}V(0,T) - \int_0^T\left[\frac{\sigma}{a}(1-e^{-a(T-u)}) + \rho\frac{\eta}{b}(1-e^{-b(T-u)})\right]d\widetilde{W}_1(u) \\ &- \frac{\eta}{b}\sqrt{1-\rho^2}\int_0^T(1-e^{-b(T-u)})d\widetilde{W}_2(u)\bigg\}.\end{aligned}$$

Since

$$\int_0^T\left[\frac{\sigma}{a}(1-e^{-a(T-u)}) + \rho\frac{\eta}{b}(1-e^{-b(T-u)})\right]^2 du \\ - \frac{\eta^2}{b^2}(1-\rho^2)\int_0^T(1-e^{-b(T-u)})^2 du = V(0,T),$$

the Girsanov theorem implies that the two processes $\widetilde{W}_1^T$ and $\widetilde{W}_2^T$ defined by

$$\begin{aligned}d\widetilde{W}_1^T(t) &= d\widetilde{W}_1(t) + \left[\frac{\sigma}{a}(1-e^{-a(T-t)}) + \rho\frac{\eta}{b}(1-e^{-b(T-t)})\right]dt \\ d\widetilde{W}_2^T(t) &= d\widetilde{W}_2(t) + \frac{\eta}{b}\sqrt{1-\rho^2}(1-e^{-b(T-t)})dt\end{aligned} \quad (4.37)$$

are two independent Brownian motions under the measure Q^T. Then defining W_1^T and W_2^T by

$$dW_1^T(t) = d\widetilde{W}_1^T(t),$$
$$dW_2^T(t) = \rho d\widetilde{W}_1^T(t) + \sqrt{1-\rho^2}d\widetilde{W}_2^T(t),$$

and combining (4.8) with (4.37) we obtain (4.18).

Formulas (4.19) follow from integration of (4.18) where the following equalities are being used:

$$\int_s^t \left[\frac{\sigma^2}{a}(1-e^{-a(T-u)}) + \rho\frac{\sigma\eta}{b}(1-e^{-b(T-u)})\right]e^{-a(t-u)}du = M_x^T(s,t)$$

$$\int_s^t \left[\frac{\eta^2}{b}(1-e^{-b(T-u)}) + \rho\frac{\sigma\eta}{a}(1-e^{-a(T-u)})\right]e^{-b(t-u)}du = M_y^T(s,t).$$

Formulas (4.20) are straightforward.

Appendix C: Proof of Theorem 4.2.2

The change-of-numeraire technique, which basically consists in changing the probability measure from Q to Q^T, implies that

$$\mathbf{ZBC}(t,T,\tau,K) = P(t,T)E^{Q^T}\{(P(T,\tau) - K)^+|\mathcal{F}_t\}$$

(see formula 2.21). Since

$$P(T,\tau) = \frac{P^M(0,\tau)}{P^M(0,T)}\exp\left\{\frac{1}{2}[V(T,\tau) - V(0,\tau) + V(0,T)]\right.$$
$$\left. -\frac{1-e^{-a(\tau-T)}}{a}x(T) - \frac{1-e^{-b(\tau-T)}}{b}y(T)\right\},$$

under Q^T the logarithm of $P(T,\tau)$ conditional on $\mathcal{F}_t$ is normally distributed with mean

$$M_p = \ln\frac{P^M(0,\tau)}{P^M(0,T)} + \frac{1}{2}[V(T,\tau) - V(0,\tau) + V(0,T)]$$
$$-\frac{1-e^{-a(\tau-T)}}{a}E^{Q^T}\{x(T)|\mathcal{F}_t\} - \frac{1-e^{-b(\tau-T)}}{b}E^{Q^T}\{y(T)|\mathcal{F}_t\}$$

and variance

$$(V_p)^2 = \frac{\sigma^2}{2a^3}(1-e^{-a(\tau-T)})^2(1-e^{-2a(T-t)})$$
$$+ \frac{\eta^2}{2b^3}(1-e^{-b(\tau-T)})^2(1-e^{-2b(T-t)})$$
$$+ 2\rho\frac{\sigma\eta}{ab(a+b)}(1-e^{-a(\tau-T)})(1-e^{-b(\tau-T)})(1-e^{-(a+b)(T-t)}).$$

Since (see also formula (B.1) in Appendix B at the end of the book)

$$\int_{-\infty}^{+\infty} \frac{1}{\sqrt{2\pi V_p}} (e^z - K)^+ e^{-\frac{1}{2}\frac{(z-M_p)^2}{V_p^2}} dz$$

$$= e^{M_p + \frac{1}{2}V_p^2} \Phi\left(\frac{M_p - \ln K + V_p^2}{V_p}\right) - K\Phi\left(\frac{M_p - \ln K}{V_p}\right),$$

we have that

$$\mathbf{ZBC}(t,T,\tau,K)$$
$$= P(t,T) \left[e^{M_p + \frac{1}{2}V_p^2} \Phi\left(\frac{M_p - \ln K + V_p^2}{V_p}\right) - K\Phi\left(\frac{M_p - \ln K}{V_p}\right) \right].$$

Noting that $\Sigma(t,T,\tau)^2 = V_p^2$, to retrieve (4.21) we have to use the equality

$$M_p = \ln \frac{P(t,\tau)}{P(t,T)} - \frac{1}{2}\Sigma(t,T,\tau)^2,$$

that can be proved by noting that $\frac{P(t,\tau)}{P(t,T)}$ is a martingale under Q^T, hence

$$\frac{P(t,\tau)}{P(t,T)} = E^{Q^T}\{P(T,\tau)|\mathcal{F}_t\} = e^{M_p + \frac{1}{2}V_p^2}.$$

Formula (4.22) immediately follows from the put-call parity:

$$\mathbf{ZBC}(t,T,\tau,K) + KP(t,T) = \mathbf{ZBP}(t,T,\tau,K) + P(t,\tau).$$

Appendix D: Proof of Theorem 4.2.3

By the general formula (2.21), the arbitrage-free price of the above European swaption is

$$\mathbf{ES}(0,T,\mathcal{T},N,X,\omega)$$
$$= NP(0,T)E^T\left\{\left[\omega\left(1 - \sum_{i=1}^n c_i P(T,t_i)\right)\right]^+\right\}$$
$$= NP(0,T)\int_{\mathbb{R}^2}\left[\omega\left(1 - \sum_{i=1}^n c_i A(T,t_i)e^{-B(a,T,t_i)x - B(b,T,t_i)y}\right)\right]^+ f(x,y)dydx,$$

where f is the density of the random vector $(x(T), y(T))$, i.e.,

$$f(x,y) := \frac{\exp\left\{-\frac{1}{2(1-\rho_{xy}^2)}\left[\left(\frac{x-\mu_x}{\sigma_x}\right)^2 - 2\rho_{xy}\frac{(x-\mu_x)(y-\mu_y)}{\sigma_x \sigma_y} + \left(\frac{y-\mu_y}{\sigma_y}\right)^2\right]\right\}}{2\pi\sigma_x\sigma_y\sqrt{1-\rho_{xy}^2}}.$$

164 4. Two-Factor Short-Rate Models

Freezing x in the integrand and integrating over y, from $-\infty$ to $+\infty$, we obtain

$$\int_{\bar{y}(x)}^{+\infty \cdot \omega} \left(1 - \sum_{i=1}^{n} \lambda_i e^{-B(b,T,t_i)y}\right) \gamma e^{E+F(y-\mu_y)-G(y-\mu_y)^2} dy,$$

where

$$\gamma := \frac{1}{2\pi \sigma_x \sigma_y \sqrt{1-\rho_{xy}^2}},$$

$$E := -\frac{1}{2(1-\rho_{xy}^2)} \left(\frac{x-\mu_x}{\sigma_x}\right)^2$$

$$F := \frac{\rho_{xy}}{1-\rho_{xy}^2} \frac{x-\mu_x}{\sigma_x \sigma_y}$$

$$G := \frac{1}{2(1-\rho_{xy}^2)\sigma_y^2},$$

and the dependence on x has been omitted for simplicity. Using the general formula

$$\int_a^b e^{-Ax^2+Bx} dx = \frac{\sqrt{\pi}}{\sqrt{A}} e^{\frac{B^2}{4A}} \left[\Phi\left(b\sqrt{2A} - \frac{B}{\sqrt{2A}}\right) - \Phi\left(a\sqrt{2A} - \frac{B}{\sqrt{2A}}\right)\right],$$

with $A > 0$, a, b and B real constants, the integral becomes

$$\gamma \frac{\sqrt{\pi}}{\sqrt{G}} e^{E+\frac{F^2}{4G}} \left[\Phi\left(+\infty \cdot \omega\sqrt{2G} - \frac{F}{\sqrt{2G}}\right) - \Phi\left((\bar{y}-\mu_y)\sqrt{2G} - \frac{F}{\sqrt{2G}}\right)\right]$$

$$- \gamma \frac{\sqrt{\pi}}{\sqrt{G}} e^E \sum_{i=1}^{n} \lambda_i e^{-B(b,T,t_i)\mu_y + \frac{[F-B(b,T,t_i)]^2}{4G}}$$

$$\cdot \left[\Phi\left(+\infty \cdot \omega\sqrt{2G} - \frac{F-B(b,T,t_i)}{\sqrt{2G}}\right) - \Phi\left((\bar{y}-\mu_y)\sqrt{2G} - \frac{F-B(b,T,t_i)}{\sqrt{2G}}\right)\right].$$

A little algebra then leads to

$$\gamma \frac{\sqrt{\pi}}{\sqrt{G}} e^{E+\frac{F^2}{4G}} \left\{\frac{\omega+1}{2} - \Phi\left((\bar{y}-\mu_y)\sqrt{2G} - \frac{F}{\sqrt{2G}}\right)\right.$$

$$- \sum_{i=1}^{n} \lambda_i e^{-B(b,T,t_i)\mu_y + \frac{B(b,T,t_i)[B(b,T,t_i)-2F]}{4G}}$$

$$\left.\cdot \left[\frac{\omega+1}{2} - \Phi\left((\bar{y}-\mu_y)\sqrt{2G} - \frac{F-B(b,T,t_i)}{\sqrt{2G}}\right)\right]\right\},$$

Since $(\omega+1)/2 - \Phi(z) = \omega\Phi(-\omega z)$, for each real constant z, we obtain (4.31) by noting that

$$\gamma \frac{\sqrt{\pi}}{\sqrt{G}} = \frac{1}{\sigma_x \sqrt{2\pi}},$$

$$E + \frac{F^2}{4G} = -\frac{1}{2}\left(\frac{x - \mu_x}{\sigma_x}\right)^2,$$

$$\frac{F}{\sqrt{2G}} = \frac{\rho_{\tau y}}{\sqrt{1 - \rho_{xy}^2}} \frac{x - \mu_\tau}{\sigma_x},$$

$$\sqrt{2G} = \frac{1}{\sigma_y \sqrt{1 - \rho_{xy}^2}}.$$

4.3 The Two-Additive-Factor Extended CIR/LS Model CIR2++

In this section, we propose an alternative two-factor short-rate model, which is based on adding a deterministic shift to a sum of two independent square-root processes. This model, which we will refer to as CIR2++, can be viewed as the natural two-factor generalization of the CIR++ model of Section 3.9. As for the CIR++ model, also in this two-factor formulation the deterministic shift is added so as to exactly retrieve the term structure of zero-coupon rates at the initial time.

We start by studying the non-shifted (time-homogeneous) two-factor CIR model without requiring an exact fitting of the initial term structure, and observe that it is equivalent to the Longstaff and Schwartz's (1992b) model (LS). We then derive the model formulas by using the one-factor case as a guide. The interested reader is therefore advised to acquire at least a basic knowledge of the CIR and CIR++ models as illustrated in the previous chapter. Indeed, the model presented in this section is largely built upon its one-factor version, using in particular its explicit formulas.

As stated earlier, a good feature of the CIR2++ model is that it can maintain positive rates through reasonable restrictions on the parameters. We do not explicitly consider such restrictions here, and just remember that a flavor of the type of restrictions involved in maintaining positive rates can be found in the one-factor CIR++ case in Section 3.9.2.

Distributionally, the short rate is a sum of two independent non-central chi-square variables, and as such it has fat tails, which is a desirable distributional property improving on the G2++ case. However, the volatility structures allowed here are poorer than in the Gaussian case. Consider once again the curve

$$T \mapsto \sqrt{\operatorname{Var}[d\, f(t,T)]/dt}.$$

It can be easily checked that in the CIR2++ model this curve can only be decreasing, as hinted at below, after formula (4.44). As a consequence, no hump can be allowed in the absolute volatilities of instantaneous forward rates. It follows that the caplet-volatility term structure implied by the model may not reproduce large humps, see the discussion in Section 4.2.3 and especially the discussion towards the end of Section 4.1. Hence, if the market features a largely-humped term structure of cap or caplet volatilities, the CIR2++ model may yield an unsatisfactory calibration to such market data.

4.3.1 The Basic Two-Factor CIR2 Model

In the two-factor CIR model the instantaneous interest rate is obtained by adding two independent processes like x in (3.21) under the risk-neutral measure:

$$dx(t) = k_1(\theta_1 - x(t))dt + \sigma_1\sqrt{x(t)}\,dW_1(t),$$
$$dy(t) = k_2(\theta_2 - y(t))dt + \sigma_2\sqrt{y(t)}\,dW_2(t),$$

where W_1 and W_2 are independent Brownian motions under the risk-neutral measure, and $k_1, \theta_1, \sigma_1, k_2, \theta_2$ and σ_2 are positive constants such that $2k_1\theta_1 > \sigma_1^2$ and $2k_2\theta_2 > \sigma_2^2$.

The short rate is then defined as

$$\xi_t^\alpha := x(t) + y(t), \quad \alpha = (\alpha_1, \alpha_2),$$

with $\alpha_1 = (k_1, \theta_1, \sigma_1)$ and $\alpha_2 = (k_2, \theta_2, \sigma_2)$.

Due to independence of the factors, it is immediate to derive the following formulas (based on the notation for the one-factor case).

The price at time t of a zero-coupon bond with maturity T is explicitly given by

$$P^\xi(t, T; x(t), y(t), \alpha) = P^1(t, T; x(t), \alpha_1)\, P^1(t, T; y(t), \alpha_2), \quad (4.38)$$

where P^1 denotes the bond-price formula for the one-factor CIR model as a function of the one-factor instantaneous short rate and of the parameters, which has been given in (3.24). For example, $P^1(t, T; x(t), \alpha_1)$ is formula (3.24) with $r(t)$ replaced by $x(t)$ and k, θ, σ replaced by $(k_1, \theta_1, \sigma_1) = \alpha_1$.

The continuously-compounded spot rate at time t for the maturity T is given by

$$R^\xi(t, T; x(t), y(t), \alpha) = R^1(t, T; x(t), \alpha_1) + R^1(t, T; y(t), \alpha_2), \quad (4.39)$$

where R^1 denotes the spot rate for the one-factor CIR model, whose formula is obtained immediately from that of P^1. From the structure of the R's we see that we have an affine term structure in two dimensions.

Under the risk-neutral measure the bond price dynamics can be easily obtained via Itô's formula:

$$dP^\xi(t, T; \alpha) = P^\xi(t, T; \alpha)[\xi_t^\alpha\, dt - B(t, T; \alpha_1)\sigma_1\sqrt{x(t)}\,dW_1(t)$$
$$-B(t, T; \alpha_2)\sigma_2\sqrt{y(t)}\,dW_2(t)],$$

where the deterministic function B is defined as in (3.25).

4.3.2 Relationship with the Longstaff and Schwartz Model (LS)

Longstaff and Schwartz (1992b) consider an interest-rate model where the short rate ξ_t, under the risk-neutral measure, is obtained as a linear combination of two basic processes X and Y as follows:

$$dX_t = a(b - X_t)dt + \sqrt{X_t}\, dW_1(t),$$
$$dY_t = c(e - Y_t)dt + \sqrt{Y_t}\, dW_2(t),$$
$$\xi_t = \mu_x X_t + \mu_y Y_t,$$

where all parameters have positive values with $\mu_x \neq \mu_y$, and again W_1 and W_2 are independent Brownian motions under the risk-neutral measure.

Longstaff and Schwartz give an equilibrium derivation of their model and show how it can be expressed as a stochastic-volatility model via a change of variable. Here our purpose is to show that the model is essentially a two-factor CIR model. Set

$$x(t) = \mu_x X_t, \quad y(t) = \mu_y Y_t,$$

so that

$$\xi_t = x(t) + y(t).$$

It is immediate to check that

$$dx(t) = a(\mu_x b - x(t))dt + \sqrt{\mu_x}\sqrt{x(t)}\, dW_1(t)$$
$$=: k_x(\theta_x - x(t))dt + \sigma_x \sqrt{x(t)}\, dW_1(t),$$
$$dy(t) = c(\mu_y e - y(t))dt + \sqrt{\mu_y}\sqrt{y(t)}\, dW_2(t)$$
$$=: k_y(\theta_y - y(t))dt + \sigma_y \sqrt{y(t)}\, dW_2(t).$$

Since both x and y describe a one-factor CIR model, we see that the LS model can be interpreted as a two-factor CIR model. The parameters are linked by the relationships

$$k_x = a,$$
$$\theta_x = \mu_x b,$$
$$\sigma_x = \sqrt{\mu_x},$$
$$k_y = c,$$
$$\theta_y = \mu_y e,$$
$$\sigma_y = \sqrt{\mu_y}.$$

Therefore, when extending a two-factor CIR model in order to exactly fit the observed term structure, we are implicitly extending also the Longstaff and Schwartz (1992b) model. The only requirement is that in the basic CIR model $\sigma_x \neq \sigma_y$, so that in the LS formulation the condition $\mu_x \neq \mu_y$ is satisfied.

4.3.3 Forward-Measure Dynamics and Option Pricing for CIR2

Since the two factors $x(t)$ and $y(t)$ are independent, the T-forward measure dynamics for each factor is exactly the same as in (3.27)

$$dx(t) = [k_1\theta_1 - (k_1 + B(t,T;\alpha_1)\sigma_1^2)x(t)]dt + \sigma_1\sqrt{x(t)}dW_1^T(t),$$
$$dy(t) = [k_2\theta_2 - (k_2 + B(t,T;\alpha_2)\sigma_2^2)y(t)]dt + \sigma_2\sqrt{y(t)}dW_2^T(t),$$

where W_1^T and W_2^T are independent standard Brownian motions under the T-forward adjusted measure. Pricing a zero-coupon-bond option can be carried out as follows. From the general formula for a call-option price at time t with maturity $T > t$ for an underlying zero-coupon bond with maturity $S > T$ and notional amount N and with strike price K we obtain

$$C^\xi(t,T,S,N,K;x(t),y(t),\alpha)$$
$$= E_t\left\{\exp\left[-\int_t^T \xi_u^\alpha du\right][NP^\xi(T,S;x(T),y(T),\alpha) - K]^+\right\}$$
$$= P^\xi(t,T;x(t),y(t),\alpha) E_t^T\{[NP^\xi(T,S;x(T),y(T),\alpha) - K]^+\}.$$

We recall from (3.28) that we know the densities of each x_T and y_T conditional on x_t and y_t respectively, under the T-forward measure. Therefore, we obtain

$$C^\xi(t,T,S,N,K;x(t),y(t),\alpha)$$
$$= P^\xi(t,T;x(t),y(t),\alpha)\int_0^{+\infty}\int_0^{+\infty}[NP^1(T,S;x_1,\alpha_1)P^1(T,S;x_2,\alpha_2) - K]^+$$
$$\cdot p_{x(T)|x(t)}^T(x_1)p_{y(T)|y(t)}^T(x_2)dx_1 dx_2,$$
(4.40)

which is an integral against the product of two non-central chi-square densities. Such a formula is the one being proposed by Longstaff and Schwartz (1992b). We need to remark that Chen and Scott (1992) develop a specific method for computing this integral through a suitable change of variable, reducing the calculation to that of a one-dimensional integral.

4.3.4 The CIR2++ Model and Option Pricing

In perfect analogy with the general method developed in Section 3.8, used in Section 3.9 for the one-factor case, let us define the instantaneous spot rate, under the risk-neutral measure, by

$$r_t = \varphi(t;\alpha) + \xi_t^\alpha = \varphi(t;\alpha) + x(t) + y(t),\tag{4.41}$$

with $x(0) = x_0$, $y(0) = y_0$, and where $\varphi(t;\alpha)$ is a deterministic function of time, depending on the augmented parameter vector

4.3 The Two-Additive-Factor Extended CIR/LS Model CIR2++

$$\alpha := (x_0, y_0, k_1, \theta_1, \sigma_1, k_2, \theta_2, \sigma_2),$$

which is chosen so as to exactly retrieve the initial zero-coupon curve. In particular, $\varphi(0; \alpha) = r_0 - x_0 - y_0$.

It is easy to see that in order to fit exactly the zero-coupon curve observed in the market, it suffices to set

$$\varphi(t; \alpha) = f^M(0, t) - f^1(0, t; x_0, \alpha_1) - f^1(0, t; y_0, \alpha_2), \quad (4.42)$$

where f^1 is the one-factor instantaneous forward rate f^{CIR} as given in formula (3.77) and f^M is the market instantaneous forward rate.

In what follows, it is helpful to define the quantity

$$\Phi^\xi(u, v; \alpha) := \exp\left[-\int_u^v \varphi(s; \alpha) ds\right]$$
$$= \frac{P^M(0, v)}{P^M(0, u)} \frac{P^\xi(0, u; \alpha)}{P^\xi(0, v; \alpha)}$$
$$= \exp\left\{[R^\xi(0, v; \alpha) - R^M(0, v)]v - [R^\xi(0, u; \alpha) - R^M(0, u)]u\right\}, \quad (4.43)$$

which is known based on the initial market term structures of discount factors $T \to P^M(0, T)$ or spot rates $T \to R^M(0, T)$, on the initial instantaneous short rate r_0 and on the analytical expressions (4.38) or (4.39).

Zero-Coupon Bonds

As extensively explained in Section 3.8, adding a deterministic shift to the process ξ^α that admits explicit bond-price formulas, preserves the original analytical tractability in that also the new process implies explicit bond-price formulas.

Indeed, it is straightforward to verify that the price at time t of a zero-coupon bond maturing at time T and with unit face value is given by

$$P(t, T; x(t), y(t), \alpha) = \Phi^\xi(t, T; \alpha) P^\xi(t, T; x(t), y(t), \alpha).$$

The derivation of this formula is perfectly analogous to that of (3.70) in the general one-factor case.

Instantaneous Forward Rates

It is immediate to check that, from the definition of the instantaneous forward rate

$$f(t, T) = -\frac{\partial}{\partial T} \ln P(t, T),$$

we have

170   4. Two-Factor Short-Rate Models

$$\text{Var}[df(t,T;x(t),y(t),\alpha)] = \left(\frac{\partial B}{\partial T}(t,T;\alpha_1)\sigma_1\sqrt{x(t)}\right)^2 dt \quad (4.44)$$
$$+ \left(\frac{\partial B}{\partial T}(t,T;\alpha_2)\sigma_2\sqrt{y(t)}\right)^2 dt,$$

where again the deterministic function B is that figuring in the bond-price formulas for the one-factor CIR model, being defined as in (3.25).

It is a little laborious, though not difficult, to check analytically, by substituting the B's expressions in (3.25), that this is a monotonically-decreasing function which allows no humped shape for the curve

$$T \mapsto \text{Var}[df(t,T;x(t),y(t),\alpha)],$$

with consequences discussed in Section 4.1.

Zero-Coupon-Bond Options and Caps and Floors

The price at time t of an European call option with maturity $T > t$ and strike price K on a zero-coupon bond with face value N maturing at $S > T$ is given by the following straightforward modification of the CIR2 formula:

$$\mathbf{ZBC}(t,T,S,N,K;x(t),y(t),\alpha)$$
$$= N\Phi^\xi(t,S;\alpha)C^\xi(t,T,S,N,K/\Phi^\xi(T,S);x(t),y(t),\alpha)$$

where C^ξ is given by (4.40). This formula is analogous to (3.73), as derived in the general one-factor case, and can be proven exactly in the same manner.

The analogous put option price is obtained from the put-call parity (3.3) and from the call-option-price formula:

$$\mathbf{ZBP}(t,T,S,N,K;x(t),y(t),\alpha) = \mathbf{ZBC}(t,T,S,N,K;x(t),y(t),\alpha)$$
$$-NP(t,S;x(t),y(t),\alpha)$$
$$+ KP(t,T;x(t),y(t),\alpha).$$

Since caps and floors can be viewed as portfolios of options on zero-coupon bonds, the general formulas (2.26), see also (3.80) and (3.81), remain valid in the two-factor case provided one considers them as expressed in terms of the two factors $x(t)$ and $y(t)$.

European Swaptions and Path-Dependent Payoffs

In the two-factor case Jamshidian's decomposition for coupon-bearing-bond options and European swaptions (see for instance Section 3.11.1) is not applicable. Therefore, such products need to be priced via alternative methods, such as numerical integration of the payoff or Monte Carlo simulation of scenarios. Alternatively, trees might be used as in the G2++ model. Since the

4.3 The Two-Additive-Factor Extended CIR/LS Model CIR2++

two factors x and y are independent, conceptually the tree implementation follows from the implementation of two different one-dimensional CIR-like trees, whose construction has been explained in Section 3.9.1. Once the model has been calibrated to a set of market data, pricing path-dependent payoffs can be achieved through Monte Carlo simulation in complete analogy with the one-factor case. The only difference is that now we need to simulate a two-dimensional process x, y. However, since the two components are independent processes, the generation of paths can take place exactly in the same way as in the one-factor case, with the chosen method applied independently to each factor. Then the pricing of exotic interest-rate derivatives can be carried out exactly as in the one-factor case, by simulating the two factors separately, each as explained in Section 3.11.2. All other considerations of that section remain valid once the single factor process used there is replaced by the two factors considered here.

5. The Heath-Jarrow-Morton (HJM) Framework

I decided I'd spent too much time philosophizing.
It is, unfortunately, one of my character flaws.
J'onn J'onnz in "Martian Manhunter annual" 2, 1999, DC Comics.

Modeling the interest-rate evolution through the instantaneous short rate has some advantages, mostly the large liberty one has in choosing the related dynamics. For example, for one-factor short-rate models one is free to choose the drift and instantaneous volatility coefficient in the related diffusion dynamics as one deems fit, with no general restrictions. We have seen several examples of possible choices in Chapter 3. However, short-rate models have also some clear drawbacks. For example, an exact calibration to the initial curve of discount factors and a clear understanding of the covariance structure of forward rates are both difficult to achieve, especially for models that are not analytically tractable.

The first historically important alternative to short-rate models has been proposed by Ho and Lee (1986), who modeled the evolution of the entire yield curve in a binomial-tree setting. Their basic intuition was then translated in continuous time by Heath, Jarrow and Morton (HJM) (1992) who developed a quite general framework for the modeling of interest-rate dynamics. Precisely, by choosing the instantaneous forward rates as fundamental quantities to model, they derived an arbitrage-free framework for the stochastic evolution of the entire yield curve, where the forward-rates dynamics are fully specified through their instantaneous volatility structures. This is a major difference with arbitrage free one-factor short-rate dynamics, where the volatility of the short rate alone does not suffice to characterize the relevant interest-rate model. But in order to clarify the matter, let us consider the Merton (1973) toy short-rate model.

Assume we take the following equation for the short rate under the risk-neutral measure:

$$dr_t = \theta dt + \sigma dW_t, \quad r_0.$$

If you wish, this is a very particular toy version of the Hull-White model (3.33) seen in Chapter 3 with constant coefficient θ. Now, for this (affine) model, one can easily compute the bond price,

$$P(t,T) = \exp\left[\frac{\sigma^2}{6}(T-t)^3 - \frac{\theta}{2}(T-t)^2 - (T-t)r_t\right],$$

and the instantaneous forward rate

$$f(t,T) = -\frac{\partial \ln P(t,T)}{\partial T} = -\frac{\sigma^2}{2}(T-t)^2 + \theta(T-t) + r_t.$$

Differentiate this and substitute the short-rate dynamics to obtain

$$df(t,T) = (\sigma^2(T-t) - \theta)dt + \theta dt + \sigma dW_t,$$

or

$$df(t,T) = \sigma^2(T-t)dt + \sigma dW_t.$$

Look at this last equation. *The drift $\sigma^2(T-t)$ in the f's dynamics is determined as a suitable transformation of the diffusion coefficient σ in the same dynamics.* This is no mere coincidence due to the simplicity of our toy model, but a general fact that Heath, Jarrow and Morton (1992) expressed in full generality.

Clearly, if one wishes to directly model this instantaneous forward rate, there is no liberty in selecting the drift of its process, as it is completely determined by the chosen volatility coefficient. This is essentially due to the fact that we are modeling a derived quantity f and not the fundamental quantity r. Indeed, f is expressed in terms of the more fundamental r by

$$f(t,T) = -\frac{\partial \ln E_t\left[\exp\left(-\int_t^T r(s)ds\right)\right]}{\partial T},$$

and, as you see, an expectation has already acted in the definition of f, adding structure and taking away freedom, so to say.

More generally, under the HJM framework, one assumes that, for each T, the forward rate $f(t,T)$ evolves according to

$$df(t,T) = \alpha(t,T)dt + \sigma(t,T)dW(t),$$

where W is a (possibly multi-dimensional) Brownian motion. As we have just seen in the above example, contrary to the short-rate modeling case, where one is free to specify the drift of the considered diffusion, here the function α is completely determined by the choice of the (vector) diffusion coefficient σ.

The importance of the HJM theory lies in the fact that virtually any (exogenous term-structure) interest-rate model can be derived within such a framework.[1] However, only a restricted class of volatilities is known to imply a Markovian short-rate process. This means that, in general, burdensome procedures, like those based on non-recombining lattices, are needed to price interest-rate derivatives. Substantially, the problem remains of defining

---

[1] Even the celebrated LIBOR market model was developed starting from instantaneous-forward-rate dynamics in Brace, Gatarek and Musiela (1997), although it is possible to obtain it also through the change-of-numeraire approach, as we will see in the next chapter.

a suitable volatility function for practical purposes. This is the reason why in this book we do not devote too much attention to the HJM theory, preferring to deal with explicitly formulated models.

In this chapter, we briefly review the HJM framework and explicitly write the HJM no-arbitrage condition. We then describe some analogies with instantaneous short-rate models. We show, in particular, that a one-factor HJM model with deterministic volatility is equivalent to the Hull and White (1990b) short-rate model (3.32). We then introduce the Ritchken and Sankarasubramanian (1995) (RS) framework and briefly illustrate the Li, Ritchken and Sankarasubramanian (1995a, 1995b) algorithm for pricing derivatives. We finally mention the Mercurio and Moraleda (2000a) humped-volatility model as a specific example of a one-factor Gaussian model within the HJM framework.

5.1 The HJM Forward-Rate Dynamics

Heath, Jarrow and Morton (1992) assumed that, for a fixed a maturity T, the instantaneous forward rate $f(t,T)$ evolves, under a given measure, according the following diffusion process:

$$df(t,T) = \alpha(t,T)dt + \sigma(t,T)dW(t),$$
$$f(0,T) = f^M(0,T), \qquad (5.1)$$

with $T \mapsto f^M(0,T)$ the market instantaneous-forward curve at time $t = 0$, and where $W = (W_1,\ldots,W_N)$ is an N-dimensional Brownian motion, $\sigma(t,T) = (\sigma_1(t,T),\ldots,\sigma_N(t,T))$ is a vector of adapted processes and $\alpha(t,T)$ is itself an adapted process. The product $\sigma(t,T)dW(t)$ is intended to be the scalar product between the two vectors $\sigma(t,T)$ and $dW(t)$.

The advantage of modeling forward rates as in (5.1) is that the current term structure of rates is, by construction, an input of the selected model. Remember, in fact, that the following relations between zero-bond prices and forward rates hold (see (1.23)):

$$f(t,T) = -\frac{\partial \ln P(t,T)}{\partial T},$$
$$P(t,T) = e^{-\int_t^T f(t,u)du}.$$

The dynamics in (5.1) is not necessarily arbitrage-free. Following an approach similar to that of Harrison and Pliska (1981) (see Chapter 2) Heath, Jarrow and Morton proved that, in order for a unique equivalent martingale measure to exist, the function α cannot be arbitrarily chosen, but it must equal a quantity depending on the vector volatility σ and on the drift rates in the dynamics of N selected zero-coupon bond prices. In particular, if the dynamics (5.1) are under the risk-neutral measure, then we must have

$$\alpha(t,T) = \sigma(t,T) \int_t^T \sigma(t,s)ds = \sum_{i=1}^{N} \sigma_i(t,T) \int_t^T \sigma_i(t,s)ds, \quad (5.2)$$

so that the integrated dynamics of $f(t,T)$ under the risk-neutral measure are

$$f(t,T) = f(0,T) + \int_0^t \sigma(u,T) \int_u^T \sigma(u,s)ds\, du + \int_0^t \sigma(s,T)dW(s)$$

$$= f(0,T) + \sum_{i=1}^{N} \int_0^t \sigma_i(u,T) \int_u^T \sigma_i(u,s)ds\, du + \sum_{i=1}^{N} \int_0^t \sigma_i(s,T)dW_i(s)$$

and are fully specified once the vector volatility function σ is provided. Given this dynamics of the instantaneous forward rate $f(t,T)$, application of Ito's lemma gives the following dynamics of the zero-coupon bond price $P(t,T)$:

$$dP(t,T) = P(t,T)\left[r(t)dt - \left(\int_t^T \sigma(t,s)ds\right)dW(t)\right],$$

where $r(t)$ is the instantaneous short term interest rate at time t, that is

$$r(t) = f(t,t) = f(0,t) + \int_0^t \sigma(u,t) \int_u^t \sigma(u,s)ds\, du + \int_0^t \sigma(s,t)dW(s)$$

$$= f(0,t) + \sum_{i=1}^{N} \int_0^t \sigma_i(u,t) \int_u^t \sigma_i(u,s)ds\, du + \sum_{i=1}^{N} \int_0^t \sigma_i(s,t)dW_i(s).$$

$$(5.3)$$

5.2 Markovianity of the Short-Rate Process

The short-rate process (5.3) is not a Markov process in general. Notice in fact that the time t appears in the stochastic integral both as extreme of integration and inside the integrand function. However, there are suitable specifications of σ for which r is indeed a Markov process. As proven by Carverhill (1994), this happens, for example, if we can write, for each $i = 1, \ldots, N$,

$$\sigma_i(t,T) = \xi_i(t)\psi_i(T), \quad (5.4)$$

with ξ_i and ψ_i strictly positive and deterministic functions of time. Under such a separable specification, the short-rate process becomes

$$r(t) = f(0,t) + \sum_{i=1}^{N} \int_0^t \xi_i(u)\psi_i(t) \int_u^t \xi_i(u)\psi_i(s)ds\, du + \sum_{i=1}^{N} \int_0^t \xi_i(s)\psi_i(t)dW_i(s)$$

$$= f(0,t) + \sum_{i=1}^{N} \psi_i(t) \int_0^t \xi_i^2(u) \int_u^t \psi_i(s)ds\, du + \sum_{i=1}^{N} \psi_i(t) \int_0^t \xi_i(s)dW_i(s).$$

Notice that in the one-factor case ($N = 1$), if we define the (strictly-positive) deterministic function A by

$$A(t) := f(0,t) + \psi_1(t) \int_0^t \xi_1^2(u) \int_u^t \psi_1(s) ds \, du,$$

and assume its differentiability, we can write

$$dr(t) = A'(t)dt + \psi_1'(t) \int_0^t \xi_1(s)dW_1(s) + \psi_1(t)\xi_1(t)dW_1(t)$$

$$= \left[A'(t) + \psi_1'(t) \frac{r(t) - A(t)}{\psi_1(t)}\right] dt + \psi_1(t)\xi_1(t)dW_1(t)$$

$$= [a(t) + b(t)r(t)] \, dt + c(t)dW_1(t),$$

with obvious definition of the coefficients a, b and c.

We therefore end up with the general short-rate dynamics proposed by Hull and White (1990b), see (3.32), thus establishing an equivalence between the HJM one-factor model for which (5.4) holds and the general formulation of the Gaussian one-factor short-rate model of Hull and White (1990b). In particular, we can easily derive the HJM forward-rate dynamics that is equivalent to the short-rate dynamics (3.33). To this end, let us set

$$\sigma_1(t,T) = \sigma e^{-a(T-t)},$$

where a and σ are now real constants, so that

$$\xi_1(t) = \sigma e^{at},$$
$$\psi_1(T) = e^{-aT},$$
$$A(t) = f(0,t) + \frac{\sigma^2}{2a^2} \left(1 - e^{-at}\right)^2.$$

The resulting short-rate dynamics is then given by

$$dr(t) = \left[\frac{\partial f}{\partial T}(0,t) + \frac{\sigma^2}{a}\left(e^{-at} - e^{-2at}\right)\right.$$
$$\left. -a\left(r(t) - f(0,t) - \frac{\sigma^2}{2a^2}\left(1 - e^{-at}\right)^2\right)\right] dt + \sigma dW_1(t)$$
$$= \left[\frac{\partial f}{\partial T}(0,t) + af(0,t) + \frac{\sigma^2}{2a}\left(1 - e^{-2at}\right) - ar(t)\right] dt + \sigma dW_1(t),$$

which is equivalent to (3.33) when combined with (3.34).

5.3 The Ritchken and Sankarasubramanian Framework

It is now clear that an arbitrary specification of the forward-rate volatility will likely lead to a non-Markovian instantaneous short-rate process. In such a case, we would soon encounter major computational problems when discretizing the dynamics (5.3) for the pricing of a general derivative. In fact, the approximating lattice will not be recombining, and the number of nodes in the tree will grow exponentially with the number of steps. This will make the numerical procedure quite difficult to handle, especially as far as execution time (combined with a pricing accuracy) is concerned.

These pricing problems can be addressed by noting that, even though the short-rate process is not Markovian, there may yet exist a higher-dimensional Markov process having the short rate as one of its components. Exploiting such intuition, Ritchken and Sankarasubramanian (1995) have identified necessary and sufficient conditions on the volatility structure of forward rates for capturing the path dependence of r through a single sufficient statistic. Precisely, they proved the following.

Proposition 5.3.1. *Consider a one-factor HJM model. If the volatility function $\sigma(t,T)$ is differentiable with respect to T, a necessary and sufficient condition for the price of any (interest-rate) derivative to be completely determined by a two-state Markov process $\chi(\cdot) = (r(\cdot), \phi(\cdot))$ is that the following condition holds:*

$$\sigma(t,T) = \sigma_{RS}(t,T) := \eta(t) e^{-\int_t^T \kappa(x)dx}, \tag{5.5}$$

where η is an adapted process and κ is a deterministic (integrable) function. In such a case, the second component of the process χ is defined by

$$\phi(t) = \int_0^t \sigma_{RS}^2(s,t)ds.$$

Accordingly, zero-coupon-bond prices are explicitly given by

$$P(t,T) = \frac{P(0,T)}{P(0,t)} \exp\left\{-\frac{1}{2}\Lambda^2(t,T)\phi(t) + \Lambda(t,T)[f(0,t) - r(t)]\right\},$$

where

$$\Lambda(t,T) = \int_t^T e^{-\int_t^u \kappa(x)dx} du.$$

Differentiation of equation (5.3) shows that, under the RS class of volatilities (5.5), the process χ, and hence the instantaneous short-rate r, evolve according to

$$d\chi(t) = \begin{pmatrix} dr(t) \\ d\phi(t) \end{pmatrix} = \begin{pmatrix} \mu(r,t)dt + \eta(t)dW(t) \\ [\eta^2(t) - 2\kappa(t)\phi(t)]\,dt \end{pmatrix} \tag{5.6}$$

with

$$\mu(r,t) = \kappa(t)[f(0,t) - r(t)] + \phi(t) + \frac{\partial}{\partial t}f(0,t).$$

We can now see that η is nothing but the instantaneous short-rate volatility process.

5.3 The Ritchken and Sankarasubramanian Framework

The yield curve dynamics described by (5.6) can be, therefore, discretized in a Markovian (recombining) lattice in terms of the two variables r and ϕ. This was suggested by Li, Ritchken and Sankarasubramanian (1995a, 1995b) (LRS), who developed an efficient lattice to approximate the processes (5.6). Their tree construction procedure is briefly outlined in the following, under the particular case where

$$\eta(t) = \hat{\sigma}(r(t))$$
$$\hat{\sigma}(x) := vx^\rho$$

with v and ρ positive constants, $\rho \in [0,1]$, so that

$$\sigma_{RS}(t,T) = v[r(t)]^\rho e^{-\int_t^T \kappa(x)dx}.$$

Li, Ritchken and Sankarasubramanian considered the following transformation, which yields a process with constant volatility

$$Y(t) = \int \frac{1}{\hat{\sigma}(x)} dx \bigg|_{x=r(t)}$$

where the right-hand side denotes a primitive of $1/\hat{\sigma}(x)$ calculated in $x = r(t)$ and where the constant in the primitive is set to zero. This is exactly the transformation of r needed to have a unit diffusion coefficient when applying Ito's formula to compute its differential, as we shall see in a moment. Notice that, by substituting the expression for $\hat{\sigma}$ and by integrating, we have

$$Y(t) = \bar{Y}(r(t)) := \begin{cases} \frac{1}{v}\ln(r(t)) & \text{if } \rho = 1 \\ \frac{1}{v(1-\rho)}[r(t)]^{1-\rho} & \text{if } 0 \leq \rho < 1 \end{cases} \quad (5.7)$$

with the function Y defined in a suitable domain $\mathcal{D}_\rho$ depending on the value of ρ. Denoting by $x = \varphi(y)$ the inverse function of $y = \bar{Y}(x)$ on $\mathcal{D}_\rho$, we have

$$\varphi(y) = \begin{cases} e^{vy} & \text{if } \rho = 1 \\ (v(1-\rho)y)^{1/(1-\rho)} & \text{if } 0 \leq \rho < 1 \end{cases} \quad (5.8)$$

Application of Ito's lemma and straightforward algebra show that

$$dY(t) = m(Y, \phi, t)dt + dW(t)$$
$$d\phi(t) = [\hat{\sigma}(r(t)) - 2\kappa(t)\phi(t)] dt$$

where indeed the diffusion coefficient in the Y dynamics is one, and where

$$m(Y, \phi, t) = \frac{\kappa(t)[f(0,t) - \varphi(Y(t))] + \phi(t) + \frac{\partial f}{\partial t}(0,t)}{v[\varphi(Y(t))]^\rho} - \frac{v\rho}{2[\varphi(Y(t))]^{1-\rho}} \quad (5.9)$$

for any $0 \leq \rho \leq 1$.

If, for instance, $\rho = 0.5$, we have that, for $r(t) > 0$, $Y(t) = 2\sqrt{r(t)}/v$, $\varphi(y) = \frac{v^2 y^2}{4}$, and $m(Y, \phi, t)$ in (5.9) becomes

$$m(Y, \phi, t) = \frac{\kappa(t)[f(0, t) - \frac{1}{4}v^2(Y(t))^2] + \phi(t) + \frac{\partial f}{\partial t}(0, t)}{\frac{1}{2}v^2 Y(t)} - \frac{1}{2Y(t)},$$

whereas, if $\rho = 1$, we have that, for $r(t) > 0$, $Y(t) = \ln[r(t)]/v$, $\varphi(y) = e^{vy}$ and

$$m(Y, \phi, t) = \frac{\kappa(t)[f(0, t) - e^{vY(t)}] + \phi(t) + \frac{\partial f}{\partial t}(0, t)}{v e^{vY(t)}} - \frac{1}{2}v.$$

Building a lattice for Y is eased by the presence of a unit diffusion coefficient, which is the reason for adopting such a transformation in the first place. Also, since ϕ's dynamics (5.6) has no diffusion part, the related lattice component need not branch. The approximating lattice is then constructed as follows. Divide the given time horizon into intervals of equal length Δt, and suppose that, at the beginning t of some time interval, the state variables are y and ϕ, and that, in the next time period, they move either to (y^+, ϕ^*) or to (y^-, ϕ^*) where

$$y^+ = y + (J(y, \phi) + 1)\sqrt{\Delta t},$$
$$y^- = y + (J(y, \phi) - 1)\sqrt{\Delta t},$$
$$\phi^* = \phi + [\hat{\sigma}^2(\varphi(y)) - 2\kappa(t)\phi]\Delta t$$

Setting $Z(y, \phi) = \text{int}\left[m(y, \phi, t)\sqrt{\Delta t}\right]$, where $\text{int}[x]$ denotes the largest integer smaller or equal than the real x, the function J is defined by

$$J(y, \phi) = \begin{cases} |Z(y, \phi)| & \text{if } Z(y, \phi) \text{ is even} \\ Z(y, \phi) + 1 & \text{otherwise.} \end{cases}$$

The branching probabilities, p for an up-move and $1 - p$ for a down-move, are then derived by solving

$$p(y^+ - y) + (1 - p)(y^- - y) = m(y, \phi, t)\Delta t,$$

thus obtaining

$$p = \frac{m(y, \phi, t)\Delta t + y - y^-}{y^+ - y^-}.$$

The above choices ensure that

$$y^+ \geq y + m(y, \phi, t)\Delta t \geq y^-,$$

and hence that the probabilities of moving from one node to another in the lattice lie always in $[0, 1]$.[2]

Once the LRS tree has been built, derivatives prices can then be calculated in a quite standard way.

---

[2] We refer to Li, Ritchken and Sankarasubramanian (1995a) for a detailed description of all the calculations above.

5.4 The Mercurio and Moraleda Model

We conclude the chapter by briefly reviewing the Mercurio and Moraleda (2000a) model, which explicitly assumes a humped volatility structure in the instantaneous-forward-rate dynamics. Such assumption is motivated by the fact that the volatility structure of forward rates, as implied by market quotes, is commonly humped. We also refer to Section 3.6 for a detailed explanation of the opportunity and practical relevance of such an assumption.

Mercurio and Moraleda (2000a) proposed a one-factor Gaussian model in the HJM framework by considering the following form for the volatility of instantaneous forward rates:

$$\sigma(t,T) = \sigma[\gamma(T-t) + 1]e^{-\frac{\lambda}{2}(T-t)}, \qquad (5.10)$$

where σ, γ and λ are non-negative constants. Under such specification, (5.1) becomes

$$df(t,T) = \bar{\alpha}(T-t)dt + \sigma[\gamma(T-t) + 1]e^{-\frac{\lambda}{2}(T-t)}dW(t),$$

$$\bar{\alpha}(\tau) = -\frac{2\sigma^2}{\lambda}[\gamma\tau + 1]e^{-\lambda\tau}\left[\gamma\tau + \left(\frac{2\gamma}{\lambda} + 1\right)\left(1 - e^{\frac{\lambda}{2}\tau}\right)\right],$$

which implies that instantaneous (forward and spot) rates are normally distributed.

The choice of the volatility function (5.10) is motivated by the following features.

(a) It provides a humped volatility structure for strictly positive σ, γ and λ, and $2\gamma > \lambda$;
(b) It depends only on the "time to maturity" $T - t$ rather than on time t and maturity T separately;
(c) It leads to analytical formulas for European options on discount bonds;
(d) It generalizes the volatility specification of the Hull and White model (3.33), in that for $\gamma = 0$ and $\lambda = 2a$ we get the volatility of forward rates as implied by (3.33).[3]

It is obvious that (b) and (d) hold. Basic calculus shows that (a) is also true. In fact, for strictly positive σ, γ and λ, the function $f(x) = \sigma[\gamma x + 1]\exp\{-\frac{\lambda}{2}x\}$ has the following features: (i) it is strictly positive for $x \geq 0$; (ii) it is increasing and concave in the interval $[0, (2\gamma - \lambda)/(\gamma\lambda)]$; (iii) it has a maximum in $(2\gamma - \lambda)/(\gamma\lambda)$ whose value is $(2\sigma\gamma/\lambda)\exp\{(-2\gamma+\lambda)/(2\gamma)\}$; (iv) it is decreasing and concave in the interval $[(2\gamma - \lambda)/(\gamma\lambda), (4\gamma - \lambda)/(\gamma\lambda)]$;

---
[3] When modeling humped volatility structures, many other specifications can of course be considered. For example the term between square brackets in (5.10) can be generalized to be any polynomial in $(T-t)$. It is disputable however whether there exists a simpler characterization than (5.10) and for which (a), (b), (c) and (d) hold.

(v) it is decreasing and convex from $(4\gamma - \lambda)/(\gamma\lambda)$ onwards; (vi) it tends asymptotically to zero.

As for property (c), Mercurio and Moraleda (2000a) used the results of Merton (1973) to prove that the time t-price of a European call option with maturity T and strike price X on a pure discount bond with maturity S is given by

$$\mathbf{ZBC}(t,T,S,X) = P(t,S)\Phi(d_1(t)) - XP(t,T)\Phi(d_2(t)),$$

where

$$d_1(t) := \frac{\ln(P(t,S)/(XP(t,T)) + \frac{1}{2}v_t^2}{v_t},$$

$$d_2(t) := d_1(t) - v_t,$$

and

$$v_t^2 = \frac{4\sigma^2}{\lambda^7}(A^2\lambda^2 + 2AB\lambda + 2B^2)\left(e^{\lambda T} - e^{\lambda t}\right) - \frac{8\sigma^2 B}{\lambda^6}(A\lambda + B)\left(Te^{\lambda T} - te^{\lambda t}\right)$$
$$+ \frac{4\sigma^2 B^2}{\lambda^5}\left(T^2 e^{\lambda T} - t^2 e^{\lambda t}\right),$$

where

$$A := (\lambda + 2\gamma)\left(e^{-\frac{\lambda}{2}S} - e^{-\frac{\lambda}{2}T}\right) + \gamma\lambda\left(Se^{-\frac{\lambda}{2}S} - Te^{-\frac{\lambda}{2}T}\right),$$

$$B := \gamma\lambda\left(e^{-\frac{\lambda}{2}S} - e^{-\frac{\lambda}{2}T}\right).$$

However, no analytical formula for pure discount bonds can be derived. Notice, in fact, that the instantaneous-short-rate process is not Markovian since (5.10) does not belong to the RS class (5.5).

Mercurio and Moraleda also tested empirically their model. They considered a time series of market cap prices and compared the fitting quality of their model with that implied by the Hull and White model (3.33). Using the Schwarz-Information-Criterion test, they concluded that, in most situations, their humped-volatility model is indeed preferable. We refer to Mercurio and Moraleda (2000a) for a detailed description of the calibration results.

6. The LIBOR and Swap Market Models (LFM and LSM)

There is so much they don't tell. Is that why I became a cop?
To learn what they don't tell?
John Jones in "Martian Manhunter: American Secrets" 1, DC Comics, 1992

6.1 Introduction

In this chapter we consider one of the most popular and promising families of interest-rate models: The market models.

Why are such models so popular? The main reason lies in the agreement between such models and well-established market formulas for two basic derivative products. Indeed, the lognormal forward-LIBOR model (LFM) prices caps with Black's cap formula, which is the standard formula employed in the cap market. Moreover, the lognormal forward-swap model (LSM) prices swaptions with Black's swaption formula, which again is the standard formula employed in the swaption market. Since the caps and swaptions markets are the two main markets in the interest-rate-options world, it is important for a model to be compatible with such market formulas.

Before market models were introduced, there was no interest-rate dynamics compatible with either Black's formula for caps or Black's formula for swaptions. These formulas were actually based on mimicking the Black and Scholes model for stock options under some simplifying and inexact assumptions on the interest-rates distributions. The introduction of market models provided a new derivation of Black's formulas based on rigorous interest-rate dynamics.

However, even with full rigor given separately to the caps and swaptions classic formulas, we point out the now classic problem of this setup: The LFM and the LSM are not compatible. Roughly speaking, if forward LIBOR rates are lognormal each under its measure, as assumed by the LFM, forward swap rates cannot be lognormal at the same time under their measure, as assumed by the LSM. Although forward swap rates obtained from lognormal forward LIBOR rates are not far from being lognormal themselves under the relevant measure (as shown in some empirical works and also in Chapter 8

184   6. The LIBOR and Swap Market Models (LFM and LSM)

of the present book), the problem still stands and reduces one's enthusiasm for the theoretical setup of market models, if not for the practical one.

In this chapter, we will derive the LFM dynamics under different measures by resorting to the change-of-numeraire technique. We will show how caps are priced in agreement with Black's cap formula, and explain how swaptions can be priced through a Monte Carlo method in general. Analytical approximations leading to swaption-pricing formulas are also presented, as well as closed-form formulas for terminal correlations based on similar approximations.

We will suggest parametric forms for the instantaneous covariance structure (volatilities and correlations) in the LFM. Part of the parameters in this structure can be obtained directly from market-quoted cap volatilities, whereas other parameters can be obtained by calibrating the model to swaption prices. The calibration to swaption prices can be made computationally efficient through the analytical approximations mentioned above.

We will derive results and approximations connecting semi-annual caplet volatilities to volatilities of swaptions whose underlying swap is one-year long. We will also show how one can obtain forward rates over non-standard periods from standard forward rates either through drift interpolation or via a bridging technique.

We will then introduce the LSM and show how swaptions are priced in agreement with Black's swaptions formula, although Black's formula for caps does not hold under this model.

We will finally consider the "smile problem" for the cap market, and introduce some possible extensions of the basic LFM that are analytically tractable and allow for a volatility smile.

6.2 Market Models: a Guided Tour

Before market models were introduced, short-rate models used to be the main choice for pricing and hedging interest-rate derivatives. Short-rate models are still chosen for many applications and are based on modeling the instantaneous spot interest rate ("short rate") via a (possibly multi-dimensional) diffusion process. This diffusion process characterizes the evolution of the complete yield curve in time. We have seen examples of such models in Chapters 3 and 4.

To fix ideas, let us consider the time-0 price of a T_2-maturity caplet resetting at time T_1 ($0 < T_1 < T_2$) with strike X and a notional amount of 1. Caplets and caps have been defined in Chapter 1 and will be described more generally in Section 6.4. Let τ denote the year fraction between T_1 and T_2. Such a contract pays out the amount

$$\tau (L(T_1, T_2) - X)^+$$

at time T_2, where in general $L(u,s)$ is the LIBOR rate at time u for maturity s.

Again to fix ideas, let us choose a specific short-rate model and assume we are using the shifted two-factor Vasicek model G2++ given in (4.4). The parameters of this two-factor Gaussian additive short-rate model are here denoted by $\theta = (a, \sigma, b, \eta, \rho)$. Then the short rate r_t is obtained as the sum of two linear diffusion processes x_t and y_t, plus a deterministic shift φ that is used for fitting the initially observed yield curve at time 0:

$$r_t = x_t + y_t + \varphi(t; \theta).$$

Such model allows for an analytical formula for forward LIBOR rates F,

$$F(t; T_1, T_2) = F(t; T_1, T_2; x_t, y_t, \theta),$$
$$L(T_1, T_2) = F(T_1; T_1, T_2; x_{T_1}, y_{T_1}, \theta).$$

At this point one can try and price a caplet. To this end, one can compute the risk-neutral expectation of the payoff discounted with respect to the bank account numeraire $\exp\left(\int_0^{T_2} r_s ds\right)$ so that one has

$$E\left[\exp\left(-\int_0^{T_2} r_s ds\right) \tau (F(T_1; T_1, T_2, x_{T_1}, y_{T_1}, \theta) - X)^+\right].$$

This too turns out to be feasible, and leads to a function

$$U_C(0, T_1, T_2, X, \theta).$$

On the other hand, the market has been pricing caplets (actually caps) with Black's formula for years. One possible derivation of Black's formula for caplets is based on the following approximation. When pricing the discounted payoff

$$E\left[\exp\left(-\int_0^{T_2} r_s ds\right) \tau (L(T_1, T_2) - X)^+\right] = \cdots$$

one first assumes the discount factor $\exp\left(-\int_0^{T_2} r_s ds\right)$ to be deterministic and identifies it with the corresponding bond price $P(0, T_2)$. Then one factors out the discount factor to obtain:

$$\cdots \approx P(0, T_2)\tau E\left[(L(T_1, T_2) - X)^+\right] = P(0, T_2)\tau E\left[(F(T_1; T_1, T_2) - X)^+\right].$$

Now, inconsistently with the previous approximation, one goes back to assuming rates to be stochastic, and models the forward LIBOR rate $F(t; T_1, T_2)$ as in the classical Black and Scholes option pricing setup, i.e as a (driftless) geometric Brownian motion:

$$dF(t; T_1, T_2) = vF(t; T_1, T_2)dW_t , \qquad (6.1)$$

186   6. The LIBOR and Swap Market Models (LFM and LSM)

where v is the instantaneous volatility, assumed here to be constant for simplicity, and W is a standard Brownian motion under the risk-neutral measure Q.

Then the expectation

$$E\left[(F(T_1;T_1,T_2) - X)^+\right]$$

can be viewed simply as a T_1-maturity call-option price with strike X and whose underlying asset has volatility v, in a market with zero risk-free rate.

We therefore obtain:

$$\begin{aligned}
\mathbf{Cpl}(0,T_1,T_2,X) &:= P(0,T_2)\tau E(F(T_1;T_1,T_2) - X)^+ \\
&= P(0,T_2)\tau[F(0;T_1,T_2)\Phi(d_1(X,F(0;T_1,T_2),v\sqrt{T_1})) \\
&\quad - X\Phi(d_2(X,F(0;T_1,T_2),v\sqrt{T_1}))],
\end{aligned}$$

$$d_1(X,F,u) = \frac{\ln(F/X) + u^2/2}{u},$$

$$d_2(X,F,u) = \frac{\ln(F/X) - u^2/2}{u},$$

where Φ is the standard Gaussian cumulative distribution function.

From the way we just introduced it, this formula seems to be partly based on inconsistencies. However, within the change-of-numeraire setup, the formula can be given full mathematical rigor as follows. Denote by Q^2 the measure associated with the T_2-bond-price numeraire $P(\cdot,T_2)$ (T_2-forward measure) and by E^2 the corresponding expectation. Then, by the change-of-numeraire approach, we can switch from the bank-account numeraire $\exp\left(\int_0^t r_s ds\right)$ associated with risk-neutral measure Q to the bond-price numeraire $P(t,T_2)$ and obtain

$$E\left[\exp\left(-\int_0^{T_2} r_s ds\right)\tau(L(T_1,T_2) - X)^+\right] = P(0,T_2)E^2[\tau(L(T_1,T_2)-X)^+].$$

What has just been done, rather than assuming deterministic discount factors, is a change of measure. We have "factored out" the stochastic discount factor and replaced it with the related bond price, but in order to do so we had to change the probability measure under which the expectation is taken. Now the last expectation is no longer taken under the risk-neutral measure but rather under the T_2-forward measure. Since by definition $F(t;T_1,T_2)$ can be written as the price of a tradable asset divided by $P(t,T_2)$, it needs follow a martingale under the measure associated with the numeraire $P(t,T_2)$, i.e. under Q^2. As we have hinted at in Appendix A, martingale means "driftless" when dealing with diffusion process. Therefore, the dynamics of $F(t;T_1,T_2)$ under Q^2 is driftless, so that the dynamics

6.2 Market Models: a Guided Tour

$$dF(t; T_1, T_2) = vF(t; T_1, T_2)dW_t \tag{6.2}$$

is correct under the measure Q^2, where W is a standard Brownian motion under Q^2. Notice that the driftless (lognormal) dynamics above is precisely the dynamics we need in order to recover *exactly* Black's formula, without approximation. *We can say that the choice of the numeraire $P(\cdot, T_2)$ is based on this fact: It makes the dynamics (6.2) of F driftless under the related Q^2 measure, thus replacing rigorously the earlier arbitrary assumption on the F dynamics (6.1) under the risk-neutral measure Q.* Following this rigorous approach we indeed obtain Black's formula, since the process F has the same distribution as in the approximated case above, and hence the expected value has the same value as before.

The example just introduced is a simple case of what is known as "lognormal forward-LIBOR model". It is known also as Brace-Gatarek-Musiela (1997) model, from the name of the authors of one of the first papers where it was introduced rigorously. This model was also introduced by Miltersen, Sandmann and Sondermann (1997). Jamshidian (1997) also contributed significantly to its development. At times in the literature and in conversations, especially in Europe, the LFM is referred to as "BGM" model, from the initials of the three above authors. In other cases, colleagues in the U.S. called it simply an "HJM model", related perhaps to the fact that the BGM derivation was based on the HJM framework rather than on the change-of-numeraire technique. However, a common terminology is now emerging and the model is generally known as "LIBOR Market Model". We will stick to "Lognormal Forward-LIBOR Model" (LFM), since this is more informative on the properties of the model: Modeling forward LIBOR rates through a lognormal distribution (under the relevant measures).

Let us now go back to our short-rate model formula U_C and ask ourselves whether this formula can be compatible with the above reported Black's market formula. It is well known that the two formulas are not compatible. Indeed, by the two-dimensional version of Ito's formula we may derive the Q^2-dynamics of the forward LIBOR rate between T_1 and T_2 under the short-rate model,

$$dF(t; T_1, T_2; x_t, y_t, \theta) = \frac{\partial F}{\partial(t, x, y)} \, d[t \ x_t \ y_t]' + \tfrac{1}{2} \, d[x_t \ y_t] \, \frac{\partial^2 F}{\partial^2(x, y)} \, d[x_t \ y_t]', \ Q^2, \tag{6.3}$$

where the Jacobian vector and the Hessian matrix have been denoted by their partial derivative notation. This dynamics clearly depends on the linear-Gaussian dynamics of x and y under the T_2-forward measure. The thus obtained dynamics is easily seen to be incompatible with the lognormal dynamics leading to Black's formula. More specifically, for no choice of the parameters θ does the distribution of the forward rate F in (6.3) produced by the short-rate model coincide with the distribution of the "Black"-like forward rate F following (6.2). In general, no known short-rate model can lead to Black's formula for caplets (and more generally for caps).

What is then done with short-rate models is the following. After setting the deterministic shift φ so as to obtain a perfect fit of the initial term structure, one looks for the parameters θ that produce caplet (actually cap) prices U_C that are closest to a number of observed market cap prices. The model is thus calibrated to (part of) the cap market and should reproduce well the observed prices. Still, the prices U_C are complicated nonlinear functions of the parameters θ, and this renders the parameters themselves difficult to interpret. On the contrary, the parameter v in the above "market model" for F has a clear interpretation as a lognormal percentage (instantaneous) volatility, and traders feel confident in handling such kind of parameters.

When dealing with several caplets involving different forward rates, different structures of instantaneous volatilities can be employed. One can select a different v for each forward rate by assuming each forward rate to have a constant instantaneous volatility. Alternatively, one can select piecewise-constant instantaneous volatilities for each forward rate. Moreover, different forward rates can be modeled as each having different random sources W that are instantaneously correlated. Modeling correlation is necessary for pricing payoffs depending on more than a single rate at a given time, such as swaptions. Possible volatility and correlation structures are discussed in Sections 6.3.1 and 6.9. The implications of such structures as far as caplets and caps are concerned are discussed in Section 6.4, and their consequences on the term structure of volatilities as a whole are discussed in Section 6.5.

As hinted at above, the model we briefly introduced is the market model for "half" of the interest-rate-derivatives world, i.e. the cap market. But what happens when dealing with basic products from the other "half" of this world, such as swaptions? Swaptions are options on interest-rate swaps. Interest rate swaps and swaptions have been defined in Chapter 1 and will be again described in Section 6.7. Swaptions are priced by a Black-like market formula that is, in many respects, similar to the cap formula. This market formula can be given full rigor as in the case of the caps formula. However, doing so involves choosing a numeraire under which the relevant forward *swap* rate (rather than a particular forward LIBOR rate) is driftless and lognormal. This numeraire is indeed different from any of the bond-price numeraires used in the derivation of Black's formula for caps according to the LFM. The obtained model is known as "lognormal (forward) swap model" (LSM) and is also referred to as the Jamshidian (1997) market model or "swap market model".

One may wonder whether the two models are distributionally compatible or not, similarly to our previous comparison of the LFM with the shifted two-factor Vasicek model G2++. As before, the two models are incompatible. If we adopt the LFM for caps we cannot recover the market formula given by the LSM for swaptions. The two models (LFM and LSM) collide. We will point out this incompatibility in Section 6.8.

There are some recent works investigating the "size" of the discrepancy between these two models, and we will address this issue in Chapter 8 by comparing swap-rates distributions under the two models. Results seem to suggest that the difference is not large in most cases. However, the problem remains of choosing either of the two models for the whole market.

When the choice is made, the half market consistent with the model is calibrated almost automatically, see for example the cap calibration with the LFM in Section 6.4. But one has still the problem of calibrating the chosen model to the remaining half, e.g. the swaption market in case the LFM is adopted for both markets.

Indeed, Brace, Dun and Barton (1998) suggest to adopt the LFM as central model for the two markets, mainly for its mathematical tractability. We will stick to their suggestion, also because of the fact that forward rates are somehow more natural and more representative coordinates of the yield-curve than swap rates. Indeed, it is more natural to express forward swap rates in terms of a suitable preselected family of LIBOR forward rates, rather than doing the converse.

We are now left with the problem of finding a way to compute swaption prices with the LFM. In order to understand the difficulties of this task, let us consider a very simple swaption. Assume we are at time 0. The underlying interest-rate swap (IRS) starts at T_1 and pays at T_2 and T_3. All times are equally spaced by a year fraction denoted by τ, and we take a unit notional amount. The (payer) swaption payoff can be written as

$$[P(T_1, T_2)\tau(F_2(T_1) - K) + P(T_1, T_3)\tau(F_3(T_2) - K)]^+ ,$$

where in general we set $F_k(t) = F(t; T_{k-1}, T_k)$. Recall that the discount factors $P(T_1, T_2)$ and $P(T_1, T_3)$ can be expressed in terms of $F_2(T_1)$ and $F_3(T_2)$, so that this payoff actually depends only on the rates $F_2(T_1)$ and $F_3(T_2)$. The key point, however, is the following. The payoff is not additively "separable" with respect to the different rates. As a consequence, when you take expectation of such a payoff, the *joint* distribution of the two rates F_2 and F_3 is involved in the calculation, so that the correlation between the two rates F_2 and F_3 has an impact on the value of the contract. This does not happen with caps. Indeed, let us go back to caps for a moment and consider a cap consisting of the T_2- and T_3-caplets. The cap payoff as seen from time 0 would be

$$\exp\left(-\int_0^{T_2} r_u du\right) \tau (F_2(T_1) - K)^+ + \exp\left(-\int_0^{T_3} r_u du\right) \tau (F_3(T_2) - K)^+ .$$

This time, the payoff is additively separated with respect to different rates. Indeed, we can compute

$$E\left[\exp\left(-\int_0^{T_2} r_u du\right) \tau(F_2(T_1) - K)^+\right.$$
$$\left. + \exp\left(-\int_0^{T_3} r_u du\right) \tau(F_3(T_2) - K)^+\right]$$
$$= E\left[\exp\left(-\int_0^{T_2} r_u du\right) \tau(F_2(T_1) - K)^+\right]$$
$$+ E\left[\exp\left(-\int_0^{T_3} r_u du\right) \tau(F_3(T_2) - K)^+\right]$$
$$= P(0,T_2)\tau E^2\left[(F_2(T_1) - K)^+\right] + P(0,T_3)\tau E^3\left[(F_3(T_2) - K)^+\right].$$

In this last expression we have two expectations, each one involving a *single* rate. The joint distribution of the two rates F_2 and F_3 is not involved, therefore, in the calculation of this last expression, since it is enough to know the marginal distributions of F_2 and F_3 separately. Accordingly, the terminal correlation between rates F_2 and F_3 does not affect this payoff.

As a consequence of this simple example, it is clear that adequately modeling correlation can be important in defining a model that can be effectively calibrated to swaption prices. When the number of swaptions to which the model has to be calibrated is large, correlation becomes definitely relevant. If a short-rate model has to be chosen, it is better to choose a multi-factor model. Multi-dimensional instantaneous sources of randomness guarantee correlation patterns among terminal rates F that are clearly more general than in the one-factor case. However, producing a realistic correlation pattern with, for instance, a two-factor short-rate model is not always possible.

As far as the LFM is concerned, as we hinted at above, the solution is usually to assign a different Brownian motion to each forward rate and to assume such Brownian motions to be instantaneously correlated. Manipulating instantaneous correlation leads to manipulation of correlation of simple rates (terminal correlation), although terminal correlation is also influenced by the way in which average volatility is distributed among instantaneous volatilities, see Section 6.6 on this point.

Choosing an instantaneous-correlation structure flexible enough to express a large number of swaption prices and, at the same time, parsimonious enough to be tractable is a delicate task. The integrated covariance matrix need not have full rank. Usually ranks two or three are sufficient, see Brace Dun and Barton (1998) on this point. We will suggest some parametric forms for the covariance structure of instantaneous forward rates in Section 6.9.

Brace (1996) proposed an approximated formula to evaluate analytically swaptions in the LFM. The formula is based on a rank-one approximation of the integrated covariance matrix plus a drift approximation and works well in specific contexts and in non-pathological situations. More generally, a rank-r approximation is considered in Brace (1997). See again Brace, Dun and Barton (1998) for numerical experiments and results.

The rank-one approximated formula is reviewed in Section 6.11, while the rank-r approximation is reviewed in Section 6.12. There are also analytical swaption-pricing formulas that are simpler and still accurate enough for most practical purposes. Such formulas are based on expressing forward swap rates as linear combinations of forward LIBOR rates, to then take variance on both sides and integrate while freezing some coefficients. This "freezing the drift and collapsing all measures" approximation is reviewed in Section 6.13 and has also been tested by Brace, Dun and Barton (1998). Given its importance, we performed numerical tests of our own. These are reported in Section 8.2 of Chapter 8, and confirm that the formula works well.

More generally, to evaluate swaptions and other payoffs with the LFM one has usually to resort to Monte Carlo simulation. Once the numeraire is chosen, one simulates all forward rates involved in the payoff by discretizing their joint dynamics with a numerical scheme for stochastic differential equations (SDEs).

Notice that each forward rate is driftless only under its associated measure. Indeed, for example, while F_2 is driftless under Q^2 (recall (6.2)), it is not driftless under Q^1. The dynamics of F_2 under Q^1 is derived below and leads to a process that we need to discretize in order to obtain simulations, whereas in the driftless case (6.2) the transition distribution of F is known to be lognormal and no numerical scheme is needed. This point is discussed in detail again in Section 6.10. As an introductory illustration, consider again the above swaption payoff, now discounted at time 0, and take its risk-neutral expectation:

$$E\left\{\exp\left(-\int_0^{T_1} r_u du\right)[P(T_1,T_2)\tau(F_2(T_1)-K) \right.$$
$$\left. +P(T_1,T_3)\tau(F_3(T_2)-K)]^+\right\} = \cdots$$

Take $P(\cdot, T_1)$ as numeraire, which corresponds to choose the T_1-forward-adjusted measure Q^1. One then obtains:

$$\cdots = P(0,T_1)E^1\left[P(T_1,T_2)\tau(F_2(T_1)-K) + P(T_1,T_3)\tau(F_3(T_2)-K)\right]^+.$$

It is known that F_1 is driftless under the chosen measure Q^1, while F_2 and F_3 are not. For example, take for simplicity the one-factor case, so that

$$dF_2(t) = v_2 F_2(t) dW_t, \qquad (6.4)$$

where W is now a standard Brownian motion under Q^2. Then if Z is a standard Brownian motion under Q^1, it can be shown that the change-of-numeraire technique leads to

192   6. The LIBOR and Swap Market Models (LFM and LSM)

$$dF_2(t) = \frac{v_2^2 F_2^2(t)\tau}{1+\tau F_2(t)}dt + v_2 F_2(t)dZ_t. \tag{6.5}$$

Now it is clear that, as we stated above, the no-arbitrage dynamics (6.5) of F_2 under Q^1 is not driftless, and moreover its transition distribution (which is necessary to perform exact simulations) is not known, contrary to the Q^2 driftless case (6.4).

In order to price the swaption, it is the Q^1 dynamics of F_2 that matters, so that we need to discretize the related equation (6.5) in order to be able to simulate it. Numerical schemes for SDEs are available, like the Euler or Milstein scheme, see also Appendix A. Alternative schemes that guarantee the (weak) no-arbitrage condition to be maintained in discrete time and not just in the continuous-time limit have been proposed by Glasserman and Zhao (2000).

Whichever scheme is chosen, Monte Carlo pricing is to be performed by simulating the relevant forward LIBOR rates. While path-dependent derivatives can be priced by this approach in general, as far as early-exercise (e.g. American-style or Bermudan-style) products are concerned, the situation is delicate, since the joint dynamics of the LFM usually does not lead to a recombining lattice for the short rate, so that it is not immediately clear how to evaluate a Bermudan- or American-style product with a tree in the LFM.

Usually, ad-hoc techniques are needed, such as Carr and Yang's (1997) who provide a method for simulating Bermudan-style derivatives with the LFM via a Markov chain approximation, or Andersen's (1999) who approximates the early exercise boundary as a function of intrinsic value and "still-alive" nested European swaptions. However, a general method for combining backward induction with Monte Carlo simulation has been proposed recently by Longstaff and Schwartz (2000), and this method is rather promising, especially because of its generality. We briefly review all these methods in Chapter 10.

Several other problems remain, like for example the possibility of including a volatility smile in the model. We will address the smile problem for the cap market by proposing both the CEV dynamics suggested by Andersen and Andreasen (2000) with possibly shifted distributions, and the shifted "lognormal-mixture" dynamics model illustrated in Brigo and Mercurio (2000a, 2000b, 2001b).

6.3 The Lognormal Forward-LIBOR Model (LFM)

Let $t=0$ be the current time. Consider a set $\mathcal{E} = \{T_0, \ldots, T_M\}$ from which expiry-maturity pairs of dates (T_{i-1}, T_i) for a family of spanning forward rates are taken. We shall denote by $\{\tau_0, \ldots, \tau_M\}$ the corresponding year fractions, meaning that τ_i is the year fraction associated with the expiry-maturity pair

6.3 The Lognormal Forward-LIBOR Model (LFM)

(T_{i-1}, T_i) for $i > 0$, and τ_0 is the year fraction from settlement to T_0. Times T_i will be usually expressed in years from the current time. We set $T_{-1} := 0$.

Consider the generic forward rate $F_k(t) = F(t; T_{k-1}, T_k)$, $k = 1, \ldots, M$, which is "alive" up to time T_{k-1}, where it coincides with the simply-compounded spot rate $F_k(T_{k-1}) = L(T_{k-1}, T_k)$. In general $L(S, T)$ is the simply compounded spot rate prevailing at time S for the maturity T.

Consider now the probability measure Q^k associated with the numeraire $P(\cdot, T_k)$, i.e. to the price of the bond whose maturity coincides with the maturity of the forward rate. Q^k is often called the *forward (adjusted) measure for the maturity* T_k. Under simple compounding, it follows immediately by definition that

$$F_k(t) P(t, T_k) = [P(t, T_{k-1}) - P(t, T_k)]/\tau_k.$$

Therefore, $F_k(t)P(t, T_k)$ is the price of a tradable asset (difference between two discount bonds with notional amounts $1/\tau_k$). As such, when its price is expressed with respect to the numeraire $P(\cdot, T_k)$, it has to be a martingale under the measure Q^k associated with that numeraire (by definition of measure associated with a numeraire). But the price $F_k(t)P(t, T_k)$ of our tradable asset divided by this numeraire is simply $F_k(t)$ itself. Therefore, F_k follows a martingale under Q^k. It follows that if F_k is modeled according to a diffusion process, it needs to be driftless under Q^k.

We assume the following driftless dynamics for F_k under Q^k:

$$dF_k(t) = \underline{\sigma}_k(t) F_k(t) dZ^k(t), \quad t \leq T_{k-1}, \tag{6.6}$$

where $Z^k(t)$ is an M-dimensional Brownian motion (under Q^k) with instantaneous covariance $\rho = (\rho_{i,j})_{i,j=1,\ldots,M}$,

$$dZ^k(t) \, dZ^k(t)' = \rho \, dt,$$

and where $\underline{\sigma}_k(t)$ is the horizontal M-vector volatility coefficient for the forward rate $F_k(t)$.

Unless differently stated, from now on we will assume that

$$\underline{\sigma}_j(t) = [0 \ 0 \ \ldots \ \sigma_j(t) \ \ldots \ 0 \ 0],$$

with the only non-zero entry $\sigma_j(t)$ occurring at the j-th position in the vector $\underline{\sigma}_j(t)$.

We hope the reader is not disoriented by our notation. The point of switching from vector to scalar volatility is that in some computations the former will be more convenient, whereas in other cases the latter will be useful. Just notice that the above equation for F_k can be rewritten under Q^k as

$$dF_k(t) = \sigma_k(t) F_k(t) dZ_k^k(t), \quad t \leq T_{k-1},$$

where $Z_k^k(t)$ is the k-th component of the vector Brownian motion Z^k and is thus a standard Brownian motion. Lower indices indicate the component we are considering in a vector, whereas upper indices show under which measure we are working. Upper indices are usually omitted when the context is clear, so that we write

$$dF_k(t) = \sigma_k(t)F_k(t)dZ_k(t), \quad t \leq T_{k-1}. \tag{6.7}$$

With this scalar notation, $\sigma_k(t)$ now bears the usual interpretation of instantaneous volatility at time t for the forward LIBOR rate F_k.

Notice that in case σ is bounded (as in all of our volatility parameterizations below), this last equation has a unique strong solution, since it describes a geometric Brownian motion. This can be checked immediately also by writing, through Ito's formula,

$$d\ln F_k(t) = -\frac{\sigma_k(t)^2}{2}dt + \sigma_k(t)dZ_k(t), \quad t \leq T_{k-1},$$

and by observing that the coefficients of this last equation are bounded, so that there exists trivially a unique strong solution. Indeed, we have immediately the unique solution

$$\ln F_k(T) = \ln F_k(0) - \int_0^T \frac{\sigma_k(t)^2}{2}dt + \int_0^T \sigma_k(t)dZ_k(t).$$

We will often consider piecewise-constant instantaneous volatilities:

$$\sigma_k(t) = \sigma_{k,\beta(t)},$$

(with $\sigma_k(0) = \sigma_{k,1}$) where in general $\beta(t) = m$ if $T_{m-2} < t \leq T_{m-1}$, so that

$$t \in (T_{\beta(t)-2}, T_{\beta(t)-1}].$$

At times we will use the notation $Z_t = Z(t)$.

As is clear from our assumption on the vector "noise" dZ above, the single "noises" in the dynamics of different forward rates are assumed to be instantaneously correlated according to

$$dZ_i(t)\,dZ_j(t) = d\langle Z_i, Z_j\rangle_t = \rho_{i,j}dt.$$

Remark 6.3.1. (**Achieving decorrelation**). Historical one-factor short-rate models, such as for instance Hull and White's (1990b) or Black and Karasinski's (1991), imply forward-rate dynamics that are perfectly instantaneously correlated. This means that for such models we would have $\rho_{i,j} = 1$ for all i,j. As a consequence, we can say that forward rates in these models are usually too correlated.

When trying to improve these models, one of the objectives is to lower the correlation of the forward rates implied by the model. Some authors refer to this objective as to achieving *decorrelation*. As we shall see later on in Section 6.6, decorrelation can be achieved not only by "lowering" instantaneous correlations, but also by carefully redistributing integrated variances of forward rates over time.

6.3.1 Some Specifications of the Instantaneous Volatility of Forward Rates

A few general remarks are now in order. First, as we said before, we will often assume that the forward rate $F_k(t)$ has a piecewise-constant instantaneous volatility. In particular, the instantaneous volatility of $F_k(t)$ is constant in each "expiry-maturity" time interval (associated with any other forward rate) $T_{m-2} < t \leq T_{m-1}$ where it is "alive".

Under this assumption, it is possible to organize instantaneous volatilities in the following matrix, where "Instant. Vols" and "Fwd" are abbreviations for Instantaneous Volatility and Forward respectively:

TABLE 1

Instant. Vols	Time: $t \in (0, T_0]$	$(T_0, T_1]$	$(T_1, T_2]$	...	$(T_{M-2}, T_{M-1}]$
Fwd Rate: $F_1(t)$	$\sigma_{1,1}$	Dead	Dead	...	Dead
$F_2(t)$	$\sigma_{2,1}$	$\sigma_{2,2}$	Dead	...	Dead
$\vdots$	...	...	...	...	...
$F_M(t)$	$\sigma_{M,1}$	$\sigma_{M,2}$	$\sigma_{M,3}$	...	$\sigma_{M,M}$

A first assumption can be made on the entries of TABLE 1 so as to reduce the number of volatility parameters. We can assume that volatilities depend only on the time-to-maturity $T_k - T_{\beta(t)-1}$ of a forward rate rather than on time t and maturity T_k separately. In such a case, by assuming

$$\sigma_k(t) = \sigma_{k,\beta(t)} =: \eta_{k-(\beta(t)-1)}, \tag{6.8}$$

the above matrix looks like:

TABLE 2

Instant. Vols	Time: $t \in (0, T_0]$	$(T_0, T_1]$	$(T_1, T_2]$	...	$(T_{M-2}, T_{M-1}]$
Fwd Rate: $F_1(t)$	η_1	Dead	Dead	...	Dead
$F_2(t)$	η_2	η_1	Dead	...	Dead
$\vdots$	...	...	...	...	...
$F_M(t)$	η_M	η_{M-1}	η_{M-2}	...	η_1

A second alternative assumption can be made on the entries of TABLE 1 so as to reduce the number of parameters. We can assume

$$\sigma_k(t) = \sigma_{k,\beta(t)} := s_k \tag{6.9}$$

for all t, which amounts to assuming that a forward rate $F_k(t)$ has constant instantaneous volatility s_k regardless of t.

This leads to the following table:

TABLE 3

Instant. Vols	Time: $t \in (0, T_0]$	$(T_0, T_1]$	$(T_1, T_2]$	...	$(T_{M-2}, T_{M-1}]$
Fwd Rate: $F_1(t)$	s_1	Dead	Dead	...	Dead
$F_2(t)$	s_2	s_2	Dead	...	Dead
$\vdots$	...	...	...	...	...
$F_M(t)$	s_M	s_M	s_M	...	s_M

A third alternative assumption can be made on the entries of TABLE 1 so as to reduce the number of parameters with respect to the general case. We can assume

$$\sigma_k(t) = \sigma_{k,\beta(t)} := \Phi_k \Psi_{\beta(t)} \qquad (6.10)$$

for all t. This leads to the following table:

TABLE 4

Instant. Vols	Time: $t \in (0, T_0]$	$(T_0, T_1]$	$(T_1, T_2]$	...	$(T_{M-2}, T_{M-1}]$
Fwd Rate: $F_1(t)$	$\Phi_1 \Psi_1$	Dead	Dead	...	Dead
$F_2(t)$	$\Phi_2 \Psi_1$	$\Phi_2 \Psi_2$	Dead	...	Dead
$\vdots$	...	...	...	...	...
$F_M(t)$	$\Phi_M \Psi_1$	$\Phi_M \Psi_2$	$\Phi_M \Psi_3$	...	$\Phi_M \Psi_M$

This last formulation includes formula (6.9) as a special case, for example by taking all Ψ's equal to one and the Φ's equal to the s's. It is the most general formulation leading to a rank-one integrated covariance matrix in case one takes all instantaneous correlations ρ equal to one (one-factor model), see for example Brace, Dun and Barton (1998) for a proof.

A fourth (final) alternative assumption can be made on the entries of TABLE 1 so as to reduce the number of parameters. We can assume

$$\sigma_k(t) = \sigma_{k,\beta(t)} := \Phi_k \psi_{k-(\beta(t)-1)} \qquad (6.11)$$

for all t. This leads to the following table:

TABLE 5

Instant. Vols	Time: $t \in (0, T_0]$	$(T_0, T_1]$	$(T_1, T_2]$	...	$(T_{M-2}, T_{M-1}]$
Fwd Rate: $F_1(t)$	$\Phi_1 \psi_1$	Dead	Dead	...	Dead
$F_2(t)$	$\Phi_2 \psi_2$	$\Phi_2 \psi_1$	Dead	...	Dead
$\vdots$	...	...	...	...	...
$F_M(t)$	$\Phi_M \psi_M$	$\Phi_M \psi_{M-1}$	$\Phi_M \psi_{M-2}$	...	$\Phi_M \psi_1$

6.3 The Lognormal Forward-LIBOR Model (LFM)

This formulation includes formula (6.8) as a special case, for example by taking all Φ's equal to one and the ψ's equal to the η's. It is the product of a structure dependent only on the time to maturity (the ψ's) by a structure dependent only on the maturity (the Φ's). It does not lead to a rank-one integrated covariance matrix in general, not even in the case of perfect instantaneous correlations ρ all equal to one. We will clarify the peculiarities of this last formulation later on. At the moment we just anticipate that this form has the potential for maintaining the qualitative shape of the term structure of volatilities in time through its ψ part, as we will explain in Section 6.5, while being rich enough to allow a satisfactory calibration to market data.

The five tables conclude the piecewise-constant models for instantaneous volatilities we planned to present. Now we introduce instead some parametric forms for the same quantities. A first possibility is the following:

FORMULATION 6

$$\boxed{\sigma_i(t) = \psi(T_{i-1} - t; a, b, c, d) := [a(T_{i-1} - t) + d]e^{-b(T_{i-1}-t)} + c} \quad (6.12)$$

This form allows a humped shape in the graph of the instantaneous-volatility of the generic forward rate F_i as a function of the time to maturity,

$$T_{i-1} - t \mapsto \sigma_i(t).$$

We will see later on that this parametric form possesses some desirable and undesirable qualitative features, and that its flexibility is not sufficient for practical purposes when jointly calibrating to the caps and swaptions markets. Dependence on the time to maturity, rather than time and maturity separately, renders this form a parametric analogue of the piecewise-constant case (6.8) of TABLE 2, the function ψ being an analogue of the η's.

The above formulation can be perfected into a richer parametric form. Indeed consider

FORMULATION 7

$$\boxed{\sigma_i(t) = \Phi_i\, \psi(T_{i-1} - t; a, b, c, d) := \Phi_i \left([a(T_{i-1} - t) + d]e^{-b(T_{i-1}-t)} + c\right)}$$
(6.13)

This parametric form reduces to the previous one when all the Φ's are set to one. This form can be seen to have a parametric core ψ that is locally altered for each maturity T_i by the Φ's. These local modifications, if small, do not destroy the essential dependence on the time to maturity, so as to maintain the desirable qualitative properties we will describe in Section 6.5; at the same time, such modifications add flexibility to the parametric form, flexibility that can be used in order to improve the joint calibration of the model to the caps and swaptions markets, as we will observe later on. When seen as a local modification of a structure depending only on the time to maturity,

198   6. The LIBOR and Swap Market Models (LFM and LSM)

this parametric form is the analogue of the piecewise-constant case (6.11) of TABLE 5.

We will discuss the benefits of this parametric assumption in Section 6.5, when describing the associated term structure of volatilities.

We will come back to matters concerning the assumptions on instantaneous volatilities later on.

6.3.2 Forward-Rate Dynamics under Different Numeraires

For reasons that will be clear afterwards, we are interested in finding the dynamics of $F_k(t)$ under a measure Q^i different from Q^k, for $t \leq \min(T_i, T_{k-1})$ (both the chosen numeraire and the forward rate being modeled have to be alive in t). This can be done by writing down the general rule under which asset price dynamics changes when one changes the numeraire (Chapter 2), and then by applying this tool to the specific case of forward-rates dynamics.

Proposition 6.3.1. (Forward-measure dynamics in the LFM). *Under the lognormal assumption, we obtain that the dynamics of F_k under the forward-adjusted measure Q^i in the three cases $i < k$, $i = k$ and $i > k$ are, respectively,*

$$i < k, \ t \leq T_i : \ dF_k(t) = \sigma_k(t) F_k(t) \sum_{j=i+1}^{k} \frac{\rho_{k,j} \tau_j \sigma_j(t) F_j(t)}{1 + \tau_j F_j(t)} dt$$
$$+ \sigma_k(t) F_k(t) \, dZ_k(t), \qquad (6.14)$$

$$i = k, \ t \leq T_{k-1} : \ dF_k(t) = \sigma_k(t) F_k(t) \, dZ_k(t),$$

$$i > k, \ t \leq T_{k-1} : \ dF_k(t) = -\sigma_k(t) F_k(t) \sum_{j=k+1}^{i} \frac{\rho_{k,j} \tau_j \sigma_j(t) F_j(t)}{1 + \tau_j F_j(t)} dt$$
$$+ \sigma_k(t) F_k(t) \, dZ_k(t),$$

where, as explained before, $Z = Z^i$ is a standard Brownian motion under Q^i. All of the above equations admit a unique strong solution if the coefficients $\sigma(\cdot)$ are bounded.

Proof. Consider the forward rate $F_k(t) = F(t, T_{k-1}, T_k)$ and suppose we wish to derive its dynamics first under the T_i-forward measure Q^i with $i < k$. We know that the dynamics under the T_k-forward measure Q^k has null drift. From this dynamics, we propose to recover the dynamics under Q^i. Let us apply (2.15) to $X = F_k = F(\cdot, T_{k-1}, T_k)$, where we set $S = P(\cdot, T_k)$ and $U = P(\cdot, T_i)$, and the dynamics of X under Q^k is given by (6.6).

We obtain the percentage drift as

6.3 The Lognormal Forward-LIBOR Model (LFM)

$$m_t^i = -\frac{d\langle \ln X, \ln(P(\cdot,T_k)/P(\cdot,T_i))\rangle_t}{dt}.$$

Now notice that

$$\ln(P(t,T_k)/P(t,T_i)) = \ln\left(1/\left[\prod_{j=i+1}^{k}(1+\tau_j F_j(t))\right]\right)$$

$$= -\sum_{j=i+1}^{k}\ln(1+\tau_j F_j(t))$$

from which

$$m_t^i = -\frac{d\langle \ln F_k, \ln(P(\cdot,T_k)/P(\cdot,T_i))\rangle_t}{dt} = \sum_{j=i+1}^{k}\frac{d\langle \ln F_k, \ln(1+\tau_j F_j(t))\rangle_t}{dt}$$

$$= \sum_{j=i+1}^{k}\frac{\tau_j}{1+\tau_j F_j(t)}\frac{d\langle \ln F_k, F_j\rangle_t}{dt} = \sum_{j=i+1}^{k}\frac{\rho_{j,k}\tau_j\sigma_k(t)\sigma_j(t)F_j(t)}{1+\tau_j F_j(t)}.$$

Secondly, consider the case $i > k$, where the numeraire is a bond whose maturity is longer than the maturity of the forward rate being modeled. In such a case, the derivation is similar, but now

$$\ln(P(t,T_k)/P(t,T_i)) = \ln\left(\left[\prod_{j=k+1}^{i}(1+\tau_j F_j(t))\right]\right) = \sum_{j=k+1}^{i}\ln(1+\tau_j F_j(t))$$

from which

$$m_t^i = -\sum_{j=k+1}^{i}\frac{\rho_{j,k}\,\tau_j\,\sigma_k(t)\,\sigma_j(t)\,F_j(t)}{1+\tau_j\,F_j(t)}.$$

As for existence and uniqueness of the solution, we have treated earlier the trivial case $i = k$. In the case $i < k$, compute through Ito's formula

$$d\ln F_k(t) = \sigma_k(t)\sum_{j=i+1}^{k}\frac{\rho_{k,j}\tau_j\sigma_j(t)F_j(t)}{1+\tau_j F_j(t)}dt - \frac{\sigma_k(t)^2}{2}dt + \sigma_k(t)dZ_k(t).$$

Now notice that the diffusion coefficient of this equation is both deterministic and bounded. Moreover, since

$$0 < \frac{\tau_j F_j(t)}{1+\tau_j F_j(t)} < 1,$$

also the drift is bounded, besides being smooth in the F's (that are positive). This ensures existence and uniqueness of a strong solution for the above SDE. The case $i > k$ is analogous. □

Remark 6.3.2. (**Relationship between Brownian motions under different numeraires**). By applying formula (2.13) (the change-of-numeraire toolkit) to the above situation for changing from $C dW^k$ to $C dW^{k+1}$ (i.e. with $S = P(\cdot, T_{k+1})$ and $U = P(\cdot, T_k)$) we obtain, after straightforward calculations similar to the ones in the above proof:

$$dZ^{k+1} = dZ^k + \frac{\tau_{k+1} F_{k+1}(t)}{1 + \tau_{k+1} F_{k+1}(t)} \rho \underline{\sigma}_{k+1}(t)' dt. \tag{6.15}$$

This relationship connects Brownian motions under two adjacent forward measures, and it is easily generalizable to any two forward measures in our family. It will be useful in the sequel when in need to recover non-standard expiry-maturity forward rates from our original family $F_1, \ldots, F_M$. Also, Brace, Musiela and Schlögl (1998), for example, point out the importance of this relationship for simulations.

The above dynamics constitute the **lognormal forward-LIBOR model** (**LFM**). The name is due to the fact that for $i = k$ the distribution of F_k is lognormal. The above dynamics (except in the case $i = k$) do not feature known transition densities, in that knowledge of σ's and ρ's above does not allow one to write an analytical expression for the density of forward rates in a future instant given the forward rates in a past instant. As a consequence, no analytical formula or simple numeric integration can be used in order to price contingent claims depending on the joint dynamics. Moreover, when generating paths for the forward rates, they cannot be generated *one-shot*, but rather have to be obtained from a time-discretization of equation (6.14). In other words, if we need for example one million realizations of

$$\varphi(t) := [F(t; T_0, T_1) \ F(t; T_1, T_2) \ \ldots \ F(t; T_{i-1}, T_i)]$$

for a certain $t \leq T_0$ under Q^0, we cannot generate them directly, because the distribution of $\varphi(t)$ is not known. One then has to discretize equations (6.14) between 0 and t with a sufficiently (but not too) small time step Δt, and generate the distributionally-known Gaussian shocks $Z_{t+\Delta t} - Z_t$.

We are not yet done with numeraire changes, since some products (typically Eurodollar futures) require the *risk-neutral* dynamics of forward rates. Recall that the risk-neutral measure Q is the measure associated with the bank-account numeraire $B(t) = 1/D(0, t)$, where $D(0, t)$ is the stochastic discount factor for maturity t.

The problem with deriving the risk-neutral dynamics with the LFM is that the numeraire $B(t)$ is not natural for forward rates with preassigned tenor and maturities, and thus leads to a "residual" in the change of measure, which is somehow awkward and mixes discrete tenor and continuous tenor quantities. Since the LFM has as a cornerstone feature the idea of modeling discrete tenor quantities, the reappearance of continuous-tenor terms à la HJM in the dynamics is an undesirable ingredient.

6.3 The Lognormal Forward-LIBOR Model (LFM)

There is a rather obvious way to avoid this undesirable mixture of continuous and discrete features, and we will come to that below. For now, let us derive the risk-neutral dynamics of the LFM.

Recall from Chapter 1 that typically, in terms of the instantaneous spot rate r,

$$D(0,T) = \exp\left(-\int_0^T r_t dt\right), \quad B(T) = \exp\left(\int_0^T r_t dt\right),$$

$$P(0,T) = E\left[\exp\left(-\int_0^T r_t dt\right)\right].$$

By remembering that B has dynamics

$$dB(t) = r_t B(t) dt$$

with zero volatility, one derives the risk-neutral dynamics of F_k by applying our change-of-numeraire toolkit to move from the numeraire $P(\cdot, T_k)$ to the numeraire $B(t)$. Indeed, by using formula (2.12) with $U = B$, $S = P(\cdot, T_k)$ and $X = F_k$, one obtains after straightforward calculations:

$$dF_k(t) = -\underline{\sigma}_{P(\cdot,T_k)} \rho \underline{\sigma}_k(t)' F_k(t) dt + \underline{\sigma}_k(t) F_k(t) d\widetilde{Z}(t)$$

where $\underline{\sigma}_{P(\cdot,T_k)}$ is the vector percentage volatility of the bond price (diffusion coefficient of the log-price). Since the bond price can be expressed in terms of spanning forward rates as

$$P(t, T_k) = P(t, T_{\beta(t)-1}) \prod_{j=\beta(t)}^{k} \frac{1}{1 + \tau_j F_j(t)}$$

we can compute the above drift as ("diff_coeff" is an abbreviation for diffusion coefficient)

$$\widetilde{\mu}_k(t) := -\underline{\sigma}_{P(t,T_k)} \bar{\rho} \underline{\sigma}_k(t)'$$
$$= -\text{diff\_coeff}[d\ln P(t,T_k)] \rho \underline{\sigma}_k(t)'$$
$$= -\text{diff\_coeff}\left[d\ln P(t,T_{\beta(t)-1}) + \sum_{j=\beta(t)}^{k} d\ln\left(\frac{1}{1+\tau_j F_j(t)}\right)\right] \rho \underline{\sigma}_k(t)'$$

Now notice that since in general

$$\ln P(t,T) = -\int_t^T f(t,u)\, du,$$

we have easily that

202   6. The LIBOR and Swap Market Models (LFM and LSM)

$$\text{diff\_coeff}[d\ln P(t,T)] = -\int_t^T \underline{\sigma}_f(t,u)\,du\,,$$

where $\underline{\sigma}_f(t,u)$ is the absolute instantaneous vector volatility of the instantaneous forward rate $f(t,u)$. We therefore have the following.

Proposition 6.3.2. (Risk-neutral dynamics in the LFM). *The risk-neutral dynamics of forward LIBOR rates in the LFM is:*

$$dF_k(t) = \widetilde{\mu}_k(t)F_k(t)\,dt + \sigma_k(t)F_k(t)\,d\widetilde{Z}_k(t)\,, \qquad (6.16)$$

where

$$\widetilde{\mu}_k(t) = \sum_{j=\beta(t)}^{k} \frac{\tau_j\,\rho_{j,k}\,\sigma_j(t)\,\sigma_k(t)\,F_j(t)}{1+\tau_j F_j(t)} + \underline{\sigma}_k(t)\,\rho\int_t^{T_{\beta(t)-1}} \underline{\sigma}_f(t,u)'\,du$$

$$= \sum_{j=\beta(t)}^{k} \frac{\tau_j\,\rho_{j,k}\,\sigma_j(t)\,\sigma_k(t)\,F_j(t)}{1+\tau_j F_j(t)} + \sum_{j=\beta(t)}^{k} \rho_{k,j}\,\sigma_k(t) \int_t^{T_{\beta(t)-1}} (\underline{\sigma}_f)_j(t,u)'\,du.$$

As we had anticipated, the drift in (6.16) looks awkward, because of the second summation. The second summation originates from the continuous-tenor part coming from the evolution of $P(t,T_{\beta(t)-1})$, which cannot be deduced from forward rates in our family. We are thus forced to model the instantaneous forward rate volatility in order to close the equation, but this is something *external* to the discrete-tenor setting we have adopted.

We can partly improve this situation by considering a discretely rebalanced bank-account numeraire as an alternative to the continuously rebalanced bank account $B(t)$, whose value, at any time t, changes according to $dB(t) = r_t B(t)dt$. We then introduce a bank account that is rebalanced only on the times in our discrete-tenor structure. To this end, consider the numeraire asset

$$B_d(t) = \frac{P(t,T_{\beta(t)-1})}{\prod_{j=1}^{\beta(t)-1} P(T_{j-1},T_j)} = \prod_{j=1}^{\beta(t)-1} (1+\tau_j F_j(T_{j-1}))\,P(t,T_{\beta(t)-1})\,.$$

The interpretation of $B_d(t)$ is that of the value at time t of a portfolio defined as follows. The portfolio starts with one unit of currency at the initial time $t=0$, exactly as in the continuous-bank-account case ($B(0)=1$), but this unit amount is now invested in a quantity X_0 of T_0 zero-coupon bonds. Such X_0 is readily found by noticing that, since we invested one unit of currency, the present value of the bonds needs to be one, so that $X_0 P(0,T_0) = 1$, and hence $X_0 = 1/P(0,T_0)$. At time T_0, we cash the bonds payoff X_0 and invest it in a quantity $X_1 = X_0/P(T_0,T_1) = 1/(P(0,T_0)P(T_0,T_1))$ of T_1 zero-coupon bonds. We continue this procedure until we reach the last tenor date $T_{\beta(t)-2}$ preceding the current time t, where we invest

$$X_{\beta(t)-1} = 1/ \prod_{j=1}^{\beta(t)-1} P(T_{j-1}, T_j)$$

in $T_{\beta(t)-1}$ zero-coupon bonds. The present value at the current time t of this investment is $X_{\beta(t)-1} P(t, T_{\beta(t)-1})$, i.e. our $B_d(t)$ above. Thus, $B_d(t)$ is obtained by starting from one unit of currency and reinvesting at each tenor date in zero-coupon bonds for the next tenor. This gives a discrete-tenor counterpart of the continuous bank account B, and the subscript "d" in B_d stands for "discrete".

Now choose B_d as numeraire and apply the change-of-numeraire technique starting from the dynamics (6.6) under Q^k, to obtain the dynamics under B_d. If you go through the (change-of-numeraire toolkit) calculations, you will notice that now the "awkward" contributions of the term $P(t, T_{\beta(t)-1})$ cancel out, since this term appears both in $P(t, T_k)$ and in $B_d(t)$. The measure Q^d associated with B_d is called *spot LIBOR measure*. We then have the following.

Proposition 6.3.3. (Spot-LIBOR-measure dynamics in the LFM).
The spot LIBOR measure dynamics of forward LIBOR rates in the LFM is:

$$dF_k(t) = \sigma_k(t) F_k(t) \sum_{j=\beta(t)}^{k} \frac{\tau_j \rho_{j,k} \sigma_j(t) F_j(t)}{1 + \tau_j F_j(t)} dt + \sigma_k(t) F_k(t) dZ_k^d(t). \quad (6.17)$$

Notice, indeed, that, under this last numeraire, no continuous-tenor term appears in the drift.

Remark 6.3.3. Both the spot-measure dynamics and the risk-neutral dynamics admit no known transition densities, so that the related equations need to be discretized in order to perform simulations.

We finally notice that a full no-arbitrage derivation of the LFM dynamics above, based on the change-of-numeraire technique, can be found in Musiela and Rutkowski (1997).

6.4 Calibration of the LFM to Caps and Floors Prices

The calibration to caps and floors prices for the LFM is almost automatic, since one can simply input in the model volatilities σ given by the market in form of Black-like implied volatilities for cap prices.

Indeed, we have already hinted in Section 6.2 that the LFM cap pricing formula coincides with that used in the market for pricing caps, namely the Black formula (1.26), which can be then re-derived under a fully consistent theoretical apparatus.

We now shortly review this alternative derivation of Black's cap formula, in order also to clarify some points on the calibration.

204   6. The LIBOR and Swap Market Models (LFM and LSM)

The discounted payoff at time 0 of a cap with first reset date T_α and payment dates $T_{\alpha+1}, \ldots, T_\beta$ is given by

$$\sum_{i=\alpha+1}^{\beta} \tau_i D(0, T_i)(F(T_{i-1}, T_{i-1}, T_i) - K)^+,$$

where we assume a unit nominal amount.

We remind that the caps whose implied volatilities are quoted by the market typically have either T_0 equal to three months, $\alpha = 0$ and all other T's equally three-months spaced, or T_0 equal to six months, $\alpha = 0$ and all other T's equally six-months spaced.

The pricing of the cap can be obtained by considering the risk-neutral expectation E of its discounted payoff:

$$E\left\{ \sum_{i=\alpha+1}^{\beta} \tau_i D(0, T_i)(F_i(T_{i-1}) - K)^+ \right\}$$

$$= \sum_{i=\alpha+1}^{\beta} \tau_i P(0, T_i) E^i \left[(F_i(T_{i-1}) - K)^+ \right]$$

where we used the forward-adjusted measure Q^i. The above payoff is therefore reduced to a sum of payoffs. The single additive term

$$(F_i(T_{i-1}) - K)^+$$

in the above payoff is associated with a contract called T_{i-1}-*caplet*, and we will often identify the contract with its payoff, thus naming "caplet" the payoff itself. Caps and caplets were introduced earlier in Chapter 1. Recall that a T_{i-1}-caplet is a contract paying at time T_i the difference between the T_i-maturity spot rate reset at time T_{i-i} and a strike rate K, if this difference is positive, and zero otherwise. The price at the initial time 0 of the T_{i-1}-caplet is then given by

$$P(0, T_i)\ E^i(F_i(T_{i-1}) - K)^+ .$$

Remark 6.4.1. (**Correlations have no impact on caps**). As we have remarked earlier in the "guided tour" to market models, notice carefully that the joint dynamics of forward rates is not involved in this payoff. As a consequence, the correlation between different rates does not afflict the payoff, since marginal distributions of the single F's are enough to compute the expectations appearing in the payoff. Indeed, there are no expectations involving two or more forward rates at the same time, so that correlations are not relevant.

Recall also that the T_{i-1}-caplet is said to be "at-the-money" when its strike price $K = K_i$ equals the current value $F_i(0)$ of the underlying forward rate. The caplet is said to be "in-the-money" when $F_i(0) > K_i$, reflecting

6.4 Calibration of the LFM to Caps and Floors Prices

the fact that if rates were deterministic, we would get money (amounting to $F_i(0) - K_i$) from the caplet contract at maturity. The caplet is said to be "out-of-the-money" when $F_i(0) < K_i$.

It is advisable to keep in mind also the notion of "at-the-money" (ATM) cap. As we just recalled, a cap is a collection of caplets with a common strike K. Each caplet, however, would be ATM for a different strike K_i. We cannot therefore define the ATM cap strike K in terms of the single ATM caplets strikes K_i, since they are different in general. We can however select a single rate that takes into account all the forward rates of the underlying caplets. This is chosen to be the forward swap rate $K = S_{\alpha,\beta}(0)$, which we will reintroduce in Section 6.7.

Going back to a general strike K, in order to compute

$$E^i[(F_i(T_{i-1}) - K)^+],$$

remember that under Q^i the process F_i follows the martingale

$$dF_i(t) = \sigma_i(t) F_i(t) dZ_i(t), \quad Q^i, \quad t \leq T_{i-1}.$$

Given the lognormal distribution for F_i, the above expectation is easily computed as a Black and Scholes price for a stock call option whose underlying "stock" is F_i, struck at K, with maturity T_{i-1}, with zero constant "risk-free rate" and instantaneous percentage volatility $\sigma_i(t)$. We thus obtain the following.

Proposition 6.4.1. (Equivalence between LFM and Black's caplet prices). *The price of the T_{i-1}-caplet implied by the LFM coincides with that given by the corresponding Black caplet formula, i.e.*

$$\mathbf{Cpl}^{LFM}(0, T_{i-1}, T_i, K) = \mathbf{Cpl}^{Black}(0, T_{i-1}, T_i, K, v_i)$$
$$= P(0, T_i)\tau_i \operatorname{Bl}(K, F_i(0), v_i),$$

$$\operatorname{Bl}(K, F_i(0), v_i) = E^i(F_i(T_{i-1}) - K)^+$$
$$= F_i(0)\Phi(d_1(K, F_i(0), v_i)) - K\Phi(d_2(K, F_i(0), v_i)),$$

$$d_1(K, F, v) = \frac{\ln(F/K) + v^2/2}{v},$$

$$d_2(K, F, v) = \frac{\ln(F/K) - v^2/2}{v},$$

where

$$v_i^2 = T_{i-1} v_{T_{i-1}-caplet}^2, \qquad (6.18)$$

$$v_{T_{i-1}-caplet}^2 := \frac{1}{T_{i-1}} \int_0^{T_{i-1}} \sigma_i(t)^2 dt.$$

Notice that in our notation v_i^2 will always denote the integrated instantaneous variance not divided by the time amount, contrary to $v_{T_{i-1}-\text{caplet}}^2$, which will be always standardized with respect to time (average instantaneous variance over time). The quantity $v_{T_{i-1}-\text{caplet}}$ is termed T_{i-1}-*caplet volatility* and has thus been (implicitly) defined as the square root of the average percentage variance of the forward rate $F_i(t)$ for $t \in [0, T_{i-1})$.

We now review the implications of the different volatility structures introduced in Section 6.3 as far as calibration is concerned.

6.4.1 Piecewise-Constant Instantaneous-Volatility Structures

These are the structures described in TABLEs 1–5. In general, we have:

$$v_{T_{i-1}-\text{caplet}}^2 = \frac{1}{T_{i-1}} \int_0^{T_{i-1}} \sigma_{i,\beta(t)}^2 dt = \frac{1}{T_{i-1}} \sum_{j=1}^{i} \tau_{j-2,j-1}\, \sigma_{i,j}^2, \qquad (6.19)$$

where we denote by $\tau_{i,j} = T_j - T_i$ the time between T_i and T_j in years.

Formulation of TABLE 2. If we assume that the piecewise-constant instantaneous volatilities $\sigma_{i,\beta(t)}$ depend only on the time to maturity, according to the formulation of TABLE 2, where $\sigma_{i,\beta(t)} = \eta_{i-\beta(t)+1}$, we then obtain:

$$v_i^2 = \sum_{j=1}^{i} \tau_{j-2,j-1}\, \eta_{i-j+1}^2. \qquad (6.20)$$

In this case the parameters η can be used to exactly fit the (squares of the) market caplet volatilities (multiplied by time) v_i^2. Although calibration to swaptions will be dealt with later on, we already hint at the fact that, with this formulation, the only parameters left to tackle swaptions calibration are the instantaneous correlations of forward rates.

Formulation of TABLE 3. If we instead assume that the piecewise-constant instantaneous volatilities $\sigma_{i,\beta(t)}$ depend only on the maturity T_i of the considered forward rate, according to the formulation leading to TABLE 3 where $\sigma_{i,\beta(t)} = s_i$, we obtain:

$$v_i^2 = T_{i-1} s_i^2, \quad v_{T_{i-1}-\text{caplet}} = s_i. \qquad (6.21)$$

Again, the parameters s can be used to exactly fit the (squares of the) market caplet volatilities (multiplied by time) v_i^2, and the only parameters left to tackle swaptions calibration are again the instantaneous correlations of forward rates.

Formulation of TABLE 4. On the other hand, if we assume that the

6.4 Calibration of the LFM to Caps and Floors Prices

piecewise-constant instantaneous volatilities $\sigma_{i,\beta(t)}$ follow the separable structure of the formulation leading to TABLE 4, where $\sigma_{i,\beta(t)} = \Phi_i \Psi_{\beta(t)}$, we obtain:

$$v_i^2 = \Phi_i^2 \sum_{j=1}^{i} \tau_{j-2,j-1} \Psi_j^2. \quad (6.22)$$

If the squares of the caplet volatilities (multiplied by time) v_i^2 are read from the market, $(v_i^{\text{MKT}})^2$, the parameters Φ can be given in terms of the parameters Ψ as

$$\Phi_i^2 = \frac{(v_i^{\text{MKT}})^2}{\sum_{j=1}^{i} \tau_{j-2,j-1} \Psi_j^2}.$$

Therefore the caplet prices are incorporated in the model by determining the Φ's in term of the Ψ's. The parameters Ψ, together with the instantaneous correlation of forward rates, can be then used in the calibration to swaption prices.

Formulation of TABLE 5. Finally, if we assume that the piecewise-constant instantaneous volatilities $\sigma_{i,\beta(t)}$ follow the separable structure of the formulation leading to TABLE 5, where $\sigma_{i,\beta(t)} = \Phi_i \psi_{i-(\beta(t)-1)}$, we obtain:

$$v_i^2 = \Phi_i^2 \sum_{j=1}^{i} \tau_{j-2,j-1} \psi_{i-j+1}^2. \quad (6.23)$$

If the squares of the caplet volatilities (multiplied by time) v_i^2 are read from the market, the parameters Φ can now be given in terms of the parameters ψ as

$$\Phi_i^2 = \frac{(v_i^{\text{MKT}})^2}{\sum_{j=1}^{i} \tau_{j-2,j-1} \psi_{i-j+1}^2}. \quad (6.24)$$

Therefore, the caplet prices are incorporated in the model by determining the Φ's in terms of the ψ's. As in the previous case, the parameters ψ, together with the instantaneous correlation of forward rates, can then be used in the calibration to swaption prices.

6.4.2 Parametric Volatility Structures

As far as the parametric structures of Formulations 6 and 7 are concerned, we observe the following.

Formulation 6. If we assume Formulation 6 for instantaneous volatilities, as from (6.12), we obtain the following expression for the squares of caplet volatilities (multiplied by time):

208 6. The LIBOR and Swap Market Models (LFM and LSM)

$$v_i^2 = \int_0^{T_{i-1}} \left([a(T_{i-1}-t)+d]e^{-b(T_{i-1}-t)}+c\right)^2 dt =: I^2(T_{i-1};a,b,c,d).$$
(6.25)

In this case the parameters a, b, c, d can be used to fit the (squares of the) market caplet volatilities (multiplied by time) v_i^2. Therefore, with this formulation the only parameters left to tackle swaptions calibration are the instantaneous correlations of forward rates.

The function I can be computed also through a software for formal manipulations, with a command line such as

```
int(((a*(T-t)+d)*exp(-b*(T-t))+c)
   *((a*(T-t)+d)*exp(-b*(T-t))+c),t=0..T);
```

Formulation 7. If we assume Formulation 7 for instantaneous volatilities, as from (6.13), we obtain the following expression for the squares of caplet volatilities (multiplied by time):

$$v_i^2 = \Phi_i^2 \int_0^{T_{i-1}} \left([a(T_{i-1}-t)+d]e^{-b(T_{i-1}-t)}+c\right)^2 dt = \Phi_i^2\, I^2(T_{i-1};a,b,c,d).$$
(6.26)

Now the Φ's parameters can be used to calibrate automatically caplet volatilities. Indeed, if the v_i are inferred from market data, we may set

$$\Phi_i^2 = \frac{(v_i^{\mathrm{MKT}})^2}{I^2(T_{i-1};a,b,c,d)}.$$
(6.27)

Thus caplet volatilities are incorporated by expressing the parameters Φ as functions of the parameters a, b, c, d, which are still free. In this way, for swaptions calibration we can rely upon the parameters a, b, c, d and upon the instantaneous correlations between forward rates.

6.4.3 Cap Quotes in the Market

As already pointed out, the market typically quotes volatilities for caps with first reset date either in three months (T_0 equal to three months, $\alpha = 0$ and all other T's equally three-months spaced) or in six months (T_0 equal to six months, $\alpha = 0$ and all other T's equally six-months spaced), and progressively increasing maturities. However, in this chapter we will essentially consider examples of caps with a six-month reset period. We set $\mathcal{T}_j = [T_0, \ldots, T_j]$ for all j.

An equation is considered between the market price $\mathbf{Cap}^{\mathrm{MKT}}(0, \mathcal{T}_j, K)$ of the cap with $\alpha = 0$ and $\beta = j$ and the sum of the first j caplets prices:

$$\mathbf{Cap}^{\mathrm{MKT}}(0, \mathcal{T}_j, K) = \sum_{i=1}^{j} \tau_i P(0, T_i)\, \mathrm{Bl}(K, F_i(0), \sqrt{T_{i-1}}\, v_{T_j-\mathrm{cap}})$$

6.4 Calibration of the LFM to Caps and Floors Prices

where a same average-volatility value $v_{T_j-\text{cap}}$ has been put in all caplets up to j. The quantities $v_{T_j-\text{cap}}$ are called sometimes *forward volatilities*. The market solves the above equation in $v_{T_j-\text{cap}}$ and quotes $v_{T_j-\text{cap}}$, annualized and in percentages.

Remark 6.4.2. (One kind of inconsistency in cap volatilities). Notice carefully that *the same* average volatility $v_{T_j-\text{cap}}$ is assumed for all caplets concurring to the T_j-maturity cap. However, when the same caplets concur to a different cap, say a T_{j+1}-maturity cap, their average volatility is changed. This appears to be somehow inconsistent. In the cap volatility system, the same caplet is linked to different volatilities when concurring to different caps.

Clearly, to recover correctly cap prices with our forward-rate dynamics we need to have

$$\sum_{i=1}^{j} \tau_i P(0,T_i) \operatorname{Bl}(K, F_i(0), \sqrt{T_{i-1}}\, v_{T_j-\text{cap}})$$
$$= \sum_{i=1}^{j} \tau_i P(0,T_i) \operatorname{Bl}(K, F_i(0), \sqrt{T_{i-1}}\, v_{T_{i-1}-\text{caplet}}).$$

The quantities $v_{T_{i-1}-\text{caplet}}$ are called sometimes *forward forward volatilities*. Notice that *different* average volatilities $v_{T_{i-1}-\text{caplet}}$ are assumed for different caplets concurring to the T_j-maturity cap.

A stripping algorithm can be used for recovering the v_{caplet}'s from the market quoted v_{cap}'s based on the last equality applied to $j = 1, 2, 3, \ldots$. This can be done off-line, before any model-calibration/pricing procedure is tackled. In turn, from the v_{caplet}'s we can go back to the σ's via (6.19), to the η's via (6.20), or to the s's via (6.21). Notice that with the general formulation (6.19) deriving from TABLE 1, or with the last reduced formulations (6.22) and (6.23) of TABLE 4 and TABLE 5, we cannot recover the whole table, since we have more unknown than equations. However, having more parameters can be helpful at later stages, when calibration to swaptions is considered, especially when choosing few instantaneous correlation parameters. On the contrary, formulations (6.8) and (6.9), by reducing the number of unknowns, allow the complete determination of TABLE 2 and TABLE 3, respectively, based on cap prices.

Therefore, modulo some trivial algebra and numerical non-linear equation solving, the calibration to cap prices with these formulations, leading to (6.20) and (6.21), is automatic since the instantaneous volatilities parameters η's and s's can be deduced directly from the market via an algebraic (numerical) procedure.

Finally, for the purpose of jointly calibrating caps and swaptions, we notice that the formulation leading to (6.22) would be particularly handy when using Brace's analytical approximation, reviewed in Sections 6.11 and 6.12,

210 6. The LIBOR and Swap Market Models (LFM and LSM)

although the formulation leading to (6.23), or its parametric analogue leading to (6.26), turn out to better combine both flexibility for a good initial fitting and "controllability" for a qualitatively acceptable evolution of the term structure of volatilities, as we are about to see.

6.5 The Term Structure of Volatility

Consider again the set of times $\mathcal{E} = \{T_0, \ldots, T_M\}$ representing adjacent expiry-maturity pairs for a family of spanning forward rates. The term structure of volatility at time T_j is a graph of expiry times T_{h-1} against average volatilities $V(T_j, T_{h-1})$ of the forward rates $F_h(t)$ up to that expiry time itself, i.e. for $t \in (T_j, T_{h-1})$. In other terms, at time $t = T_j$, the volatility term structure is the graph of points

$$\{(T_{j+1}, V(T_j, T_{j+1})), (T_{j+2}, V(T_j, T_{j+2})), \ldots, (T_{M-1}, V(T_j, T_{M-1}))\}$$

where

$$V^2(T_j, T_{h-1}) = \frac{1}{\tau_{j,h-1}} \int_{T_j}^{T_{h-1}} \frac{dF_h(t)\, dF_h(t)}{F_h(t) F_h(t)} = \frac{1}{\tau_{j,h-1}} \int_{T_j}^{T_{h-1}} \sigma_h^2(t) dt,$$

for $h > j+1$ (we remind that $\tau_{i,j} = T_j - T_i$). The term structure of volatilities at time 0 is

$$\{(T_0, V(0, T_0)), \ldots, (T_{M-1}, V(0, T_{M-1}))\}$$
$$= \{(T_0, v_{T_0-\text{caplet}}), \ldots, (T_{M-1}, v_{T_{M-1}-\text{caplet}})\}$$

These are simply (pairs of expiries with corresponding) forward forward volatilities at time 0, and a typical example (with annualized versions of the caplet volatilities, as from Section 6.16) from the Euro market is displayed in Figure 6.1. Notice the humped structure, starting from about 15% for the one-year caplet, up to about 16% for the two-year caplet, and afterwards always decreasing, down to less than 10% for the nineteen-year caplet.

Different assumptions on the behaviour of instantaneous volatilities imply different evolutions for the term structure of volatilities. We now examine the impact of different formulations of instantaneous volatilities on the evolution of the term structure.

6.5.1 Piecewise-Constant Instantaneous Volatility Structures

We first examine the two cases corresponding to TABLE 2 and TABLE 3 respectively.

Formulation of TABLE 2. Formulation (6.8) of TABLE 2 gives

6.5 The Term Structure of Volatility

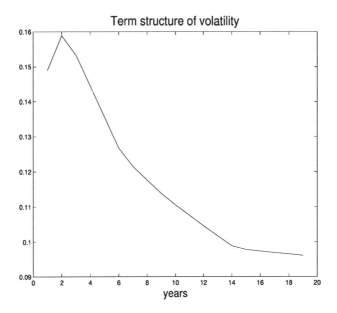

Fig. 6.1. Example of term structure of volatility, $T \mapsto v_{T-\text{caplet}}$, from the Euro market.

$$V^2(T_j, T_{h-1}) = \frac{1}{\tau_{j,h-1}} \sum_{k=j+1}^{h-1} \tau_{k-1,k} \eta_{h-k}^2.$$

Assume for simplicity that $\tau_{i-1,i} = \tau > 0$, so that $\tau_{j,h-1} = (h-j-1)\tau$. It follows easily that, in such a case,

$$V^2(T_j, T_{h-1}) = \frac{1}{h-j-1} \sum_{k=j+1}^{h-1} \eta_{h-k}^2, \tag{6.28}$$

which immediately implies that

$$V(T_j, T_{h-1}) = V(T_{j+1}, T_h).$$

Therefore, when time moves from T_j to T_{j+1}, the volatility term structure remains the same, except that now it is shorter, since the number of possible expiries has decreased by one. We move in fact from

$$\{(T_{j+1}, V(T_j, T_{j+1})), (T_{j+2}, V(T_j, T_{j+2})), (T_{j+3}, V(T_j, T_{j+3})), \ldots$$
$$\ldots, (T_{M-1}, V(T_j, T_{M-1}))\}$$

to

$$\{(T_{j+2}, V(T_j, T_{j+1})), (T_{j+3}, V(T_j, T_{j+2})), \ldots, (T_{M-1}, V(T_j, T_{M-2}))\}.$$

Summarizing, with instantaneous volatilities of forward rates like in TABLE 2, as time passes from an expiry date to the next, the volatility term structure remains the same, except that the "tail" of the graph is cut away. In particular, the volatility term structure does not change over time. It has been often observed that *qualitatively* this is a desirable property, since the actual shape of the market term structure (typically humped) does not change too much over time (see also Rebonato (1998)). We present an example of such a desirable evolution in the first part of Figure 6.2.

Still, there remains the fundamental question of whether this formulation allows for a humped structure to be fitted at the initial time. What we have just seen, in fact, is that, under this volatility parameterization, *if* the initial term structure is humped then it remains humped at future times. But does this volatility specification allow for a humped-volatility structure in the first place?

Consider formula (6.28) for $j = -1$, i.e. the volatility term structure at time 0. It is easy to see that the map

$$T_{h-1} \mapsto \sqrt{T_{h-1}} V(0, T_{h-1}) = \sqrt{\tau \sum_{k=0}^{h-1} \eta_{h-k}^2}$$

is increasing. Therefore, if the market data imply an initial volatility term structure such that $\sqrt{T}V(0,T)$ is humped as a function of T, the instantaneous-volatility parameterization of TABLE 2 cannot be used since it implies, instead, an increasing behaviour.

However, one is usually interested in $T \mapsto V(0,T)$ to be humped, rather than $T \mapsto \sqrt{T}V(0,T)$. This gives us further hope. In fact, we want a decreasing structure following the hump, namely

$$V(0, T_{h-1}) \geq V(0, T_h)$$

for T_h (typically) larger than three (or four-five) years. On the other hand, we have just seen that the parameterization obtained through the η's imposes that

$$\sqrt{T_{h-1}}\, V(0, T_{h-1}) \leq \sqrt{T_h}\, V(0, T_h) \, .$$

Putting these two constraints together yields:

$$1 \leq \frac{V(0, T_{h-1})}{V(0, T_h)} \leq \sqrt{\frac{h}{h-1}}.$$

We observe that, for large h, the last term is close to one, meaning that the term structure gets almost flat at the end. Indeed, the above constraints imply $V(0, T_h) \approx V(0, T_{h-1})$ for large h. In case the term structure is steeply decreasing also for large maturities, the parameterization of TABLE 2 is thus to be avoided.

6.5 The Term Structure of Volatility

Formulation of TABLE 3. Consider now formulation (6.9) for instantaneous volatilities, which leads to TABLE 3. In such a case

$$V^2(T_j, T_{h-1}) = s_h^2.$$

This is the simplest case of volatility term structure. When moving from expiry time T_j to T_{j+1}, the term structure changes from

$$\{(T_{j+1}, s_{j+2}), (T_{j+2}, s_{j+3}), (T_{j+3}, s_{j+4}), \ldots, (T_{M-1}, s_M)\}$$

to

$$\{(T_{j+2}, s_{j+3}), (T_{j+3}, s_{j+4}), \ldots, (T_{M-1}, s_M)\}.$$

With instantaneous volatilities of forward rates like in TABLE 3, as time passes from an expiry date to the next, the volatility term structure is obtained from the previous one by "cutting off" the head instead of the tail. This is less desirable than in the previous formulation, since now the qualitative behaviour of the term structure can be altered over time. Typically, if the term structure features a hump around two years, this hump will be absent in the term structure occurring in three years. See the second part of Figure 6.2 for an example.

Formulations of TABLES 4 and 5. Now consider the separable case of TABLE 4, as from (6.10). In such a case

$$V^2(T_j, T_{h-1}) = \frac{\Phi_h^2}{\tau_{j,h-1}} \sum_{k=j+2}^{h} \tau_{k-2,k-1} \Psi_k^2.$$

Here the qualitative behaviour depends on the particular specification of both the Φ's and the Ψ's, and there is no a priori clear qualitative pattern for the evolution of the term structure of volatilities as a whole. Moreover, it is hard to control the future term structure of volatilities by acting on the parameters with this formulation.

This task results to be easier under the last formulation (6.11), which yields

$$V^2(T_j, T_{h-1}) = \frac{\Phi_h^2}{\tau_{j,h-1}} \sum_{k=j+1}^{h-1} \tau_{k-1,k} \psi_{h-k}^2.$$

By assuming again all adjacent τ's to be equal, this reduces to

$$V^2(T_j, T_{h-1}) = \frac{\Phi_h^2}{h-j-1} \sum_{k=j+1}^{h-1} \psi_{h-k}^2.$$

If the weights Φ are all equal, this formulation is equivalent to that of TABLE 2. Therefore, in such a case the term structure of volatility remains

214   6. The LIBOR and Swap Market Models (LFM and LSM)

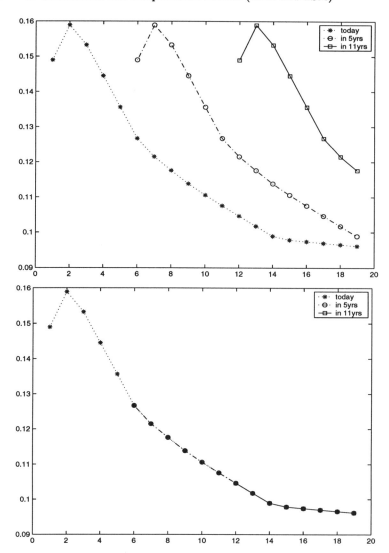

Fig. 6.2. Examples of evolution of the term structure of volatilities under Formulations 2 and 3.

unchanged as time passes and in particular it maintains the hump over time. The presence of the Φ terms, when they are not all equal, can change this situation, but if we make sure that the Φ's remain sufficiently close to each other, the qualitative behaviour will not be affected and the hump in the evolution of the term structure in time will be preserved. An example of such a desirable behaviour is given in Figure 6.3 for the similar parametric formulation 7 below.

6.5 The Term Structure of Volatility   215

The abundance of parameters and the "controllability" of the future term structure make this parameterization particularly appealing among the piecewise-constant ones.

6.5.2 Parametric Volatility Structures

We now consider the evolution of the term structure of volatilities in time as implied by Formulations 6 and 7.

Evolution under Formulation 6. We have already noticed that Formulation 6 is the analogue of the piecewise-constant formulation of TABLE 2. As such, the same qualitative behaviour can be expected. The term structure remains the same as time passes, and in particular it maintains its humped shape if initially humped. Nevertheless, the problem of a non-decreasing structure is still present. If the initial term structure is decreasing for large maturities, it cannot be calibrated with this formulation, as observed for the case of TABLE 2.

Evolution under Formulation 7. We have also already noticed that Formulation 7 is the analogue of the piecewise-constant formulation of TABLE 5. Again, the same qualitative behaviour can be expected. The term structure remains the same as time passes, and in particular it can maintain its humped shape if initially humped and if all Φ's are not too different from one another. This form has been suggested, among others, by Rebonato (1999d).

An example of the evolution of the term structure of volatilities, as implied by this formulation, is given in Figure 6.3. We start with the same initial term structure of volatilities as in the previous Figure 6.2, which is given here by setting $a = 0.19085664, b = 0.97462314, c = 0.08089168, d = 0.01344948$, and with the Φ vector plotted in the first graph of this figure. Clearly, this short example is just a sample of the kind of tests and games one has to play with a formulation before feeling confident about it. In particular, we give it as an example of the kind of questions that are posed by traders and practitioners, in general, when a model is presented to them.

A final remark is that, compared to other models, the LFM allows for an immediate calculation of the future term structures of volatilities, which is a rather important advantage. This calculation requires no simulation and leads to a deterministic evolution. For other models this is not the case. Usually the instantaneous percentage volatility of forward rates (i.e. the diffusion coefficient of $d \ln F_k(t)$ or of $dF_k(t)/F_k(t)$) is not deterministic, so that the integrated instantaneous variance leading to future volatilities is a random variable. Displaying scenarios of future term structures of volatility implies, therefore, simulation of trajectories for the underlying processes.

Indeed, if we were using, for example, a one-factor short-rate model, say Vasicek's $dr_t = k(\theta - r_t)dt + \sigma dW_t$, we would have $\ln F_k(t) = \phi_k(r_t)$ for some

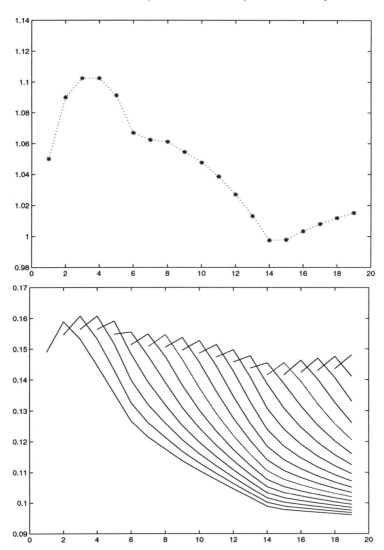

Fig. 6.3. An example of Φ (first graph) and the evolution of the term structure of volatilities (second graph) under Formulation 7.

function ϕ_k. Then by Ito's formula, the instantaneous percentage volatility would be given by the diffusion coefficient of $d\ln F_k(t)$, which is equal to $\phi'_k(r_t)\sigma$ (where $'$ denotes differentiation). Now take $T_1 = 2y$, $T_2 = 3y$ and compute, for instance,

$$V^2(T_1, T_2) = \int_{T_1}^{T_2} d\ln F_3(t) d\ln F_3(t) = \int_{T_1}^{T_2} \phi'_3(r_t)^2 \sigma^2 dt.$$

This is a random variable depending on the trajectory of r between T_1 and T_2. If $T_1 > 0$, there is no escape from this situation, and we need to cope with a random future term structure of volatilities. However, if we are considering the term structure of volatilities at time 0, there is a way out. We can in fact obtain a deterministic volatility by pricing at-the-money caplets with the model and then by inverting Black's formula for caplets to derive an implied Black-like caplet volatility for the Vasicek model. With the LFM this is not necessary, and there is no difference between the model "intrinsic" caplet volatility and the model "implied" caplet volatility, as in the Vasicek case, for example (for a discussion of such matters in the short-rate world see Sections 3.6 and 4.2.3).

To sum up, the LFM has the advantage over other models of allowing one to display (deterministic) future term structures of volatilities once the model has been calibrated. With other models such structures would be stochastic, except for the structure at time 0 when its definition can be modified in "implied" terms.

6.6 Instantaneous Correlation and Terminal Correlation

In the following we shall consider payoffs whose evaluation will be based on expected values of quantities involving several rates at the same time, typically swaptions. As a consequence, prices will depend on the terminal correlation between different forward rates. The point of this section is showing that the terminal correlation between different forward rates depends not only on the instantaneous correlations of forward-rate dynamics, but also on the instantaneous volatility specification.

Indeed, we have assumed earlier each forward rate F_k to be driven by a different Brownian motion Z_k, and we assumed also different Brownian motions to be instantaneously correlated. Now it is important to understand how this instantaneous correlation in the forward-rate dynamics translates into a terminal correlation of simple rates.

Recall that the *instantaneous* correlation is a quantity summarizing the degree of "dependence" between *changes* of different forward rates. Roughly speaking, for example,

$$\rho_{2,3} = \frac{dF_2(t)\, dF_3(t)}{\text{Std}(dF_2(t))\, \text{Std}(dF_3(t))},$$

where "Std" denotes the standard deviation conditional on the information available at time t at which the change occurs. From this formula it is clear that indeed the instantaneous correlation ρ is related to changes dF in forward rates. Instead, the *terminal* correlation is a quantity summarizing the degree of "dependence" between two different forward rates at a given "terminal" time-instant. Typically, the T_1 terminal correlation between F_2 and

218   6. The LIBOR and Swap Market Models (LFM and LSM)

F_3 is the correlation between $F_2(T_1)$ and $F_3(T_1)$. Is this terminal correlation completely determined by the instantaneous correlations $\rho_{2,3}$ between Z_2 and Z_3?

The answer is negative in general. The terminal correlation between different forward rates certainly depends on the introduced instantaneous correlations, but not only on them. We will show, in fact, that such a terminal correlation depends also on the way the "total" average volatility of each forward rate (caplet volatility) is "disintegrated" or "distributed" in instantaneous volatility. Precisely, the correlation between $F_2(T)$ and $F_3(T)$ depends also on the particular functions of time $\sigma_{2,\beta(t)}$ and $\sigma_{3,\beta(t)}$ that are used to recover the average volatilities v_2 and v_3 in $[0, T_1]$ through integration. Keeping the same instantaneous correlations and decomposing v_2 and v_3 in two different ways "a" ($\sigma^a_{2,\beta(t)}, \sigma^a_{3,\beta(t)}$) and "b" ($\sigma^b_{2,\beta(t)}, \sigma^b_{3,\beta(t)}$), leads to different correlations between $F_2(T_1)$ and $F_3(T_1)$ in the two cases. This elementary and fundamental feature of terminal correlation was noticed and pointed out in Rebonato (1998, 1999d).

We notice also that, while the instantaneous correlation does not depend on the particular probability measure (or numeraire asset) under which we are working, the terminal correlation does. Indeed, the Girsanov theorem establishes that the instantaneous covariance structure is the same for all the equivalent measures under which a process can be expressed, so that the particular measure under which we work makes no difference. This is not the case for terminal correlation. However, we will see further on in Section 6.14 that approximated terminal correlations based on "freeze part of the drift"-like approximations in the dynamics will not depend on the particular probability measure being chosen.

To clarify dependence of the terminal correlation on the volatility decomposition we now consider two examples, the first of which, though not addressing completely this issue, has the advantage of being based on exact calculations.

As first example, consider a payoff depending on the two forward rates $F_2(t), F_3(t)$ for $t \leq T_1$. Take $T_0 > 0$, $T_1 = 2T_0$, and take as underlying measure Q^2, under which F_2 is a martingale. We thus have:

$$dF_2(t) = \sigma_{2,\beta(t)} F_2(t) dZ_2(t)$$
$$dF_3(t) = \frac{\tau_3 \sigma^2_{3,\beta(t)} F_3^2(t)}{1 + \tau_3 F_3(t)} dt + \sigma_{3,\beta(t)} F_3(t) dZ_3(t), \quad t \leq T_1 \ .$$

Assume the payoff depends on $\ln F_3(T_1) - \ln F_3(0)$ and on $F_2(T_1) - F_2(0)$ jointly, and consider the product of these two variables as an academic example of a quantity whose expectation depends on the T_1-terminal correlation between F_2 and F_3. Ito's isometry plus some straightforward algebra give

6.6 Instantaneous Correlation and Terminal Correlation

$$E^2\left[(F_2(T_1) - F_2(0))(\ln F_3(T_1) - \ln F_3(0))\right] = F_2(0)\rho_{2,3}\int_0^{T_1}\sigma_{2,\beta(t)}\sigma_{3,\beta(t)}dt$$

$$= F_2(0)\rho_{2,3}(\sigma_{2,1}\sigma_{3,1}T_0 + \sigma_{2,2}\sigma_{3,2}\tau_{0,1}). \tag{6.29}$$

Now notice that any caplet price involving the above rates would simply depend on the average volatilities

$$v_2^2 = T_0\,\sigma_{2,1}^2 + \tau_{0,1}\,\sigma_{2,2}^2,$$

and

$$v_3^2 = T_0\,\sigma_{3,1}^2 + \tau_{0,1}\,\sigma_{3,2}^2 + \tau_{1,2}\,\sigma_{3,3}^2$$

no matter how the total variances v_2^2 and v_3^2 are distributed between F_2 and F_3, i.e. no matter the particular values of $\sigma_{2,1}, \sigma_{2,2}, \sigma_{3,1}, \sigma_{3,2}$ are chosen, as long as v_2 and v_3 are preserved.

On the contrary, the correlation-dependent term (6.29) does depend on this choice. Consider for example $\tau_{i-1,i} = T_i - T_{i-1} = 0.5$ for all i, $F_2(0) = F_3(0) = 0.05$, $\rho_{2,3} = 0.75$, $\sigma_{3,3} = 0.1$, and finally take the two cases

a) $\sigma_{2,1} = 0.5$, $\sigma_{3,1} = 0.1$, $\sigma_{2,2} = 0.1$, $\sigma_{3,2} = 0.5$

and

b) $\sigma_{2,1} = 0.1$, $\sigma_{3,1} = 0.1$, $\sigma_{2,2} = 0.5$, $\sigma_{3,2} = 0.5$.

In cases a) and b) v_2 and v_3 have the same values, whereas the correlation-dependent quantity (6.29) gives $1.875E - 3$ in case a) and $2.34375E - 4$ in case b). This example shows that the value of a payoff depending on the terminal correlation between different rates may depend not only on the instantaneous correlations ρ of the forward-rate dynamics, but also on the way the instantaneous volatilities σ are modeled from average volatilities v coming from cap prices.

A more extreme and direct example is as follows. As stated earlier, we will derive in Section 6.14 an approximated formula for computing the terminal correlation of forward rates. The formula, in this particular example, would read

$$\mathrm{Corr}(F_2(T_1), F_3(T_1)) \approx \rho_{2,3}\frac{\int_0^{T_1}\sigma_2(t)\sigma_3(t)dt}{\sqrt{\int_0^{T_1}\sigma_2^2(t)dt}\sqrt{\int_0^{T_1}\sigma_3^2(t)dt}}.$$

Take $T_0 = 1y$, $T_1 = 2y$, $T_2 = 3y$.

Given the piecewise-constant volatility formulation, and the fact that T_1 is exactly the expiry time for the forward rate F_2, we can write

$$\mathrm{Corr}(F_2(T_1), F_3(T_1)) \approx \rho_{2,3}\frac{\sigma_{2,1}\sigma_{3,1} + \sigma_{2,2}\sigma_{3,2}}{v_2\sqrt{\sigma_{3,1}^2 + \sigma_{3,2}^2}}.$$

Let us fix the instantaneous correlation $\rho_{2,3} = 1$ and, as usual, require that our piecewise-constant functions $\sigma_2(t)$ and $\sigma_3(t)$ be compatible with T_1- and T_2-caplet volatilities. We therefore require

220   6. The LIBOR and Swap Market Models (LFM and LSM)

$$v_2^2 = \sigma_{2,1}^2 + \sigma_{2,2}^2,$$

and

$$v_3^2 = \sigma_{3,1}^2 + \sigma_{3,2}^2 + \sigma_{3,3}^2 \ .$$

There are two extreme cases of interest. The first is obtained as

$$\sigma_{2,1} = v_2, \ \sigma_{2,2} = 0;$$

$$\sigma_{3,1} = v_3, \sigma_{3,2} = 0, \ \sigma_{3,3} = 0\,.$$

In this case, the above formula yields easily

$$\text{Corr}(F_2(T_1), F_3(T_1)) \approx \rho_{2,3} = 1\,.$$

The second extreme case is obtained as

$$\sigma_{2,1} = 0, \ \sigma_{2,2} = v_2;$$

$$\sigma_{3,1} = v_3, \sigma_{3,2} = 0, \ \sigma_{3,3} = 0\,.$$

In this second case, the above formula yields immediately

$$\text{Corr}(F_2(T_1), F_3(T_1)) \approx 0 \, \rho_{2,3} = 0\,.$$

So we have seen two cases with the same instantaneous correlation and caplet volatilities, where a different rebalancing of instantaneous volatilities leads respectively to terminal correlations of 0 and 1.

Whichever choice of the instantaneous volatilities structure is considered, say for example (6.8), (6.9), or (6.10), the point of this section is making clear that this choice influences also the terminal correlation between different rates.

6.7 Swaptions and the Lognormal Forward-Swap Model (LSM)

Assume a unit notional amount. Recall that a (prototypical) interest-rate swap (IRS) is a contract that exchanges payments between two differently indexed legs. At every instant T_j in a prespecified set of dates $T_{\alpha+1},...,T_\beta$ the fixed leg pays out an amount corresponding to a fixed interest rate K,

$$\tau_j K\,,$$

where τ_j is the year fraction from T_{j-1} to T_j, whereas the floating leg pays an amount corresponding to the interest rate $F_j(T_{j-1})$ set at the previous instant T_{j-1} for the maturity given by the current payment instant T_j.

Clearly, the floating-leg rate is reset at dates $T_\alpha, T_{\alpha+1}, \ldots, T_{\beta-1}$ and paid at dates $T_{\alpha+1}, \ldots, T_\beta$.

6.7 Swaptions and the Lognormal Forward-Swap Model (LSM)

The payoff at time T_α for the payer party of such an IRS can be expressed as

$$\sum_{i=\alpha+1}^{\beta} D(T_\alpha, T_i) \tau_i (F_i(T_{i-1}) - K).$$

Accordingly, the *discounted* payoff at a time $t < T_\alpha$ is

$$\sum_{i=\alpha+1}^{\beta} D(t, T_i) \tau_i (F_i(T_{i-1}) - K). \tag{6.30}$$

The value of such a contract is easily computed as

$$\begin{aligned}
\mathbf{PFS}(t, [T_\alpha, \ldots, T_\beta], K) &= E_t \left\{ \sum_{i=\alpha+1}^{\beta} D(t, T_i) \tau_i (F_i(T_{i-1}) - K) \right\} \\
&= \sum_{i=\alpha+1}^{\beta} P(t, T_i) \tau_i E_t^i (F_i(T_{i-1}) - K) \\
&= \sum_{i=\alpha+1}^{\beta} P(t, T_i) \tau_i (F_i(t) - K) \\
&= \sum_{i=\alpha+1}^{\beta} [P(t, T_{i-1}) - (1 + \tau_i K) P(t, T_i)].
\end{aligned} \tag{6.31}$$

Notice that the same value is obtained by defining as T_α-payoff

$$\sum_{i=\alpha+1}^{\beta} P(T_\alpha, T_i) \tau_i (F_i(T_\alpha) - K),$$

leading to the t-discounted payoff

$$D(t, T_\alpha) \sum_{i=\alpha+1}^{\beta} P(T_\alpha, T_i) \tau_i (F_i(T_\alpha) - K),$$

and by taking the risk-neutral expectation.

From the last formula we can also notice that, as is well known, neither volatility nor correlation of rates affect the pricing of this financial product.

The *forward swap rate* corresponding to the above IRS is the particular value of the fixed-leg rate K that makes the contract fair, i.e. that makes its present value equal to zero. The forward swap rate associated with the above IRS is therefore obtained by equating to zero the last expression in (6.31) and by solving it in K. We obtain

$$S_{\alpha,\beta}(t) = \frac{P(t, T_\alpha) - P(t, T_\beta)}{\sum_{i=\alpha+1}^{\beta} \tau_i P(t, T_i)} = \frac{1 - \text{FP}(t; T_\alpha, T_\beta)}{\sum_{i=\alpha+1}^{\beta} \tau_i \text{FP}(t; T_\alpha, T_i)}$$

$$=: \exp(\psi(F_{\alpha+1}(t), F_{\alpha+2}(t), \ldots, F_\beta(t)))$$

$$\text{FP}(t; T_\alpha, T_i) = \frac{P(t, T_i)}{P(t, T_\alpha)} = \prod_{j=\alpha+1}^{i} \text{FP}_j(t), \quad \text{FP}_j(t) = \frac{1}{1 + \tau_j F_j(t)},$$

where FP denotes the "forward discount factor". The expression in terms of an exponential of a function of the underlying forward rates, which is implicitly defined by the last equality, will be useful in the following and is written also to point out that a forward swap rate is actually a (nonlinear) function of the underlying forward LIBOR rates. Indeed, we can rewrite the above formula directly as

$$S_{\alpha,\beta}(t) = \frac{1 - \prod_{j=\alpha+1}^{\beta} \frac{1}{1 + \tau_j F_j(t)}}{\sum_{i=\alpha+1}^{\beta} \tau_i \prod_{j=\alpha+1}^{i} \frac{1}{1 + \tau_j F_j(t)}} \quad (6.32)$$

We may also derive an alternative expression for the forward swap rate by equating to zero the expression (6.31) given in terms of F's, and obtain

$$S_{\alpha,\beta}(t) = \sum_{i=\alpha+1}^{\beta} w_i(t) F_i(t)$$
$$w_i(t) = \frac{\tau_i \text{FP}(t, T_\alpha, T_i)}{\sum_{k=\alpha+1}^{\beta} \tau_k \text{FP}(t, T_\alpha, T_k)} = \frac{\tau_i P(t, T_i)}{\sum_{k=\alpha+1}^{\beta} \tau_k P(t, T_k)}. \quad (6.33)$$

This last expression for S is important because it can lead to useful approximations as follows. It looks like a weighted average, so that forward swap rates can be interpreted as weighted averages of spanning forward rates. However, notice carefully that the weights w's depend on the F's, so that we do not have properly a weighted average. Based on empirical studies showing the variability of the w's to be small compared to the variability of the F's, one can approximate the w's by their (deterministic) initial values $w(0)$ and obtain

$$S_{\alpha,\beta}(t) \approx \sum_{i=\alpha+1}^{\beta} w_i(0) F_i(t).$$

This can be helpful for example in estimating the absolute volatility of swap rates from the absolute volatility of forward rates.

6.7 Swaptions and the Lognormal Forward-Swap Model (LSM) 223

Finally, notice that the IRS discounted payoff (6.30) for a K different from the swap rate can be expressed, at time $t = 0$, also in terms of swap rates as

$$D(0, T_\alpha) \left(S_{\alpha,\beta}(T_\alpha) - K \right) \sum_{i=\alpha+1}^{\beta} \tau_i P(T_\alpha, T_i), \tag{6.34}$$

We can now move to define a swaption. A swaption is a contract that gives its holder the right (but not the obligation) to enter at a future time $T_\alpha > 0$ an IRS, whose first reset time usually coincides with T_α, with payments occurring at dates $T_{\alpha+1}, T_{\alpha+2}, \ldots, T_\beta$. The fixed rate K of the underlying swap is usually called the swaption strike. As explained in Chapter 1, for the payer swaption this right will be exercised only when the swap rate at the exercise time T_α is larger than the IRS fixed rate K (the resulting IRS has positive value). Consequently, if we assume unit notional amount, the (payer) swaption payoff can be written as

$$D(0, T_\alpha) \left(S_{\alpha,\beta}(T_\alpha) - K \right)^+ \sum_{i=\alpha+1}^{\beta} \tau_i P(T_\alpha, T_i). \tag{6.35}$$

Recall that a swaption is said to be "at-the-money" when its strike price K equals the current value $S_{\alpha,\beta}(0)$ of the underlying forward swap rate. The (payer) swaption is said to be "in-the-money" when $S_{\alpha,\beta}(0) > K$, reflecting the fact that if rates were deterministic, we would get money (a positive multiple of $S_{\alpha,\beta}(0) - K$) from the (payer) swaption contract. The (payer) swaption is said to be "out-of-the-money" when $S_{\alpha,\beta}(0) < K$. Clearly, moneyness for a receiver swaption is defined in the opposite way. A numeraire under which the above forward swap rate $S_{\alpha,\beta}$ follows a martingale, is

$$C_{\alpha,\beta}(t) = \sum_{i=\alpha+1}^{\beta} \tau_i P(t, T_i).$$

Indeed, the product $C_{\alpha,\beta}(t) S_{\alpha,\beta}(t) = P(t, T_\alpha) - P(t, T_\beta)$ gives the price of a tradable asset that, expressed in $C_{\alpha,\beta}$ units, coincides with our forward swap rate. Therefore, when choosing the swap's "present value for basis point" $C_{\alpha,\beta}$ as numeraire, the forward swap rate $S_{\alpha,\beta}(t)$ evolves according to a martingale under the measure $Q^{\alpha,\beta}$ associated with the numeraire $C_{\alpha,\beta}$. The measure $Q^{\alpha,\beta}$ is called *forward-swap measure* or simply swap measure.

By assuming a lognormal dynamics, we obtain

$$dS_{\alpha,\beta}(t) = \sigma^{(\alpha,\beta)}(t) S_{\alpha,\beta}(t) \, dW_t^{\alpha,\beta}, \tag{6.36}$$

where the instantaneous percentage volatility $\sigma^{(\alpha,\beta)}(t)$ is deterministic and $W^{\alpha,\beta}$ is a standard Brownian motion under $Q^{\alpha,\beta}$.

We denote by $v_{\alpha,\beta}^2(T)$ the average percentage variance of the forward swap rate in the interval $[0, T]$ times the interval length:

224   6. The LIBOR and Swap Market Models (LFM and LSM)

$$v_{\alpha,\beta}^2(T) = \int_0^T (\sigma^{(\alpha,\beta)}(t))^2 \, dt.$$

This model for the evolution of forward swap rates is known as **lognormal forward-swap model (LSM)**, since each swap rate $S_{\alpha,\beta}$ has a lognormal distribution under its swap measure $Q^{\alpha,\beta}$.

When pricing a swaption, this model is particularly convenient, since it yields the well-known Black formula for swaptions, as we state in the following.

Proposition 6.7.1. (Equivalence between LSM and Black's swaption prices). *The price of the above payer swaption, as implied by the LSM, coincides with that given by Black's formula for swaptions, i.e.*

$$\mathbf{PS}^{LSM}(0, T_\alpha, [T_\alpha, \ldots, T_\beta], K) = \mathbf{PS}^{Black}(0, T_\alpha, [T_\alpha, \ldots, T_\beta], K)$$
$$= C_{\alpha,\beta}(0) \, \mathrm{Bl}(K, S_{\alpha,\beta}(0), v_{\alpha,\beta}(T_\alpha)),$$

where $\mathrm{Bl}(\cdot, \cdot, \cdot)$ *was defined in Section (6.4).*

Proof. The swaption price is the risk-neutral expectation of the discounted payoff (6.35),

$$E\left(D(0, T_\alpha)\left(S_{\alpha,\beta}(T_\alpha) - K\right)^+ C_{\alpha,\beta}(T_\alpha)\right) = C_{\alpha,\beta}(0) \, E^{\alpha,\beta}\left\{(S_{\alpha,\beta}(T_\alpha) - K)^+\right\}, \tag{6.37}$$

as follows immediately from the definition of measure associated with a numeraire (formula (2.6) with $Z(T) = (S_{\alpha,\beta}(T_\alpha) - K)^+ C_{\alpha,\beta}(T_\alpha)$, $U = B$, $N = C_{\alpha,\beta}$). Now notice that, given the lognormal distribution of S, computing the last expectation in the above formula with the dynamics (6.36) leads to Black's formula for swaptions. Indeed, the above expectation is the classical Black and Scholes price for a call option whose underlying "asset" is $S_{\alpha,\beta}$, struck at K, with maturity T_α, with 0 constant "risk-free rate" and instantaneous percentage volatility $\sigma^{(\alpha,\beta)}(t)$. □

6.7.1 Swaptions Hedging

The knowledge of the self-financing strategy replicating a given (attainable) derivative is extremely useful when in need of managing the risk of short/long positions in the claim itself. For instance, when selling a stock option, the resulting exposure can be hedged by (continuously) trading in the underlying stock and in bonds, if some simplified assumptions are introduced as in Black and Scholes (1973). Another classical example concerns options on zero-coupon bonds, which can be dynamically hedged with two bonds: the underlying one and the zero-coupon bond expiring at the option maturity, see Musiela and Rutkowski (1998).

In general, it is rather difficult to come up with a suitable hedging strategy associated to a given claim. However, in case of European swaptions,

6.7 Swaptions and the Lognormal Forward-Swap Model (LSM) 225

and hence caps and floors (which can be viewed as one-period swaptions), a replicating strategy can be easily derived.

Exploiting the similarities between Black's formula for swaptions and the classical Black-Scholes option-pricing formula, Jamshidian (1996) constructed a self-financing strategy exactly replicating the swaption payoff at maturity. His derivation is outlined in the following.

Black's formula for a (payer) swaption at a time $t \geq 0$ is

$$C_{\alpha,\beta}(t)\operatorname{Bl}(K, S_{\alpha,\beta}(t), v_{\alpha,\beta}(t, T_\alpha)),$$

where we set

$$v_{\alpha,\beta}^2(t,T) := \int_t^T (\sigma^{(\alpha,\beta)}(u))^2 \, du.$$

Simple algebra shows that such a formula can be rewritten as:

$$\alpha_+(t)[P(t, T_\alpha) - P(t, T_\beta)] - K\alpha_-(t)\, C_{\alpha,\beta}(t),$$

where

$$\alpha_+(t) = \Phi\left(\frac{\ln(S_{\alpha,\beta}(t)/K) + v_{\alpha,\beta}^2(t, T_\alpha)/2}{v_{\alpha,\beta}(t, T_\alpha)}\right),$$

$$\alpha_-(t) = \Phi\left(\frac{\ln(S_{\alpha,\beta}(t)/K) - v_{\alpha,\beta}^2(t, T_\alpha)/2}{v_{\alpha,\beta}(t, T_\alpha)}\right).$$

Jamshidian then considered the portfolio that at any time t is made of a long position in $\alpha_+(t)$ bonds $P(t, T_\alpha)$, a short position in $\alpha_+(t)$ bonds $P(t, T_\beta)$ and short positions in $K\alpha_-(t)\tau_i$ bonds $P(t, T_i)$, $i = \alpha+1, \ldots, \beta$. This portfolio has a value $V(t)$ that coincides, by construction, with the Black swaption price at every time t, and at time T_α in particular. Moreover, it is self-financing in the sense of Chapter 2, since

$$dV(t) = \alpha_+(t)\, dP(t, T_\alpha) - \alpha_+(t)\, dP(t, T_\beta) - \sum_{i=\alpha+1}^{\beta} K\alpha_-(t)\tau_i\, dP(t, T_i).$$

This has been proved by Jamshidian (1996) under the assumption of a deterministic $v_{\alpha,\beta}(t, T_\alpha)$, which is trivially true in our case given the deterministic nature of the instantaneous percentage volatility $\sigma^{(\alpha,\beta)}(t)$.

The just stated result shows that we can exactly replicate a European-swaption payoff by trading a portfolio of zero-coupon bonds, which must be continuously rebalanced so as to hold the proper amounts of bonds at each time until maturity. However, a due care is needed. Indeed, the self-financing nature of the above replicating strategy deeply relies on the assumption of a deterministic swap-rate volatility, which basically implies the knowledge today of all future implied volatility structures. As a result, the replicating

226   6. The LIBOR and Swap Market Models (LFM and LSM)

portfolio is just made of bonds, a clear sign that we are only hedging the risk due to fluctuations of interest rates.

In practice, a trader will always try and hedge her exposure to volatility risk, which can even be the largest portion of the risk involved in a deal. However, there is no universal recipe for this and hedging strategies are usually constructed so as to be insensitive to local variations of the risk parameters (interest rates, volatilities, ...). With this respect, the trader's experience and sensibility is invaluable and cannot be replaced with the sheer output of any quantitative model.

6.7.2 Cash-Settled Swaptions

The swaptions that are actually traded in the Euro market are usually defined with a different payoff at maturity T_α, namely

$$(S_{\alpha,\beta}(T_\alpha) - K)^+ \sum_{i=\alpha+1}^{\beta} \tau_i \frac{1}{(1 + S_{\alpha,\beta}(T_\alpha))^{\tau_{\alpha,i}}}.$$

Such swaptions are said to be "cash settled". Here $\tau_{\alpha,i}$ denotes the year fraction between T_α and T_i. At maturity, the relevant swap rate in the payoff comes from the average of swap rates quoted by a number of major banks operating in the Euro market.

In this formulation, instead of the proper discount factors contributing to the usual numeraire

$$\sum_{i=\alpha+1}^{\beta} \tau_i P(t, T_i),$$

a flat yield curve at the swap-rate level at maturity is used to discount the swap payments at maturity. In the Euro market, this is done in order to drastically simplify the determination of the cash settlement. This settlement simplification is not used in the US market, where the classic swaption payoff is kept. The "flat-curve" payoff amounts to choosing the numeraire

$$G_{\alpha,\beta}(t) := \sum_{i=\alpha+1}^{\beta} \tau_i \frac{1}{(1 + S_{\alpha,\beta}(t))^{\tau(t,T_i)}},$$

where we denote by $\tau(t, T_i)$ the year fraction between t and T_i. The correct price of the related cash-settled swaption would be, by risk-neutral valuation:

$$E\left[D(0, T_\alpha)(S_{\alpha,\beta}(T_\alpha) - K)^+ G_{\alpha,\beta}(T_\alpha)\right]$$
$$= G_{\alpha,\beta}(0) E^{G_{\alpha,\beta}}\left[(S_{\alpha,\beta}(T_\alpha) - K)^+\right].$$

However, we cannot impose a suitable martingale-dynamics distribution of $S_{\alpha,\beta}$ under the numeraire $G_{\alpha,\beta}$, since $S_{\alpha,\beta} G_{\alpha,\beta}$ is not a tradable asset. In order to derive an analytical formula one can use instead the LSM numeraire in the expectation:

$$E\left[D(0,T_\alpha)(S_{\alpha,\beta}(T_\alpha) - K)^+ G_{\alpha,\beta}(T_\alpha)\right]$$
$$\approx G_{\alpha,\beta}(0)\, E^{\alpha,\beta}\left[(S_{\alpha,\beta}(T_\alpha) - K)^+\right]$$
$$= G_{\alpha,\beta}(0)\, \text{Bl}(K, S_{\alpha,\beta}(0), v_{\alpha,\beta}(T_\alpha)).$$

The expectation can now be computed as the classical Black-like formula, as we have recalled after (6.37) above.

This last formula could be used to connect swaptions volatilities to swaptions prices. In such a case, when reading volatilities from the Euro market, we should recover the related swaptions prices through this last formula, instead of the LSM Black formula given in Proposition 6.7.1. Nonetheless, we consider classic swaption payoffs and keep the original LSM as swaptions pricing model. We are therefore assuming that prices associated with the two different numeraires are close enough, as seems to happen in most situations. We will therefore ignore from now on the "cash-settled" feature of swaptions in the Euro market and treat them as swaptions in the US market, as suggested by experienced practitioners working in the market (Castagna (2001)).

6.8 Incompatibility between the LFM and the LSM

Recall the LSM dynamics introduced in Section 6.7. When pricing a swaption, this model is particularly convenient, since it yields the well-known Black formula for swaptions. However, for different products involving swap rates there are no analytical formulae in general. Since we selected forward LIBOR rates, rather than forward swap rates, for a basic description of the yield curve, we will rarely consider the dynamics of forward swap rates. The only computation we shall consider here concerns the dynamics of the forward swap rates under the numeraire $P(\cdot, T_\alpha)$, which is a possible numeraire for the forward-rate dynamics given earlier. We can thus express forward-rates and forward-swap-rates dynamics under the unique measure $Q^{P(\cdot,T_\alpha)}$. This unique-measure setup will be useful in the following.

By applying the change-of-numeraire toolkit (formula (2.15)), we obtain the percentage drift $m^\alpha(t)$ for $S_{\alpha,\beta}(t)$ under $Q^{P(\cdot,T_\alpha)}$ as follows:

$$m^\alpha dt = 0\, dt\ - d\ln(S_{\alpha,\beta}(t))\, d\ln\left(\frac{C_{\alpha,\beta}(t)}{P(t,T_\alpha)}\right).$$

The covariation term can be computed as follows. Notice that

$$\ln\left(\frac{C_{\alpha,\beta}(t)}{P(t,T_\alpha)}\right) = \ln\left(\sum_{i=\alpha+1}^{\beta} \tau_i \text{FP}(t;T_\alpha,T_i)\right) =: \chi(F_{\alpha+1}(t), F_{\alpha+2}(t), \ldots, F_\beta(t)),$$

228   6. The LIBOR and Swap Market Models (LFM and LSM)

and recall the function ψ defined earlier by

$$S_{\alpha,\beta}(t) = \exp\left(\psi(F_{\alpha+1}(t), F_{\alpha+2}(t), \ldots, F_\beta(t))\right).$$

It follows that

$$m^\alpha dt = \sum_{i,j=\alpha+1}^{\beta} \frac{\partial \psi}{\partial F_i} \frac{\partial \chi}{\partial F_j} dF_i(t) dF_j(t).$$

After straightforward but lengthy computations we obtain the following.

Proposition 6.8.1. (Forward-swap-rate dynamics under the numeraire $P(\cdot, T_\alpha)$). *The dynamics of the forward swap rate $S_{\alpha,\beta}$ under the numeraire $P(\cdot, T_\alpha)$ is given by*

$$dS_{\alpha,\beta}(t) = m^\alpha(t) S_{\alpha,\beta}(t) dt + \sigma^S(t) S_{\alpha,\beta}(t) dW_t,$$

$$m^\alpha(t) = \frac{\sum_{h,k=\alpha+1}^{\beta} \mu_{h,k}(t) \tau_h \tau_k FP_h(t) FP_k(t) \rho_{h,k} \sigma_h(t) \sigma_k(t) F_h(t) F_k(t)}{1 - FP(t; T_\alpha, T_\beta)}$$

$$\mu_{h,k}(t) = \frac{\left[FP(t;T_\alpha,T_\beta) \sum_{i=\alpha+1}^{h-1} \tau_i FP(t;T_\alpha,T_i) + \sum_{i=h}^{\beta} \tau_i FP(t;T_\alpha,T_i)\right]}{\left(\sum_{i=\alpha+1}^{\beta} \tau_i FP(t;T_\alpha,T_i)\right)^2}$$

$$\cdot \sum_{i=k}^{\beta} \tau_i FP(t;T_\alpha,T_i)$$

(6.38)

where W is a $Q^{P(\cdot,T_\alpha)}$ standard Brownian motion and where we set $FP_k(t) = FP(t;T_\alpha,T_k)$ for all k for brevity.

Symmetrically, it is possible to work out the dynamics of forward LIBOR rates F under the LSM numeraire $C_{\alpha,\beta}$. Again by applying the change-of-numeraire technique, we have the following.

Proposition 6.8.2. (Forward-rate dynamics under $Q^{\alpha,\beta}$). *The forward-rate dynamics under the forward-swap measure $Q^{\alpha,\beta}$ is given by*

$$dF_k(t) = \sigma_k(t) F_k(t) \left(\mu_k^{\alpha,\beta}(t) dt + dZ_k(t)\right),$$ (6.39)

$$\mu_k^{\alpha,\beta}(t) = \sum_{j=\alpha+1}^{\beta} (2\,1_{(j \le k)} - 1) \tau_j \frac{P(t,T_j)}{C_{\alpha,\beta}(t)} \sum_{i=\min(k+1,j+1)}^{\max(k,j)} \frac{\tau_i \rho_{k,i} \sigma_i(t) F_i(t)}{1 + \tau_i F_i(t)},$$

where the Z's are now Brownian motions under $Q^{\alpha,\beta}$.

Notice that (6.39) is a closed set of SDEs when k ranges from $\alpha+1$ to β, since the terms

$$\frac{P(t,T_j)}{C_{\alpha,\beta}(t)}$$

6.8 Incompatibility between the LFM and the LSM

can be easily expressed as suitable functions of the spanning forward rates

$$F_{\alpha+1}(t), \ldots, F_\beta(t).$$

We are now able to appreciate the theoretical incompatibility of the two market models, the LFM and the LSM. Lest this appreciation be lost because of an excess of details and formulas, we now recap the situation. When computing the swaption price as

$$E\left(D(0,T_\alpha)\left(S_{\alpha,\beta}(T_\alpha) - K\right)^+ C_{\alpha,\beta}(T_\alpha)\right) = C_{\alpha,\beta}(0) E^{\alpha,\beta}[(S_{\alpha,\beta}(T_\alpha) - K)^+]$$

we have two possibilities. We can either do this with the LFM where the swap rate (6.32) is expressed in terms of the LFM dynamics (6.39), or with the LSM (6.36). Once we have selected the swap measure as our basic measure for pricing, corresponding to choosing $C_{\alpha,\beta}$ as numeraire, we can compare the situations induced by the two different models. Indeed, for a distributional comparison to make sense, we need to work with both models under the same measure, the swap measure in this case.

Pricing the swaption through the above expectation with the LSM means we are basing our computations directly on swap-rate dynamics such as (6.36). In this case the swap-rate distribution is exactly lognormal and the expectation reduces to the well-known Black formula.

Instead, pricing the swaption through the above expectation with the LFM means we are basing our computations on the dynamics of forward LIBOR rates (6.39). These forward LIBOR rates define a swap rate through the algebraic relationship (6.32), and there is no reason for the distribution of the swap rate thus obtained to be lognormal. This therefore explains the incompatibility issue from a theoretical point of view.

However, one may wonder whether this incompatibility holds also from a practical point of view. The answer seems to be negative in general. Indeed, Brace, Dun and Barton (1998) argue that the distribution of swap rates, as implied by the LFM, is not far from being lognormal *in practice*.

We will test ourselves lognormality of the swap rates $S_{\alpha,\beta}(T_\alpha)$, computed as functions of the forward rates, in several situations in Section 8.2, where we will take the forward measure Q^α (instead of the swap measure $Q^{\alpha,\beta}$) as reference measure. In Section 8.2, we will also present several plots of the LFM Q^α-density of the swap rate versus a related lognormal density, and we will show that the two agree well most of times.

We can finally resume the difference between the LFM and the LSM in the following remark.

Remark 6.8.1. (**Distributional Incompatibility of LFM and LSM**). While the swap rate coming from the LSM dynamics (6.36) is lognormally distributed, the swap rate coming from the LFM dynamics (6.32, 6.39) is not lognormal. This results in the two models being theoretically incompatible. However, this incompatibility is mostly theoretical, since in practice we have that the LFM distribution for $S_{\alpha,\beta}$ is almost lognormal.

6.9 The Structure of Instantaneous Correlations

> "How did you get these correlations? Did you roll the dice?"
> [pulls the lever of an imaginary slot machine]
>
> *Head of the Interest-Rate-Derivatives Desk of Banca IMI*

In the previous sections we discussed several possible volatility structures for the LFM and considered the possible consequences of each formulation as far as some fundamental market structures are concerned. We have also seen that both the instantaneous-volatility formulation and the chosen instantaneous correlation can contribute to terminal correlations, which are not determined by instantaneous correlations alone. Now we focus on instantaneous correlations and present a possible parametric formulation for them.

In general, the full instantaneous-correlation matrix features $M(M-1)/2$ parameters, which can be a lot, where $M = \beta - \alpha$ is the number of forward rates. Therefore, a parsimonious parametric form has to be found for ρ, based on a reduced number of parameters.

We start by reducing the number of noise factors. Suppose we are given a full rank $M \times M$ instantaneous-correlation matrix ρ. We know that, being ρ a positive-definite symmetric matrix, it can be written as

$$\rho = PHP',$$

where P is a real orthogonal matrix, $P'P = PP' = I_M$, and H is a diagonal matrix whose entries are the (positive) eigenvalues of ρ. Let Λ be the diagonal matrix whose entries are the square roots of the corresponding entries of H, so that if we set $A := P\Lambda$ we have

$$AA' = \rho, \quad A'A = H.$$

We can try and mimic the decomposition $\rho = AA'$ by means of a suitable n-rank $M \times n$ matrix B such that BB' is an n-rank correlation matrix, with $n < M$ (typically $n << M$).

The advantage of doing so is that we may take as new noise an n-dimensional Brownian motion W and replace the original random shocks $dZ(t)$ with $BdW(t)$. In other terms, we move from a noise correlation structure

$$dZ\,dZ' = \rho dt$$

to

$$BdW(BdW)' = BdW\,dW'B' = BB'dt.$$

Therefore, with a noise given by BdW, our new instantaneous noise-correlation matrix is BB' whose rank is $n << M$, and the dimension of our random shocks has decreased to n. We set

$$\rho^B = BB'.$$

6.9 The Structure of Instantaneous Correlations

The problem remains of choosing a suitable parametric form for the matrix B, such that BB' is a possible correlation matrix. Rebonato (1999a, 1999d) suggests the following general form for the i-th row of B:

$$b_{i,1} = \cos\theta_{i,1} \tag{6.40}$$
$$b_{i,k} = \cos\theta_{i,k}\sin\theta_{i,1}\cdots\sin\theta_{i,k-1}, \quad 1 < k < n,$$
$$b_{i,n} = \sin\theta_{i,1}\cdots\sin\theta_{i,n-1},$$

for $i = 1, 2, \ldots, M$. Notice that with this parameterization ρ^B is clearly positive semidefinite and its diagonal terms are ones. It follows that ρ^B is a possible correlation matrix. The number of parameters in this case is $M \times (n-1)$.

The first attempt is a simple two-factor structure, $n = 2$, consisting of M parameters. This is obtained as

$$b_{i,1} = \cos\theta_{i,1}, \quad b_{i,2} = \sin\theta_{i,1}. \tag{6.41}$$

Dropping the second subscript for θ, we have

$$\rho^B_{i,j} = b_{i,1}b_{j,1} + b_{i,2}b_{j,2} = \cos(\theta_i - \theta_j). \tag{6.42}$$

This structure consists of M parameters $\theta_1, \ldots, \theta_M$. If M is large (typically twenty), we can still have troubles. We can then select a subparameterization for the θ's of the type

$$\theta_k = \vartheta(k),$$

where $\vartheta(\cdot)$ is a function depending on a small (say four or five) number of parameters. Such a function could be, for example, a linear-exponential combination similar to the one proposed for the volatility Formulation 6.

However, in our experience, keeping all θ's free (no subparameterization) can be necessary when in need of calibrating to a large number of swaptions. A justification of this statement can be found in Section 8.3.1, where we will consider a case of nineteen forward rates and show two examples involving both positive and negative correlations. See Figure 8.1 in Section 8.3.1 for a typical case involving positive correlations. In that section, we will also see that, with a rank-two structure, it is difficult to obtain a matrix where decorrelation occurs quickly. This is evident from the example with positive correlations, but is even clearer in the second example, see Figure 8.6 of Section 8.3.1, where we allow for negative correlations. In fact, take for instance the first column of the matrix in Figure 8.6. It starts with 1, but it is difficult that the next entries immediately jump to values close to 0 and stay there for the rest of the column. Indeed, entries remain close to one for a while and then jump quickly to negative values, to become stable again around the final negative entry. If we plot the first column against its indices, we obtain a typical sigmoid-like shape, and an analogous remark holds for all other columns. This is shown in Figure 6.4, where the first and last columns,

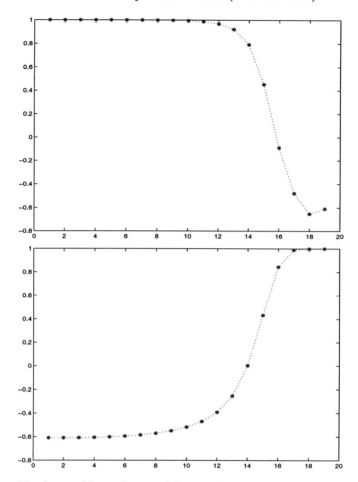

Fig. 6.4. The first and last columns of the correlation matrix given in the table of Figure 8.6.

respectively, are plotted against their indices. Rebonato (1998) explains that this is due to the low rank (two) of the correlation matrix and that there is no way to circumvent this when retaining only two factors. Only when the number of factors approaches the number of forward rates (19 in our case) the sigmoid-like shape can be changed in a shape closer to a straight line. Rebonato (1998) devotes a whole chapter to the statistical approach to yield curve models, where matters such as the sigmoid shape are discussed at length.

A three factor version ($n = 3$) of the correlation structure can be attempted to improve the decorrelation, but the improvement is not significant in general. We remark, anyway, that what finally matters is the terminal correlation of forward rates, so that, even for sigmoid-like instantaneous correlations, we can play with instantaneous volatilities to try and attain more desirable terminal-correlation structures.

A three-factor structure is given by

$$b_{i,1} = \cos\theta_{i,1}, \quad b_{i,2} = \cos\theta_{i,2}\sin\theta_{i,1}, \quad b_{i,3} = \sin\theta_{i,1}\sin\theta_{i,2}, \quad (6.43)$$

so that it follows easily

$$\begin{aligned}\rho^B_{i,j} &= b_{i,1}b_{j,1} + b_{i,2}b_{j,2} + b_{i,3}b_{j,3} \\ &= \cos\theta_{i,1}\cos\theta_{j,1} + \sin\theta_{i,1}\sin\theta_{j,1}\cos(\theta_{i,2} - \theta_{j,2}).\end{aligned}$$

Clearly, if $\theta_{k,2}$ is almost constant in k we are back to the two-factor case.

Other parametric forms are possible. For example, one can retain all M noise factors but impose the following structure on the θ's and therefore on ρ^B:

$$\theta_{i,i} = \bar{\theta}, \quad \theta_{M-i+1,k} = \pi/2 - a_k i, \quad k \neq i, \; k = 1,\ldots,M-1, \quad (6.44)$$

for $i = 1,\ldots,M$. Here we have M parameters $a_1,\ldots,a_{M-1},\bar{\theta}$. The number of parameters is M but can be further reduced by choosing a parametric form in k for a_k. This parameterization allows for a large variety of matrices. For instance, for $M = 19$ factors, the case $a_1 = \ldots = a_{18}, \bar{\theta} = 0$ yields the identity matrix (no instantaneous correlation).

Whichever volatility and correlation parameterizations are chosen, the following step consists of calibrating the model to swaption prices. A set of European swaptions, whose prices are quoted in the market, is thus considered. These swaptions are priced with one of the methods we will see later on, and the model prices will depend on the correlation parameters appearing in ρ^B. Such functions of ρ^B, and possibly of some remaining instantaneous volatility parameters, are then forced to match as much as possible the corresponding market swaptions prices, so that the parameters values implied by the market, $\rho^B = \rho^B_{\text{MKT}}$, are found. In the two-factor case for example, one obtains the values of $\theta_1,\ldots,\theta_M$ (and of the volatility parameters not determined by the calibration to caps) that are implied by the market.

6.10 Monte Carlo Pricing of Swaptions with the LFM

Consider again the swaption price

$$E\left(D(0,T_\alpha)\,(S_{\alpha,\beta} - K)^+ \sum_{i=\alpha+1}^{\beta} \tau_i P(T_\alpha, T_i)\right)$$

$$= P(0,T_\alpha)E^\alpha\left[(S_{\alpha,\beta} - K)^+ \sum_{i=\alpha+1}^{\beta} \tau_i P(T_\alpha, T_i)\right],$$

where, this time, we take the LFM numeraire $P(\cdot, T_\alpha)$ rather than the LSM numeraire $C_{\alpha,\beta}$. The above expectation is "closed" within the LFM because, based on (6.32), the swap rate can be expressed in terms of spanning forward rates at time T_α.

Now, while keeping in mind (6.32), notice carefully that this last expectation depends on the *joint* distribution of spanning forward rates

$$F_{\alpha+1}(T_\alpha), F_{\alpha+2}(T_\alpha), \ldots, F_\beta(T_\alpha).$$

Therefore, as already mentioned, correlations between different forward rates *do matter* in pricing swaptions.

We now consider the Monte Carlo pricing of swaptions. Recalling the dynamics of forward rates under Q^α:

$$dF_k(t) = \sigma_k(t) F_k(t) \sum_{j=\alpha+1}^{k} \frac{\rho_{k,j} \tau_j \sigma_j(t) F_j(t)}{1 + \tau_j F_j(t)} dt + \sigma_k(t) F_k(t) dZ_k(t), \quad (6.45)$$

for $k = \alpha+1, \ldots, \beta$, we need to generate, according to such dynamics, m realizations of

$$F_{\alpha+1}(T_\alpha), F_{\alpha+2}(T_\alpha), \ldots, F_\beta(T_\alpha).$$

Subsequently, we can evaluate the payoff

$$(S_{\alpha,\beta}(T_\alpha) - K)^+ \sum_{i=\alpha+1}^{\beta} \tau_i P(T_\alpha, T_i)$$

in each realization, and average. This leads to the Monte Carlo price of the swaption.

We now explain how to generate m realizations of the $M = \beta - \alpha$ forward rates

$$F_{\alpha+1}(T_\alpha), F_{\alpha+2}(T_\alpha), \ldots, F_\beta(T_\alpha),$$

which are consistent with the dynamics (6.45). Since the dynamics (6.45) does not lead to a distributionally known process, we need to discretize it with a sufficiently (but not too) small time step Δt, in order to reduce the random inputs to the distributionally known (independent Gaussian) shocks $Z_{t+\Delta t} - Z_t$.

In doing so, "taking logs" can be helpful. By Ito's formula,

$$d\ln F_k(t) = \sigma_k(t) \sum_{j=\alpha+1}^{k} \frac{\rho_{k,j} \tau_j \sigma_j(t) F_j(t)}{1 + \tau_j F_j(t)} dt - \frac{\sigma_k(t)^2}{2} dt + \sigma_k(t) dZ_k(t). \quad (6.46)$$

This last equation has the advantage that the diffusion coefficient is deterministic. As a consequence, the naive Euler scheme coincides with the more sophisticated Milstein scheme, see for example Klöden and Platen (1995) or Appendix A, so that the discretization

6.10 Monte Carlo Pricing of Swaptions with the LFM

$$\ln F_k^{\Delta t}(t + \Delta t) = \ln F_k^{\Delta t}(t) + \sigma_k(t) \sum_{j=\alpha+1}^{k} \frac{\rho_{k,j}\,\tau_j\,\sigma_j(t)\,F_j^{\Delta t}(t)}{1 + \tau_j F_j^{\Delta t}(t)} \Delta t$$
$$- \frac{\sigma_k(t)^2}{2}\Delta t + \sigma_k(t)(Z_k(t + \Delta t) - Z_k(t)), \quad (6.47)$$

leads to an approximation of the true process such that there exists a δ_0 with

$$E^\alpha\{|\ln F_k^{\Delta t}(T_\alpha) - \ln F_k(T_\alpha)|\} \leq c(T_\alpha)\Delta t \quad \text{for all} \quad \Delta t \leq \delta_0$$

where $c(T_\alpha)$ is a positive constant (strong convergence of order 1, from the exponent of Δt on the right-hand side). Recall that $Z_{t+\Delta t} - Z_t \sim \sqrt{\Delta t}\,\mathcal{N}(0,\rho)$, and that $Z_{t_3} - Z_{t_2}$ is independent of $Z_{t_2} - Z_{t_1}$ for all $t_1 < t_2 < t_3$, so that the joint shocks can be taken as independent draws from a multivariate normal distribution $\mathcal{N}(0,\rho)$.

Remark 6.10.1. (A refined variance for simulating the shocks). Notice that in integrating equation (6.46) between t and $t + \Delta t$, the resulting Brownian-motion part, in vector notation, is

$$\Delta \zeta_t := \int_t^{t+\Delta t} \underline{\sigma}(s)\,dZ(s) \sim \mathcal{N}(0, \text{COV}_t)$$

(here the product of vectors acts component by component), where the matrix COV_t is given by

$$(\text{COV}_t)_{h,k} = \int_t^{t+\Delta t} \rho_{h,k} \sigma_h(s) \sigma_k(s)\,ds.$$

Therefore, in principle we have no need to approximate this term by

$$\underline{\sigma}(t)(Z(t + \Delta t) - Z(t)) \sim \mathcal{N}(0, \Delta t\,\underline{\sigma}(t)\,\rho\,\underline{\sigma}(t)')$$

as is done in the classical general scheme (6.47). Indeed, we may consider a more refined scheme coming from (6.47) where the following substitution occurs:

$$\underline{\sigma}(t)(Z(t + \Delta t) - Z(t)) \longrightarrow \Delta \zeta_t.$$

The new shocks vector $\Delta \zeta_t$ can be simulated easily through its Gaussian distribution given above. This is the technique we have employed ourselves when implementing numerically the LFM.

Remark 6.10.2. (The Glasserman and Zhao (2000) no-arbitrage discretization scheme). Discretizing the continuous-time exact dynamics does not lead to discrete-time interest-rate processes that are compatible with discrete-time no-arbitrage. In other terms, the discretized process leads to bond prices that are not martingales when expressed with respect to the relevant numeraires. Alternatively, one can introduce a discretization scheme that maintains the martingale property required by no-arbitrage in discrete time. This matter is addressed in Glasserman and Zhao (2000). We do not address it here, since we assume that the violation of the no-arbitrage condition due to the time-discretization of no-arbitrage continuous-time processes is negligible enough when choosing sufficiently small discretization steps (as is known to happen in most practical situations).

6.11 Rank-One Analytical Swaption Prices

In this section we present approximations that lead to analytical formulae for swaption prices. This can be helpful for calibration and hedging purposes. The presentation is based on Brace (1996). Consider the same swaption as in the previous section, whose underlying is $S_{\alpha,\beta}$. In order to obtain an analytical formula for swaption prices, some simplifications are needed. Indeed, the swaption price can be also written

$$E\left[D(0,T_\alpha)\left(\sum_{i=\alpha+1}^{\beta} P(T_\alpha,T_i)\tau_i(F_i(T_\alpha)-K)\right)^+\right]$$

$$= \sum_{i=\alpha+1}^{\beta} \tau_i P(0,T_i) E^i\left[(F_i(T_\alpha)-K)1_A\right],$$

where

$$A := \left\{\left(\sum_{i=\alpha+1}^{\beta} P(T_\alpha,T_i)\tau_i(F_i(T_\alpha)-K)\right) > 0\right\} = \{S_{\alpha,\beta}(T_\alpha) > K\},$$

and where 1_A denotes the indicator function of the set A. The problem consists of finding approximations leading to analytical formulas for

$$E^i\left[(F_i(T_\alpha)-K)1_A\right].$$

We can approach the problem as follows. Consider the vector-dynamics for forward rates $F_{\alpha+1},\ldots,F_\beta$. Denote by

$$Z_t = [Z_{\alpha+1}(t),\ldots,Z_\beta(t)]'$$

the vector-collected Brownian motions of the forward rates, which are still "alive" at time $t < T_\alpha$, and by F the corresponding vector of forward rates

$$F(t) = [F_{\alpha+1}(t),\ldots,F_\beta(t)]'.$$

We know that the dynamics under Q^γ, with $\gamma \geq \alpha$, is

6.11 Rank-One Analytical Swaption Prices

$$dF_k(t) = -\sigma_k(t)F_k(t) \sum_{j=k+1}^{\gamma} \frac{\rho_{k,j}\tau_j\sigma_j(t)F_j(t)}{1+\tau_j F_j(t)} dt + \sigma_k(t)F_k(t)dZ_k(t),$$

for $k = \alpha+1, \ldots, \gamma-1$,

$$dF_k(t) = \sigma_k(t)F_k(t)dZ_k(t), \quad \text{for } k = \gamma,$$

$$dF_k(t) = \sigma_k(t)F_k(t) \sum_{j=\gamma+1}^{k} \frac{\rho_{k,j}\tau_j\sigma_j(t)F_j(t)}{1+\tau_j F_j(t)} dt + \sigma_k(t)F_k(t)dZ_k(t),$$

for $k = \gamma+1, \ldots, \beta$.

A first approximation consists in replacing the drift with deterministic coefficients:

$$-\sum_{j=k+1}^{\gamma} \frac{\rho_{k,j}\tau_j\sigma_j(t)F_j(t)}{1+\tau_j F_j(t)} \approx -\sum_{j=k+1}^{\gamma} \frac{\rho_{k,j}\tau_j\sigma_j(t)F_j(0)}{1+\tau_j F_j(0)} =: \mu_{\gamma,k}(t), \ k < \gamma,$$

$$0 =: \mu_{\gamma,\gamma}(t), \ k = \gamma,$$

$$\sum_{j=\gamma+1}^{k} \frac{\rho_{k,j}\,\tau_j\sigma_j(t)F_j(t)}{1+\tau_j F_j(t)} \approx \sum_{j=\gamma+1}^{k} \frac{\rho_{k,j}\tau_j\sigma_j(t)\,F_j(0)}{1+\tau_j F_j(0)} =: \mu_{\gamma,k}(t), \ k > \gamma.$$

The forward-rate dynamics is now

$$dF_k(t) = \sigma_k(t)\mu_{\gamma,k}(t)F_k(t)dt + \sigma_k(t)F_k(t)dZ_k(t)$$

under Q^γ. This equation describes a geometric Brownian motion and can be easily integrated. It turns out, by doing so, that

$$F(T_\alpha) = F(0)\exp\left(\int_0^{T_\alpha} \sigma(t)\mu_{\gamma,\cdot}(t)dt - \tfrac{1}{2}\int_0^{T_\alpha} \sigma(t)^2 dt\right)\exp(X^\gamma) \quad (6.48)$$

where operators are supposed to act componentwise and

$$X^\gamma \sim N(0,V), \quad V_{i,j} = \int_0^{T_\alpha} \sigma_i(t)\sigma_j(t)\rho_{i,j}dt.$$

It is easy to check that

$$X^\gamma = X^\alpha - \int_0^{T_\alpha} \sigma(t)[\mu_{\gamma,\cdot}(t) - \mu_{\alpha,\cdot}(t)]dt.$$

Now a second approximation can be introduced. We approximate the (generally $(\beta - \alpha)$-rank) matrix V by the rank-one matrix obtained through V's dominant eigenvalue. To do so, *we need to assume $\rho > 0$*. Indeed, in such a case, all components of V are strictly positive. It follows that V is an irreducible matrix, and as such (by the Perron-Frobenius theorem) it admits a unique dominant eigenvalue whose associate dominant eigenvector has positive components. Define

238   6. The LIBOR and Swap Market Models (LFM and LSM)

$$\Gamma := \sqrt{\lambda_1(V)}\, e_1(V),$$

where $\lambda_1(V)$ denotes the largest eigenvalue of V and $e_1(V)$ the corresponding eigenvector. Positivity of Γ, ensured by positivity of ρ, will be needed in the following.

More generally, denote by $\lambda_k(V)$ the k-th largest eigenvalue and by $e_k(V)$ the corresponding eigenvector.

We approximate V by the rank-one matrix

$$V^1 := \Gamma\Gamma',$$

and replace V with V^1 in the previous dynamics. Therefore, we still consider (6.48) but now we take

$$X^\gamma \sim N(0, V^1),$$

which, distributionally, amounts to setting

$$X_i^\gamma = \Gamma_i U^\gamma,$$

with U^γ a scalar standard Gaussian random variable under Q^γ. The advantage is that now the whole vector $F(t)$ has all its randomness condensed in the scalar random variable U^γ.

By backward substitution we see that

$$\int_0^{T_\alpha} \sigma_k(t)\mu_{\gamma,k}(t)dt = -\Gamma_k \sum_{j=k+1}^{\gamma} \frac{\tau_j \Gamma_j F_j(0)}{1+\tau_j F_j(0)} =: \Gamma_k\, q_{\gamma,k},\ k<\gamma,$$

$$\int_0^{T_\alpha} \sigma_\gamma(t)\mu_{\gamma,\gamma}(t)dt = -\Gamma_\gamma 0 =: \Gamma_\gamma\, q_{\gamma,\gamma},\ k=\gamma, \qquad (6.49)$$

$$\int_0^{T_\alpha} \sigma_k(t)\mu_{\gamma,k}(t)dt = \Gamma_k \sum_{j=\gamma+1}^{k} \frac{\tau_j \Gamma_j F_j(0)}{1+\tau_j F_j(0)} =: \Gamma_k\, q_{\gamma,k},\ k>\gamma.$$

The corresponding forward-rate vector is then

$$F(T_\alpha) = F(0)\, \exp(\Gamma q_{\gamma,\cdot} - \tfrac{1}{2}\Gamma^2)\exp(X^\gamma). \qquad (6.50)$$

Notice that

$$X^\gamma = X^g - \Gamma(q_{\gamma,\cdot} - q_{g,\cdot}),\ \gamma \geq g,\ g = \alpha,\ldots,\beta,$$

which can be rewritten in terms of U's as

$$X^\gamma = \Gamma(U^g + q_{g,\cdot} - q_{\gamma,\cdot}).$$

Since in our treatment below the maturity T_α will play a central role, we specialize our notation for the case $g = \alpha$. We set

$$p = q_{\alpha,\cdot},\quad U = U^\alpha,$$

6.11 Rank-One Analytical Swaption Prices

so that
$$X^\gamma = \Gamma(U + p - q_{\gamma,\cdot}), \quad \gamma = \alpha, \ldots, \beta.$$

Notice also that the q's can be easily expressed in terms of p's as
$$q_{\gamma,i} = p_i - p_\gamma.$$

We are now able to express the event A in a convenient way. Denote by $F_j(T_\alpha; U^j)$ the forward rate obtained according to the above approximation when $\gamma = j$:

$$F_j(T_\alpha; U^j) := F_j(0) \exp(-\tfrac{1}{2}\Gamma_j^2) \exp(\Gamma_j U^j) \qquad (6.51)$$
$$= F_j(0) \exp(-\tfrac{1}{2}\Gamma_j^2) \exp(\Gamma_j(U + p_j)) =: F_j(T_\alpha; U),$$

from which we deduce that
$$U^j = U + p_j.$$

Then rewrite A as

$$\sum_{j=\alpha+1}^{\beta} \left(\prod_{k=\alpha+1}^{j} \frac{1}{1 + \tau_k F_k(T_\alpha; U)} \right) \tau_j (F_j(T_\alpha; U) - K) \qquad (6.52)$$
$$=: \sum_{j=\alpha+1}^{\beta} P(T_\alpha, T_j; U) \tau_j (F_j(T_\alpha; U) - K) > 0,$$

or, by expressing the forward rates in terms of bond prices,

$$1 - P(T_\alpha, T_\beta; U) - \sum_{j=\alpha+1}^{\beta} \tau_j K P(T_\alpha, T_j; U) > 0.$$

It is possible to show that the partial derivative with respect to U of the left hand side of this inequality is always positive, *provided that* $\Gamma > 0$ (and this is why we need irreducibility of V, guaranteed by $\rho > 0$). Indeed, it is enough to show that (the logarithm of) each term in the above summation has positive partial derivative with respect to U. This amounts to showing that
$$\frac{\partial}{\partial U} \ln P(T_\alpha, T_j; U) < 0, \quad j = \alpha+1, \ldots, \beta.$$

This can be rewritten as

$$\frac{\partial}{\partial U} \left[\sum_{k=\alpha+1}^{j} \ln(1 + \tau_k F_k(T_\alpha; U)) \right] > 0,$$

and, again, it is enough to prove that each term in the summation has partial derivative with respect to U with the right sign. In doing so we can take away logarithms. Eventually, a sufficient condition for the partial derivative of the left-hand side of (6.52) to be positive is

$$\frac{\partial}{\partial U} F_k(T_\alpha; U) > 0, \quad k = \alpha+1, \ldots, \beta.$$

A quick investigation shows this to be the case, provided that $\Gamma > 0$. As a consequence, the left-hand side of (6.52) is an increasing function of U, and therefore the equation

$$\sum_{j=\alpha+1}^{\beta} \left(\prod_{k=\alpha+1}^{j} \frac{1}{1 + \tau_k F_k(T_\alpha; U_*)} \right) \tau_j (F_j(T_\alpha; U_*) - K) = 0$$

has a unique solution U_*. Moreover, from monotonicity, inequality (6.52) is equivalent to

$$U > U_*,$$

or

$$U^j > U_* + p_j.$$

We can finally compute the above expectation with the following approximation

$$E^i \left[(F_i(T_\alpha) - K) 1_A \right] = E^i \left[(F_i(T_\alpha) - K) 1_{(U^i > U_* + p_i)} \right],$$

where $U^i \sim \mathcal{N}(0,1)$. More explicitly,

$$E^i \left[(F_i(T_\alpha) - K) 1_A \right]$$
$$= E^i \left[(F_i(T_\alpha) - K) 1_{(U^i > U_* + p_i)} \right]$$
$$= E^i \left[(F_i(0) \exp(-\tfrac{1}{2}\Gamma_i^2 + \Gamma_i U^i) - K) 1_{(U^i > U_* + p_i)} \right]$$
$$= E^i \left[(F_i(0) \exp(-\tfrac{1}{2}\Gamma_i^2 + \Gamma_i U^i) - K) 1_{(F_i(0) \exp(-\tfrac{1}{2}\Gamma_i^2 + \Gamma_i U^i) > F_i^*)} \right]$$
$$= E^i \left[(F_i(0) \exp(-\tfrac{1}{2}\Gamma_i^2 + \Gamma_i U^i) - F_i^* + F_i^* - K) 1_{(F_i(0) \exp(-\tfrac{1}{2}\Gamma_i^2 + \Gamma_i U^i) > F_i^*)} \right]$$
$$= E^i \left[F_i(0) \exp(-\Gamma_i^2 + \Gamma_i U^i) - F_i^* \right]^+$$
$$\quad + (F_i^* - K) E^i \left[1_{(F_i(0) \exp(-\tfrac{1}{2}\Gamma_i^2 + \Gamma_i U^i) > F_i^*)} \right]$$
$$= F_i(0) \Phi(d_1(F_i^*, F_i(0), \Gamma_i)) - K \Phi(d_2(F_i^*, F_i(0), \Gamma_i))$$

where the last expression is obtained by adding a call price to a cash-or-nothing-call price, and

$$F_i^* := F_i(0) \exp(-\tfrac{1}{2}\Gamma_i^2 + \Gamma_i(U_* + p_i)).$$

We then have the following.

6.11 Rank-One Analytical Swaption Prices 241

Proposition 6.11.1. (Brace's rank-one formula). *The approximated price of the above swaption is given by*

$$\sum_{i=\alpha+1}^{\beta} \tau_i P(0, T_i) \left[F_i(0) \Phi(\Gamma_i - U_* - p_i) - K \Phi(-U_* - p_i) \right]. \quad (6.53)$$

This formula is analytical and just involves a root-searching procedure to find U_*.

The formulation we presented here works for any possible choice of the instantaneous volatilities $(\sigma_k(t))_k$ and correlation ρ structures. However, in applying this apparatus, if one starts from a parametric form for volatilities and correlations, one might as well choose a parameterization that renders the computation of the matrix Γ as simple as possible. Another desirable feature is the approximation between the real integrated covariance matrix V and its Γ-based approximation V^1 to be as good as possible.

Now suppose to choose a say rank-two instantaneous-correlation matrix ρ^B and then apply the formula given here. The problem with this setup is that, since the integrated covariance matrix V will be approximated via a rank-one matrix, we may as well start from a rank-one instantaneous covariance matrix. This is obtained by taking a one-factor model, i.e. by setting all ρ's components equal to one. If one does this, the approximation needed to obtain a rank one matrix V^1 from V will be minimal.

But we can do even better. We can check whether there is a volatility formulation among (6.8), (6.9), (6.10) and (6.11) leading to an integrated covariance matrix V that has automatically rank one. It turns out that this is possible. Indeed, when taking a one-factor model ($\rho_{i,j} = 1$ for all i, j), the integrated covariance matrix has rank one if (and only if) the volatility structure is separable as in (6.10), (and in particular as in (6.9)), see Brace, Dun and Barton (1998) for the "only if" part. Indeed, under formulation (6.10) we have

$$V_{i,j} = \Phi_i \Phi_j \sum_{k=0}^{\alpha} \tau_{j-1,j} \Psi_{j+1}^2,$$

which gives a rank-one matrix, so that it suffices to set

$$\Gamma = \sqrt{\sum_{k=0}^{\alpha} \tau_{j-1,j} \Psi_{j+1}^2} \ [\Phi_{\alpha+1}, \ldots, \Phi_\beta]',$$

and no eigenvalue/eigenvector analysis or calculations are needed.

A simpler formulation is obtained with the volatility structure (6.9) (and $\rho_{i,j} = 1$ for all i, j as before). Indeed, in such a case, $V_{i,j} = s_i s_j \tau_{0,\alpha}$, so that we can directly take $\Gamma_i = s_i \sqrt{\tau_{0,\alpha}}$.

Finally notice that even in the one-factor case ($\rho_{i,j} = 1$ for all i, j), formulations (6.8) and more generally (6.11) do not lead to a rank-one integrated covariance matrix V in general, so that if we choose (6.8) or (6.11) we actually need to include an eigenvector/eigenvalue calculation in order to apply Brace's formula (6.53). The integrated covariance matrix with formulation (6.11) is given by

$$V_{i,j} = \Phi_i \Phi_j \sum_{k=0}^{\alpha} \tau_{j-1,j} \psi_{j-k} \psi_{i-k},$$

and Γ has to be derived through the dominant eigenvector of this matrix. This last approach has the disadvantage of requiring an explicit eigenvalues/eigenvectors calculation. However, as stated in Section 6.5, it has the advantage of giving a possibly good qualitative behaviour for the term structure of volatilities over time, and is also more controllable than the previous one.

6.12 Rank-r Analytical Swaption Prices

The rank-one analytical approximation of Brace (1996) has been extended to the rank-r case by Brace (1997) himself as follows. The derivation is the same as in Section 6.11 up to the definition of Γ. This time we select Γ to be the following matrix

$$\Gamma := \left[\sqrt{\lambda_1(V)}\, e_1(V) \quad \sqrt{\lambda_2(V)}\, e_2(V) \quad \ldots \quad \sqrt{\lambda_r(V)}\, e_r(V) \right],$$

leading to the following rank-r approximation of V,

$$V^r := \Gamma \Gamma'.$$

Again, we substitute backwards in the dynamics. Therefore, we still consider (6.48) but now we take

$$X^\gamma \sim N(0, V^r),$$

which amounts to setting

$$X_i^\gamma = \Gamma_{(i)} U^\gamma,$$

where $\Gamma_{(i)}$ denotes the i-th row of the Γ matrix, and with U^γ a r-th dimensional standard Gaussian random vector under Q^γ. Now the whole $(\beta - \alpha)$-dimensional vector $F(t)$ has all its randomness condensed in the r-dimensional random vector U^γ, with typically $r << \beta - \alpha$.

Formulas (6.49) still hold when replacing Γ_k with $\Gamma_{(k)}$ and Γ_j with $\Gamma'_{(j)}$. Now the q's are r-vectors, and so are the p_j's. Under the same adjustments, (6.51) still holds, and the "exercise set" A can still be expressed as (6.52). However, this time the situation is more delicate since U is a vector. We can deal with this more general case as follows.

Brace (1997) observed that close numerical examination of the U-surface

$$\sum_{j=\alpha+1}^{\beta}\left(\prod_{k=\alpha+1}^{j}\frac{1}{1+\tau_k F_k(T_\alpha;U)}\right)\tau_j(F_j(T_\alpha;U)-K) \tag{6.54}$$

$$=:\sum_{j=\alpha+1}^{\beta} P(T_\alpha,T_j;U)\tau_j(F_j(T_\alpha;U)-K)=0$$

inside the hypercube $[-5,5]^r$ (where almost all of the probability density of U is concentrated) reveals a slightly curved surface close to be a hyperplane, which is almost perpendicular to the U_1 axis. We therefore assume the solution of (6.54) to be a hyperplane:

$$U_1 = s_1 + \sum_{k=2}^{r} s_k U_k. \tag{6.55}$$

Under this assumption, if the partial derivative with respect to U_1 of the left-hand side $f(U)$ of (6.54) is positive, we can deduce that

$$A = \left\{[U_1, U_2, \ldots, U_r] : U_1 > s_1 + \sum_{k=2}^{r} s_k U_k\right\}. \tag{6.56}$$

Indeed, in such a case,

$$f(U_1, U_2, \ldots, U_k) > f\left(s_1 + \sum_{k=2}^{r} s_k U_k, U_2, \ldots, U_k\right) = 0,$$

which was the inequality to be solved.

The partial derivative with respect to U_1 of the left-hand side of (6.54) can be shown to be positive in exactly the same way as in the rank-one case. This is guaranteed by positivity of the first column of Γ, which, in turn, is guaranteed by positivity of ρ (and therefore of V) through the Perron-Frobenius theorem.

At this point we need to determine $s_1, s_2, \ldots, s_r$. To do so, we need to perform $2r - 1$ root searches similar to the ones of the rank-one case. We proceed as follows.

- Set $U = [\alpha_1, 0, \ldots, 0]$ and solve numerically (6.54) thus finding α_1. As a solution, $U = [\alpha_1, 0, \ldots, 0]$ satisfies (6.55) and therefore

$$s_1 = \alpha_1.$$

- Set $U = [\alpha_2^-, -1/2, 0, \ldots, 0]$ and solve numerically (6.54), thus finding α_2^-. Set $U = [\alpha_2^+, 1/2, 0, \ldots, 0]$ and solve numerically (6.54), thus finding α_2^+. Being solutions, both $[\alpha_2^-, -1/2, 0, \ldots, 0]$ and $[\alpha_2^+, 1/2, 0, \ldots, 0]$ satisfy (6.55), from which it follows easily

$$s_2 = \alpha_2^+ - \alpha_2^-.$$

- For all other k, set $U = [\alpha_k^-, 0, \ldots, 0, -1/2, 0, \ldots, 0]$, with $1/2$ as k-th component of the vector U, and solve numerically (6.54), thus finding α_k^-. Set $U = [\alpha_k^+, 0, \ldots, 0, 1/2, 0, \ldots, 0]$ and solve numerically (6.54), thus finding α_k^+. Being solutions, these two vectors U both satisfy (6.55), from which it follows easily

$$s_k = \alpha_k^+ - \alpha_k^-.$$

All the above equations admit a unique solution, since the partial derivative with respect to U_1 of the left hand side of our equation (6.54) is positive.

Once the s's have been determined, we can proceed and compute the swaption price. To do so, notice that we can express A in terms of U_i by combining (6.56) with $U^i = U + p_i$, thus obtaining

$$A = \left\{ U^i : U_1^i - \sum_{k=2}^r s_k U_k^i > s_1 + (p_i)_1 - \sum_{k=2}^r s_k (p_i)_k \right\}.$$

To ease notation, set

$$w := [1, -s_2, \ldots, -s_r], \quad s_i^* := s_1 + (p_i)_1 - \sum_{k=2}^r s_k (p_i)_k,$$

so that

$$A = \{U^i : wU^i > s_i^*\}.$$

This time we compute expectations as follows:

$$E^i \left[(F_i(T_\alpha) - K) \mathbf{1}_A \right]$$
$$= E^i \left[(F_i(T_\alpha; U^i) - K) \mathbf{1}_{(wU^i > s_i^*)} \right], \quad U^i \sim \mathcal{N}(0, I_r)$$
$$= E^i \left[(F_i(0) \exp(\Gamma_{(i)} U^i - \tfrac{1}{2} |\Gamma_{(i)}|^2) - K) \mathbf{1}_{(wU^i > s_i^*)} \right]$$
$$= F_i(0) E^i \left[\exp(\Gamma_{(i)} U^i - \tfrac{1}{2} |\Gamma_{(i)}|^2) \mathbf{1}_{(wU^i > s_i^*)} \right] - K E^i \mathbf{1}_{(wU^i > s_i^*)}$$
$$= F_i(0) E^i \left[\mathbf{1}_{(w[U^i + \Gamma_{(i)}'] > s_i^*)} \right] - K \Phi\left(-\frac{s_i^*}{|w|} \right)$$
$$= F_i(0) \Phi\left(-\frac{s_i^* - w\Gamma_{(i)}'}{|w|} \right) - K \Phi\left(-\frac{s_i^*}{|w|} \right)$$

where use has been made of the general property

$$E^i \left\{ \exp(b' U^i - \tfrac{1}{2} b'b) g(U^i) \right\} = E^i \{ g(U^i + b) \}, \quad b = [b_1, \ldots, b_r]'.$$

We then have the following.

Proposition 6.12.1. (Brace's rank-r formula). *The swaption price can be approximated as*

6.12 Rank-r Analytical Swaption Prices

$$\sum_{i=\alpha+1}^{\beta} \tau_i P(0,T_i) \left[F_i(0) \Phi\left(-\frac{s_i^* - w\Gamma'_{(i)}}{|w|} \right) - K\Phi\left(-\frac{s_i^*}{|w|} \right) \right]. \quad (6.57)$$

This formula is analytical and involves $2r - 1$ root-searching procedures to find

$$\alpha_1, \alpha_2^-, \alpha_2^+, \ldots, \alpha_r^-, \alpha_r^+,$$

needed to determine $s_1, \ldots, s_r$ and, therefore, w and s^* appearing in the formula.

Brace (1997) argued that the rank-two approximation ($r = 2$) is enough for obtaining sufficient accuracy, based on numerical examination of the U-surface.

Let us therefore take $r = 2$ and let us see which volatility and correlation parametric structures are handy when one tries to apply formula (6.57) with $r = 2$.

Exactly as in the rank-one case, the formulation we presented here works for any possible choice of the instantaneous volatilities $(\sigma_k(t))_k$ and correlation-ρ structures. As before, in applying this apparatus, one may choose a parameterization that renders the Γ matrix computation as simple as possible and the approximation of the real integrated covariance matrix V, with its Γ-based approximation V^2, as good as possible. In particular, since the integrated covariance matrix V will be approximated via a rank-two matrix, we may as well start from a rank-two instantaneous covariance matrix. This is obtained, for example, by taking a two-factor model, i.e. by taking $\rho = \rho^B$ as in (6.42),

$$\rho_{i,j}^B = b_{i,1}b_{j,1} + b_{i,2}b_{j,2} = \cos(\theta_i - \theta_j),$$

and the separable instantaneous-volatility structure (6.10). Notice carefully that the above formula holds under the assumption $\rho_{i,j} > 0$ for all i,j, as required in order to apply the Perron-Frobenius theorem. This is guaranteed by

$$|\theta_i - \theta_j| < \pi/2,$$

which in turn is guaranteed by taking all θ's in an interval whose length is less than $\pi/2$, say $(-\pi/4, \pi/4)$. With this choice, the resulting V has automatically rank 2 and we can set

$$\Gamma = \sqrt{\sum_{k=0}^{\alpha} \tau_{j-1,j} \Psi_{j+1}^2} \; \text{Diag}(\Phi_{\alpha+1}, \ldots, \Phi_\beta) B$$

where $\text{Diag}(x_a, \ldots, x_b)$ denotes a diagonal matrix whose diagonal entries are $x_a, x_{a+1}, \ldots, x_b$, and where $B = (b_{i,j})$ is the matrix defined in (6.41),

$$b_{i,1} = \cos\theta_i, \quad b_{i,2} = \sin\theta_i.$$

Finally notice again that the alternative formulations (6.8) and (6.11) do not lead to a rank-2 integrated covariance matrix V in general, so that if we choose (6.11) we actually need to include an eigenvector/eigenvalue calculation in order to apply Brace's formula (6.57). The integrated covariance matrix with formulation (6.11) is now given by

$$V_{i,j} = \Phi_i \Phi_j \cos(\theta_i - \theta_j) \sum_{k=0}^{\alpha} \tau_{j-1,j} \psi_{j-k} \psi_{i-k},$$

and $\varGamma$ has to be derived through the first two dominant eigenvectors of this matrix.

As observed before, this last approach has the disadvantage of requiring an explicit eigenvalues/eigenvectors calculation. But once again, as stated in Section 6.5, it has the advantage of giving a possibly good qualitative behaviour for the term structure of volatilities over time.

6.13 A Simpler LFM Formula for Swaptions Volatilities

There is a further approximation method to compute swaption prices with the LFM without resorting to Monte Carlo simulation. This method is rather simple and its quality has been tested, for example, by Brace, Dun, and Barton (1999). We have tested the method ourselves and will present the related numerical results in Chapter 8.

Recall from Section 6.7 the forward-swap-rate dynamics underlying the LSM (leading to Black's formula for swaptions):

$$dS_{\alpha,\beta}(t) = \sigma^{(\alpha,\beta)}(t) S_{\alpha,\beta}(t) dW_t^{\alpha,\beta}, \quad Q^{\alpha,\beta}.$$

A crucial role in the LSM is played by the (squared) Black swaption volatility (multiplied by T_α)

$$(v_{\alpha,\beta}(T_\alpha))^2 := \int_0^{T_\alpha} \sigma_{\alpha,\beta}^2(t) dt = \int_0^{T_\alpha} (d\ln S_{\alpha,\beta}(t))(d\ln S_{\alpha,\beta}(t))$$

entering Black's formula for swaptions. We plan to compute, under a number of approximations, an analogous quantity $v_{\alpha,\beta}^{\text{LFM}}$ in the LFM.

Recall from formula (6.33) that forward swap rates can be thought of as algebraic transformations of forward rates, according to

$$S_{\alpha,\beta}(t) = \sum_{i=\alpha+1}^{\beta} w_i(t) F_i(t),$$

where

6.13 A Simpler LFM Formula for Swaptions Volatilities

$$w_i(t) = w_i(F_{\alpha+1}(t), F_{\alpha+2}(t), \ldots, F_\beta(t)) = \frac{\tau_i \mathrm{FP}(t, T_\alpha, T_i)}{\sum_{k=\alpha+1}^{\beta} \tau_k \mathrm{FP}(t, T_\alpha, T_k)}$$

$$= \frac{\tau_i \prod_{j=\alpha+1}^{i} \frac{1}{1+\tau_j F_j(t)}}{\sum_{k=\alpha+1}^{\beta} \tau_k \prod_{j=\alpha+1}^{k} \frac{1}{1+\tau_j F_j(t)}}.$$

A first approximation is derived as follows. Start by freezing the w's at time 0, so as to obtain

$$S_{\alpha,\beta}(t) \approx \sum_{i=\alpha+1}^{\beta} w_i(0) F_i(t).$$

This approximation is justified by the fact that the variability of the w's is much smaller than the variability of the F's. This can be tested both historically and through simulation of the F's (and therefore of the w's) via a Monte Carlo method.

Then differentiate both sides

$$dS_{\alpha,\beta}(t) \approx \sum_{i=\alpha+1}^{\beta} w_i(0) dF_i(t) = (\ldots) dt + \sum_{i=\alpha+1}^{\beta} w_i(0) \sigma_i(t) F_i(t)\, dZ_i(t),$$

under any of the forward-adjusted measures, and compute the quadratic variation

$$dS_{\alpha,\beta}(t)\, dS_{\alpha,\beta}(t) \approx \sum_{i,j=\alpha+1}^{\beta} w_i(0) w_j(0) F_i(t) F_j(t) \rho_{i,j} \sigma_i(t) \sigma_j(t)\, dt.$$

The percentage quadratic variation is

$$\left(\frac{dS_{\alpha,\beta}(t)}{S_{\alpha,\beta}(t)}\right)\left(\frac{dS_{\alpha,\beta}(t)}{S_{\alpha,\beta}(t)}\right) = (d\ln S_{\alpha,\beta}(t))(d\ln S_{\alpha,\beta}(t))$$

$$\approx \frac{\sum_{i,j=\alpha+1}^{\beta} w_i(0) w_j(0) F_i(t) F_j(t) \rho_{i,j} \sigma_i(t) \sigma_j(t)}{S_{\alpha,\beta}(t)^2}\, dt.$$

Introduce now a further approximation by freezing all F's in the above formula (as was done earlier for the w's) to their time-zero value:

$$(d\ln S_{\alpha,\beta}(t))(d\ln S_{\alpha,\beta}(t)) \approx \sum_{i,j=\alpha+1}^{\beta} \frac{w_i(0) w_j(0) F_i(0) F_j(0) \rho_{i,j}}{S_{\alpha,\beta}(0)^2} \sigma_i(t) \sigma_j(t)\, dt.$$

Using this last formula, finally compute an approximation $(v_{\alpha,\beta}^{\mathrm{LFM}})^2$ of the integrated percentage variance of S as

$$\int_0^{T_\alpha} (d\ln S_{\alpha,\beta}(t))(d\ln S_{\alpha,\beta}(t))$$

$$\approx \sum_{i,j=\alpha+1}^{\beta} \frac{w_i(0) w_j(0) F_i(0) F_j(0) \rho_{i,j}}{S_{\alpha,\beta}(0)^2} \int_0^{T_\alpha} \sigma_i(t) \sigma_j(t)\, dt =: (v_{\alpha,\beta}^{\mathrm{LFM}})^2,$$

so that we have the following.

Proposition 6.13.1. (Rebonato's formula). *The LFM Black-like (squared) swaption volatility (multiplied by T_α) can be approximated by*

$$\boxed{(v_{\alpha,\beta}^{\text{LFM}})^2 = \sum_{i,j=\alpha+1}^{\beta} \frac{w_i(0)w_j(0)F_i(0)F_j(0)\rho_{i,j}}{S_{\alpha,\beta}(0)^2} \int_0^{T_\alpha} \sigma_i(t)\sigma_j(t)\,dt}. \quad (6.58)$$

We refer to formula (6.58) as to "Rebonato's formula" for brevity, since a sketchy version of it can be found in Rebonato's (1998) book, and since Rebonato was one of the first to explicitly point out the interpretation of swap rates as linear combinations of forward rates.

The quantity $v_{\alpha,\beta}^{\text{LFM}}$ can be used as a proxy for the Black volatility $v_{\alpha,\beta}(T_\alpha)$ of the swap rate $S_{\alpha,\beta}$. Putting this quantity in Black's formula for swaptions allows one to compute approximated swaptions prices with the LFM. Formula (6.58) is obtained under a number of assumptions, and at first one would imagine its quality to be rather poor. However, it turns out that the approximation is not at all bad, as also pointed out by Brace, Dun and Barton (1998). In Chapter 8, we will present numerical investigations of our own and confirm that the approximation is indeed satisfactory in general.

A slightly more sophisticated version of the above procedure has been proposed, for example, by Hull and White (1999). Hull and White differentiate $S_{\alpha,\beta}(t)$ without freezing the w's, thus obtaining

$$dS_{\alpha,\beta}(t) = \sum_{i=\alpha+1}^{\beta} (w_i(t)\,dF_i(t) + F_i(t)\,dw_i(t)) + (\ldots)\,dt$$

$$= \sum_{i,h=\alpha+1}^{\beta} \left(w_h(t)\delta_{\{h,i\}} + F_i(t)\frac{\partial w_i(t)}{\partial F_h}\right) dF_h(t) + (\ldots)\,dt,$$

where $\delta_{\{i,i\}} = 1$ and $\delta_{\{i,h\}} = 0$ for $h \neq i$. By straightforward calculations one obtains

$$\frac{\partial w_i(t)}{\partial F_h} = \frac{w_i(t)\tau_h}{1+\tau_h F_h(t)} \left[\frac{\sum_{k=h}^{\beta} \tau_k \prod_{j=\alpha+1}^{k} \frac{1}{1+\tau_j F_j(t)}}{\sum_{k=\alpha+1}^{\beta} \tau_k \prod_{j=\alpha+1}^{k} \frac{1}{1+\tau_j F_j(t)}} - 1_{\{i\geq h\}}\right],$$

so that by setting

$$\bar{w}_h(t) := w_h(t) + \sum_{i=\alpha+1}^{\beta} F_i(t)\frac{\partial w_i(t)}{\partial F_h},$$

we have

$$dS_{\alpha,\beta}(t) = \sum_{h=\alpha+1}^{\beta} \bar{w}_h(t)\,dF_h(t) + (\ldots)\,dt,$$

where the $\bar{w}(t)$'s are completely determined in terms of the forward rates $F(t)$'s. Now we may freeze all F's (and therefore $\bar{w}$'s) at time 0 and obtain

$$dS_{\alpha,\beta}(t) \approx \sum_{h=\alpha+1}^{\beta} \bar{w}_h(0)\, dF_h(t) + (\ldots)\, dt\ .$$

Now, in order to derive an approximated Black swaption volatility $\bar{v}_{\alpha,\beta}^{\text{LFM}}$, based on the weights $\bar{w}$ rather than on the weights w, we may reason as in the previous derivation so as to obtain the following.

Proposition 6.13.2. (Hull and White's formula). *The LFM Black-like (squared) swaption volatility (multiplied by T_α) can be better approximated by*

$$\boxed{(\bar{v}_{\alpha,\beta}^{\text{LFM}})^2 := \sum_{i,j=\alpha+1}^{\beta} \frac{\bar{w}_i(0)\bar{w}_j(0)F_i(0)F_j(0)\rho_{i,j}}{S_{\alpha,\beta}(0)^2} \int_0^{T_\alpha} \sigma_i(t)\sigma_j(t)\, dt}\ . \quad (6.59)$$

We have investigated, among other features, the difference between the outputs of formulas (6.58) and (6.59) in a number of situation, and will describe the results in Section 8.2. We anticipate that the difference between the two formulas is practically negligible in most situations.

6.14 A Formula for Terminal Correlations of Forward Rates

We have seen earlier in Section 6.6 that the terminal correlation between forward rates depends not only on the instantaneous correlations of the forward-rate dynamics, but also on the way the instantaneous volatilities are modeled from average volatilities coming from caps, and possibly swaptions, prices.

In general, if one is interested in terminal correlations of forward rates at a future time instant, as implied by the LFM, the computation has to be based on a Monte Carlo simulation technique. Indeed, assume we are interested in computing the terminal correlation between the forward rates $F_i = F(\cdot; T_{i-1}, T_i)$ and $F_j = F(\cdot; T_{j-1}, T_j)$ at time T_α, $\alpha \leq i-1 < j$, say under the measure Q^γ, $\gamma \geq \alpha$. Then we need to compute

$$\text{Corr}^\gamma(F_i(T_\alpha), F_j(T_\alpha)) \qquad (6.60)$$
$$= \frac{E^\gamma[(F_i(T_\alpha) - E^\gamma F_i(T_\alpha))(F_j(T_\alpha) - E^\gamma F_j(T_\alpha))]}{\sqrt{E^\gamma[(F_i(T_\alpha) - E^\gamma F_i(T_\alpha))^2]}\sqrt{E^\gamma[(F_j(T_\alpha) - E^\gamma F_j(T_\alpha))^2]}}\ .$$

Recall the dynamics of F_i and F_j under Q^γ:

$$dF_k(t) = -\sigma_k(t)F_k(t) \sum_{j=k+1}^{\gamma} \frac{\rho_{k,j}\tau_j\sigma_j(t)F_j(t)}{1+\tau_j F_j(t)} dt + \sigma_k(t)F_k(t)\, dZ_k(t),$$
$$k = \alpha+1,\ldots,\gamma-1,$$
$$dF_k(t) = \sigma_k(t)F_k(t)\, dZ_k(t), \quad k = \gamma,$$
$$dF_k(t) = \sigma_k(t)F_k(t) \sum_{j=\gamma+1}^{k} \frac{\rho_{k,j}\tau_j\sigma_j(t)F_j(t)}{1+\tau_j F_j(t)} dt + \sigma_k(t)F_k(t)\, dZ_k(t),$$
$$k = \gamma+1,\ldots,\beta,$$

where Z is a Brownian motion under Q^γ. The expected values appearing in the expression (6.60) for the terminal correlation can be obtained by simulating the above dynamics for $k = i$ and $k = j$ respectively, thus simulating F_i and F_j up to time T_α. The simulation can be based on a discretized Milstein dynamics analogous to (6.47), which was introduced to Monte Carlo price swaptions.

However, at times, traders may need to quickly check reliability of the model's terminal correlations, so that there could be no time to run a Monte Carlo simulation. Fortunately, there does exist an approximated formula, in the spirit of the approximated formulas for Black's swaption volatilities, which allows us to compute terminal correlations algebraically from the LFM parameters ρ and $\sigma(\cdot)$. We now derive such an approximated formula.

The first approximation we introduce is a partial freezing of the drift in the dynamics, analogous to what was done for Brace's rank-one formula:

$$-\sum_{j=k+1}^{\gamma} \frac{\rho_{k,j}\tau_j\sigma_j(t)F_j(t)}{1+\tau_j F_j(t)} \approx -\sum_{j=k+1}^{\gamma} \frac{\rho_{k,j}\tau_j\sigma_j(t)F_j(0)}{1+\tau_j F_j(0)} =: \mu_{\gamma,k}(t), \, k < \gamma,$$

$$0 =: \mu_{\gamma,\gamma}(t), \, k = \gamma,$$

$$\sum_{j=\gamma+1}^{k} \frac{\rho_{k,j}\tau_j\sigma_j(t)F_j(t)}{1+\tau_j F_j(t)} \approx \sum_{j=\gamma+1}^{k} \frac{\rho_{k,j}\tau_j\sigma_j(t)\, F_j(0)}{1+\tau_j F_j(0)} =: \mu_{\gamma,k}(t), \, k > \gamma.$$

Under this approximation, the forward-rate dynamics, under Q^γ, is

$$dF_k(t) = \bar{\mu}_{\gamma,k}(t)F_k(t)\, dt + \sigma_k(t)F_k(t)\, dZ_k(t), \quad \bar{\mu}_{\gamma,k}(t) := \sigma_k(t)\mu_{\gamma,k}(t),$$

which describes a geometric Brownian motion and can thus be easily integrated. If considered for $k = i$ and $k = j$, this equation leads to jointly normally distributed variables $\ln F_i(T_\alpha)$ and $\ln F_j(T_\alpha)$ under the measure Q^γ. This allows for an exact evaluation of the expected value in the numerator of (6.60). Indeed, we obtain easily

$$F_k(T_\alpha) = F_k(0) \exp\left[\int_0^{T_\alpha} \left(\bar{\mu}_{\gamma,k}(t) - \frac{\sigma_k^2(t)}{2}\right) dt + \int_0^{T_\alpha} \sigma_k(t)\, dZ_k(t)\right],$$

for $k \in \{i,j\}$, and

6.14 A Formula for Terminal Correlations of Forward Rates

$$F_i(T_\alpha)F_j(T_\alpha) = F_i(0)F_j(0)\exp\left[\int_0^{T_\alpha}\left(\bar{\mu}_{\gamma,i}(t)+\bar{\mu}_{\gamma,j}(t)-\frac{\sigma_i^2(t)+\sigma_j^2(t)}{2}\right)dt\right.$$
$$\left.+\int_0^{T_\alpha}\sigma_i(t)\,dZ_i(t)+\int_0^{T_\alpha}\sigma_j(t)\,dZ_j(t)\right].$$

Recalling that, by a trivial application of Ito's isometry, the two-dimensional random vector

$$\left[\int_0^{T_\alpha}\sigma_i(t)\,dZ_i(t),\ \int_0^{T_\alpha}\sigma_j(t)\,dZ_j(t)\right]'$$

is jointly normally distributed with mean $[0,0]'$ and covariance matrix

$$\begin{bmatrix}\int_0^{T_\alpha}\sigma_i^2(t)dt & \rho_{i,j}\int_0^{T_\alpha}\sigma_i(t)\sigma_j(t)dt \\ \rho_{i,j}\int_0^{T_\alpha}\sigma_i(t)\sigma_j(t)dt & \int_0^{T_\alpha}\sigma_j^2(t)dt\end{bmatrix},$$

we see that the correlation approximated according to this distribution is easily obtained. We present the related formula in the following

Proposition 6.14.1. (Analytical terminal-correlation formula). *The terminal correlation between the forward rates F_i and F_j at time T_α, $\alpha \leq i-1 < j$, under the measure Q^γ, $\gamma \geq \alpha$, can be approximated as follows:*

$$\boxed{\mathrm{Corr}^\gamma(F_i(T_\alpha),F_j(T_\alpha))\approx\frac{\exp\left(\int_0^{T_\alpha}\sigma_i(t)\sigma_j(t)\rho_{i,j}\,dt\right)-1}{\sqrt{\exp\left(\int_0^{T_\alpha}\sigma_i^2(t)\,dt\right)-1}\sqrt{\exp\left(\int_0^{T_\alpha}\sigma_j^2(t)\,dt\right)-1}}.}$$

(6.61)

Notice that a first order expansion of the exponentials appearing in formula (6.61) yields

Rebonato's terminal-correlation formula:

$$\boxed{\mathrm{Corr}^{\mathrm{REB}}(F_i(T_\alpha),F_j(T_\alpha))=\rho_{i,j}\frac{\int_0^{T_\alpha}\sigma_i(t)\sigma_j(t)\,dt}{\sqrt{\int_0^{T_\alpha}\sigma_i^2(t)\,dt}\sqrt{\int_0^{T_\alpha}\sigma_j^2(t)\,dt}}.} \quad (6.62)$$

This formula shows, through the Schwartz inequality, that

$$\mathrm{Corr}^{\mathrm{REB}}(F_i(T_\alpha),F_j(T_\alpha))\leq\rho_{i,j}\quad\text{if}\quad\rho_{i,j}\geq 0, \quad (6.63)$$
$$\mathrm{Corr}^{\mathrm{REB}}(F_i(T_\alpha),F_j(T_\alpha))\geq\rho_{i,j}\quad\text{if}\quad\rho_{i,j}<0. \quad (6.64)$$

In words, we can say that *terminal correlations are, in absolute value, always smaller than or equal to instantaneous correlations*. In agreement with this general observation, recall that through a clever repartition of integrated volatilities in instantaneous volatilities $\sigma_i(t)$ and $\sigma_j(t)$ we can make the terminal correlation $\mathrm{Corr}^{\mathrm{REB}}(F_i(T_\alpha), F_j(T_\alpha))$ arbitrarily close to zero, even when the instantaneous correlation $\rho_{i,j}$ is one (see Section 6.6).

The approximate formulas above are particularly appealing. They do not depend on the measure Q^γ under which the correlation has been computed. Moreover, they can be computed algebraically based on the volatility and correlation parameters σ and ρ of the LFM. Therefore, once the LFM parameters determining σ and ρ have been obtained through calibration to the cap and swaption markets, we can immediately check the terminal-correlation structure at any future date. There remains the problem of actually checking that the above formulas lead to acceptable approximation errors. This can be done by rigorously evaluating the correlation Corr^γ through a Monte Carlo method, as we will do in Chapter 8.

6.15 Calibration to Swaptions Prices

Errors using inadequate data are much less than those using no data at all.
Charles Babbage (1791-1871)

We have seen several ways to compute swaption prices with the LFM. However, as already mentioned, computing market (plain-vanilla) swaption prices is not the purpose of an interest-rate model. In fact, it is common practice in the market to compute such prices through a Black-like formula. At most, the LFM can be used to determine the price of illiquid swaptions or of standard swaptions for which the Black volatility is not quoted or is judged to be not completely reliable. But in general, as far as standard (plain-vanilla) swaptions are concerned, the market is happy with Black's formula. Black's formula, indeed, is a "metric" by which traders translate prices into implied volatilities, and it is both pointless and hopeless to expect traders to give up such a formula.

Since traders already know standard-swaptions prices from the market, they wish a chosen model to incorporate as many such prices as possible. In case we are adopting the LFM, we need to find the instantaneous-volatility and correlation parameters in the LFM dynamics that reflect the swaptions prices observed in the market.

First, let us quickly see how the market organizes swaption prices in a table. To simplify ideas, assume we are interested only in swaptions with maturity and underlying-swap length (tenor) given by multiples of one year. Traders typically consider a matrix of at-the-money Black's swaption volatilities, where each row is indexed by the swaption maturity T_α, whereas each column is indexed in terms of the underlying swap length, $T_\beta - T_\alpha$. The $x \times y$-swaption is then the swaption in the table whose maturity is x years and whose underlying swap is y years long. Thus a 2×10 swaption is a swaption maturing in two years and giving then the right to enter a ten-year swap. Here, we consider maturities of $1, 2, 3, 4, 5, 7, 10$ years and underlying-swap lengths of $1, 2, 3, 4, 5, 6, 7, 8, 9, 10$ years.

A typical example of table of swaption volatilities is shown below.

Black's implied volatilities of at-the-money swaptions on May 16, 2000.

	1y	2y	3y	4y	5y	6y	7y	8y	9y	10y
1y	16.4	15.8	14.6	13.8	13.3	12.9	12.6	12.3	12.0	11.7
2y	17.7	15.6	14.1	13.1	12.7	12.4	12.2	11.9	11.7	11.4
3y	17.6	15.5	13.9	12.7	12.3	12.1	11.9	11.7	11.5	11.3
4y	16.9	14.6	12.9	11.9	11.6	11.4	11.3	11.1	11.0	10.8
5y	15.8	13.9	12.4	11.5	11.1	10.9	10.8	10.7	10.5	10.4
7y	14.5	12.9	11.6	10.8	10.4	10.3	10.1	9.9	9.8	9.6
10y	13.5	11.5	10.4	9.8	9.4	9.3	9.1	8.8	8.6	8.4

This is a submatrix of the complete table provided by a broker on May 16, 2000. Its entries are the implied volatilities obtained by inverting the related at-the-money swaption prices through Black's formula for swaptions. A remark at this point is in order.

Remark 6.15.1. **(Possible misalignments in the swaption matrix).** Usually, one should not completely rely on the swaption matrix provided by a single broker, and, in any case, one should not take it for granted. The problem is that the matrix is not necessarily uniformly updated. The most liquid swaptions are updated regularly, whereas some entries of the matrix refer to older market situations. This "temporal misalignment" in the swaptions matrix can cause troubles, since, when we try a calibration, the model parameters might reflect this misalignment by assuming "weird" values (we will see in Chapter 7 examples leading to imaginary and complex forward-rate volatilities). If one trusts the model, this can indeed be used to detect such misalignments. The model can then be calibrated to the liquid swaptions, and used to price the remaining swaptions, looking at the values that most differ from the corresponding market ones. This may give an indication of which swaptions can cause troubles.

What is one to do with the above table when presented with the problem of calibrating the LFM? The problem is incorporating as much information as possible from such a table into the LFM parameters. To focus ideas, consider the LFM with (volatility) Formulation 7 (given by formula (6.13)) and rank-two correlations expressed by the angles θ. We have seen earlier several ways to compute swaption prices with the LFM: Monte Carlo simulation, Brace's rank-one and rank-two formulas, Rebonato's formula and Hull and White's formula. Whichever method is chosen, for given values of a, b, c, d and θ we can price all the swaptions in the table, thus obtaining a table of LFM prices for swaptions corresponding to the chosen values of the LFM parameters.

254   6. The LIBOR and Swap Market Models (LFM and LSM)

Indeed, each price of an at-the-money European (payer) swaption with underlying swap $S_{\alpha,\beta}$ with the LFM is a function of the LFM parameters a, b, c, d and θ's (the Φ's being determined through caplet volatilities via (6.27)). We can then try and change the parameters a, b, c, d and θ and consider the related Φ's (6.27) in such a way that the LFM table approaches as much as possible, in some sense, the market swaptions table. We can, for instance, minimize the sum of the squares of the differences of the corresponding swaption prices in the two tables. Such a sum will be a function of a, b, c, d and θ, and when we find the parameters that minimize it, we can say we have calibrated the LFM to the swaption market.

We will approach this problem by using Rebonato's formula to price swaptions under the LFM. As we will see in Section 8.2, we have tested this formula in a number of situations and we have found it to be sufficiently accurate. The formula gives directly the swaption volatility as a simple function of the parameters. Neither simulations (as in the Monte Carlo method) nor root searches (as in Brace's formulas) are needed, so that this method is ideally suited to calibrate a large table with a contained computational effort.

We need also to point out that recently some attention has moved on a pre-selected instantaneous-correlation matrix. Typically, one estimates the instantaneous correlation ρ historically from time series of zero rates at a given set of maturities, and then approximates it by a lower rank matrix, thus finding, for example, the θ's. With this approach, the parameters that are left for the swaptions calibration are the free parameters in the volatility structure. Take Formulation 7 as an example. The Φ's are determined by the cap market as functions of a, b, c and d, which are parameters to be used in the calibration to swaptions prices. However, if the number of swaptions is large, four parameters are not sufficient for practical purposes. One then needs to consider richer parametric forms for the instantaneous volatility, such as for example the formulation of TABLE 5, although in this case care must be taken for the resulting market structures to be regular enough.

6.16 Connecting Caplet and $S \times 1$-Swaption Volatilities

We now present a result that may facilitate the joint calibration of the LFM to the caps and swaptions markets. Indeed, an important problem is that in the cap market forward rates are mostly semi-annual, whereas those entering the forward-swap-rate expressions are typically annual rates. Therefore, when considering both markets at the same time, we may have to reconcile volatilities of semi-annual forward rates and volatilities of annual forward rates. In this section we address this problem.

Consider three instants $0 < S < T < U$, all six-months spaced. To fix ideas, assume we are dealing with an $S \times 1$ swaption and with S and T-expiry six-month caplets. For instance, we might have $S = 5$ years, $T = 5.5$

6.16 Connecting Caplet and $S \times 1$-Swaption Volatilities

years and $U = 6$ years. We aim at deriving a relationship between the Black swaption volatility and the two Black caplet volatilities.

Consider the three following forward rates at a generic instant $t < S$:

$$F_1(t) := F(t; S, T), \quad F_2(t) := F(t; T, U), \quad F(t) := F(t; S, U).$$

The first two, F_1 and F_2, are semi-annual forward rates, whereas the third one is the annual forward rate in which the two previous forward rates are "nested". We assume year fractions of 0.5 for F_1 and F_2 and of 1 for F.

The algebraic relationship between F, F_1 and F_2 is easily derived by expressing all forward rates in terms of zero-coupon-bond prices. Start from

$$F_1(t) = \frac{1}{0.5}\left[\frac{P(t,S)}{P(t,T)} - 1\right], \quad F_2(t) = \frac{1}{0.5}\left[\frac{P(t,T)}{P(t,U)} - 1\right]$$

and

$$F(t) = \frac{1}{1}\left[\frac{P(t,S)}{P(t,U)} - 1\right].$$

Now observe that

$$F(t) = \frac{P(t,S)}{P(t,T)}\frac{P(t,T)}{P(t,U)} - 1,$$

and then substitute the expression for the two inner fractions of discount factors from the above expressions of F_1 and F_2. One easily obtains

$$\boxed{F(t) = \frac{F_1(t) + F_2(t)}{2} + \frac{F_1(t)F_2(t)}{4}} \qquad (6.65)$$

so that if F_1 and F_2 are lognormal, F cannot be exactly lognormal at the same time.

Assume now the following dynamics

$$dF_1(t) = (\ldots)\, dt + \sigma_1(t) F_1(t)\, dZ_1(t),$$
$$dF_2(t) = (\ldots)\, dt + \sigma_2(t) F_2(t)\, dZ_2(t),$$
$$dZ_1\, dZ_2 = \rho\, dt$$

for the two semi-annual rates. The quantity ρ is the "infra-correlation" between the "inner rates" F_1 and F_2. We obtain easily by differentiation

$$dF(t) = (\ldots)\, dt + \sigma_1(t)\left(\frac{F_1(t)}{2} + \frac{F_1(t)F_2(t)}{4}\right) dZ_1(t)$$
$$+ \sigma_2(t)\left(\frac{F_2(t)}{2} + \frac{F_1(t)F_2(t)}{4}\right) dZ_2(t).$$

By taking variance on both sides, conditional on the information available at time t, and by calling $\sigma(t)$ the percentage volatility of F we have

6. The LIBOR and Swap Market Models (LFM and LSM)

$$\sigma^2(t)F^2(t) = \sigma_1(t)^2 \left(\frac{F_1(t)}{2} + \frac{F_1(t)F_2(t)}{4} \right)^2$$

$$+ \sigma_2(t)^2 \left(\frac{F_2(t)}{2} + \frac{F_1(t)F_2(t)}{4} \right)^2$$

$$+ 2\rho\sigma_1(t)\sigma_2(t) \left(\frac{F_1(t)}{2} + \frac{F_1(t)F_2(t)}{4} \right) \left(\frac{F_2(t)}{2} + \frac{F_1(t)F_2(t)}{4} \right).$$

Set

$$\boxed{u_1(t) := \frac{1}{F(t)} \left(\frac{F_1(t)}{2} + \frac{F_1(t)F_2(t)}{4} \right)}$$

$$\boxed{u_2(t) := \frac{1}{F(t)} \left(\frac{F_2(t)}{2} + \frac{F_1(t)F_2(t)}{4} \right)}$$

so that

$$\sigma^2(t) = u_1^2(t)\sigma_1(t)^2 + u_2^2(t)\sigma_2(t)^2 + 2\rho\sigma_1(t)\sigma_2(t)u_1(t)u_2(t).$$

Let us introduce a first (deterministic) approximation by freezing all F's (and therefore u's) at their time-zero value:

$$\boxed{\sigma_{\text{appr}}^2(t) = u_1^2(0)\sigma_1(t)^2 + u_2^2(0)\sigma_2(t)^2 + 2\rho\sigma_1(t)\sigma_2(t)u_1(0)u_2(0)}.$$

Now recall that F is the particular (one-period) swap rate underlying the $S \times 1$ swaption, whose (squared) Black's swaption volatility is therefore

$$v_{\text{Black}}^2 \approx \frac{1}{S} \int_0^S \sigma_{\text{appr}}^2(t)\, dt = \frac{1}{S} \left[u_1^2(0) \int_0^S \sigma_1(t)^2\, dt + u_2^2(0) \int_0^S \sigma_2(t)^2\, dt \right.$$

$$\left. + 2\rho u_1(0) u_2(0) \int_0^S \sigma_1(t)\sigma_2(t)\, dt \right]. \qquad (6.66)$$

The problem is evaluating the last three integrals. Consider the first one:

$$\frac{1}{S} \int_0^S \sigma_1(t)^2\, dt = v_{S-\text{caplet}}^2,$$

since S is exactly the expiry of the semi-annual rate F_1. Therefore the first integral can be inputed directly as a market caplet volatility. Not so for the second and third integrals. They both require some parametric assumption on the instantaneous volatility structure of rates in order to be computed. The simplest solution is to assume that forward rates have constant volatilities. In such a case, we compute immediately the second integral as

$$\frac{1}{S} \int_0^S \sigma_2(t)^2\, dt = \frac{1}{S} \int_0^S v_{T-\text{caplet}}^2\, dt = v_{T-\text{caplet}}^2,$$

6.16 Connecting Caplet and $S \times 1$-Swaption Volatilities

while the third one is

$$\frac{1}{S}\int_0^S \sigma_1(t)\sigma_2(t)\,dt = \frac{1}{S}\int_0^S v_{T-\text{caplet}} v_{S-\text{caplet}}\,dt = v_{T-\text{caplet}} v_{S-\text{caplet}}.$$

Under this assumption of constant volatility affecting the second and third integral we obtain

$$\boxed{v_{\text{Black}}^2 \approx u_1^2(0) v_{S-\text{caplet}}^2 + u_2^2(0) v_{T-\text{caplet}}^2 + 2\rho u_1(0) u_2(0) v_{S-\text{caplet}} v_{T-\text{caplet}}}.$$
(6.67)

Other assumptions on the volatility term structure are possible, but the simple one above is one of the few rendering the approximated formula self-sufficient on the basis of direct market quantities. When dealing with swaptions, De Jong, Driessen and Pelsser (1999) noticed that flat volatilities in forward rates tend to overprice swaptions, so that we can argue that the approximated formula above can give volatilities that are slightly larger than the actual ones. This is partly confirmed analytically as follows. As far as the second integral approximation is concerned, we may note that when S is large and the instantaneous volatility $\sigma_2(t)$ does not differ largely in $[S,T]$ from its average value in $[0,S]$, we can approximate the integral as

$$\frac{1}{S}\int_0^S \sigma_2(t)^2\,dt \approx \frac{1}{T}\int_0^T \sigma_2(t)^2\,dt = v_{T-\text{caplet}}^2.$$

As far as the third integral is concerned, by the Schwartz inequality we have:

$$\int_0^S \sigma_1(t)\sigma_2(t)\,dt \leq \sqrt{\int_0^S \sigma_1(t)^2\,dt}\sqrt{\int_0^S \sigma_2(t)^2\,dt} \approx S\, v_{S-\text{caplet}}\, v_{T-\text{caplet}},$$

which, in case of positive correlation, shows that (6.67) overestimates volatility with respect to (6.66).

The above formula (6.67) can also be used to back out an implied correlation between adjacent semi-annual rates starting from the $S \times 1$-Black swaption volatility and from the two Black caplet volatilities for expiries S and T. Indeed, one reads from the swaption market v_{Black} and from the cap market one calculates $v_{S-\text{caplet}}$ and $v_{T-\text{caplet}}$. Then one inverts formula (6.67) and obtains ρ. One has to keep in mind, however, that this correlation depends on the constant-volatility assumption above.

A fundamental use for formula (6.67) is easing the joint calibration of the LFM to the caps and swaptions markets. Indeed, the problem (ignored by us in earlier sections and chapters) is that in the cap market forward rates are semi-annual, whereas the forward rates concurring to a swap rate are annual forward rates. Therefore, when calibrating the LFM, we have volatilities of semi-annual forward rates, such as F_1 and F_2, from the cap market but we need to put volatilities of annual forward rates, such as F, in the swaption

258   6. The LIBOR and Swap Market Models (LFM and LSM)

calibration apparatus. What can be helpful here is formula (6.66). Whichever structure is assumed for the instantaneous volatilities of semi-annual rates, this formula allows us to compute the corresponding integrated volatility for annual forward rates. Hence, we can treat caplets as if they were on annual forward rates, and the joint calibration can now be based on the same family of (annual) forward rates. Notice that the ρ's between adjacent semi-annual forward rates are further parameters that can be used in the calibration, even though they should not be allowed to take small or negative values, since instantaneous correlations between adjacent forward rates are usually close to one.

We close this section by presenting some numerical tests of our formulas above.

We adopt the piecewise-constant instantaneous-volatility formulation (6.8) that through the η's preserves the humped shape of the term structure of volatilities through time.

The η's have been deduced by a typical Euro-market caplet-volatility table and are reported below in Table 6.1. In the first column we have the expiry of the relevant semi-annual forward rate, in the second column the related caplet volatility, in the third column the time interval where the instantaneous volatility of the forward rate, whose expiry is the terminal point of the interval, is equal to the value of η reported in the corresponding position of the last column. So, for example, in the third row we read that the caplet resetting at 1.5 years (and paying at 2 years) has a caplet volatility of 19.78%. The related forward rate is $F(\cdot; 1.5, 2)$ and has piecewise-constant instantaneous volatility given by 0.225627879 for $t \in [0, 0.5)$, with subsequent values in $[0.5, 1)$ and $[1, 1.5)$ obtained by moving up along the column (amounting respectively to 0.208017523 and 0.1523).

We plan to test formula (6.66) (approximation 1, shortly "a1") versus formula (6.67) ("a2"), and finally compare them to the real implied volatility obtained by inverting the Monte Carlo price of the related annual caplet through Black's formula. In doing this, we shall always assume for each pair of adjacent rows the first forward rate to be $F_1(0) = 0.04$ and the second to be $F_2(0) = 0.05$, while always taking as caplet strike $K = 0.0455 = F(0)$.

The "true" Monte Carlo annual caplet volatility is obtained as follows. Given the instantaneous volatilities η, one simulates the discretized Milstein dynamics of F_1 and F_2 up to the expiry of F_1, under the "canonical" forward measure for F_2. Then one calculates the annual forward rate F along each path based on the simulated F_1 and F_2 through Formula (6.65).

At this point one can evaluate the annual caplet, whose underlying rate is F, through a Monte Carlo pricing, by averaging the discounted payoff in F along all paths. By including 1.65 times the standard deviation of the simulated payoff divided by the square root of the number of paths we can obtain a 90% window for the true annual caplet price, and we can check

6.16 Connecting Caplet and $S \times 1$-Swaption Volatilities

Expiries	Caplet volatilities	Time intervals	η's
0.5	0.1523	[0, 0.5]	0.1523
1	0.1823	[0.5, 1]	0.208017523
1.5	0.1978	[1, 1.5]	0.225627879
2	0.1985	[1.5, 2]	0.200585343
2.5	0.1999	[2, 2.5]	0.205404601
3	0.1928		0.152417158
3.5	0.1868		0.145700515
4	0.1807		0.130231486
4.5	0.1754		0.12516597
5	0.1715		0.131286176
5.5	0.1677	[5, 5.5]	0.123424835
6	0.1638		0.112290204
6.5	0.16		0.104089961
7	0.1576		0.122182814
7.5	0.1552		0.116520213
8	0.1528		0.110724162
8.5	0.1504		0.104772515
9	0.148		0.098637113
9.5	0.1456		0.092281309
10	0.145		0.133087039
10.5	0.1444	[10, 10.5]	0.131827766
11	0.1439		0.132966387
11.5	0.1433		0.129398029
12	0.1427		0.128126851
12.5	0.1422		0.129622683
13	0.1416		0.125672431
13.5	0.141		0.124388263
14	0.1405		0.126253713
14.5	0.1399		0.121906891
15	0.1403		0.151441111
15.5	0.1406	[15, 15.5]	0.149319992
16	0.141		0.152881784
16.5	0.1414		0.153651163
17	0.1418		0.154419817
17.5	0.1421		0.151947985
18	0.1425	[17.5, 18]	0.15585458
18.5	0.1429	[18, 18.5]	0.156621103
19	0.1433	[18.5, 19]	0.157386944
19.5	0.1436	[19, 19.5]	0.154569143
20	0.1436	[19.5, 20]	0.1436

Table 6.1. Testing "a1" and "a2" approximations: inputs (expiries and time intervals are in years, caplets are semi-annual).

whether the approximated analytical formulas (6.66) and (6.67) are inside this window ("in") or not ("out"). Our results are reported in Table 6.2.

In this table, we can see what happens in three different cases of infra-instantaneous correlations ρ between F_1 and F_2. The upper table concerns the case $\rho = 1$, which should be close to the appropriate value of correlation between adjacent rates. In the middle table, we see the case with $\rho = 0.5$, where adjacent rates are decorrelated to a large extent, whereas in the final table, we see the case of complete decorrelation $\rho = 0$, which sounds rather unrealistic for close rates.

In each case we reported the annual caplet volatilities under both approximation "a1" and "a2", the "true" Monte Carlo (MC) implied volatility and the related 95% window. We have written "in" or "out" to signal when

260 6. The LIBOR and Swap Market Models (LFM and LSM)

$\rho=1$					95% Window		Errors (%)			Vols			
mat.	a1		a2		MC	inf	sup	a1-a2	a1-MC	a2-MC	F_1 a1	F_2 a2	F_2 a1
1y	20.34	in	19.30	out	20.40	20.23	20.58	-5.14	-0.30	-5.42	18.23	19.78	21.70
2y	20.67	in	20.15	out	20.78	20.59	20.97	-2.51	-0.55	-3.05	19.85	19.99	21.01
5y	17.15	in	17.13	in	17.19	17.02	17.36	-0.17	-0.20	-0.37	17.15	16.77	16.92
10y	14.56	in	14.63	out	14.46	14.31	14.61	0.44	0.68	1.12	14.5	14.44	14.40
19y	14.47	in	14.50	in	14.39	14.21	14.56	0.26	0.54	0.81	14.33	14.36	14.34

$\rho=.5$					95% Window		Errors (%)			Vols			
mat.	a1		a2		MC	inf	sup	a1-a2	a1-MC	a2-MC	F_1 a1	F_2 a2	F_2 a1
1y	17.74	in	16.78	out	17.64	17.49	17.79	-5.43	0.57	-4.89	18.23	19.78	21.70
2y	17.97	in	17.48	in	17.89	17.72	18.05	-2.70	0.46	-2.25	19.85	19.99	21.01
5y	14.90	in	14.85	in	14.84	14.70	14.98	-0.29	0.40	0.11	17.15	16.77	16.92
10y	12.65	out	12.69	in	12.79	12.66	12.91	0.35	-1.10	-0.76	14.5	14.44	14.40
19y	12.56	out	12.59	out	12.73	12.59	12.87	0.21	-1.34	-1.14	14.33	14.36	14.34

$\rho=0$					95% Window		Errors (%)			Vols			
mat.	a1		a2		MC	inf	sup	a1-a2	a1-MC	a2-MC	F_1 a1	F_2 a2	F_2 a1
1y	14.68	out	13.80	out	14.89	14.77	15.02	-5.99	-1.47	-7.37	18.23	19.78	21.70
2y	14.79	in	14.34	out	14.91	14.78	15.04	-3.06	-0.80	-3.83	19.85	19.99	21.01
5y	12.23	out	12.17	out	12.41	12.29	12.52	-0.52	-1.44	-1.95	17.15	16.77	16.92
10y	10.38	out	10.40	out	10.57	10.46	10.67	0.17	-1.73	-1.56	14.5	14.44	14.40
19y	10.31	out	10.32	out	10.62	10.51	10.73	0.10	-2.94	-2.84	14.33	14.36	14.34

Table 6.2. Testing "a1" and "a2" approximations: results.

an approximation is inside or outside the window. The related percentage differences are also reported.

Finally, if we call T_{row} the expiry in the first column of the table, in the last three columns we have reported the following.

- "F_1 a1" is the average volatility of the first forward rate, F_1, up to its expiry T_{row}, and is therefore the F_1 caplet volatility for that row:

$$\sqrt{\frac{1}{T_{\text{row}}} \int_0^{T_{\text{row}}} \sigma(t; T_{\text{row}}, T_{\text{row}} + 6m)^2 \, dt} = v_{T_{\text{row}} - \text{caplet}}$$

- "F_2 a1" is the average volatility of the second forward rate, F_2, up to the expiry T_{row} of F_1, which, under "a1", is computed exactly:

$$\sqrt{\frac{1}{T_{\text{row}}} \int_0^{T_{\text{row}}} \sigma(t; T_{\text{row}} + 6m, T_{\text{row}} + 1y)^2 \, dt}$$

- "F_2 a2" is the average volatility of the second forward rate, F_2, up to the expiry T_{row}, which, under "a2", is approximated by the $T_{\text{row}} + 6m$-semi-annual caplet volatility for F_2:

$$\sqrt{\frac{1}{T_{\text{row}}} \int_0^{T_{\text{row}}} \sigma(t; T_{\text{row}} + 6m, T_{\text{row}} + 1y)^2 \, dt} \approx v_{T_{\text{row}} + 6m - \text{caplet}}.$$

In this way we can see the difference between the approximations "a1" and "a2" for the second integral. This is important because "a2" is obtained directly as a caplet volatility, and requires no explicit knowledge of instantaneous volatilities.

Results seem to show that the differences between "a1" and "a2" are small for large expiries, say $T_{\text{row}} \geq 5y$. This is intuitive: "a2" is based on assuming the averages of $\sigma_2^2(\cdot)$ in $[0, T_{\text{row}}]$ and $[0, T_{\text{row}} + 6m]$, respectively, to be close, which is more likely to happen for large T_{row}.

Moreover, "a1" results to be acceptable compared to the true volatility in most situations, and especially when $\rho = 1$. Instead, in such a comparison, "a2" is in trouble for short maturities, exactly as in its comparison with "a1".

6.17 Forward and Spot Rates over Non-Standard Periods

Assume that, in order to price a financial product, we need to know spot LIBOR rates $L(S) := L(S, S + \delta)$ at certain dates $S = s_1, \ldots, S = s_n$, where δ is the common time-to-maturity of the considered rates (typically half a year or one year). This set of rates can be obtained through a suitable family of forward rates defined over expiry-maturity pairs that are non-standard, i.e. that are not contained in our starting set $\{T_0, \ldots, T_M\}$. This can help when pricing products such as trigger or accrual swaps, depending on the daily or weekly evolution of the spot LIBOR rate. Specifically, suppose we are interested in propagating the forward rates

$$F(\cdot; s_1, U_1), F(\cdot; s_2, U_2), \ldots, F(\cdot; s_n, U_n),$$

with $s_1 < s_2 < \ldots < s_n$ and $U_1 < U_2 < \ldots < U_n$, from time 0 to time s_n. Typically, as we said above, $U_i = s_i + \delta$. Notice that this propagation provides us also with the spot LIBOR rates

$$L(s_1, U_1), L(s_2, U_2), \ldots, L(s_n, U_n).$$

How can we obtain the dynamics of the above forward rates from the dynamics of the original family $F_1, \ldots, F_M$? Let us examine two possible methods.

6.17.1 Drift Interpolation

Marry, sir, here's my drift   Polonius in Hamlet, II.1

To fix ideas, take a generic maturity U and assume it to be included in $[T_k, T_{k+1}]$, and assume the related expiry S to be not too far back from U (i.e. $S - U$ should not be too much larger than the average τ_k).

The key idea here is to use formula (6.15) as a guide to write

$$dZ^U \approx dZ^k + \frac{(U - T_k)F_{k+1}(t)}{1 + \tau_{k+1}F_{k+1}(t)} \rho \underline{\sigma}_{k+1}(t)' dt. \tag{6.68}$$

Notice that for $U = T_{k+1}$ we obtain formula (6.15) by assuming $\tau_{k+1} = T_{k+1} - T_k$. Also, for $U = T_k$ we obtain an identity, as should be. We need also to define instantaneous correlations between these shocks. We set

$$dZ_t^{U_a} dZ_t^{U_b} = \rho_{k,j} dt \quad \text{if} \quad k - 1 < U_a \leq k, \ j - 1 < U_b \leq j,$$

so as to extend the original instantaneous correlations to non-standard maturities in the simplest possible way, that is by taking the same instantaneous correlations of the original rates where the considered non-standard rates are "nested". Now use the last approximated shock above in the martingale dynamics

$$dF(t; S, U) = \underline{\sigma}(t; S, U) F(t; S, U) dZ^U(t)$$

to deduce the following.

Proposition 6.17.1. (Interpolated-drift dynamics for non-standard rates in the LFM). *The dynamics of the non-standard forward rate $F(t; S, U)$, derived through the above drift interpolation, is given by*

$$dF(t; S, U) = \frac{(U - T_k)F_{k+1}(t)}{1 + \tau_{k+1}F_{k+1}(t)} F(t; S, U) \underline{\sigma}(t; S, U) \rho \underline{\sigma}_{k+1}'(t) dt \tag{6.69}$$

$$+ \underline{\sigma}(t; S, U) F(t; S, U) dZ^k(t), \quad T_k < U < T_{k+1}.$$

6.17 Forward and Spot Rates over Non-Standard Periods

The (deterministic) volatility $\underline{\sigma}(t;S,U)$ is readily obtained if we have calibrated a certain functional form of instantaneous volatilities of the LFM, such as for example Formulations 6 or 7. As for volatility parameters depending on maturities, such as the Φ's of Formulation 7 or the volatilities themselves in the piecewise-constant formulations, the parameters for $\underline{\sigma}(t;S,U)$ can be obtained by suitably interpolating the corresponding parameters of $\underline{\sigma}_k(t)$ and $\underline{\sigma}_{k+1}(t)$, i.e. by interpolating the instantaneous volatilities of the closest adjacent rates in the original family. We may as well set directly $\underline{\sigma}(t;S,U) \approx \underline{\sigma}_{k+1}(t)$.

For Monte Carlo simulation, the obtained dynamics can be discretized first by taking logarithms and then by applying the Milstein scheme to the obtained dynamics:

$$\ln F(t+\Delta t;S,U) = \ln F(t;S,U) + \frac{(U-T_k)F_{k+1}(t)}{1+\tau_{k+1}F_{k+1}(t)}\underline{\sigma}(t;S,U)\rho\underline{\sigma}'_{k+1}(t)\Delta t$$
$$-\tfrac{1}{2}|\underline{\sigma}(t;S,U)|^2\Delta t + \underline{\sigma}(t;S,U)(Z^k(t+\Delta t)-Z^k(t)).$$

Now, to obtain the required family of forward rates, we just set $U = U_1$, $U = U_2$ up to $U = U_n$ and use the above scheme from time 0 to time s_n, making sure that we choose a time step Δt such that all the dates $s_1, s_2, \ldots, s_n$ are encountered in the discretization. Notice that here too the analogous of Remark 6.10.1 applies for the shocks' simulation.

There is still a possible problem. Assume we need the daily evolution of the one-year spot LIBOR rate up to 5 years for valuing an accrual swap with daily accruing and yearly payment period. Let us focus on the second payment, and assume that the swap leg we are to price pays out at the second year the amount given by the one-year LIBOR rate resetting after one year, times the number of days the spot rate has remained in-between two given levels in the period from one to two years, divided by 365. More formally, the discounted payoff reads

$$D(0,2)L(1,2)\frac{\sum_{t=1y}^{2y-1d}1_{\{B_1<L(t+1y,t+2y)<B_2\}}}{365},$$

B_1 and B_2 being the levels, $d = 1/365y$ being a shorthand notation for "day" and the step in the summation being one day.

To evaluate this payment of the swap leg at time 0, we would need to set $s_1 = 1y$ up to $s_{365} = 2y - 1d$, and $U_1 = 2y$ up to $U_{365} = 3y - 1d$. However, it is clearly undesirable to simulate jointly a 365 dimensional vector. It is true that, as time passes, the vector length diminishes by one each day and the "alive" forward rates reduce in number. Yet, the task can still be too demanding. What can be done is simulating say weekly or three-monthly forward rates, and then at each simulated instant one can interpolate directly the forward rates themselves. A naive way to do this, again with reference to the considered example, is as follows.

Suppose we simulate monthly rates by choosing $s_1 = 1y,\ldots,s_n = 1y+12m$ and the corresponding U's one-year shifted, thus having $\delta = 1y$. Suppose further that we are at day $1y + 35d = 1y + 1m + 5d$ and we need the one-year spot LIBOR rate $L(1y + 35d, 2y + 35d) = F(1y + 35d; 1y + 35d, 2y + 35d)$ to be put in the second-payment payoff. Our numerical Monte Carlo scheme provides us with $F(1y+35d; 1y+1m, 2y+1m) := F(1y+1m; 1y+1m, 2y+1m)$ (by extending forward rates after their "death" in the trivial "frozen" way) and $F(1y + 35d; 1y + 2m, 2y + 2m)$. It is now easy to interpolate between the points $(1y + 1m, F(1y + 35d; 1y + 1m, 2y + 1m))$ and $(1y + 2m, F(1y + 35d; 1y + 2m, 2y + 2m))$ to obtain $F(1y + 35d; 1y + 35d, 2y + 35d)$. There are several classical interpolation methods that can be used.

Of course, the above "drift interpolation" technique, possibly partially combined with the above "directly interpolate rates" reductions, is not the only possibility. We propose below a technique based on a "Brownian bridge"-like method.

6.17.2 The Bridging Technique

We propose a different solution to the problem of simulating spot-rate values $L(t) := L(t, t + \delta)$ in-between the two spot-rate values $L(T_{\beta(t)-2})$ and $L(T_{\beta(t)-1})$ obtained from the original family of spanning forward rates as $F_{\beta(t)-1}(T_{\beta(t)-2})$ and $F_{\beta(t)}(T_{\beta(t)-1})$, respectively. We here assume that the T_i's are equally δ-spaced.

To fix ideas, assume the current time t to be between T_{i-1} and T_i,

$$t \in [T_{i-1}, T_i] = [T_{\beta(t)-2}, T_{\beta(t)-1}],$$

and that we need the LIBOR rate $L(t)$ at times $s_1 = T_{i-1}, s_2, \ldots, s_{l-1}, s_l = T_i$. This second method is based on ideas mutuated from the notion of Brownian bridge. It works according to the following steps:

a) Through the discretized forward-rate dynamics analogous to (6.47) we have generated m realizations of $L(T_{i-1}) = F_i(T_{i-1})$ and m corresponding realizations of $L(T_i) = F_{i+1}(T_i)$. We denote the j-th realization by a superscript j, so that, for example, $L^j(T_{i-1})$ denotes the realization of $L(T_{i-1})$ obtained under the j-th scenario.

b) Assume a "geometric-Brownian-motion"-like dynamics for the spot rate $L(t)$ in-between the two already-generated values $L(T_{i-1})$ and $L(T_i)$,

$$dL(t) = \mu_i L(t)\, dt + v_i L(t)\, dZ(t - T_{i-1}),\ t \in [T_{i-1}, T_i], \quad (6.70)$$

where Z is a standard Brownian motion under the same measure used for generating the forward-rate dynamics.

c) Consider

$$\ln \frac{L(T_i)}{L(T_{i-1})} = (\mu_i - v_i^2/2)(T_i - T_{i-1}) + v_i Z(T_i - T_{i-1}) \quad (6.71)$$

$$\sim \mathcal{N}\left((\mu_i - v_i^2/2)(T_i - T_{i-1}), v_i^2(T_i - T_{i-1})\right).$$

6.17 Forward and Spot Rates over Non-Standard Periods

Since we know the maximum-likelihood estimators for mean and variance of Gaussian random variables, we can use such estimators to estimate μ_i and v_i as implied from the m values generated for $L(T_{i-1})$ and $L(T_i)$. The maximum-likelihood estimators consist simply on the sample mean and variance,

$$(T_i - T_{i-1})(\hat{\mu}_i - \hat{v}_i^2/2) = \frac{1}{m}\sum_{j=1}^{m} \ln \frac{L^j(T_i)}{L^j(T_{i-1})},$$

$$(T_i - T_{i-1})\hat{v}_i^2 = \left(\frac{1}{m}\sum_{j=1}^{m} \ln^2 \frac{L^j(T_i)}{L^j(T_{i-1})} - (T_i - T_{i-1})^2(\hat{\mu}_i - \hat{v}_i^2/2)^2\right)$$

which can be easily inverted to obtain $\hat{\mu}_i$ and $\hat{v}_i$. The superscript j denotes the scenario under which the considered variables have been generated.

d) Consider again (6.71), replace μ_i and v_i with their estimates $\hat{\mu}_i$ and $\hat{v}_i$, and solve for $Z(T_i - T_{i-1})$. We find

$$Z(T_i - T_{i-1}) = \ln \frac{L(T_i)}{L(T_{i-1})} - (\hat{\mu}_i - \hat{v}_i^2/2)(T_i - T_{i-1}).$$

The final value of Z is therefore known in advance. In particular, under the j-th scenario, we have the final value of Z given by

$$Z^j(T_i - T_{i-1}) = \ln \frac{L^j(T_i)}{L^j(T_{i-1})} - (\hat{\mu}_i - \hat{v}_i^2/2)(T_i - T_{i-1}).$$

e) For each scenario j, we have now the initial and final values $L^j(T_{i-1})$ and $L^j(T_i)$, and the final value of the Brownian motion Z in the L-dynamics (6.70). We now need to connect the initial and final values by finding a path

$$L(s_1) = L^j(T_{i-1}), L(s_2), L(s_3), \ldots, L(s_{l-2}), L(s_{l-1}), L(s_l) = L^j(T_i)$$

describing all the spot-rate values needed under the j-th scenario.

Now assume we are in scenario j. All we need to do is simulating the shocks dZ consistently with the known final value $Z^j(T_i - T_{i-1})$. In order to do so, we replace $Z(t - T_{i-1})$ with a process with the same initial and final known values and whose increments are as close as possible to the increments of a Brownian motion. The process we consider is

$$\zeta(t - T_{i-1}) := V(t - T_{i-1}) - \frac{t - T_{i-1}}{T_i - T_{i-1}}\left(V(T_i - T_{i-1}) - Z^j(T_i - T_{i-1})\right), \tag{6.72}$$

where V is a new standard Brownian motion, independent of the previous one. By construction, $\zeta(0) = Z(0) = 0$, $\zeta(T_i - T_{i-1}) = Z^j(T_i - T_{i-1})$. Clearly,

$$d\zeta(t - T_{i-1}) = dV(t - T_{i-1}) - \frac{V(T_i - T_{i-1}) - Z^j(T_i - T_{i-1})}{T_i - T_{i-1}} dt,$$

so that this is a Brownian motion with constant drift. Of course

$$d\zeta(t - T_{i-1}) d\zeta(t - T_{i-1}) = dt$$

as for a standard Brownian motion.

f) At this point, we simulate the intermediate values of the process (6.70) under the j-th scenario through a single path from the following modified version of its exact discrete dynamics:

$$L^j(s_{h+1}) = L^j(s_h) \exp\left[(\hat{\mu}_i - \hat{v}_i^2/2)(s_{h+1} - s_h) + \hat{v}_i(\zeta(s_{h+1}) - \zeta(s_h))\right],$$

for $h = 1, \ldots, l-1$. This is perfectly feasible at this point, since all the quantities above are known. In particular, one path for ζ through $s_1, \ldots, s_k$ can be obtained by simulating a priori one path for the independent standard Brownian motion V and subsequently using (6.72).

The above "bridging" technique, coupled with the considered forward-rate dynamics, is a possible tool for evaluating derivatives whose payoff depends on the daily or weekly evolution of spot rates. The only limitation lies in the geometric Brownian motion assumption, which features no mean reversion for the spot rate. However, notice that the final value of the considered rates is consistent with forward-rate dynamics by construction, since the drift $\hat{\mu}$ and volatility $\hat{v}$ of the in-between geometric Brownian motion reflect the correct initial and final values for the spot-rate process as implied from the correct forward-rates dynamics. Moreover, the fact that usually the geometric Brownian motion is reset at each payment date T_i to the "true" spot rate –as implied by forward-rate dynamics– renders the absence of mean reversion quite bearable, since mean reversion is usually intended to act on longer periods than the typically six-month long $[T_{i-1}, T_i]$. Finally, one can combine this method with "directly interpolating rates" so as to propagate a much smaller number of forward rates, similarly to what was done for the previous method. However, both the feasibility and the quality of this method need to be tested with simulations and numerical investigations before drawing any conclusion about its performances.

6.18 Including the Caplet Smile in the LFM

6.18.1 A Mini-tour on the Smile Problem

We have seen earlier that Black's formula for caplets is the standard in the cap market. This formula is consistent with the LFM, in that it comes as the expected value of the discounted caplet payoff under the related forward measure when the forward-rate dynamics is given by the LFM.

6.18 Including the Caplet Smile in the LFM

To fix ideas, let us consider again the time-0 price of a T_2-maturity caplet resetting at time T_1 $(0 < T_1 < T_2)$ with strike K and a notional amount of 1. Caplets and caps have been described more generally in Section 6.4. Let τ denote the year fraction between T_1 and T_2. Such a contract pays out the amount

$$\tau(F(T_1;T_1,T_2) - K)^+,$$

at time T_2, so that its value at time 0 is

$$P(0,T_2)\tau E_0^2[(F(T_1;T_1,T_2) - K)^+].$$

The dynamics for F in the above expectation under the T_2-forward measure is the lognormal LFM dynamics

$$dF(t;T_1,T_2) = \sigma_2(t)F(t;T_1,T_2)\,dW_t. \tag{6.73}$$

Lognormality of the T_1-marginal distribution of this dynamics implies that the above expectation results in Black's formula

$$\mathbf{Cpl}^{\text{Black}}(0,T_1,T_2,K) = P(0,T_2)\tau\text{Bl}(K,F_2(0),v_2(T_1)),$$

$$v_2(T_1)^2 = \int_0^{T_1} \sigma_2^2(t)dt.$$

It is clear that in this derivation, the average volatility of the forward rate in $[0,T_1]$, i.e. $v_2(T_1)/\sqrt{T_1}$, does not depend on the strike K of the option. Indeed, in this formulation, volatility is a characteristic of the forward rate underlying the contract, and has nothing to do with the nature of the contract itself. In particular, it has nothing to do with the strike K of the contract.

Now take two different strikes K_1 and K_2. Suppose that the market provides us with the prices of the two related caplets $\mathbf{Cpl}^{\text{MKT}}(0,T_1,T_2,K_1)$ and $\mathbf{Cpl}^{\text{MKT}}(0,T_1,T_2,K_2)$. Both caplets have the same underlying forward rates and the same maturity.

Life would be simple if the market followed Black's formula in a consistent way. But is this the case? Does there exist a *single* volatility parameter $v_2(T_1)$ such that both

$$\mathbf{Cpl}^{\text{MKT}}(0,T_1,T_2,K_1) = P(0,T_2)\tau\text{Bl}(K_1,F_2(0),v_2(T_1))$$

and

$$\mathbf{Cpl}^{\text{MKT}}(0,T_1,T_2,K_2) = P(0,T_2)\tau\text{Bl}(K_2,F_2(0),v_2(T_1))$$

hold? The answer is a resounding "no". In general, market caplet prices do not behave like this. What one sees when looking at the market is that two *different* volatilities $v_2(T_1,K_1)$ and $v_2(T_1,K_2)$ are required to match the observed market prices if one is to use Black's formula:

$$\mathbf{Cpl}^{\text{MKT}}(0,T_1,T_2,K_1) = P(0,T_2)\tau\text{Bl}(K_1,F_2(0),v_2^{\text{MKT}}(T_1,K_1)),$$
$$\mathbf{Cpl}^{\text{MKT}}(0,T_1,T_2,K_2) = P(0,T_2)\tau\text{Bl}(K_2,F_2(0),v_2^{\text{MKT}}(T_1,K_2)).$$

In other terms, each caplet market price requires its own Black volatility $v_2^{\text{MKT}}(T_1, K)$ depending on the caplet strike K.

The market therefore uses Black's formula simply as a metric to express caplet prices as volatilities. The curve $K \mapsto v_2^{\text{MKT}}(T_1, K)/\sqrt{T_1}$ is the so called volatility smile of the T_1-expiry caplet. If Black's formula were consistent along different strikes, this curve would be flat, since volatility should not depend on the strike K. Instead, this curve is commonly seen to exhibit "smiley" or "skewed" shapes. The term skew is generally used for those structures where, for a fixed maturity, low-strikes implied volatilities are higher than high-strikes implied volatilities. The term smile is used instead to denote those structures where, again for a fixed maturity, the volatility has a minimum value around the current value of underlying forward rate.

Clearly, only some strikes $K = K_i$ are quoted by the market, so that usually the remaining points have to be determined through interpolation or through an alternative model. Interpolation in K, for a fixed expiry T_1, can be easy but it does not give any insight as to the underlying forward-rate dynamics compatible with such prices.

Indeed, suppose that we have a few market caplet prices for expiry T_1 and for a set of strikes $K = K_i$. By interpolation, we can obtain the price for every other possible K, i.e. we can build a function $K \mapsto \mathbf{Cpl}^{\text{MKT}}(0, T_1, T_2, K)$. Now, if this strike-$K$ price corresponds really to an expectation, we have

$$\mathbf{Cpl}^{\text{MKT}}(0, T_1, T_2, K) = P(0, T_2)\tau E_0^2(F(T_1; T_1, T_2) - K)^+ \quad (6.74)$$

$$= P(0, T_2)\tau \int (x - K)^+ p_2(x)\, dx, \quad (6.75)$$

where p_2 is the probability density function of $F_2(T_1)$ under the T_2-forward measure. If Black's formula were consistent, this density would be the lognormal density, coming for example from a dynamics such as (6.73). We have seen that this is not the case in the market. However, by differentiating the above integral twice with respect to K we see that, see also Breeden and Litzenberger (1978),

$$\frac{\partial^2 \mathbf{Cpl}^{\text{MKT}}(0, T_1, T_2, K)}{\partial K^2} = P(0, T_2)\tau p_2(K),$$

so that by differentiating the interpolated-prices curve we can find the density p_2 of the forward rate at time T_1 that is compatible with the given interpolated prices. However, the method of interpolation may interfere with the recovery of the density, since a second derivative of the interpolated curve is involved. Moreover, what kind of dynamics, alternative to (6.73), does the density p_2 come from?

A partial answer to these issues can be given the other way around, by starting from an alternative dynamics. Indeed, assume that

$$dF(t; T_1, T_2) = \nu(t, F(t; T_1, T_2))\, dW_t \quad (6.76)$$

under the T_2-forward measure, where ν can be either a deterministic or a stochastic function of $F(t;T_1,T_2)$. In the latter case we would be using a so called "stochastic-volatility model", where for example $\nu(t,F) = \xi(t)F$, with ξ following a second stochastic differential equation.

In this section, instead, we will concentrate on a deterministic $\nu(t,\cdot)$, leading to a so called "local-volatility model", such as for example $\nu(t,F) = \sigma_2(t)F^\gamma$ (CEV model), where $0 \leq \gamma \leq 1$ and where σ_2 is itself deterministic. We will then propose a new $\nu(t,\cdot)$ of our own, flexible enough for practical purposes.

We have seen above how the "true" forward-adjusted density p_2 of the underlying forward rate is linked to market prices through second-order differentiation. The problem we will face is finding a dynamics alternative to (6.73) and as compatible as possible with the density p_2 ideally associated with market prices. This will be done by fitting directly the prices implied by our alternative model to the market prices $\mathbf{Cpl}^{\mathrm{MKT}}(0,T_1,T_2,K_i)$ for the considered strikes K_i, or equivalently by fitting the model implied volatilities to the implied volatilities $v_2^{\mathrm{MKT}}(T_1,K_i)$ for the observed strikes. To understand this point, it may be helpful to explicitly clarify how an alternative dynamics such as (6.76) leads to a volatility smile to be fitted to the market smile. The alternative dynamics generates a smile, which is obtained as follows.

1. Set K to a starting value;
2. Compute the model caplet price

$$\Pi(K) = P(0,T_2)\tau E_0^2(F(T_1;T_1,T_2) - K)^+$$

with F obtained through the alternative dynamics (6.76).
3. Invert Black's formula for this strike, i.e. solve

$$\Pi(K) = P(0,T_2)\tau \mathrm{Bl}(K_1, F_2(0), v(K)\sqrt{T_1})$$

in $v(K)$, thus obtaining the (average) model implied volatility $v(K)$.
4. Change K and restart from point 2.

The fact that the alternative dynamics is not lognormal implies that we obtain a curve $K \mapsto v(K)$ that is not flat. Clearly, one needs to choose $\nu(t,\cdot)$ flexible enough for this curve to be able to resemble or even match the corresponding volatility curves coming from the market. Indeed, the model implied volatilities $v(K_i)$ corresponding to the observed strikes have to be made as close as possible to the corresponding market volatilities $v_2^{\mathrm{MKT}}(T_1,K_i)/\sqrt{T_1}$, by acting on the coefficient $\nu(\cdot,F)$ in the alternative dynamics. We will address this problem in the following sections.

We finally point out that one has to deal, in general, with an implied-volatility surface, since we have a caplet-volatility curve for each considered expiry. The calibration issues, however, are essentially unchanged, apart from the obviously larger computational effort required when trying to fit a bigger set of data.

6.18.2 Modeling the Smile

We start with a little history and a few references.

Similarly to what happens in the equity or foreign-exchange markets, also in the interest-rate market non-flat structures are normally observed when plotting implied volatilities against strikes and maturities. For example, you may plot the curve $K \mapsto v_2^{\text{MKT}}(T_1, K)/\sqrt{T_1}$ given in the earlier section and you may find it to have a smile-like shape in several cases.

As we have just seen in the mini-tour, modeling the dynamics of forward LIBOR rates as in the LFM leads to implied caplet (and hence cap) volatilities that are constant for each fixed caplet (cap) maturity. The LFM, therefore, can be used to exactly retrieve ATM cap prices, but it fails to reproduce non-flat volatility surfaces when a whole range of strikes is considered.

The above considerations suggest the need for an alternative model that is capable of suitably fitting the larger set of prices that is usually available to a trader. Several approaches can then be followed. A first approach is based on assuming (alternative) explicit dynamics for the forward-rate processes that immediately lead to volatility smiles or skews. Examples are the general CEV process of Cox (1975) and Cox and Ross (1976) or the displaced diffusion of Rubinstein (1983). A second approach is based on the assumption of a continuum of traded strikes and goes back to Breeden and Litzenberger (1978). Successive developments are due to Dupire (1994, 1997) and Derman and Kani (1994, 1998). However, the former approach does not provide sufficient flexibility to properly calibrate the whole volatility surface, whereas the latter has the major drawback that one needs to smoothly interpolate option prices between consecutive strikes in order to be able to differentiate them twice with respect to the strike.

In general the problem of finding a distribution that consistently prices all quoted options is largely undetermined, since there are infinitely many curves connecting (smoothly) finitely many points. A possible solution is then given by assuming a particular parametric distribution depending on several, time-dependent parameters. But the question remains of finding forward-rate dynamics consistent with the chosen parametric density. A possible solution to this issue is given by Brigo and Mercurio (2000a, 2000b) who find dynamics leading to a parametric distribution that may be flexible enough for practical purposes. Their resulting forward-rate process combines the parametric risk-neutral distribution approach with the alternative-dynamics approach.

In the following sections, we first introduce the forward-LIBOR model that can be obtained by displacing a given lognormal diffusion. We then describe the CEV model used by Andersen and Andreasen (2000) to model the evolution of the forward-rate process. We finally illustrate the lognormal-mixture approach proposed by Brigo and Mercurio (2000a, 2000b, 2001b).

6.18.3 The Shifted-Lognormal Case

A very simple way of constructing forward-rate dynamics that implies non-flat volatility structures is by shifting the generic lognormal dynamics analogous to (6.73). Indeed, let us assume that the forward rate F_j evolves, under its associated T_j-forward measure, according to

$$F_j(t) = X_j(t) + \alpha,$$
$$dX_j(t) = \beta(t) X_j(t) \, dW_t, \tag{6.77}$$

where α is a real constant, β is a deterministic function of time and W is a standard Brownian motion. We immediately have that

$$dF_j(t) = \beta(t)(F_j(t) - \alpha) \, dW_t, \tag{6.78}$$

so that, for $t < T \leq T_{j-1}$, the forward rate F_j can be explicitly written as

$$F_j(T) = \alpha + (F_j(t) - \alpha) e^{-\frac{1}{2} \int_t^T \beta^2(u) \, du + \int_t^T \beta(u) \, dW_u}. \tag{6.79}$$

The distribution of $F_j(T)$, conditional on $F_j(t)$, $t < T \leq T_{j-1}$, is then a shifted lognormal distribution with density

$$p_{F_j(T)|F_j(t)}(x) = \frac{1}{(x-\alpha) U(t,T) \sqrt{2\pi}} \exp\left\{ -\frac{1}{2} \left(\frac{\ln \frac{x-\alpha}{F_j(t) - \alpha} + \frac{1}{2} U^2(t,T)}{U(t,T)} \right)^2 \right\}, \tag{6.80}$$

for $x > \alpha$, where

$$U(t,T) := \sqrt{\int_t^T \beta^2(u) \, du}. \tag{6.81}$$

The resulting model for F_j preserves the analytical tractability of the geometric Brownian motion X. Notice indeed that

$$P(t, T_j) E^j \{ [F_j(T_{j-1}) - K]^+ | \mathcal{F}_t \} = P(t, T_j) E^j \{ [X_j(T_{j-1}) - (K - \alpha)]^+ | \mathcal{F}_t \},$$

so that, for $\alpha < K$, the caplet price $\mathbf{Cpl}(t, T_{j-1}, T_j, \tau, N, K)$ associated with (6.78) is simply given by

$$\mathbf{Cpl}(t, T_{j-1}, T_j, \tau, N, K) = \tau N P(t, T_j) \mathrm{Bl}(K - \alpha, F_j(t) - \alpha, U(t, T_{j-1})). \tag{6.82}$$

The implied Black volatility $\hat{\sigma} = \hat{\sigma}(K, \alpha)$ corresponding to a given strike K and to a chosen α is obtained by backing out the volatility parameter $\hat{\sigma}$ in Black's formula that matches the model price:

$$\tau N P(t, T_j) \mathrm{Bl}(K, F_j(t), \hat{\sigma}(K, \alpha) \sqrt{T_{j-1} - t})$$
$$= \tau N P(t, T_j) \mathrm{Bl}(K - \alpha, F_j(t) - \alpha, U(t, T_{j-1})).$$

272   6. The LIBOR and Swap Market Models (LFM and LSM)

We can now understand why the simple affine transformation (6.77) can be useful in practice. The resulting forward-rate process, in fact, besides having explicit dynamics and known marginal density, immediately leads to closed-form formulas for caplet prices that allow for skews in the caplet implied volatility. An example of the skewed volatility structure $K \mapsto \hat{\sigma}(K, \alpha)$ that is implied by such a model is shown in Figure 6.5.[1]

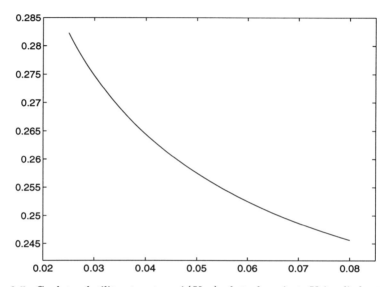

Fig. 6.5. Caplet volatility structure $\hat{\sigma}(K, \alpha)$ plotted against K implied, at time $t = 0$, by the forward-rate dynamics (6.78), where we set $T_{j-1} = 1$, $T_j = 1.5$, $\alpha = -0.015$, $\beta(t) = 0.2$ for all t and $F_j(0) = 0.055$.

Introducing a non-zero parameter α has two effects on the implied caplet volatility structure, which for $\alpha = 0$ is flat at the constant level $U(0, T_{j-1})$. First, it leads to a strictly decreasing ($\alpha < 0$) or increasing ($\alpha > 0$) curve. Second, it moves the curve upwards ($\alpha < 0$) or downwards ($\alpha > 0$). More generally, ceteris paribus, increasing α shifts the volatility curve $K \mapsto \hat{\sigma}(K, \alpha)$ down, whereas decreasing α shifts the curve up. The formal proof of these properties is straightforward. Notice, for example, that at time $t = 0$ the implied at-the-money ($K = F_j(0)$) caplet volatility $\hat{\sigma}$ satisfies

$$\text{Bl}(F_j(0), F_j(0), \sqrt{T_{j-1}}\hat{\sigma}(F_j(0), \alpha)) = \text{Bl}(F_j(0) - \alpha, F_j(0) - \alpha, U(0, T_{j-1})),$$

which reads

---

[1] Such a figure shows a decreasing caplet-volatility curve. In real markets, however, different structures can be encountered too (smile-shaped, skewed to the right,...).

$$(F_j(0) - \alpha)\left[2\Phi\left(\frac{1}{2}U(0, T_{j-1})\right) - 1\right] = F_j(0)\left[2\Phi\left(\frac{1}{2}\sqrt{T_{j-1}}\hat{\sigma}(F_j(0), \alpha)\right) - 1\right].$$

When increasing α the left hand side of this equation decreases, thus decreasing the $\hat{\sigma}$ in the right-hand side that is needed to match the decreased left hand side. Moreover, when differentiating (6.82) with respect to α we obtain a quantity that is always negative.

Shifting a lognormal diffusion can then help in recovering skewed volatility structures. However, such structures are often too rigid, and highly negative slopes are impossible to recover. Moreover, the best fitting of market data is often achieved for decreasing implied volatility curves, which correspond to negative values of the α parameter, and hence to a support of the forward-rate density containing negative values. Even though the probability of negative rates may be negligible in practice, many people regard this drawback as an undesirable feature.

The next models we illustrate may offer the properties and flexibility required for a satisfactory fitting of market data.

6.18.4 The Constant Elasticity of Variance (CEV) Model

Another classical model leading to skews in the implied caplet-volatility structure is the CEV model of Cox (1975) and Cox and Ross (1976). Recently, Andersen and Andreasen (2000) applied the CEV dynamics as a model of the evolution of forward LIBOR rates.

Andersen and Andreasen start with a general forward-LIBOR dynamics of the following type:

$$dF_j(t) = \phi(F_j(t))\sigma_j(t)\, dZ_j^j(t),$$

where ϕ is a general function. Andersen and Andreasen suggest as a particularly tractable case in this family the CEV model, where

$$\phi(F_j(t)) = [F_j(t)]^\gamma,$$

with $0 < \gamma < 1$. Notice that the "border" cases $\gamma = 0$ and $\gamma = 1$ would lead respectively to a normal and a lognormal dynamics.

The model then reads

$$dF_j(t) = \sigma_j(t)[F_j(t)]^\gamma\, dW_t, \quad F_j = 0 \text{ absorbing boundary when } 0 < \gamma < 1/2, \tag{6.83}$$

where we set $W = Z_j^j$, a one-dimensional Brownian motion under the T_j forward measure.

For $0 < \gamma < 1/2$ equation (6.83) does not have a unique solution unless we specify a boundary condition at $F_j = 0$. This is why we take $F_j = 0$ as an absorbing boundary for the above SDE when $0 < \gamma < 1/2$.[2]

---
[2] Andersen and Andreasen (2000) also extend their treatment to the case $\gamma > 1$, while noticing that this can lead to explosion when leaving the T_j-forward measure (under which the process has null drift).

Time dependence of σ_j can be dealt with through a deterministic time change. Indeed, by first setting

$$v(\tau, T) = \int_\tau^T \sigma_j(s)^2 ds$$

and then

$$\widetilde{W}(v(0,t)) := \int_0^t \sigma_j(s) dW(s),$$

we obtain a Brownian motion $\widetilde{W}$ with time parameter v. We substitute this time change in equation (6.83) by setting $f_j(v(t)) := F_j(t)$ and obtain

$$df_j(v) = f_j(v)^\gamma d\widetilde{W}(v), \quad f_j = 0 \text{ absorbing boundary when } 0 < \gamma < 1/2. \tag{6.84}$$

This is a process that can be easily transformed into a Bessel process via a change of variable. Straightforward manipulations lead then to the transition density function of f. By also remembering our time change, we can finally go back to the transition density for the continuous part of our original forward-rate dynamics. The continuous part of the density function of $F_j(T)$ conditional on $F_j(t)$, $t < T \le T_{j-1}$, is then given by

$$p_{F_j(T)|F_j(t)}(x) = 2(1-\gamma)k^{1/(2-2\gamma)}(uw^{1-4\gamma})^{1/(4-4\gamma)}e^{-u-w}I_{1/(2-2\gamma)}(2\sqrt{uw}),$$

$$k = \frac{1}{2v(t,T)(1-\gamma)^2},$$
$$u = k[F_j(t)]^{2(1-\gamma)},$$
$$w = kx^{2(1-\gamma)},$$
(6.85)

with I_q denoting the modified Bessel function of the first kind of order q. Moreover, denoting by $g(y,z) = \frac{e^{-z}z^{y-1}}{\Gamma(y)}$ the gamma density function and by $G(y,x) = \int_x^{+\infty} g(y,z)dz$ the complementary gamma distribution, the probability that $F_j(T) = 0$ conditional on $F_j(t)$ is $G\left(\frac{1}{2(1-\gamma)}, u\right)$.

A major advantage of the model (6.83) is its analytical tractability, allowing for the above transition density function. This transition density can be useful, for example, in Monte Carlo simulations. From knowledge of the density follows also the possibility to price simple claims. In particular, the following explicit formula for a caplet price can be derived:

$$\mathbf{Cpl}(t, T_{j-1}, T_j, \tau, N, K) = \tau N P(t, T_j) \left[F_j(t) \sum_{n=0}^{+\infty} g(n+1, u) G\left(c_n, kK^{2(1-\gamma)}\right) \right.$$

$$\left. - K \sum_{n=0}^{+\infty} g(c_n, u) G\left(n+1, kK^{2(1-\gamma)}\right) \right],$$
(6.86)

where k and u are defined as in (6.85) and

$$c_n := n + 1 + \frac{1}{2(1-\gamma)}.$$

This price can be expressed also in terms of the non-central chi-squared distribution function we have encountered in the CIR model. Recall that we denote by $\chi^2(x; r, \rho)$ the cumulative distribution function for a non-central chi-squared distribution with r degrees of freedom and non-centrality parameter ρ, computed at point x. Then the above price can be rewritten as

$$\mathbf{Cpl}(t, T_{j-1}, T_j, \tau, N, K) = \tau N P(t, T_j) \left[F_j(t) \left(1 - \chi^2 \left(2K^{1-\gamma}; \frac{1}{1-\gamma} + 2, 2u \right) \right) \right.$$
$$\left. - K \chi^2 \left(2u; \frac{1}{1-\gamma}, 2kK^{1-\gamma} \right) \right].$$
(6.87)

As hinted at above, the caplet price (6.86) leads to skews in the implied volatility structure. An example of the structure that can be implied is shown in Figure 6.6. As previously done in the case of a geometric Brownian motion,

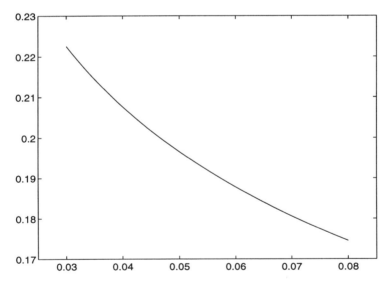

Fig. 6.6. Caplet volatility structure implied by (6.86) at time $t = 0$, where we set $T_{j-1} = 1$, $T_j = 1.5$, $\sigma_j(t) = 1.5$ for all t, $\gamma = 0.5$ and $F_j(0) = 0.055$.

an extension of the above model can be proposed based on displacing the CEV process (6.83) and defining accordingly the forward-rate dynamics. The introduction of the extra parameter determining the density shifting may improve the calibration to market data.

276   6. The LIBOR and Swap Market Models (LFM and LSM)

Finally, there is the possibly annoying feature of absorption in $F = 0$. While this does not necessarily constitute a problem for caplet pricing, it can be an undesirable feature from an empirical point of view. Also, it is not clear whether there could be some problems when pricing more exotic structures. As a remedy to this absorption problem, Andersen and Andreasen (2000) propose a "Limited" CEV (LCEV) process, where instead of $\phi(F) = F^\gamma$ they set

$$\phi(F) = F \ \min(\epsilon^{\gamma-1}, F^{\gamma-1}) \ ,$$

where ϵ is a small positive real number. This function collapses the CEV diffusion coefficient F^γ to a (lognormal) level-proportional diffusion coefficient $F\epsilon^{\gamma-1}$ when F is small enough to make little difference (smaller than ϵ itself). Andersen and Andreasen (2000) compare the LCEV and CEV models as far as cap prices are concerned and conclude that the differences are small and tend to vanish when $\epsilon \to 0$. They also investigate, to some extent, the speed of convergence. A Crank-Nicholson scheme is used to compute cap prices within the LCEV model. As for the CEV model itself, Andersen and Andreasen allow for $\gamma > 1$ also in the LCEV case, with the difference that then ϵ has to be taken very large.

As far as the calibration of the CEV model to swaptions is concerned, approximated swaption prices based on "freezing the drift" and "collapsing all measures" are also derived (analogous to Rebonato's formula in the LFM). See Andersen and Andreasen (2000) for the details.

6.18.5 A Mixture-of-Lognormals Model

I say we attack again. I can get through. I know I can
Lar Gand, "Panic in the Sky", 1993, DC Comics

A natural way to obtain more general implied volatility structures is by assuming a forward-rate distribution that depends on several, time-dependent parameters. Following this approach, Brigo and Mercurio (2000a) proposed an alternative forward-LIBOR model based on the construction of the diffusion process that is consistent with a given mixture of lognormal densities. Their derivation is reviewed in the following.

For each time t let us consider N lognormal densities

$$p_t^i(y) = \frac{1}{yV_i(t)\sqrt{2\pi}} \exp\left\{-\frac{1}{2V_i^2(t)}\left[\ln\frac{y}{F_j(0)} + \tfrac{1}{2}V_i^2(t)\right]^2\right\},$$

$$V_i(t) := \sqrt{\int_0^t \sigma_i^2(u)du},$$

(6.88)

where all σ_i's are positive and deterministic functions of time, and p_0^i is the δ-Dirac function centered in $F_j(0)$. Ideally, these densities can be thought

6.18 Including the Caplet Smile in the LFM

to be coming from the marginal laws of N instrumental geometric Brownian motions such as

$$dX_i(t) = \sigma_i(t) X_i(t)\, dW(t), \quad i = 1, \ldots, N.$$

Brigo and Mercurio (2000a) showed that it is possible to determine the *local volatility* σ in the Q^j-forward-rate dynamics:

$$dF_j(t) = \sigma(t, F_j(t)) F_j(t)\, dW_t, \tag{6.89}$$

in such a way that the SDE (6.89) admits a unique strong solution whose marginal density, at each time $t \leq T_{j-1}$, is given by the mixture of lognormals

$$p_t(y) := \frac{d}{dy} Q^j \{F_j(t) \leq y\} = \sum_{i=1}^{N} \lambda_i p_t^i(y), \tag{6.90}$$

with λ_i's strictly positive constants such that $\sum_{i=1}^{N} \lambda_i = 1$.[3] The local volatility $\sigma(t, \cdot)$ is backed out from the Fokker-Planck equation associated with the dynamics (6.89).[4] Precisely, Brigo and Mercurio (2000a) proved the following.

Proposition 6.18.1. *Let us assume that each σ_i is continuous and bounded from above and below by (strictly) positive constants, and that there exists an $\varepsilon > 0$ such that $\sigma_i(t) = \sigma_0 > 0$, for each t in $[0, \varepsilon]$ and $i = 1, \ldots, N$. Then, if we set*

$$\nu(t, y) := \sqrt{\frac{\sum_{i=1}^{N} \lambda_i \sigma_i^2(t) \frac{1}{V_i(t)} \exp\left\{-\frac{1}{2V_i^2(t)}\left[\ln \frac{y}{F_j(0)} + \frac{1}{2} V_i^2(t)\right]^2\right\}}{\sum_{i=1}^{N} \lambda_i \frac{1}{V_i(t)} \exp\left\{-\frac{1}{2V_i^2(t)}\left[\ln \frac{y}{F_j(0)} + \frac{1}{2} V_i^2(t)\right]^2\right\}}}, \tag{6.91}$$

for $(t, y) > (0, 0)$ and $\nu(t, y) = \sigma_0$ for $(t, y) = (0, F_j(0))$, the SDE

$$dF_j(t) = \nu(t, F_j(t)) F_j(t)\, dW_t \tag{6.92}$$

has a unique strong solution whose marginal density is given by the mixture of lognormals (6.90).

---

[3] Indeed, $p_t(\cdot)$ is a proper Q^j-density function since, by definition, the correct expectation is recovered:

$$\int_0^{+\infty} y p_t(y)\, dy = \sum_{i=1}^{N} \lambda_i \int_0^{+\infty} y p_t^i(y)\, dy = \sum_{i=1}^{N} \lambda_i F_j(0) = F_j(0).$$

[4] Such a problem is essentially the reverse to that of finding the marginal density function of the solution of an SDE when the coefficients are known.

278   6. The LIBOR and Swap Market Models (LFM and LSM)

The above proposition provides us with the analytical expression for the diffusion coefficient in the SDE (6.89) such that the resulting equation has a unique strong solution whose marginal density is given by (6.90). The square of the local volatility $\nu(t,y)$ can be viewed as a weighted average of the squared basic volatilities $\sigma_1^2(t), \ldots, \sigma_N^2(t)$, where the weights are all functions of the lognormal marginal densities (6.88). That is, for each $i = 1, \ldots, N$ and $(t,y) > (0,0)$, we can write

$$\nu^2(t,y) = \sum_{i=1}^{N} \Lambda_i(t,y)\sigma_i^2(t),$$

$$\Lambda_i(t,y) := \frac{\lambda_i p_t^i(y)}{\sum_{i=1}^{N} \lambda_i p_t^i(y)}.$$

As a consequence, for each $t > 0$ and $y > 0$, the function ν is bounded from below and above by (strictly) positive constants. In fact

$$\sigma_* \leq \nu(t,y) \leq \sigma^* \quad \text{for each } t, y > 0, \tag{6.93}$$

where

$$\sigma_* := \inf_{t \geq 0} \left\{ \min_{i=1,\ldots,N} \sigma_i(t) \right\} > 0,$$

$$\sigma^* := \sup_{t \geq 0} \left\{ \max_{i=1,\ldots,N} \sigma_i(t) \right\} < +\infty.$$

Remark 6.18.1. The function $\nu(t,y)$ can be extended by continuity to the semi-axes $\{(0,y) : y > 0\}$ and $\{(t,0) : t \geq 0\}$ by setting $\nu(0,y) = \sigma_0$ and $\nu(t,0) = \nu^*(t)$, where $\nu^*(t) := \sigma_{i^*}(t)$ and $i^* = i^*(t)$ is such that $V_{i^*}(t) = \max_{i=1,\ldots,N} V_i(t)$. In particular, $\nu(0,0) = \sigma_0$. Indeed, for every $\bar{y} > 0$ and every $\bar{t} \geq 0$,

$$\lim_{t \to 0} \nu(t, \bar{y}) = \sigma_0,$$

$$\lim_{y \to 0} \nu(\bar{t}, y) = \nu^*(t).$$

At time $t = 0$, the pricing of caplets under our forward-rate dynamics (6.92) is quite straightforward. Indeed,

$$P(0,T_j)E^j\{[F_j(T_{j-1}) - K]^+\} = P(0,T_j) \int_0^{+\infty} (y - K)^+ p_{T_{j-1}}(y)dy$$

$$= P(0,T_j) \sum_{i=1}^{N} \lambda_i \int_0^{+\infty} (y - K)^+ p_{T_{j-1}}^i(y)dy$$

so that, the caplet price $\mathbf{Cpl}(0, T_{j-1}, T_j, \tau, N, K)$ associated with our dynamics (6.92) is simply given by

$$\mathbf{Cpl}(0, T_{j-1}, T_j, \tau, N, K) = \tau N P(0, T_j) \sum_{i=1}^{N} \lambda_i \mathrm{Bl}(K, F_j(0), V_i(T_{j-1})). \quad (6.94)$$

The caplet price (6.94) leads to smiles in the implied volatility structure. An example of the shape that can be reproduced in shown in Figure 6.7.[5] Observe that the implied volatility curve has a minimum exactly at a strike equal to the initial forward rate $F_j(0)$. This property, which is formally proven in Brigo and Mercurio (2000a), makes the model suitable for recovering smile-shaped volatility surfaces. In fact, also skewed shapes can be retrieved, but with zero slope at the ATM level.

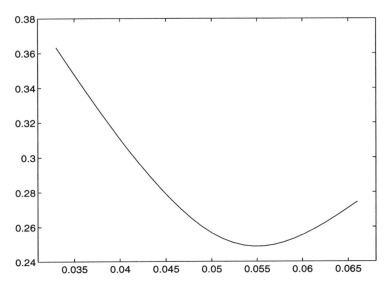

Fig. 6.7. Caplet volatility structure implied by the option prices (6.94), where we set, $T_{j-1} = 1$, $N = 3$, $(V_1(1), V_2(1), V_3(1)) = (0.6, 0.1, 0.2)$, $(\lambda_1, \lambda_2, \lambda_3) = (0.2, 0.3, 0.5)$ and $F_j(0) = 0.055$.

Given the above analytical tractability, we can easily derive an explicit approximation for the caplet implied volatility as a function of the caplet strike price. More precisely, define the moneyness m as the logarithm of the ratio between the forward rate and the strike, i.e.,

$$m := \ln \frac{F_j(0)}{K}.$$

The implied volatility $\hat{\sigma}(m)$ for the moneyness m is implicitly defined by equating the Black caplet price in $\hat{\sigma}(m)$ to the price implied by our model

---

[5] In such a figure, we consider directly the values of the V_i's. Notice that one can easily find some σ_i's satisfying our technical assumptions that are consistent with the chosen V_i's.

according to

$$\left[\Phi\left(\frac{m + \frac{1}{2}\hat{\sigma}(m)^2 T_{j-1}}{\hat{\sigma}(m)\sqrt{T_{j-1}}}\right) - e^{-m}\Phi\left(\frac{m - \frac{1}{2}\hat{\sigma}(m)^2 T_{j-1}}{\hat{\sigma}(m)\sqrt{T_{j-1}}}\right)\right]$$
$$= \sum_{i=1}^{N} \lambda_i \left[\Phi\left(\frac{m + \frac{1}{2}V_i^2(T_{j-1})}{V_i(T_{j-1})}\right) - e^{-m}\Phi\left(\frac{m - \frac{1}{2}V_i^2(T_{j-1})}{V_i(T_{j-1})}\right)\right]. \quad (6.95)$$

A repeated application of Dini's implicit function theorem and a Taylor's expansion around $m = 0$, lead to

$$\hat{\sigma}(m) = \hat{\sigma}(0) + \frac{1}{2\hat{\sigma}(0)T_{j-1}} \sum_{i=1}^{N} \lambda_i \left[\frac{\hat{\sigma}(0)\sqrt{T_{j-1}}}{V_i(T_{j-1})} e^{\frac{1}{8}(\hat{\sigma}(0)^2 T_{j-1} - V_i^2(T_{j-1}))} - 1\right] m^2$$
$$+ o(m^3) \quad (6.96)$$

where the ATM implied volatility, $\hat{\sigma}(0)$, is explicitly given by

$$\hat{\sigma}(0) = \frac{2}{\sqrt{T_{j-1}}} \Phi^{-1}\left(\sum_{i=1}^{N} \lambda_i \Phi\left(\frac{1}{2}V_i(T_{j-1})\right)\right). \quad (6.97)$$

We can finally motivate the assumption that the forward-rate density be given by the mixture of known lognormal densities. When proposing alternative dynamics, it can be quite problematic to come up with analytical formulas for caplet prices. Here, instead, such problem can be avoided from the beginning, just because the use of analytically-tractable densities $p^i_{T_{j-1}}$ immediately leads to explicit caplet prices for the process F_j. Moreover, the absence of bounds on the parameter N implies that a virtually unlimited number of parameters can be introduced in the forward-rate dynamics and used for a better calibration to market data. A last remark concerns the classic economic interpretation of a mixture of densities. We can indeed view F_j as a process whose density at time t coincides with the basic density $p^i_{T_{j-1}}$ with probability λ_i.

6.18.6 Shifting the Lognormal-Mixture Dynamics

Brigo and Mercurio (2000b) proposed a simple way to generalize the dynamics (6.92). With the main target consisting of retrieving a larger variety of volatility structures, the basic lognormal-mixture model was combined with the displaced-diffusion technique by assuming that the forward-rate process is given by

$$F_j(t) = \alpha + \bar{F}_j(t), \quad (6.98)$$

where α is a real constant and $\bar{F}_j$ evolves according to the basic "lognormal mixture" dynamics (6.92). It is easy to prove that this is actually the most

6.18 Including the Caplet Smile in the LFM

general affine transformation for which the forward-rate process is still a martingale under its canonical measure.

The analytical expression for the marginal density of such process is given by the shifted mixture of lognormals

$$p_t(y) = \sum_{i=1}^{N} \lambda_i \frac{1}{(y-\alpha)V_i(t)\sqrt{2\pi}} \exp\left\{-\frac{1}{2V_i^2(t)}\left[\ln\frac{y-\alpha}{F_j(0)-\alpha} + \frac{1}{2}V_i^2(t)\right]^2\right\},$$

with $y > \alpha$.

By Ito's formula, we obtain that the forward-rate process evolves according to

$$dF_j(t) = \nu(t, F_j(t) - \alpha)(F_j(t) - \alpha)dW_t. \tag{6.99}$$

This model for the forward-rate process preserves the analytical tractability of the original process $\bar{F}_j$. Indeed,

$$P(0,T_j)E^j\left\{[F_j(T_{j-1}) - K]^+\right\} = P(0,T_j)E^j\left\{[\bar{F}(T_{j-1}) - (K-\alpha)]^+\right\},$$

so that, for $\alpha < K$, the caplet price $\mathbf{Cpl}(0, T_{j-1}, T_j, \tau, N, K)$ associated with (6.98) is simply given by

$$\mathbf{Cpl}(t, T_{j-1}, T_j, \tau, N, K) = \tau N P(t, T_j) \sum_{i=1}^{N} \lambda_i \mathrm{Bl}(K - \alpha, F_j(0) - \alpha, V_i(T_{j-1})). \tag{6.100}$$

Moreover, the caplet implied volatility (as a function of m) can be approximated as follows:

$$\hat{\sigma}(m) = \hat{\sigma}(0) + \hat{\sigma}'(0)m + \frac{1}{2}\hat{\sigma}''(0)m^2 + o(m^2)$$

$$\hat{\sigma}'(0) = \alpha\sqrt{\frac{2\pi}{T_{j-1}}} \frac{e^{\frac{1}{8}\hat{\sigma}(0)^2 T_{j-1}}}{F_j(0)} \left(-\sum_{i=1}^{N} \lambda_i \Phi\left(\frac{1}{2}V_i(T_{j-1})\right) + \frac{1}{2}\right)$$

$$\hat{\sigma}''(0) = \frac{F_j(0)}{F_j(0) - \alpha} \sum_{i=1}^{N} \lambda_i \frac{e^{\frac{1}{8}(\hat{\sigma}(0)^2 T_{j-1} - V_i(T_{j-1})^2)}}{V_i(T_{j-1})\sqrt{T_{j-1}}} - \frac{4 - \hat{\sigma}(0)^2 \hat{\sigma}'(0)^2 T_{j-1}^2}{4\hat{\sigma}(0) T_{j-1}},$$

where the ATM implied volatility, $\hat{\sigma}(0)$, is explicitly given by

$$\hat{\sigma}(0) = \frac{2}{F_j(0)\sqrt{T_{j-1}}} \Phi^{-1}\left((F_j(0) - \alpha)\sum_{i=1}^{N} \lambda_i \Phi\left(\frac{1}{2}V_i(T_{j-1})\right) + \frac{\alpha}{2}\right). \tag{6.101}$$

For $\alpha = 0$ the process F_j obviously coincides with $\bar{F}_j$ while preserving the correct zero drift. The introduction of the new parameter α has the effect that, decreasing α, the variance of the asset price at each time increases while maintaining the correct expectation. Indeed:

282   6. The LIBOR and Swap Market Models (LFM and LSM)

$$E(F_j(t)) = F_j(0),$$

$$\text{Var}(F_j(t)) = (F_j(0) - \alpha)^2 \left(\sum_{i=1}^{N} \lambda_i e^{V_i^2(t)} - 1 \right).$$

As for the model (6.77), the parameter α affects the shape of the implied volatility curve in two ways. First, it concurs to determine the level of such curve in that changing α leads to an almost parallel shift of the curve itself. Second, it moves the strike with minimum volatility. Precisely, if $\alpha > 0$ (< 0) the minimum is attained for strikes lower (higher) than the ATM's. When varying all parameters, the parameter α can be used to add asymmetry around the ATM volatility without shifting the curve.

Finally, as far as the calibration of the above models to swaptions is concerned, once again approximated swaption prices similar to Rebonato's formula in the LFM and based on "freezing the drift" and "collapsing all measures" approaches can be attempted, although results need to be checked numerically in a sufficiently rich number of situations.

7. Cases of Calibration of the LIBOR Market Model

If I find 10,000 ways something won't work, I haven't failed. I am not discouraged, because every wrong attempt discarded is another step forward.
Thomas Edison (1847-1931)

In this chapter we present some numerical examples concerning the goodness of fit of the LFM to both the caps and swaptions markets, based on market data. We study several cases based on different instantaneous-volatility parameterizations. We will also point out a particular parameterization allowing for a closed-form-formulas calibration to swaption volatilities and establishing a one to one correspondence between swaption volatilities and LFM covariance parameters.

However, before proceeding, we would like to make ourselves clear by pointing out that the examples presented here are a first attempt at underlying the relevant choices as far as the LFM parameterization is concerned. We do not pretend to be exhaustive or even particularly systematic in these examples, and we do not employ statistical testing or econometric techniques in our analysis. We will base our considerations only on cross-sectional calibration to the market-quoted volatilities, although we will check implications of the obtained calibrations as far as the future time-evolution of key structures of the market are concerned. Therefore, to decide upon the quality of a calibration, we will look at the future starting from the present, rather than using the past to elicit information on the present. Within such cross-sectional approach we are not being totally systematic. We are aware there are several other issues concerning the number of factors, further possible parameterizations, modeling correlations implicitly as inner products of vector instantaneous volatilities, and so on. Yet, most of the available literature on interest-rate models does not deal with the questions and examples we raise here on the market model. We thought about presenting the examples below in order to let the reader appreciate what are the current problems with the LFM, especially as far as practitioners and traders are concerned. Indeed, the kind of problems we used to read about in interest-rate modeling before starting our work in a bank was rather different from the problems presented here. We hope the reader will have a clear grasp of the main issues at stake nowadays, and this is all we are trying to obtain with the examples below.

7.1 The Inputs

> On two occasions I have been asked [by members of Parliament], 'Pray, Mr. Babbage, if you put into the machine wrong figures, will the right answers come out?' I am not able rightly to apprehend the kind of confusion of ideas that could provoke such a question.
>
> Charles Babbage, Passages from the Life of a Philosopher, 1864, London.

We will try and calibrate the following data: "annualized" initial curve of forward rates, "annualized" caplet volatilities, and swaptions volatilities. Let us examine these data in more detail.

We actually take as input the vector

$$F0 = [F(0;0,0.5), F(0;0.5,1), \ldots, F(0;19.5,20)]$$

of initial semi-annual forward rates as of May 16, 2000, and the semi-annual at-the-money caplet volatilities (y stands for year/years)

$$v0 = [v_{1y-\text{caplet}}, v_{1.5y-\text{caplet}}, \ldots, v_{19.5y-\text{caplet}}],$$

with the first semi-annual caplet resetting in one year and paying at 1.5 (years), the last semi-annual caplet resetting in 19.5 (years) and paying at 20 (years), all other reset dates being six-months spaced.

These volatilities have been provided by our interest-rate traders, based on a stripping algorithm combined with personal adjustments applied to cap volatilities. All basic data are as of May 16, 2000. We transform semi-annual data in annual data through the methods highlighted in Section 6.16, and work with the annual forward rates

$$[F(\cdot; 1y, 2y), F(\cdot; 2y, 3y), \ldots, F(\cdot; 19y, 20y)]$$

and their associated annual caplet volatilities. In the transformation formula, infra-correlations are set to one. Notice that infra-correlations might be kept as further parameters to ease the calibration, see again Section 6.16 for a discussion.

In our data, the initial spot rate is $F(0;0,1y) = 0.0452$, and the other initial forward rates and caplet volatilities are shown in Table 7.1

Finally, the values of swaptions volatilities are the same as in the table given in Section 6.15.

7.2 Joint Calibration with Piecewise-Constant Volatilities as in TABLE 5

In order to satisfactorily calibrate the above data with the LFM, we first try the volatility structure of TABLE 5, with a local algorithm of minimization for finding the best-fitting parameters $\psi_1, \ldots, \psi_{19}$ and $\theta_1, \ldots, \theta_{19}$, starting

7.2 Joint Calibration with Piecewise-Constant Volatilities as in TABLE 5

initial $F0$	v_{caplet}
0.050114	0.180253
0.055973	0.191478
0.058387	0.186154
0.060027	0.177294
0.061315	0.167887
0.062779	0.158123
0.062747	0.152688
0.062926	0.148709
0.062286	0.144703
0.063009	0.141259
0.063554	0.137982
0.064257	0.134708
0.064784	0.131428
0.065312	0.128148
0.063976	0.1271
0.062997	0.126822
0.06184	0.126539
0.060682	0.126257
0.05936	0.12597

Table 7.1. Initial (annualized) forward rates and caplet volatilities.

from the initial guesses $\psi_i = 1$ and $\theta_i = \pi/2$. We thus adopt formula (6.11) for instantaneous volatilities and obtain the Φ's directly as functions of the parameters ψ by using the (annualized) caplet volatilities and formula (6.24). We compute swaptions prices as functions of the ψ's and θ's by using Rebonato's formula (6.58). Since we are using piecewise-constant instantaneous volatilities, the formula reduces to a summation of products of volatility parameters. We also impose the constraints

$$-\pi/2 < \theta_i - \theta_{i-1} < \pi/2$$

to the correlation angles, which implies that $\rho_{i,i-1} > 0$. We thus require that adjacent rates have positive correlations. As we shall see, this requirement is obviously too weak to guarantee the instantaneous correlation matrix coming from the calibration to be "reasonable".

We obtain the parameters that are shown in Table 7.2, where the Φ's have been computed through (6.24).

The fitting quality is as follows. The caplets are fitted exactly, whereas we calibrated the whole swaptions volatility matrix except for the first column of $S \times 1$-swaptions. This is left aside because of possible misalignments with the annualized caplet volatilities, since we are basically quoting twice the same volatilities. A more complete approach can be obtained by keeping semi-annual volatilities and by introducing infra-correlations as new parameters as indicated in Section 6.16. The matrix of percentage errors in the swaptions calibration,

$$100 \cdot \frac{\text{Market swaption volatility - LFM swaption volatility}}{\text{Market swaption volatility}}$$

is reported below.

286   7. Cases of Calibration of the LIBOR Market Model

Index	ψ	Φ	θ
1	1.3392	0.1346	2.3649
2	1.3597	0.1419	1.2453
3	1.1771	0.1438	0.7676
4	1.3563	0.1353	0.3877
5	1.2807	0.1287	0.2870
6	1.2696	0.1218	1.8578
7	1.1593	0.1193	1.9923
8	1.7358	0.1105	2.2133
9	0.8477	0.1114	2.1382
10	1.9028	0.1030	2.4456
11	0.6717	0.1043	1.5708
12	2.8610	0.0891	1.5708
13	0.6620	0.0897	1.5708
14	1.2649	0.0883	1.5708
15	0.1105	0.0906	1.5708
16	0.0000	0.0934	1.5708
17	0.0000	0.0961	1.5708
18	0.0000	0.0986	1.5708
19	0.6551	0.1004	1.5708

Table 7.2. Calibration results under the volatility formulation of TABLE 5: parameter values.

	2y	3y	4y	5y	6y	7y	8y	9y	10y
1y	-0.24%	1.29%	0.61%	-0.59%	-0.28%	0.97%	-0.64%	1.02%	-0.87%
2y	-1.65%	-1.29%	-1.09%	-0.33%	0.03%	-0.61%	-0.37%	-1.04%	-0.89%
3y	1.03%	1.11%	0.51%	1.45%	0.79%	1.08%	1.03%	1.30%	0.79%
4y	0.13%	-1.05%	-0.80%	-0.29%	-0.33%	-0.16%	0.49%	0.02%	0.23%
5y	0.89%	0.07%	-0.09%	-0.16%	-1.27%	-0.50%	0.00%	-0.80%	0.37%
7y	1.15%	0.53%	0.59%	0.12%	-0.33%	-0.66%	-0.58%	0.10%	0.93%
10y	-1.23%	-0.61%	0.64%	0.07%	0.46%	0.45%	-0.37%	-0.64%	-0.20%

Errors are actually small, and from this point of view, such calibration seems to be satisfactory, considering that we are trying to fit 19 caplets and 63 swaption volatilities! However, since the LFM allows for a quick check of future term structures of volatilities and terminal correlations, let us have a look at these quantities.

The first observation is that the calibrated θ's above imply quite erratic instantaneous correlations. Consider the resulting instantaneous-correlation matrix as given in Tables 7.3 and 7.4. As you can see, correlations fluctuate occasionally between positive and negative values. At the same time, from a certain index on, they become constant. This is too a weird behaviour to be trusted. Terminal correlations computed through formula (6.62) are in this case, after ten years,

	10y	11y	12y	13y	14y	15y	16y	17y	18y	19y
10y	1.000	0.574	0.625	0.525	0.602	0.500	0.505	0.445	0.418	0.413
11y	0.574	1.000	0.801	0.976	0.805	0.926	0.750	0.760	0.646	0.641
12y	0.625	0.801	1.000	0.712	0.917	0.653	0.745	0.590	0.588	0.597
13y	0.525	0.976	0.712	1.000	0.697	0.906	0.622	0.721	0.550	0.576
14y	0.602	0.805	0.917	0.697	1.000	0.655	0.866	0.570	0.666	0.540
15y	0.500	0.926	0.653	0.906	0.655	1.000	0.627	0.855	0.524	0.641
16y	0.505	0.750	0.745	0.622	0.866	0.627	1.000	0.597	0.846	0.487
17y	0.445	0.760	0.590	0.721	0.570	0.855	0.597	1.000	0.556	0.825
18y	0.418	0.646	0.588	0.550	0.666	0.524	0.846	0.556	1.000	0.516
19y	0.413	0.641	0.597	0.576	0.540	0.641	0.487	0.825	0.516	1.000

7.2 Joint Calibration with Piecewise-Constant Volatilities as in TABLE 5   287

	1y	2y	3y	4y	5y	6y	7y	8y	9y
1y	1.000	0.436	-0.026	-0.395	-0.486	0.874	0.931	0.989	0.974
2y	0.436	1.000	0.888	0.654	0.575	0.818	0.734	0.567	0.627
3y	-0.026	0.888	1.000	0.929	0.887	0.462	0.339	0.125	0.199
4y	-0.395	0.654	0.929	1.000	0.995	0.101	-0.034	-0.252	-0.179
5y	-0.486	0.575	0.887	0.995	1.000	0.000	-0.134	-0.348	-0.277
6y	0.874	0.818	0.462	0.101	0.000	1.000	0.991	0.937	0.961
7y	0.931	0.734	0.339	-0.034	-0.134	0.991	1.000	0.976	0.989
8y	0.989	0.567	0.125	-0.252	-0.348	0.937	0.976	1.000	0.997
9y	0.974	0.627	0.199	-0.179	-0.277	0.961	0.989	0.997	1.000
10y	0.997	0.362	-0.107	-0.468	-0.555	0.832	0.899	0.973	0.953
11y	0.701	0.947	0.694	0.378	0.283	0.959	0.912	0.801	0.843
12y	0.701	0.947	0.694	0.378	0.283	0.959	0.912	0.801	0.843
13y	0.701	0.947	0.694	0.378	0.283	0.959	0.912	0.801	0.843
14y	0.701	0.947	0.694	0.378	0.283	0.959	0.912	0.801	0.843
15y	0.701	0.947	0.694	0.378	0.283	0.959	0.912	0.801	0.843
16y	0.701	0.947	0.694	0.378	0.283	0.959	0.912	0.801	0.843
17y	0.701	0.947	0.694	0.378	0.283	0.959	0.912	0.801	0.843
18y	0.701	0.947	0.694	0.378	0.283	0.959	0.912	0.801	0.843
19y	0.701	0.947	0.694	0.378	0.283	0.959	0.912	0.801	0.843

Table 7.3. Calibration results under the volatility formulation of TABLE 5: instantaneous-correlation matrix (part I).

	10y	11y	12y	13y	14y	15y	16y	17y	18y	19y
1y	0.997	0.701	0.701	0.701	0.701	0.701	0.701	0.701	0.701	0.701
2y	0.362	0.947	0.947	0.947	0.947	0.947	0.947	0.947	0.947	0.947
3y	0.107	0.694	0.694	0.694	0.694	0.694	0.694	0.694	0.694	0.694
4y	0.468	0.378	0.378	0.378	0.378	0.378	0.378	0.378	0.378	0.378
5y	0.555	0.283	0.283	0.283	0.283	0.283	0.283	0.283	0.283	0.283
6y	0.832	0.959	0.959	0.959	0.959	0.959	0.959	0.959	0.959	0.959
7y	0.899	0.912	0.912	0.912	0.912	0.912	0.912	0.912	0.912	0.912
8y	0.973	0.801	0.801	0.801	0.801	0.801	0.801	0.801	0.801	0.801
9y	0.953	0.843	0.843	0.843	0.843	0.843	0.843	0.843	0.843	0.843
10y	1.000	0.641	0.641	0.641	0.641	0.641	0.641	0.641	0.641	0.641
11y	0.641	1.000	1.000	1.000	1.000	1.000	1.000	1.000	1.000	1.000
12y	0.641	1.000	1.000	1.000	1.000	1.000	1.000	1.000	1.000	1.000
13y	0.641	1.000	1.000	1.000	1.000	1.000	1.000	1.000	1.000	1.000
14y	0.641	1.000	1.000	1.000	1.000	1.000	1.000	1.000	1.000	1.000
15y	0.641	1.000	1.000	1.000	1.000	1.000	1.000	1.000	1.000	1.000
16y	0.641	1.000	1.000	1.000	1.000	1.000	1.000	1.000	1.000	1.000
17y	0.641	1.000	1.000	1.000	1.000	1.000	1.000	1.000	1.000	1.000
18y	0.641	1.000	1.000	1.000	1.000	1.000	1.000	1.000	1.000	1.000
19y	0.641	1.000	1.000	1.000	1.000	1.000	1.000	1.000	1.000	1.000

Table 7.4. Calibration results under the volatility formulation of TABLE 5: instantaneous-correlation matrix (part II).

and they still look erratic, although to a lesser degree. Notice that the corresponding entries in the instantaneous-correlation matrix are almost all equal to one, meaning that these terminal correlations are almost completely determined by instantaneous volatilities.

Finally, let us have a look at the time evolution of caplet volatilities. We know that the model reproduces exactly the initial caplet volatility structure observed in the market. However, as time passes, the above ψ and Φ parameters imply the evolution shown in Figure 7.1. This evolution shows that the

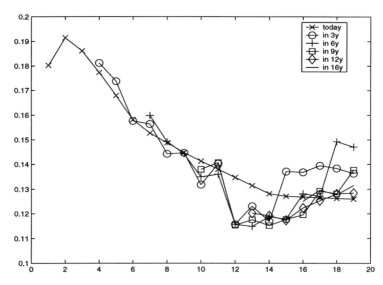

Fig. 7.1. Calibration results under the volatility formulation of TABLE 5: evolution of the term structure of caplet volatilities.

structure loses its humped shape after a short time. Moreover, it becomes somehow "noisy".

What is one to learn from such an example? The idea is that the fitting quality is not the only criterion by which a calibration session has to be judged. The so called "uncertainty principle of modeling", "the more a model fits the less it explains", has to be carefully considered here. A trader has then to decide whether he is willing to sacrifice the fitting quality for a better evolution in time of the key structures.

7.2.1 Instantaneous Correlations: Narrowing the Angles

The instantaneous correlations in the previous examples appear to be arbitrary to a certain degree. In particular, given their strange behaviour, one might suspect they are a byproduct of the particular local-optimization method used in the calibration. Do there exist more regular instantaneous correlations that imply the same results in terms of fitting quality? We try and answer this and other questions in the following sections.

A first attempt to improve the situation is based on constraining the θ's to range in smaller intervals. Let us then change the constraints on the θ's as follows:

$$-\pi/2 < \theta_i - \theta_{i-1} < \pi/2 \quad \to \quad -\pi/8 < \theta_i - \theta_{i-1} < \pi/8.$$

This implies that $\rho_{i,i-1} > 0, \ldots, \rho_{i,i-4} > 0$. We thus require also "sub-adjacent" rates up to the "fourth level" to have positive correlations. The ψ

7.2 Joint Calibration with Piecewise-Constant Volatilities as in TABLE 5

parameters now read

$$\psi_{1\div 8} = [2.5834 \ \ 1.6333 \ \ 1.1076 \ \ 1.8114 \ \ 0.2315 \ \ 2.3874 \ \ 0.4012 \ \ 1.5133],$$
$$\psi_{9\div 19} = [1.2685 \ \ 0 \ \ 3.2337 \ \ 0 \ \ 0.5089 \ \ 1.1768 \ \ 0 \ \ 0 \ \ 0 \ \ 0 \ \ 0.7236].$$

The resulting Φ's assume several values in-between 0.0098 and 0.1007.

The overall fitting quality deteriorates with respect to the previous case. Indeed, the new percentage differences, reported in the following table, are usually higher and range up to 12.1%.

	2y	3y	4y	5y	6y	7y	8y	9y	10y
1y	-0.56%	5.85%	-1.37%	12.13%	-1.43%	5.73%	1.96%	0.07%	5.68%
2y	-3.87%	-4.21%	-2.44%	1.19%	-2.00%	1.06%	-1.76%	0.89%	-1.64%
3y	-3.80%	-0.11%	-3.33%	1.89%	-0.83%	0.73%	1.49%	-0.11%	0.39%
4y	-1.43%	-2.90%	-3.85%	0.23%	-2.81%	1.08%	-1.12%	-0.27%	0.60%
5y	-0.21%	-0.65%	-2.37%	-0.57%	-1.93%	-0.86%	-0.82%	-0.51%	0.83%
7y	1.95%	-0.63%	-1.78%	-0.50%	-1.60%	-0.17%	-0.88%	0.21%	1.10%
10y	-0.65%	-0.88%	-0.77%	0.71%	-0.22%	0.76%	-0.79%	-0.67%	-0.26%

However, correlations look slightly better. We no longer observe oscillations between positive and negative values. The first ten columns of the instantaneous-correlation matrix are given in Table 7.5.

	1y	2y	3y	4y	5y	6y	7y	8y	9y	10y
1y	1.000	0.924	0.707	0.557	0.454	0.760	0.843	0.837	0.837	0.920
2y	0.924	1.000	0.924	0.833	0.760	0.951	0.985	0.983	0.983	1.000
3y	0.707	0.924	1.000	0.981	0.951	0.997	0.976	0.979	0.979	0.928
4y	0.557	0.833	0.981	1.000	0.993	0.963	0.916	0.921	0.920	0.838
5y	0.454	0.760	0.951	0.993	1.000	0.924	0.862	0.867	0.867	0.767
6y	0.760	0.951	0.997	0.963	0.924	1.000	0.990	0.992	0.992	0.954
7y	0.843	0.985	0.976	0.916	0.862	0.990	1.000	1.000	1.000	0.986
8y	0.837	0.983	0.979	0.921	0.867	0.992	1.000	1.000	1.000	0.985
9y	0.837	0.983	0.979	0.920	0.867	0.992	1.000	1.000	1.000	0.985
10y	0.920	1.000	0.928	0.838	0.767	0.954	0.986	0.985	0.985	1.000
11y	0.820	0.977	0.985	0.932	0.882	0.995	0.999	1.000	1.000	0.979
12y	0.820	0.977	0.985	0.932	0.882	0.995	0.999	1.000	1.000	0.979
13y	0.820	0.977	0.985	0.932	0.882	0.995	0.999	1.000	1.000	0.979
14y	0.820	0.977	0.985	0.932	0.882	0.995	0.999	1.000	1.000	0.979
15y	0.820	0.977	0.985	0.932	0.882	0.995	0.999	1.000	1.000	0.979
16y	0.820	0.977	0.985	0.932	0.882	0.995	0.999	1.000	1.000	0.979
17y	0.820	0.977	0.985	0.932	0.882	0.995	0.999	1.000	1.000	0.979
18y	0.820	0.977	0.985	0.932	0.882	0.995	0.999	1.000	1.000	0.979
19y	0.820	0.977	0.985	0.932	0.882	0.995	0.999	1.000	1.000	0.979

Table 7.5. Calibration results under the volatility formulation of TABLE 5, case II: first ten columns of the instantaneous-correlation matrix.

Similar remarks apply to terminal correlations, and especially to those corresponding to the portions of the instantaneous-correlation matrix where oscillations between positive and negative elements were previously present. However, even this new correlation matrix is far from being an ideal one. Columns exhibit a non-monotonic behaviour when moving away from the diagonal elements, although now we no longer have oscillations between positive

290   7. Cases of Calibration of the LIBOR Market Model

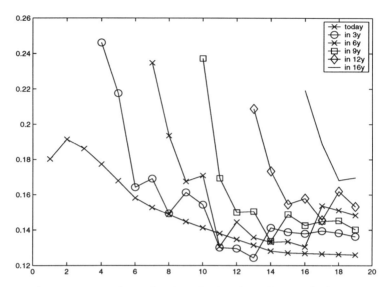

Fig. 7.2. Calibration results under the volatility formulation of TABLE 5, case II: evolution of the term structure of caplet volatilities.

and negative values. The evolution of the term structure of caplet volatilities is shown in Figure 7.2.

It is clear that we have paid our improvement on correlations in terms of fitting quality and in terms of regularity of the evolution of the term structure of volatilities.

One can try and reduce less drastically the allowed ranges for the θ's, to, say,

$$-\pi/4 < \theta_i - \theta_{i-1} < \pi/4.$$

By doing so, one obtains an intermediate situation with respect to the previous two cases. Correlations are better than in the first case and worse than in the second one, whereas the fitting quality is better than in the second case but worse than in the first. One can play with intermediate situations, but the correlation matrices are never really satisfactory. Usually accepting wild correlations allows one to obtain a good fitting even when constraining the Φ's to be close to one so as to ensure a nice evolution of the term structure of volatilities. We will see this in detail later on for Formulation 7 of instantaneous volatilities, but the same conclusions apply here.

7.2.2 Instantaneous Correlations: Fixing the Angles to Typical Values

Now we test a slightly different situation. We keep the same inputs, but fix instantaneous correlations to some reasonable values. Our purpose is to check whether the remaining parameters ψ are flexible enough to fit swaptions

7.2 Joint Calibration with Piecewise-Constant Volatilities as in TABLE 5   291

volatilities even when the θ's are fixed (for example supposing that they have been estimated historically). Suppose that the θ's are fixed at the values of Figure 8.1, where the corresponding correlations are also shown. Calibrating the given data using only the parameters ψ, we obtain

$\psi_{1 \div 10} = [1.9945\ \ 2.3007\ \ 0.2003\ \ 2.1209\ \ 0\ \ 2.2401\ \ 0\ \ 0.1513\ \ 3.5668\ \ 0]$,

$\psi_{11 \div 19} = [0\ \ 0\ \ 2.0542\ \ 0\ \ 1.4820\ \ 1.0072\ \ 0\ \ 0.6878\ \ 1.0203]$.

The related Φ's range between 0.0773 and 0.1057, and are certainly not uniform enough to guarantee a stable behaviour for the shape of the term structure of caplet volatilities over time, as we will see. What has changed with respect to the "moving angles" case? Clearly, fixing the θ's has subtracted flexibility from the model, and the calibration has worsened. Percentage differences, in fact, are given by

	2y	3y	4y	5y	6y	7y	8y	9y	10y
1y	-15.40%	12.33%	-4.41%	11.29%	-0.18%	9.92%	15.83%	2.51%	8.82%
2y	-4.18%	-0.09%	-2.76%	-1.71%	-2.99%	6.85%	2.66%	-1.54%	4.75%
3y	-2.72%	2.02%	-4.81%	-2.75%	-1.25%	0.45%	-1.37%	-2.97%	3.20%
4y	-0.81%	-2.55%	-6.52%	-2.84%	-6.45%	-2.66%	-2.91%	-3.23%	0.02%
5y	-1.46%	-2.04%	-3.77%	-2.80%	-2.14%	-0.74%	-0.31%	-1.83%	1.44%
7y	0.86%	0.54%	-6.19%	-5.28%	-3.67%	-0.36%	-0.27%	-0.88%	1.41%
10y	2.36%	3.13%	-0.46%	-1.04%	-0.91%	0.21%	-0.24%	-1.08%	0.42%

It is clear that now we have large percentage errors, especially in the first two rows, which have been penalized most by fixing instantaneous correlations. Terminal correlations after ten years look like

	10y	11y	12y	13y	14y	15y	16y	17y	18y	19y
10y	1.000	0.203	0.388	0.433	0.479	0.439	0.242	0.643	0.269	0.558
11y	0.203	1.000	0.057	0.367	0.340	0.489	0.326	0.260	0.667	0.009
12y	0.388	0.057	1.000	0.039	0.229	0.361	0.355	0.366	0.272	0.572
13y	0.433	0.367	0.039	1.000	0.022	0.320	0.431	0.327	0.390	0.465
14y	0.479	0.340	0.229	0.022	1.000	0.023	0.159	0.473	0.339	0.119
15y	0.439	0.489	0.361	0.320	0.023	1.000	0.091	0.152	0.491	0.590
16y	0.242	0.326	0.355	0.431	0.159	0.091	1.000	0.100	0.182	0.304
17y	0.643	0.260	0.366	0.327	0.473	0.152	0.100	1.000	0.099	0.276
18y	0.269	0.667	0.272	0.390	0.339	0.491	0.182	0.099	1.000	0.161
19y	0.558	0.009	0.572	0.465	0.119	0.590	0.304	0.276	0.161	1.000

Here we notice an important point. Although initial instantaneous correlations are in this case quite regular, terminal correlations after 10 years appear to be somehow erratic. Notice that there are still wide oscillations, although not involving negative values, as in some of the previous cases. This can only be due to a "noisy" behaviour of volatility, since terminal correlations depend on instantaneous correlations (which are smooth in this case) and on volatilities.

Let us then check the evolution of the term structure of volatilities, shown in Figure 7.3. Once again, the evolution is far from being regular, although it does not look "uniformly" worse than in the "narrow moving angles" case. Indeed, the initial hump is somehow preserved, although the remaining parts of the curves lose their smooth initial behaviour. This shows that narrowing

292   7. Cases of Calibration of the LIBOR Market Model

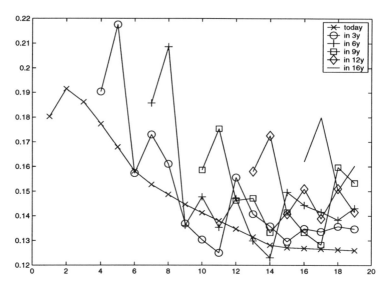

Fig. 7.3. Calibration results under the volatility formulation of TABLE 5, case III: evolution of the term structure of caplet volatilities.

the angles too much can be worse than fixing them to reasonable values, as far as the volatilities evolution is concerned, although the same cannot be said for the fitting quality to market data.

Summing up, we have gained better instantaneous correlations, terminal correlations less erratic than in some of the above cases, slightly better evolution of the term structure of volatilities but at the price of a considerable worsening of the calibration.

7.2.3 Instantaneous Correlations: Fixing the Angles to Atypical Values

Now maintain the same inputs, but fix instantaneous correlations to some rank-two values involving negatives entries in the matrix. Precisely, we fix instantaneous correlations to the values used in the test (1.c) for analytical formulas below shown in Figure 8.6. How do volatilities change to fit the same data when we impose a different fixed instantaneous correlation matrix with some negative entries? Calibrate the given data using only the parameters ψ. We obtain

$$\psi_{1 \div 10} = [2.0220 \ \ 2.3287 \ \ 0.2031 \ \ 2.1322 \ \ 0 \ \ 2.2482 \ \ 0 \ \ 0.1382 \ \ 3.5205 \ \ 0],$$
$$\psi_{11 \div 19} = [0 \ \ 0 \ \ 1.9151 \ \ 0.5790 \ \ 0.3136 \ \ 0.7184 \ \ 0.7932 \ \ 0.9221 \ \ 0.9875].$$

Notice that these values are not too different from the corresponding ones of the previous case. The related Φ's range between 0.0773 and 0.1043, and again, they are certainly not uniform enough to guarantee a stable behaviour

7.2 Joint Calibration with Piecewise-Constant Volatilities as in TABLE 5

for the shape of the term structure of caplet volatilities over time, as we will see. What has changed with respect to the previous "fixed typical angles" case? Percentage differences are given by

	2y	3y	4y	5y	6y	7y	8y	9y	10y
1y	-15.39%	12.32%	-4.37%	11.32%	-0.22%	9.89%	15.89%	2.38%	8.69%
2y	-4.19%	-0.03%	-2.70%	-1.68%	3.01%	6.88%	2.74%	-1.60%	4.69%
3y	-2.74%	2.05%	-4.77%	-2.73%	-1.26%	0.53%	-1.28%	-3.01%	3.16%
4y	-0.86%	-2.52%	-6.49%	-2.80%	-6.43%	-2.56%	-2.81%	-3.27%	0.10%
5y	-1.53%	-2.02%	-3.74%	-2.72%	-2.09%	-0.63%	-0.22%	-1.80%	1.01%
7y	0.79%	0.56%	-6.15%	-5.21%	-3.67%	-0.24%	-0.56%	-1.13%	1.96%
10y	2.30%	3.09%	-0.61%	-1.35%	-1.59%	0.43%	0.28%	-1.08%	0.33%

Also this matrix resembles a lot to the corresponding matrix of the previous case. Terminal correlations after ten years look like

	10y	11y	12y	13y	14y	15y	16y	17y	18y	19y
10y	1.000	0.207	0.393	0.427	0.403	0.301	0.002	-0.247	-0.265	-0.283
11y	0.207	1.000	0.056	0.369	0.304	0.299	0.035	-0.033	-0.347	-0.141
12y	0.393	0.056	1.000	0.038	0.216	0.254	0.059	-0.121	-0.042	-0.215
13y	0.427	0.369	0.038	1.000	0.066	0.185	0.135	-0.040	-0.163	-0.058
14y	0.403	0.304	0.216	0.066	1.000	0.075	0.037	0.078	-0.024	0.000
15y	0.301	0.299	0.254	0.185	0.075	1.000	0.089	0.046	-0.191	0.319
16y	0.002	0.035	0.059	0.135	0.037	0.089	1.000	0.136	0.088	0.218
17y	-0.247	-0.033	-0.121	-0.040	0.078	0.046	0.136	1.000	0.179	0.240
18y	-0.265	-0.347	-0.042	-0.163	-0.024	0.191	0.088	0.179	1.000	0.325
19y	-0.283	-0.141	-0.215	-0.058	0.000	0.319	0.218	0.240	0.325	1.000

As before, terminal correlations are not regular. This is due again to a "noisy" behaviour of volatility. Finally, the evolution of the term structure of volatilities is almost identical to the one shown in Figure 7.3.

7.2.4 Instantaneous Correlations: Collapsing to One Factor

Let us now push things further and fix all angles at the same value, so that the instantaneous-correlation matrix collapses to the unit matrix (all entries are one) and we basically obtain a one-factor LFM. Again, the swaptions calibration relies completely on the ψ parameters of instantaneous volatility. In this case we obtain the values

$$\psi_{1 \div 10} = [1.9961 \ 2.2976 \ 0.2001 \ 2.1047 \ 0 \ 2.2202 \ 0 \ 0.1420 \ 3.4499 \ 0],$$
$$\psi_{11 \div 19} = [0 \ 0 \ 2.0334 \ 0 \ 1.0791 \ 1.2100 \ 0 \ 0.8888 \ 1.1757],$$

which are rather close, qualitatively, to the values observed in the previous two cases.

Once again, the related Φ's range between 0.0785 and 0.1057, and are far from being uniform enough to guarantee preservation of the shape of the term structure of caplet volatilities over time. This is evident when looking at the evolution of the volatility term structure, which is almost identical to that shown in Figure 7.3.

The fitting quality is expressed by the usual percentage differences in the following table:

294   7. Cases of Calibration of the LIBOR Market Model

	2y	3y	4y	5y	6y	7y	8y	9y	10y
1y	-15.38%	12.34%	-4.36%	11.33%	-0.23%	9.88%	15.84%	2.38%	8.70%
2y	-4.18%	-0.02%	-2.70%	-1.68%	-3.01%	6.85%	2.73%	-1.58%	4.70%
3y	-2.73%	2.06%	-4.77%	-2.74%	-1.29%	0.53%	-1.26%	-2.98%	3.19%
4y	-0.85%	-2.51%	-6.49%	-2.82%	-6.45%	-2.56%	-2.80%	-3.24%	-0.08%
5y	-1.52%	-2.01%	-3.75%	-2.74%	-2.10%	-0.63%	-0.21%	-1.88%	1.31%
7y	0.80%	0.53%	-6.18%	-5.24%	-3.70%	-0.35%	-0.30%	-0.86%	1.53%
10y	2.25%	3.00%	-0.65%	-1.22%	-1.06%	0.38%	-0.07%	-1.08%	0.23%

Differences are quite close to the previous cases where we fixed the angles to produce either typical or atypical rank-two correlation matrices, as we have noticed also for the ψ values. Terminal correlations after ten years are completely determined by instantaneous volatilities, in this case where instantaneous correlations have been set to one. We obtain:

	10y	11y	12y	13y	14y	15y	16y	17y	18y	19y
10y	1.000	0.208	0.400	0.441	0.490	0.461	0.198	0.713	0.289	0.589
11y	0.208	1.000	0.057	0.377	0.343	0.511	0.339	0.212	0.737	0.011
12y	0.400	0.057	1.000	0.039	0.235	0.372	0.366	0.379	0.216	0.640
13y	0.441	0.377	0.039	1.000	0.021	0.302	0.464	0.335	0.420	0.405
14y	0.490	0.343	0.235	0.021	1.000	0.022	0.126	0.511	0.347	0.153
15y	0.461	0.511	0.372	0.302	0.022	1.000	0.088	0.123	0.533	0.588
16y	0.198	0.339	0.366	0.464	0.126	0.088	1.000	0.096	0.171	0.374
17y	0.713	0.212	0.379	0.335	0.511	0.123	0.096	1.000	0.094	0.253
18y	0.289	0.737	0.216	0.420	0.347	0.533	0.171	0.094	1.000	0.178
19y	0.589	0.011	0.640	0.405	0.153	0.588	0.374	0.253	0.178	1.000

Terminal correlations have changed, but not enough to be qualitatively different than in the previous two cases. We can conclude that in order to have a good calibration to swaptions data we need to allow for at least partially oscillating patterns in the correlation matrix. If we force a given "smooth/monotonic" correlation matrix into the calibration and rely upon volatilities, the results are the same as in the case of a one-factor LFM where correlations are all set to one. This kind of results suggests some considerations. First, one can start to suspect the following. Since by fixing rather different instantaneous correlations, the results of the calibration to the same data do not change that much, probably instantaneous correlations do not have a strong link with European swaptions prices. Therefore, one may suspect that swaptions volatilities do not always contain clear and precise information on instantaneous correlations of forward rates. This was clearly stated also in Rebonato (1999d). On the other hand, one may also suspect that this permanence of "bad results", no matter the particular "smooth" choice of fixed instantaneous correlation, reflects the impossibility of a low-rank formulation to decorrelate quickly the forward rates in a steeper initial pattern. Indeed, as we have seen in the figures of Section 6.9, instantaneous correlations of a low-rank formulation have a sigmoid-like shape that, initially, cannot decrease quickly. Instantaneous correlations in the rank-two model are then trying to mimic something like a steep initial pattern present in the market by a sigmoid like shape through an oscillating behaviour. The obvious remedy would be to increase drastically the number of factors, but this is computationally undesirable. We have tried experiments with three-factor correlation matrices, but we have obtained results analogous to the

7.3 Joint Calibration with Parameterized Volatilities as in Formulation 7

two-factor case. And resorting to a higher-dimensional model (up to nineteen factors) is not desirable in terms of implementation issues.

7.3 Joint Calibration with Parameterized Volatilities as in Formulation 7

We now adopt Formulation 7 for instantaneous volatilities. Instantaneous volatilities are now given by formula (6.13), and depend on the parameters a, b, c, d and Φ's. We again try a local algorithm of minimization for finding the fitted parameters a, b, c, d and $\theta_1, \ldots, \theta_{19}$ starting from the initial guesses $a = 0.0285$, $b = 0.20004$, $c = 0.1100$, $d = 0.0570$, and initial θ components ranging from $\theta_1 = 0$ to $\theta_{19} = 2\pi$ and equally spaced. The Φ's are obtained as functions of a, b, c, d, $\Phi = \Phi(a, b, c, d)$, through caplet volatilities according to formula (6.27), and this is the caplet calibration part. As for swaptions, we compute swaptions prices as functions of a, b, c, d and the θ's by using Rebonato's formula (6.58). To this end, the lengthy computation of terms such as

$$\int_0^T \psi(T_{i-1} - t; a, b, c, d) \, \psi(T_{j-1} - t; a, b, c, d) \, dt$$

has to be carried out. This can be done easily with a software for formal manipulations, with a command line like

```
int(((a*(S-t)+d)*exp(-b*(S-t))+c)*((a*(T-t)+d)
              *exp(-b*(T-t))+c),t=t1..t2);
```

We also impose the constraints

$$-\pi/3 < \theta_i - \theta_{i-1} < \pi/3, \quad 0 < \theta_i < \pi,$$

to the correlation angles. Putting $\pi/2$ in the first constraint would ensure that $\rho_{i,i-1} > 0$, but here we impose a stronger constraint, although not as strong as in the $\pi/4$ case. Finally, the local minimization is constrained by the requirement

$$1 - 0.1 \leq \Phi_i(a, b, c, d) \leq 1 + 0.1,$$

for all i. This constraint ensures that all Φ's will be close to one, so that the qualitative behaviour of the term structure should be preserved in time. Moreover, with this parameterization, we can expect a smooth shape for the term structure of volatilities at all instants, since with linear/exponential functions we avoid the typical erratic behaviour of piecewise-constant formulations.

For the calibration, we use only volatilities in the swaptions matrix corresponding to the 2y, 5y and 10y columns, in order to speed up the constrained optimization. The local optimization routine produced the following parameters:

$$a = 0.29342753, \quad b = 1.25080230, \quad c = 0.13145869, \quad d = 0.$$

296   7. Cases of Calibration of the LIBOR Market Model

$\theta_{1\div 6} = [1.754112 \ 0.577818 \ 1.685018 \ 0.581761 \ 1.538243 \ 2.436329]$,
$\theta_{7\div 12} = [0.880112 \ 1.896454 \ 0.486056 \ 1.280206 \ 2.440311 \ 0.944809]$,
$\theta_{13\div 19} = [1.340539 \ 2.911335 \ 1.996228 \ 0.700425 \ 0 \ 0.815189 \ 2.383766]$.

Notice that $d = 0$ has reached the lowest value allowed by the positivity constraint, meaning that possibly the optimization would have improved with a negative d. We allow d to go negative later on. The instantaneous correlations resulting from this calibration are again oscillating and non-monotonic. The first ten rows and columns of the instantaneous-correlation matrix are, for example,

	1y	2y	3y	4y	5y	6y	7y	8y	9y	10y
1y	1.000	0.384	0.998	0.388	0.977	0.776	0.642	0.990	0.298	0.890
2y	0.384	1.000	0.447	1.000	0.573	-0.284	0.955	0.249	0.996	0.763
3y	0.998	0.447	1.000	0.451	0.989	0.731	0.693	0.978	0.363	0.919
4y	0.388	1.000	0.451	1.000	0.576	-0.280	0.956	0.253	0.995	0.766
5y	0.977	0.573	0.989	0.576	1.000	0.623	0.791	0.937	0.496	0.967
6y	0.776	-0.284	0.731	-0.280	0.623	1.000	0.015	0.858	-0.370	0.403
7y	0.642	0.955	0.693	0.956	0.791	0.015	1.000	0.526	0.923	0.921
8y	0.990	0.249	0.978	0.253	0.937	0.858	0.526	1.000	0.160	0.816
9y	0.298	0.996	0.363	0.995	0.496	-0.370	0.923	0.160	1.000	0.701
10y	0.890	0.763	0.919	0.766	0.967	0.403	0.921	0.816	0.701	1.000

We find some repeated oscillations between positive and negative values that are not desirable. Terminal correlations share part of this negative behaviour. However, the evolution in time of the term structure of caplet volatilities is now reasonable, as shown in Figure 7.4.

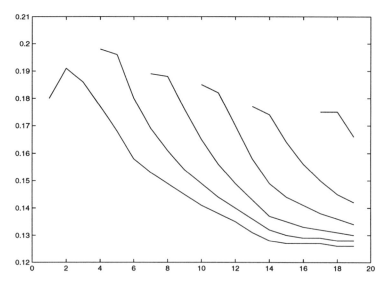

Fig. 7.4. Calibration results under Formulation 7: evolution of the term structure of caplet volatilities.

7.3 Joint Calibration with Parameterized Volatilities as in Formulation 7   297

There remains to check the fitting quality. Caplets are fitted exactly, whereas the swaption volatilities are fitted with the following percentage differences:

	2y	3y	4y	5y	6y	7y	8y	9y	10y
1y	2.28%	-3.74%	-3.19%	-4.68%	2.46%	1.50%	0.72%	1.33%	-1.42%
2y	-1.23%	-7.67%	-9.97%	2.10%	0.49%	1.33%	1.56%	-0.44%	1.88%
3y	2.23%	-6.20%	-1.30%	-1.32%	-1.43%	1.80%	-0.19%	2.42%	1.17%
4y	-2.59%	9.02%	1.70%	0.79%	3.22%	1.19%	4.85%	3.75%	1.21%
5y	-3.26%	-0.28%	-8.16%	-0.81%	-3.56%	-0.23%	-0.08%	-2.63%	2.62%
7y	0.10%	-2.59%	-10.85%	-2.00%	-3.67%	-6.84%	2.15%	1.19%	0.00%
10y	0.29%	-3.44%	-11.83%	-1.31%	-4.69%	-2.60%	4.07%	1.11%	0.00%

Recall that here we have fitted only the 2y, 5y and 10y columns. In such columns differences, in absolute value, reach at most 4.68%, and are usually much smaller. The remaining data are also reproduced with small errors, even if they have not been included in the calibration, with a number of exceptions. For example, the 10 × 4-swaption features a difference of 11.83%. However, the last columns show relatively small errors, so that the 7y, 8y and 9y columns seem to be rather aligned with the 5y and 10y columns. On the contrary, the 3y and 4y columns seem to be rather misaligned with the 2y and 5y columns, since they show larger errors.

We have performed many more experiments with this choice of volatility, and some are reported later on. We have tried rank-three correlations structures, less or more stringent constraints on the angles and on the Φ's, and so on. A variety of results have been obtained. In general, we can say that the fitting to the whole swaption matrix can be improved, but at the cost of an erratic behaviour of both correlations and of the evolution of the term structure of volatilities in time. In particular, the three-factor choice does not seem to help that much.

The above example is sufficient to let one appreciate both the potential and the disadvantages of this parameterization with respect to the piecewise-constant case. In general, this parameterization allows for an easier control of the evolution of the term structure of volatilities, but produces more erratic correlation structures, since most of the "noise" in the swaptions data now ends up in the angles, because of the fact that we have only four volatility parameters a, b, c, d that can be used to calibrate swaption volatilities. A different possible use of the model, however, is to limit the calibration to act only on swaption prices, by ignoring the cap market, or by keeping it for testing the caps/swaptions misalignment a posteriori. With this approach the Φ's become again free parameters to be used in the swaption calibration, and are no longer functions of a, b, c, d imposed by the caplet volatilities. We will partly address this possibility in a later section.

7.3.1 Formulation 7: Narrowing the Angles

Now we change the calibration constraints by imposing

$$-\pi/4 \leq \theta_i - \theta_{i-1} \leq \pi/4, \quad 0 \leq \theta_i < \pi/2$$

for all i's. These constraints imply positive instantaneous correlations. However, we allow d in the instantaneous volatility to go negative. This is not a problem in itself for the term structure of caplet volatilities, since it is the square of the instantaneous volatilities that matters. However, the partially negative structure of the ψ function can be needed to calibrate swaptions data, where integrals of products of shifted ψ's are considered. In the optimization, we also constrained all Φ's to remain in-between 0.9 and 1.1, thus ensuring regularity in the evolution of the term structure of volatilities. Notice that, since the caplet fitting forces the Φ's to be functions of a, b, c, d, we obtain, as before, that the optimization is subject to the nonlinear constraint

$$0.9 \leq \Phi_i(a, b, c, d) \leq 1.1$$

for all i. Again, we only fit the 2y, 5y and 10y columns.

Starting from typical values for the initial guesses, one has

$$a = 0.995456499, \quad b = 1.449033507, \quad c = 0.095483199, \quad d = -0.452131651.$$

The related Φ's are

$\Phi_{1 \div 10} = [1.048\ 0.996\ 1.035\ 1.067\ 1.080\ 1.073\ 1.082\ 1.091\ 1.093\ 1.093]$,
$\Phi_{11 \div 19} = [1.091\ 1.085\ 1.075\ 1.063\ 1.068\ 1.078\ 1.086\ 1.094\ 1.100]$,

and the angles are as follows:

$\theta_{1 \div 6} = [0.981285\ 1.178236\ 0.400059\ 0.888339\ 1.515577\ 0.697620]$,
$\theta_{7 \div 13} = [0.112149\ 0.970131\ 1.330376\ 1.047198\ 0\ 1.047198\ 1.570796]$,
$\theta_{14 \div 19} = [0.523599\ 1.570796\ 1.570796\ 0.523599\ 1.570796\ 1.570796]$.

The matrix of percentage differences determining the fitting quality is

	2y	3y	4y	5y	6y	7y	8y	9y	10y
1y	0.82%	2.46%	1.48%	4.60%	5.03%	7.84%	6.80%	6.10%	4.39%
2y	-3.29%	-6.98%	-3.80%	-2.16%	1.69%	2.10%	2.08%	1.52%	2.12%
3y	-4.22%	-0.90%	-4.26%	0.08%	1.04%	2.42%	2.05%	3.77%	2.82%
4y	-3.67%	-6.40%	-2.30%	-2.23%	-1.09%	-0.69%	2.24%	1.92%	2.02%
5y	-0.08%	2.98%	-1.65%	-1.63%	-2.11%	2.19%	1.77%	2.06%	1.84%
7y	0.57%	-0.44%	-4.99%	-2.28%	-1.93%	-0.06%	-1.53%	-0.34%	-0.99%
10y	0.16%	-5.42%	-3.36%	-6.60%	-4.26%	-4.85%	-8.09%	-9.73%	-11.92%

The first row exhibits positive signs, whereas the last shows negative ones. This means that the LFM swaptions volatilities are too small in the first row and too large in the last. It seems that our constraints on the correlations do not allow the model to "steepen the columns" of the fitted swaptions volatility matrix, as was instead possible in the previous case with more free angles. This might again be a consequence of the low number of factors: it might be possible to have steep "columns" but only with "wilder" rank-two correlations. The first ten rows and columns of the instantaneous-correlation matrix resulting from the calibrated angles are:

7.3 Joint Calibration with Parameterized Volatilities as in Formulation 7   299

	1y	2y	3y	4y	5y	6y	7y	8y	9y	10y
1y	1.000	0.981	0.836	0.996	0.861	0.960	0.645	1.000	0.940	0.998
2y	0.981	1.000	0.712	0.958	0.944	0.887	0.484	0.978	0.988	0.991
3y	0.836	0.712	1.000	0.883	0.440	0.956	0.959	0.842	0.598	0.798
4y	0.996	0.958	0.883	1.000	0.810	0.982	0.714	0.997	0.904	0.987
5y	0.861	0.944	0.440	0.810	1.000	0.684	0.167	0.855	0.983	0.892
6y	0.960	0.887	0.956	0.982	0.684	1.000	0.833	0.963	0.806	0.940
7y	0.645	0.484	0.959	0.714	0.167	0.833	1.000	0.654	0.345	0.594
8y	1.000	0.978	0.842	0.997	0.855	0.963	0.654	1.000	0.930	0.997
9y	0.940	0.988	0.598	0.904	0.983	0.806	0.345	0.936	1.000	0.960
10y	0.998	0.991	0.798	0.987	0.892	0.940	0.594	0.997	0.960	1.000

We have no oscillations between negative and positive values, but apart from this, we are still far from an ideal correlation matrix. A similar remark applies also to the ten-year terminal-correlation matrix, which is given by

	10y	11y	12y	13y	14y	15y	16y	17y	18y	19y
10y	1.000	0.408	0.830	0.738	0.745	0.748	0.768	0.749	0.749	0.749
11y	0.408	1.000	0.495	0.000	0.831	0.000	0.803	0.826	0.000	0.000
12y	0.830	0.495	1.000	0.861	0.857	0.855	0.876	0.854	0.854	0.854
13y	0.738	0.000	0.861	1.000	0.500	0.998	0.539	0.499	0.998	0.998
14y	0.745	0.831	0.857	0.500	1.000	0.500	0.999	1.000	0.500	0.500
15y	0.748	0.000	0.855	0.998	0.500	1.000	0.540	0.500	1.000	1.000
16y	0.768	0.803	0.876	0.539	0.999	0.540	1.000	0.999	0.540	0.540
17y	0.749	0.826	0.854	0.499	1.000	0.500	0.999	1.000	0.500	0.500
18y	0.749	0.000	0.854	0.998	0.500	1.000	0.540	0.500	1.000	1.000
19y	0.749	0.000	0.854	0.998	0.500	1.000	0.540	0.500	1.000	1.000

The evolution in time of the term structure of caplet volatilities is shown in Figure 7.5, and looks rather satisfactory in this case. However, the price

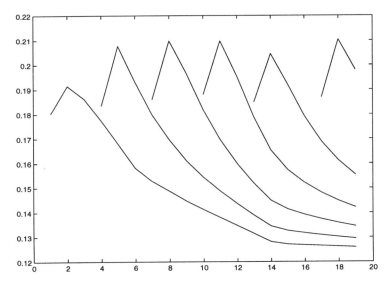

Fig. 7.5. Calibration results under Formulation 7, case II: evolution of the term structure of caplet volatilities

we have paid for this nice evolution is too high. Imposing the constraints

300   7. Cases of Calibration of the LIBOR Market Model

$0 < \theta_i < \pi/2$ seems to bring more damage than benefit. It actually prevents correlations from going negative, but the fitting quality is deteriorated and the gain in correlation smoothness is rather limited.

7.3.2 Formulation 7: Calibrating only to Swaptions

We now try the calibration to the whole swaption volatility matrix (including the first column) with parameters a, b, c, d, θ and Φ. We do not consider caplet volatilities here.

We force the Φ's to be close to one. Precisely, for each i, we impose the constraint

$$0.95 \leq \Phi_i \leq 1.05.$$

We then obtain the following parameters:

$$a = 1.0609249;\ b = 1.919928336;\ c = 0.110354707;\ d = -0.314126885;$$

$\Phi_{1 \div 10} = [0.950\ 0.967\ 1.038\ 1.050\ 1.050\ 1.050\ 1.050\ 0.950\ 1.050\ 1.038],$
$\Phi_{11 \div 19} = [0.950\ 0.950\ 0.950\ 0.950\ 1.050\ 1.050\ 0.950\ 0.951\ 0.950],$

and the angles are as follows:

$\theta_{1 \div 6} = [1.648748\ 1.275742\ 1.577678\ 2.108568\ 2.129107\ 1.343708],$
$\theta_{7 \div 12} = [1.042389\ 1.336917\ 2.114582\ 2.383600\ 1.617353\ 1.180804],$
$\theta_{13 \div 19} = [0.960047\ 0.774527\ 0.640304\ 0.406570\ 0.211008\ 0.144127\ 0].$

The fitting quality is expressed by the usual matrix of percentage differences:

	1y	2y	3y	4y	5y	6y	7y	8y	9y	10y
1y	-3.0%	0.3%	-1.4%	1.2%	2.3%	1.8%	2.4%	1.9%	1.0%	0.9%
2y	1.0%	-2.9%	-1.3%	-2.2%	-1.7%	-0.3%	0.3%	-0.2%	0.7%	-1.0%
3y	1.9%	1.9%	-1.6%	-3.8%	-0.7%	0.6%	0.6%	1.3%	0.5%	0.5%
4y	3.6%	-2.1%	-2.2%	-1.3%	-0.3%	-0.7%	0.7%	-0.1%	0.8%	1.3%
5y	2.1%	4.2%	1.4%	-2.3%	-2.8%	-0.8%	-1.1%	-0.4%	-0.2%	1.1%
7y	0.2%	0.2%	1.3%	2.0%	-1.4%	-0.8%	-0.8%	-0.6%	-0.2%	-0.2%
10y	-0.0%	-0.1%	0.0%	-0.0%	-0.1%	0.1%	0.1%	-0.0%	0.0%	0.0%

These percentage differences are partly satisfactory. In fact, one may expect smaller fitting errors when using, as in this case, 42 parameters for the calibration of 70 swaptions prices. However, we should not forget that such parameters have mainly a local influence, and that we have introduced heavy constraints so as to ensure a reasonable evolution of the term structure of volatilities.

The resulting instantaneous-correlation matrix is shown in Tables 7.6 and 7.7.

The instantaneous-correlation matrix is again far from being ideal, given the repeated oscillations between positive and negative values in the second half matrix. However, the situation changes when considering terminal correlations, which, after ten years, are as follows:

7.3 Joint Calibration with Parameterized Volatilities as in Formulation 7

	10y	11y	12y	13y	14y	15y	16y	17y	18y	19y
10y	1.000	0.668	0.338	0.138	-0.036	-0.162	-0.373	-0.534	-0.585	-0.685
11y	0.668	1.000	0.899	0.779	0.654	0.550	0.346	0.161	0.096	-0.046
12y	0.338	0.899	1.000	0.975	0.917	0.856	0.714	0.564	0.508	0.379
13y	0.138	0.779	0.975	1.000	0.983	0.949	0.851	0.732	0.685	0.573
14y	-0.036	0.654	0.917	0.983	1.000	0.991	0.933	0.845	0.808	0.715
15y	-0.162	0.550	0.856	0.949	0.991	1.000	0.973	0.909	0.879	0.802
16y	-0.373	0.346	0.714	0.851	0.933	0.973	1.000	0.981	0.966	0.918
17y	-0.534	0.161	0.564	0.732	0.845	0.909	0.981	1.000	0.998	0.978
18y	-0.585	0.096	0.508	0.685	0.808	0.879	0.966	0.998	1.000	0.990
19y	-0.685	-0.046	0.379	0.573	0.715	0.802	0.918	0.978	0.990	1.000

This matrix looks rather interesting. Indeed, we observe monotonic terminal correlation patterns when moving away from diagonal terms. This characteristic was also present in the corresponding portion of the instantaneous matrix, so that we can conclude that instantaneous volatilities behave well, in that they do not spoil a good part of the instantaneous-correlation matrix when moving to terminal correlations.

	1y	2y	3y	4y	5y	6y	7y	8y	9y	10y
1y	1.000	0.931	0.997	0.896	0.887	0.954	0.822	0.952	0.893	0.742
2y	0.931	1.000	0.955	0.673	0.657	0.998	0.973	0.998	0.668	0.447
3y	0.997	0.955	1.000	0.862	0.852	0.973	0.860	0.971	0.859	0.692
4y	0.896	0.673	0.862	1.000	1.000	0.721	0.483	0.717	1.000	0.962
5y	0.887	0.657	0.852	1.000	1.000	0.707	0.465	0.702	1.000	0.968
6y	0.954	0.998	0.973	0.721	0.707	1.000	0.955	1.000	0.717	0.506
7y	0.822	0.973	0.860	0.483	0.465	0.955	1.000	0.957	0.478	0.228
8y	0.952	0.998	0.971	0.717	0.702	1.000	0.957	1.000	0.713	0.500
9y	0.893	0.668	0.859	1.000	1.000	0.717	0.478	0.713	1.000	0.964
10y	0.742	0.447	0.692	0.962	0.968	0.506	0.228	0.500	0.964	1.000
11y	1.000	0.942	0.999	0.882	0.872	0.963	0.839	0.961	0.879	0.721
12y	0.892	0.995	0.922	0.600	0.583	0.987	0.990	0.988	0.595	0.360
13y	0.772	0.951	0.815	0.410	0.391	0.927	0.997	0.930	0.404	0.147
14y	0.642	0.877	0.694	0.235	0.215	0.842	0.964	0.846	0.229	-0.038
15y	0.533	0.805	0.592	0.102	0.082	0.763	0.920	0.767	0.096	-0.172
16y	0.323	0.645	0.389	-0.131	-0.151	0.592	0.805	0.598	-0.137	-0.395
17y	0.133	0.485	0.203	-0.321	-0.340	0.424	0.674	0.430	-0.327	-0.566
18y	0.066	0.425	0.137	-0.384	-0.402	0.363	0.623	0.369	-0.389	-0.620
19y	-0.078	0.291	-0.007	-0.512	-0.530	0.225	0.504	0.232	-0.517	-0.726

Table 7.6. Calibration results under Formulation 7, case III: instantaneous-correlation matrix (part I).

The evolution in time of the term structure of volatilities is shown in Figure 7.6. This graph looks reasonable, apart from the "zig-zag" shape of the initial term structure (and also of the following ones) around the 8y and 15y entries. It is worth noticing that the initial term structure is humped. It is also worth noticing that the $1 \times 1, 2 \times 1, 3 \times 1, 4 \times 1, 5 \times 1, 7 \times 1$ and 10×1 swaptions (first column in the table) are basically caplet volatilities, so that the initial term structure has to incorporate them somehow. It is curious to notice that the "zig-zag" effect in the fitted term structure at time 0 occurs exactly where no "caplet-like" swaptions data are available. Indeed, we have neither 8×1 nor 9×1 swaptions data, so that without imposing further constraints, the model resorts to a zig-zag shape. This choice probably renders

	11y	12y	13y	14y	15y	16y	17y	18y	19y
1y	1.000	0.892	0.772	0.642	0.533	0.323	0.133	0.066	-0.078
2y	0.942	0.995	0.951	0.877	0.805	0.645	0.485	0.425	0.291
3y	0.999	0.922	0.815	0.694	0.592	0.389	0.203	0.137	-0.007
4y	0.882	0.600	0.410	0.235	0.102	-0.131	-0.321	-0.384	-0.512
5y	0.872	0.583	0.391	0.215	0.082	-0.151	-0.340	-0.402	-0.530
6y	0.963	0.987	0.927	0.842	0.763	0.592	0.424	0.363	0.225
7y	0.839	0.990	0.997	0.964	0.920	0.805	0.674	0.623	0.504
8y	0.961	0.988	0.930	0.846	0.767	0.598	0.430	0.369	0.232
9y	0.879	0.595	0.404	0.229	0.096	-0.137	-0.327	-0.389	-0.517
10y	0.721	0.360	0.147	-0.038	-0.172	-0.395	-0.566	-0.620	-0.726
11y	1.000	0.906	0.792	0.665	0.559	0.352	0.164	0.097	-0.047
12y	0.906	1.000	0.976	0.919	0.857	0.715	0.565	0.509	0.380
13y	0.792	0.976	1.000	0.983	0.949	0.851	0.732	0.685	0.573
14y	0.665	0.919	0.983	1.000	0.991	0.933	0.845	0.808	0.715
15y	0.559	0.857	0.949	0.991	1.000	0.973	0.909	0.879	0.802
16y	0.352	0.715	0.851	0.933	0.973	1.000	0.981	0.966	0.918
17y	0.164	0.565	0.732	0.845	0.909	0.981	1.000	0.998	0.978
18y	0.097	0.509	0.685	0.808	0.879	0.966	0.998	1.000	0.990
19y	-0.047	0.380	0.573	0.715	0.802	0.918	0.978	0.990	1.000

Table 7.7. Calibration results under Formulation 7, case III: instantaneous-correlation matrix (part II).

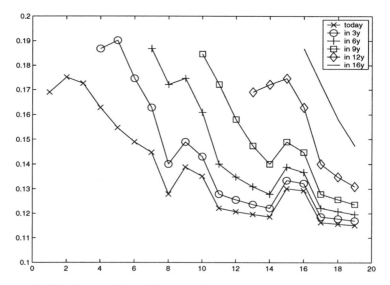

Fig. 7.6. Calibration results under Formulation 7, case III: evolution of the term structure of caplet volatilities.

the calibration of subsequent swaptions easier. It can be also appropriate to reason on the data in order to understand whether there is some "noise" (for example due to illiquidity) in the market swaption matrix causing this seeming misalignment in the data.

7.4 Exact Swaptions Calibration with Volatilities as TABLE 1

We now examine the case of the instantaneous-volatility structure summarized in TABLE 1, which is the one with the largest number of parameters. One would expect this structure to lead to a complex calibration routine, requiring optimization in a space of huge dimension. Instead, we have devised a method such that, by assuming exogenously-given instantaneous correlations ρ, the calibration can be carried out through closed-form formulas having as inputs the exogenous correlations and the swaption volatilities. Our method is illustrated in what follows.

Recall that the instantaneous-volatility structure considered here is defined by

$$\sigma_k(t) = \sigma_{k,\beta(t)},$$

where in general $T_{\beta(t)-2} < t \leq T_{\beta(t)-1}$, and TABLE 1 reads as

Instant. Vols	Time: $t \in (0,T_0]$	$(T_0,T_1]$	$(T_1,T_2]$	...	$(T_{M-2},T_{M-1}]$
Fwd Rate:$F_1(t)$	$\sigma_{1,1}$	Dead	Dead	...	Dead
$F_2(t)$	$\sigma_{2,1}$	$\sigma_{2,2}$	Dead	...	Dead
$\vdots$	...	...	...	...	...
$F_M(t)$	$\sigma_{M,1}$	$\sigma_{M,2}$	$\sigma_{M,3}$	...	$\sigma_{M,M}$

Dealing with swaptions, we denote as usual by $V_{\alpha,\beta}$ the Black volatility for the swaption whose underlying swap rate is $S_{\alpha,\beta}$ (T_α is the swap starting date and $T_{\alpha+1},\ldots,T_\beta$ are the swap payment dates). Recall also the approximated formula (6.58) for valuing swaptions volatilities $v_{\alpha,\beta}$ in the LFM. Apply this formula, dividing by time T_α, to the volatility formulation of TABLE 1 to obtain

$$(V_{\alpha,\beta})^2 \approx \sum_{i,j=\alpha+1}^{\beta} \frac{w_i(0)w_j(0)F_i(0)F_j(0)\rho_{i,j}}{T_\alpha S_{\alpha,\beta}(0)^2} \sum_{h=0}^{\alpha} \tau_{h-1,h}\sigma_{i,h+1}\sigma_{j,h+1} \quad (7.1)$$

with $\tau_{h-1,h} = T_h - T_{h-1}$, and $T_{-1} = 0$.

We remind that the weights w are specific of the swaption being considered, i.e. they depend on α and β. As usual we will omit such dependence to shorten notation, but each time we change swaption the corresponding w's change.

In order to effectively illustrate the calibration results in this case, without getting lost in notation and details, we work out an example with just six swaptions. We will then show how to generalize our procedure to an arbitrary number of contracts.

Suppose we start from the swaptions volatilities in the upper half of the swaption matrix:

304   7. Cases of Calibration of the LIBOR Market Model

Length Maturity	1y	2y	3y
$T_0 = 1\text{y}$	$V_{0,1}$	$V_{0,2}$	$V_{0,3}$
$T_1 = 2\text{y}$	$V_{1,2}$	$V_{1,3}$	-
$T_2 = 3\text{y}$	$V_{2,3}$	-	-

Let us move along this table, starting from the $(1,1)$ entry $V_{0,1}$. Using the approximating formula (7.1), we compute, after straightforward simplifications,

$$(V_{0,1})^2 \approx \sigma_{1,1}^2$$

This formula is immediately invertible and yields the volatility parameter $\sigma_{1,1}$ in the forward-rate dynamics as a function of the swaption volatility $V_{0,1}$, which we read from our matrix.

We then move on to the right, to entry $(1,2)$, containing $V_{0,2}$. The same formula gives, this time,

$$S_{0,2}(0)^2 (V_{0,2})^2 \approx w_1(0)^2 F_1(0)^2 \sigma_{1,1}^2 + w_2(0)^2 F_2(0)^2 \sigma_{2,1}^2$$
$$+ 2\rho_{1,2} w_1(0) F_1(0) w_2(0) F_2(0) \sigma_{1,1} \sigma_{2,1}.$$

Everything in this formula is known, except $\sigma_{2,1}$. We then solve the "elementary-school" algebraic second-order equation in $\sigma_{2,1}$, and, assuming existence and uniqueness of a positive solution, we analytically recover $\sigma_{2,1}$ in terms of the previously found $\sigma_{1,1}$ and of the known swaptions data. More generally, we assume for the moment that all the next algebraic second-order equations admit a unique positive solution.

We keep on moving to the right, to entry $(1,3)$, containing $V_{0,3}$. Formula (7.1) gives, this time,

$$S_{0,3}(0)^2 (V_{0,3})^2 \approx w_1(0)^2 F_1(0)^2 \sigma_{1,1}^2 + w_2(0)^2 F_2(0)^2 \sigma_{2,1}^2$$
$$+ w_3(0)^2 F_3(0)^2 \sigma_{3,1}^2 + 2\rho_{1,2} w_1(0) F_1(0) w_2(0) F_2(0) \sigma_{1,1} \sigma_{2,1}$$
$$+ 2\rho_{1,3} w_1(0) F_1(0) w_3(0) F_3(0) \sigma_{1,1} \sigma_{3,1} + 2\rho_{2,3} w_2(0) F_2(0) w_3(0) F_3(0) \sigma_{2,1} \sigma_{3,1}.$$

Similarly to the previous formula, here everything is known except for $\sigma_{3,1}$. We then solve the algebraic second-order equation in $\sigma_{3,1}$, and recover analytically $\sigma_{3,1}$ in terms of the previously found $\sigma_{1,1}$, $\sigma_{2,1}$ and of the known swaptions data.

We now move on to the second row of the swaptions matrix, entry $(2,1)$, containing $V_{1,2}$. Our formula gives:

$$T_1 V_{1,2}^2 \approx T_0 \sigma_{2,1}^2 + (T_1 - T_0) \sigma_{2,2}^2.$$

This time, everything is known except $\sigma_{2,2}$. Once again, we solve explicitly this equation for $\sigma_{2,2}$, being $\sigma_{2,1}$ known from previous passages.

We move on to the right, entry $(2,2)$, containing $V_{1,3}$. Formula (7.1) gives:

7.4 Exact Swaptions Calibration with Volatilities as TABLE 1

$$T_1 S_{1,3}(0)^2 V_{1,3}^2 \approx w_2(0)^2 F_2(0)^2 (\tau_{-1,0}\sigma_{2,1}^2 + \tau_{0,1}\sigma_{2,2}^2)$$
$$+ w_3(0)^2 F_3(0)^2 (\tau_{-1,0}\sigma_{3,1}^2 + \tau_{0,1}\sigma_{3,2}^2)$$
$$+ 2\rho_{2,3} w_2(0) F_2(0) w_3(0) F_3(0) (\tau_{-1,0}\sigma_{2,1}\sigma_{3,1} + \tau_{0,1}\sigma_{2,2}\sigma_{3,2}).$$

Here everything is known except $\sigma_{3,2}$. Once again, we solve explicitly this equation for $\sigma_{3,2}$, being $\sigma_{2,1}, \sigma_{2,2}$ and $\sigma_{3,1}$ known from previous passages.

Finally, we move to the only entry (3,1) of the third row, containing $V_{2,3}$. The usual formula gives:

$$T_2 V_{2,3}^2 \approx \tau_{-1,0}\sigma_{3,1}^2 + \tau_{0,1}\sigma_{3,2}^2 + \tau_{1,2}\sigma_{3,3}^2.$$

The only unknown entry at this point is $\sigma_{3,3}$, which can be easily found by explicitly solving this last equation.

We have thus been able to find all instantaneous volatilities

Instant. Vols	Time: $t \in (0, T_0]$	$(T_0, T_1]$	$(T_1, T_2]$
Fwd Rate: $F_1(t)$	$\sigma_{1,1}$	Dead	Dead
$F_2(t)$	$\sigma_{2,1}$	$\sigma_{2,2}$	Dead
$F_3(t)$	$\sigma_{3,1}$	$\sigma_{3,2}$	$\sigma_{3,3}$

in terms of swaptions volatilities. The following table summarizes the dependence of the swaptions volatilities V on the instantaneous forward volatilities σ.

Length Maturity	1y	2y	3y
$T_0 = 1y$	$V_{0,1}$ $\sigma_{1,1}$	$V_{0,2}$ $\sigma_{1,1}, \sigma_{2,1}$	$V_{0,3}$ $\sigma_{1,1}, \sigma_{2,1}, \sigma_{3,1}$
$T_1 = 2y$	$V_{1,2}$ $\sigma_{2,1}$ $\sigma_{2,2}$	$V_{1,3}$ $\sigma_{2,1}, \sigma_{3,1}$ $\sigma_{2,2}, \sigma_{3,2}$	-
$T_2 = 3y$	$V_{2,3}$ $\sigma_{3,1}$ $\sigma_{3,2}$ $\sigma_{3,3}$	-	-

In this table, we have put in each entry the related swaption volatility and the instantaneous volatilities upon which it depends. In reading the table left to right and top down, you realize that, each time, only one new σ appears, and this makes the relationship between the V's and the σ's invertible (analytically).

We now give the general method for calibrating our volatility formulation of TABLE 1 to the upper-diagonal part of the swaption matrix when an arbitrary number s of rows of the matrix is given. We thus generalize the just seen case, where $s = 3$, to a generic positive integer s.

Let us rewrite formula (7.1) as follows:

$$T_\alpha S_{\alpha,\beta}^2(0)V_{\alpha,\beta}^2 = \sum_{i,j=\alpha+1}^{\beta-1} w_i(0)w_j(0)F_i(0)F_j(0)\rho_{i,j} \sum_{h=0}^{\alpha} \tau_{h-1,h}\sigma_{i,h+1}\sigma_{j,h+1}$$

$$+2 \sum_{j=\alpha+1}^{\beta-1} w_\beta(0)w_j(0)F_\beta(0)F_j(0)\rho_{\beta,j} \sum_{h=0}^{\alpha-1} \tau_{h-1,h}\sigma_{\beta,h+1}\sigma_{j,h+1}$$

$$+2 \sum_{j=\alpha+1}^{\beta-1} w_\beta(0)w_j(0)F_\beta(0)F_j(0)\rho_{\beta,j}\tau_{\alpha-1,\alpha}\sigma_{j,\alpha+1}\boxed{\sigma_{\beta,\alpha+1}}$$

$$+w_\beta(0)^2 F_\beta(0)^2 \sum_{h=0}^{\alpha-1} \tau_{h-1,h}\sigma_{\beta,h+1}^2$$

$$+w_\beta(0)^2 F_\beta(0)^2 \tau_{\alpha-1,\alpha}\boxed{\sigma_{\beta,\alpha+1}^2}. \qquad (7.2)$$

In turn, by suitable definition of the coefficients A, B and C, this equation can be rewritten as:

$$A_{\alpha,\beta}\sigma_{\beta,\alpha+1}^2 + B_{\alpha,\beta}\sigma_{\beta,\alpha+1} + C_{\alpha,\beta} = 0,$$

and thus it can be solved analytically. It is important to realize (as from the above $s = 3$ example) that when one solves this equation, all quantities are indeed known with the exception of $\sigma_{\beta,\alpha+1}$ *if* the swaption matrix is visited from left to right and top down.

Our procedure can be written in algorithmic form as follows.

1. Select the number s of rows in the swaption matrix that are of interest for the calibration;
2. Set $\alpha = 0$;
3. Set $\beta = 1$;
4. Solve equation (7.2) in $\sigma_{\beta,\alpha+1}$. Since both $A_{\alpha,\beta}$ and $B_{\alpha,\beta}$ are strictly positive, (7.2) has at most one positive solution, namely

$$\sigma_{\beta,\alpha+1} = \frac{-B_{\alpha,\beta} + \sqrt{B_{\alpha,\beta}^2 - 4A_{\alpha,\beta}C_{\alpha,\beta}}}{2A_{\alpha,\beta}},$$

if and only if $C_{\alpha,\beta} < 0$.
5. Increase β by one. If β is smaller than or equal to $s - \alpha$, go back to point 4, otherwise increase α by one.
6. If $\alpha < s$ go back to 4, otherwise stop.

Of course, this recipe works if and only if the condition $C_{\alpha,\beta} < 0$ is verified every time we must solve an equation like (7.2). Our practical experience is that such condition is always met for non-pathological swaptions data. However, for sake of completeness, we will illustrate later on some examples leading to $C_{\alpha,\beta} > 0$.

7.4 Exact Swaptions Calibration with Volatilities as TABLE 1

We may wonder about what happens when in need to recover the whole matrix and not only the upper-diagonal part. By continuing our example with $s = 3$ above, we realize that the dependence of the whole set of swaptions volatilities V on the instantaneous forward volatilities σ is as follows:

Length \\ Maturity	1y	2y	3y
$T_0 = 1y$	$V_{0,1}$	$V_{0,2}$	$V_{0,3}$
	$\sigma_{1,1}$	$\sigma_{1,1}, \sigma_{2,1}$	$\sigma_{1,1}, \sigma_{2,1}, \sigma_{3,1}$
$T_1 = 2y$	$V_{1,2}$	$V_{1,3}$	$V_{1,4}$
	$\sigma_{2,1}$	$\sigma_{2,1}, \sigma_{3,1}$	$\sigma_{2,1}, \sigma_{3,1}, \sigma_{4,1}$
	$\sigma_{2,2}$	$\sigma_{2,2}, \sigma_{3,2}$	$\sigma_{2,2}, \sigma_{3,2}, \sigma_{4,2}$
$T_2 = 3y$	$V_{2,3}$	$V_{2,4}$	$V_{2,5}$
	$\sigma_{3,1}$	$\sigma_{3,1}, \sigma_{4,1}$	$\sigma_{3,1}, \sigma_{4,1}, \sigma_{5,1}$
	$\sigma_{3,2}$	$\sigma_{3,2}, \sigma_{4,2}$	$\sigma_{3,2}, \sigma_{4,2}, \sigma_{5,2}$
	$\sigma_{3,3}$	$\sigma_{3,3}, \sigma_{4,3}$	$\sigma_{3,3}, \sigma_{4,3}, \sigma_{5,3}$

To analytically determine the σ's from the V's, we proceed initially as before, thus ending up with the upper-diagonal part of the matrix:

Length \\ Maturity	1y	2y	3y
$T_0 = 1y$	$V_{0,1}$	$V_{0,2}$	$V_{0,3}$
	$\sigma_{1,1}$	$\sigma_{1,1}, \sigma_{2,1}$	$\sigma_{1,1}, \sigma_{2,1}, \sigma_{3,1}$
$T_1 = 2y$	$V_{1,2}$	$V_{1,3}$	-
	$\sigma_{2,1}$	$\sigma_{2,1}, \sigma_{3,1}$	
	$\sigma_{2,2}$	$\sigma_{2,2}, \sigma_{3,2}$	
$T_2 = 3y$	$V_{2,3}$	-	-
	$\sigma_{3,1}$		
	$\sigma_{3,2}$		
	$\sigma_{3,3}$		

We then move off the diagonal of one level, starting from the upper entry (2,3), containing $V_{1,4}$.

Length \\ Maturity	1y	2y	3y
$T_0 = 1y$	$V_{0,1}$	$V_{0,2}$	$V_{0,3}$
	$\sigma_{1,1}$	$\sigma_{1,1}, \sigma_{2,1}$	$\sigma_{1,1}, \sigma_{2,1}, \sigma_{3,1}$
$T_1 = 2y$	$V_{1,2}$	$V_{1,3}$	$V_{1,4}$
	$\sigma_{2,1}$	$\sigma_{2,1}, \sigma_{3,1}$	$\sigma_{2,1}, \sigma_{3,1}, \sigma_{4,1}$
	$\sigma_{2,2}$	$\sigma_{2,2}, \sigma_{3,2}$	$\sigma_{2,2}, \sigma_{3,2}, \sigma_{4,2}$
$T_2 = 3y$	$V_{2,3}$	-	-
	$\sigma_{3,1}$		
	$\sigma_{3,2}$		
	$\sigma_{3,3}$		

308   7. Cases of Calibration of the LIBOR Market Model

This presents us with *two* new unknowns, rather than one, and precisely $\sigma_{4,1}$ and $\sigma_{4,2}$. We can now consider again the analogue of equation (7.2), but we have two unknowns. An easy way out is to assume a relationship between them, and one of the easiest possibilities is to assume the two unknowns to be equal, $\sigma_{4,1} = \sigma_{4,2}$. By doing so, we end up again with an analytically-solvable second-order equation, we solve it (assuming existence and uniqueness of a positive solution), and we move down along the subdiagonal to entry (3,2) containing $V_{2,4}$:

Length Maturity	1y	2y	3y
$T_0 = 1y$	$V_{0,1}$	$V_{0,2}$	$V_{0,3}$
	$\sigma_{1,1}$	$\sigma_{1,1}, \sigma_{2,1}$	$\sigma_{1,1}, \sigma_{2,1}, \sigma_{3,1}$
$T_1 = 2y$	$V_{1,2}$	$V_{1,3}$	$V_{1,4}$
	$\sigma_{2,1}$	$\sigma_{2,1}, \sigma_{3,1}$	$\sigma_{2,1}, \sigma_{3,1}, \sigma_{4,1}$
	$\sigma_{2,2}$	$\sigma_{2,2}, \sigma_{3,2}$	$\sigma_{2,2}, \sigma_{3,2}, \sigma_{4,2}$
$T_2 = 3y$	$V_{2,3}$	$V_{2,4}$	-
	$\sigma_{3,1}$	$\sigma_{3,1}, \sigma_{4,1}$	
	$\sigma_{3,2}$	$\sigma_{3,2}, \sigma_{4,2}$	
	$\sigma_{3,3}$	$\sigma_{3,3}, \sigma_{4,3}$	

This time, we just have one new unknown, $\sigma_{4,3}$, so that we can solve the related second-order equation and we are done.

We then move one level further below the diagonal, i.e. to entry (3,3) containing $V_{2,5}$. Looking at the full table,

Length Maturity	1y	2y	3y
$T_0 = 1y$	$V_{0,1}$	$V_{0,2}$	$V_{0,3}$
	$\sigma_{1,1}$	$\sigma_{1,1}, \sigma_{2,1}$	$\sigma_{1,1}, \sigma_{2,1}, \sigma_{3,1}$
$T_1 = 2y$	$V_{1,2}$	$V_{1,3}$	$V_{1,4}$
	$\sigma_{2,1}$	$\sigma_{2,1}, \sigma_{3,1}$	$\sigma_{2,1}, \sigma_{3,1}, \sigma_{4,1}$
	$\sigma_{2,2}$	$\sigma_{2,2}, \sigma_{3,2}$	$\sigma_{2,2}, \sigma_{3,2}, \sigma_{4,2}$
$T_2 = 3y$	$V_{2,3}$	$V_{2,4}$	$V_{2,5}$
	$\sigma_{3,1}$	$\sigma_{3,1}, \sigma_{4,1}$	$\sigma_{3,1}, \sigma_{4,1}, \sigma_{5,1}$
	$\sigma_{3,2}$	$\sigma_{3,2}, \sigma_{4,2}$	$\sigma_{3,2}, \sigma_{4,2}, \sigma_{5,2}$
	$\sigma_{3,3}$	$\sigma_{3,3}, \sigma_{4,3}$	$\sigma_{3,3}, \sigma_{4,3}, \sigma_{5,3}$

we see that we now have three new unknowns, namely $\sigma_{5,1}$, $\sigma_{5,2}$ and $\sigma_{5,3}$. Again, if we assume them to be equal, we can solve the related second-order equation and fill the table, so that the calibration is now completed.

The generalization to cases where $s > 3$, as well as different possibilities for reducing the number of unknowns are left to the reader.

7.4 Exact Swaptions Calibration with Volatilities as TABLE 1

7.4.1 Some Numerical Results

Here, we plan to present some numerical results. We start from the simplest case of a swaption matrix with $s = 3$, and with swaptions volatilities only in the upper half of the table. Then we move to the full matrix, and, finally, we consider the case with $s = 10$.

Under the case with $s = 3$, we consider the three subcases: a) market swaption-volatility matrix as of May 16 2000; b) the same volatility matrix with modified upper corners; c) the same volatility matrix with modified upper and lower corners. These matrices are as follows:

Input: the three swaptions volatility matrices a), b) and c)

0.180	0.167	0.154	0.300	0.167	0.100	0.300	0.167	0.100
0.181	0.162		0.181	0.162		0.181	0.162	
0.178			0.178			0.100		

Calibrating to these upper-diagonal swaption matrices yields the following instantaneous-volatility parameters in the corresponding TABLE 1:

Output: instantaneous volatility parameters for a), b) and c)

0.1800	-	-	0.3000	-	-
0.1540	0.2050	-	0.0287	0.2540	-
0.1270	0.1570	0.2340	-0.0424	0.2030	0.2280

0.3000	-	-
0.0287	0.2540	-
-0.0424	0.2030	0+0.1136i

This first example helps us realize what can go wrong with our automatic and algebraic technique. While for the first swaption matrix a) we see that the parameters σ are all positive real numbers, we immediately notice that, by tilting the first row of the swaption matrix as in case b), we get a negative entry in the corresponding σ table. Even worse, by tilting also the first column, we obtain both negative and imaginary instantaneous volatilities σ, as is clear from case c). Therefore caution is in order when running a calibration. In fact, too steep matrices can yield both negative and imaginary instantaneous volatilities.

We find further confirmation of this fact by calibrating to the whole 3×3 swaptions matrices analogous to a), b) and c):

Input: the three swaptions volatility matrices d), e) and f)

0.180	0.167	0.154	0.300	0.167	0.100	0.300	0.167	0.100
0.181	0.162	0.145	0.181	0.162	0.145	0.181	0.162	0.145
0.178	0.155	0.137	0.178	0.155	0.137	0.100	0.155	0.280

310   7. Cases of Calibration of the LIBOR Market Model

We obtain the following tables of instantaneous-volatility parameters σ:

Output: instantaneous-volatility parameters for d), e) and f)

0.1800	-	-	0.300	-	-
0.1548	0.2039	-	0.041	0.253	-
0.1285	0.1559	0.2329	- 0.032	0.205	0.228
0.1105	0.1105	0.1660	0.138	0.138	0.171
0.1012	0.1012	0.1012	0.106	0.106	0.106

0.300	-	-
0.041	0.253	-
- 0.032	0.205	0 + 0.1147i
0.138	0.138	0.4053 - 0.1182i
0.5629 + 0.0001i	0.5629 + 0.0001i	0.5629 + 0.0001i

where in the equations for σ's involving more than one unknown we have assumed all unknowns to be equal. Again, we obtain first negative and then imaginary volatilities in cases e) and f).

We now move from our 3×3 toy input matrices to the full 10×10 swaption matrix, obtained from the 7×10 market data of 16 May, 2000, where the 6y, 8y, 9y lines have been added by linear interpolation. This matrix is:

	1y	2y	3y	4y	5y	6y	7y	8y	9y	10y
1y	0.180	0.167	0.154	0.145	0.138	0.134	0.130	0.126	0.124	0.122
2y	0.181	0.162	0.145	0.135	0.127	0.123	0.120	0.117	0.115	0.113
3y	0.178	0.155	0.137	0.125	0.117	0.114	0.111	0.108	0.106	0.104
4y	0.167	0.143	0.126	0.115	0.108	0.105	0.103	0.100	0.098	0.096
5y	0.154	0.132	0.118	0.109	0.104	0.104	0.099	0.096	0.094	0.092
6y	0.147	0.127	0.113	0.104	0.098	0.098	0.094	0.092	0.090	0.089
7y	0.140	0.121	0.107	0.098	0.092	0.091	0.089	0.087	0.086	0.085
8y	0.137	0.117	0.103	0.095	0.089	0.088	0.086	0.084	0.083	0.082
9y	0.133	0.114	0.100	0.091	0.086	0.085	0.083	0.082	0.081	0.080
10y	0.130	0.110	0.096	0.088	0.083	0.082	0.080	0.079	0.078	0.077

and a plot of the implied surface is shown in Figure 7.7.

Calibrating to such a matrix yields the instantaneous volatilities, collected as in TABLE 1, that are shown in Table 7.8.

This "real-market" calibration shows several negative signs in instantaneous volatilities. Recall that these undesirable negative entries might be due to "temporal misalignments" caused by illiquidity in the swaption matrix. As observed before, these misalignments can cause troubles, since after a calibration, the model parameters might reflect these misalignments. To avoid this inconvenience, we smooth the above market swaption matrix by means of the following parametric form:

$$\mathrm{vol}(S,T) = \gamma(S) + \left(\frac{\exp(f\ln(T))}{eS} + D(S)\right)\exp(-\beta\exp(p\ln(T))),$$

where

7.4 Exact Swaptions Calibration with Volatilities as TABLE 1

0.1800	-	-	-	-	-	-	-	-	-
0.1548	0.2039	-	-	-	-	-	-	-	-
0.1285	0.1559	0.2329	-	-	-	-	-	-	-
0.1178	0.1042	0.1656	0.2437	-	-	-	-	-	-
0.1091	0.0988	0.0973	0.1606	0.2483	-	-	-	-	-
0.1131	0.0734	0.0781	0.1009	0.1618	0.2627	-	-	-	-
0.1040	0.0984	0.0502	0.0737	0.1128	0.1633	0.2633	-	-	-
0.0940	0.1052	0.0938	0.0319	0.0864	0.0969	0.1684	0.2731	-	-
0.1065	0.0790	0.0857	0.0822	0.0684	0.0536	0.0921	0.1763	0.2848	-
0.1013	0.0916	0.0579	0.1030	0.1514	-0.0316	0.0389	0.0845	0.1634	0.2777
0.0916	0.0916	0.0787	0.0431	0.0299	0.2088	-0.0383	0.0746	0.0948	0.1854
0.0827	0.0827	0.0827	0.0709	0.0488	0.0624	0.1561	-0.0103	0.0731	0.0911
0.0744	0.0744	0.0744	0.0744	0.0801	0.0576	0.0941	0.1231	-0.0159	0.0610
0.0704	0.0704	0.0704	0.0704	0.0704	0.1009	0.0507	0.0817	0.1203	-0.0210
0.0725	0.0725	0.0725	0.0725	0.0725	0.0725	0.1002	0.0432	0.0619	0.1179
0.0753	0.0753	0.0753	0.0753	0.0753	0.0753	0.0753	0.0736	0.0551	0.0329
0.0719	0.0719	0.0719	0.0719	0.0719	0.0719	0.0719	0.0719	0.0708	0.0702
0.0690	0.0690	0.0690	0.0690	0.0690	0.0690	0.0690	0.0690	0.0690	0.0680
0.0663	0.0663	0.0663	0.0663	0.0663	0.0663	0.0663	0.0663	0.0663	0.0663

Table 7.8. Fitted instantaneous volatilities, collected as in TABLE 1.

$$\gamma(S) = c + (\exp(h \ln(S))a + d) \exp(-b \exp(m \ln(S))),$$
$$D(S) = (\exp(g \ln(S))q + r) \exp(-s \exp(t \ln(S))) + \delta,$$

and S and T are respectively the maturity and the tenor of the related swaption. So, for example, vol$(2, 3)$ is the volatility of the swaption whose underlying swap starts in two years and lasts three years (entry $(2, 3)$ of the swaption matrix).

We do not claim that this form has any appealing characteristic or that it always yields the precision needed by a trader, but we use it to point out the effect of smoothing.

The smoothing yields the following parameter values

$a = -0.00016$	$g = -0.10002$
$b = 0.376284$	$h = -4.18228$
$c = 0.201927$	$m = 0.875284$
$d = 0.336238$	$p = 0.241479$
$e = 5.21409$	$q = -6.37843$
$f = 0.193324$	$r = 5.817809$
$\delta = 0.809365$	$s = 0.048161$
$\beta = 0.840421$	$t = 1.293201$

which lead to the smoothed matrix:

	1y	2y	3y	4y	5y	6y	7y	8y	9y	10y
1y	0.185	0.162	0.150	0.143	0.138	0.134	0.131	0.128	0.126	0.123
2y	0.186	0.158	0.144	0.135	0.129	0.124	0.120	0.117	0.114	0.112
3y	0.176	0.149	0.135	0.126	0.120	0.115	0.112	0.108	0.106	0.103
4y	0.166	0.139	0.126	0.118	0.112	0.108	0.104	0.101	0.099	0.096
5y	0.156	0.131	0.119	0.111	0.105	0.101	0.098	0.095	0.092	0.090
6y	0.148	0.125	0.113	0.106	0.100	0.096	0.093	0.090	0.088	0.086
7y	0.141	0.119	0.107	0.100	0.095	0.091	0.088	0.085	0.083	0.081
8y	0.137	0.115	0.104	0.097	0.092	0.088	0.085	0.083	0.081	0.079
9y	0.133	0.112	0.101	0.094	0.089	0.086	0.083	0.080	0.078	0.076
10y	0.129	0.108	0.098	0.091	0.087	0.083	0.080	0.078	0.076	0.074

312   7. Cases of Calibration of the LIBOR Market Model

Moreover, the matrix of percentage differences between the market volatilities and the corresponding smoothed ones is

	1y	2y	3y	4y	5y	6y	7y	8y	9y	10y
1y	-2.64%	3.09%	2.30%	1.26%	0.07%	0.12%	-0.43%	-1.46%	-1.22%	-1.22%
2y	-2.63%	2.61%	0.71%	-0.12%	-1.44%	-0.77%	-0.01%	0.21%	0.89%	1.27%
3y	1.04%	4.14%	1.44%	-1.10%	-2.70%	-1.21%	-0.47%	-0.30%	0.38%	0.73%
4y	0.88%	2.54%	-0.39%	-2.86%	-4.06%	-2.71%	-1.16%	-1.17%	-0.60%	-0.36%
5y	-1.13%	0.79%	-0.66%	-1.84%	-1.36%	2.76%	1.34%	1.25%	1.74%	1.91%
6y	-0.99%	1.40%	-0.50%	-2.06%	-2.34%	1.35%	1.20%	1.50%	2.45%	3.08%
7y	-0.83%	2.07%	-0.34%	-2.31%	-3.43%	-0.26%	1.03%	1.77%	3.22%	4.35%
8y	-0.30%	1.93%	-0.87%	-2.72%	-3.73%	-0.57%	0.67%	1.69%	3.12%	4.22%
9y	0.18%	1.87%	-1.33%	-3.29%	-4.04%	-0.89%	0.28%	1.73%	3.13%	4.20%
10y	0.77%	1.71%	-1.94%	-3.77%	-4.37%	-1.24%	-0.14%	1.64%	3.01%	4.06%

Plots of the market volatility surface and of the corresponding smoothed surface are given in Figure 7.7.

If we run our algebraic calibration with the smoothed swaptions data as input, the resulting instantaneous-volatility values are all real and positive, as we can see from Table 7.9.

0.1848	-	-	-	-	-	-	-	-	-
0.1403	0.2221	-	-	-	-	-	-	-	-
0.1282	0.1337	0.2424	-	-	-	-	-	-	-
0.1212	0.1124	0.1329	0.2542	-	-	-	-	-	-
0.1164	0.1003	0.1082	0.1317	0.2620	-	-	-	-	-
0.1126	0.0920	0.0950	0.1049	0.1305	0.2721	-	-	-	-
0.1095	0.0853	0.0859	0.0902	0.1009	0.1332	0.2771	-	-	-
0.1065	0.0798	0.0792	0.0809	0.0858	0.1030	0.1325	0.2904	-	-
0.1046	0.0752	0.0740	0.0742	0.0760	0.0870	0.1002	0.1405	0.2977	-
0.1025	0.0716	0.0701	0.0694	0.0695	0.0774	0.0841	0.1083	0.1429	0.3017
0.0853	0.0853	0.0686	0.0664	0.0651	0.0710	0.0741	0.0916	0.1087	0.1423
0.0778	0.0778	0.0778	0.0649	0.0618	0.0662	0.0670	0.0811	0.0908	0.1072
0.0720	0.0720	0.0720	0.0720	0.0599	0.0626	0.0617	0.0737	0.0796	0.0890
0.0669	0.0669	0.0669	0.0669	0.0669	0.0607	0.0580	0.0683	0.0721	0.0777
0.0630	0.0630	0.0630	0.0630	0.0630	0.0630	0.0549	0.0637	0.0659	0.0694
0.0596	0.0596	0.0596	0.0596	0.0596	0.0596	0.0596	0.0613	0.0616	0.0636
0.0573	0.0573	0.0573	0.0573	0.0573	0.0573	0.0573	0.0573	0.0576	0.0588
0.0552	0.0552	0.0552	0.0552	0.0552	0.0552	0.0552	0.0552	0.0552	0.0552
0.0531	0.0531	0.0531	0.0531	0.0531	0.0531	0.0531	0.0531	0.0531	0.0531

Table 7.9. Fitted instantaneous volatilities, having as input the smoothed swaptions data, collected as in TABLE 1.

We thus conclude that irregularity and illiquidity in the input swaption matrix can cause negative or even imaginary values in the calibrated instantaneous volatilities. However, by smoothing the input data before calibration, usually this undesirable features can be avoided.

In the following, we show the terminal correlations and the evolution of the term structure of volatility for the two 10 × 10 matrices, recalling that, in our tests, we have considered the instantaneous-correlation parameters θ

7.4 Exact Swaptions Calibration with Volatilities as TABLE 1 313

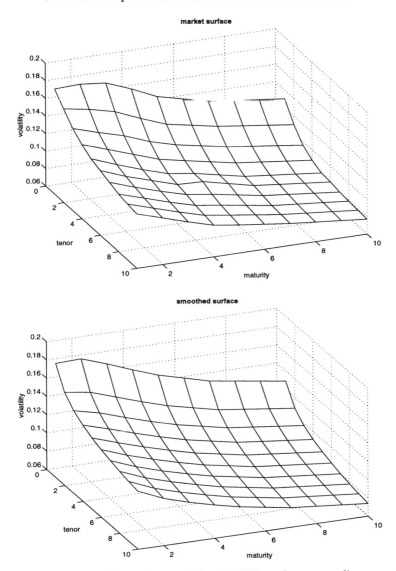

Fig. 7.7. Market volatility surface as of May 16, 2000, and corresponding smoothed surface.

that give the typical instantaneous-correlation matrix showed in figure 8.1 (with the obvious restriction in the case 3×3).

The terminal correlation matrix obtained with the market (non-smoothed) swaption-volatility data is

314   7. Cases of Calibration of the LIBOR Market Model

	10y	11y	12y	13y	14y	15y	16y	17y	18y	19y
10y	1.000	0.604	0.415	0.349	0.313	0.308	0.271	0.233	0.252	0.228
11y	0.604	1.000	0.654	0.460	0.417	0.349	0.371	0.354	0.303	0.310
12y	0.415	0.654	1.000	0.679	0.489	0.397	0.349	0.415	0.363	0.309
13y	0.349	0.460	0.679	1.000	0.682	0.486	0.391	0.348	0.404	0.398
14y	0.313	0.417	0.489	0.682	1.000	0.674	0.513	0.420	0.415	0.551
15y	0.308	0.349	0.397	0.486	0.674	1.000	0.684	0.489	0.399	0.328
16y	0.271	0.371	0.349	0.391	0.513	0.684	1.000	0.678	0.471	0.340
17y	0.233	0.354	0.415	0.348	0.420	0.489	0.678	1.000	0.670	0.423
18y	0.252	0.303	0.363	0.404	0.415	0.399	0.471	0.670	1.000	0.639
19y	0.228	0.310	0.309	0.398	0.551	0.328	0.340	0.423	0.639	1.000

whereas the terminal correlation matrix obtained with the smoothed swaption-volatility data is

	10y	11y	12y	13y	14y	15y	16y	17y	18y	19y
10y	1.000	0.534	0.418	0.363	0.330	0.304	0.283	0.263	0.248	0.233
11y	0.534	1.000	0.594	0.480	0.419	0.375	0.341	0.311	0.287	0.266
12y	0.418	0.594	1.000	0.621	0.513	0.447	0.400	0.362	0.331	0.306
13y	0.363	0.480	0.621	1.000	0.635	0.525	0.459	0.408	0.370	0.341
14y	0.330	0.419	0.513	0.635	1.000	0.636	0.527	0.456	0.407	0.371
15y	0.304	0.375	0.447	0.525	0.636	1.000	0.641	0.527	0.458	0.412
16y	0.283	0.341	0.400	0.459	0.527	0.641	1.000	0.633	0.522	0.458
17y	0.263	0.311	0.362	0.408	0.456	0.527	0.633	1.000	0.636	0.530
18y	0.248	0.287	0.331	0.370	0.407	0.458	0.522	0.636	1.000	0.642
19y	0.233	0.266	0.306	0.341	0.371	0.412	0.458	0.530	0.642	1.000

The future term structures of swaptions volatilities as implied by the two different calibrations are respectively shown in Figure 7.8 and Figure 7.9. Comparing such figures, we can easily see that, with the smoothed swaption prices in input, the evolution of the term structure of volatility is qualitatively improved, as well as the related terminal correlations that are non-negative and always decreasing when moving away from the diagonal.

7.5 Conclusions: Where Now?

Far from being conclusive or even systematic, our examples have served the purpose of pointing out many of the actual problems involved in the LFM calibration, with some discussion on possible solutions.

In conclusion, one would like a calibration of the LFM to have the following features:

1. A small calibration error, i.e. small percentage differences

$$100 \cdot \frac{\text{Market swaption volatility - LFM swaption volatility}}{\text{Market swaption volatility}}$$

2. Regular instantaneous correlations. One would like to avoid large oscillations along the columns of the correlation matrix. Indeed, one would appreciate a monotonically decreasing pattern when moving away from a diagonal term of the matrix along the related row or column.
3. Regular terminal correlations.

7.5 Conclusions: Where Now? 315

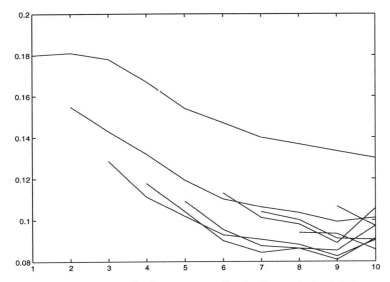

Fig. 7.8. Term-structure evolution corresponding to the market volatility swaption data

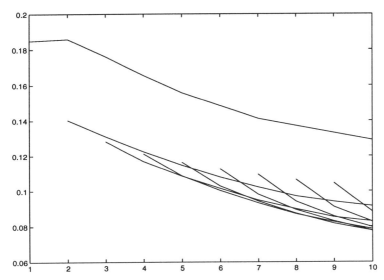

Fig. 7.9. Term-structure evolution corresponding to the smoothed volatility swaption data

4. Smooth and qualitatively stable evolution of the term structure of caplet volatilities over time.

In addition, one would like to accomplish these points, which are related to each other, by a low number of factors. However, none of the structures we proposed can perfectly meet the above requirements at the same time, although the formulation of TABLE 1 seems to go in the right direction. This shows once again that the core of the LFM is a clever choice of the covariance function. The rest are just mathematical details. One may try and combine many of the ideas presented here to come up with a different approach that might work. Research in this issue is still quite open, and probably the next months/years will show some significant evolution.

We just dare say that the formulation of TABLE 1, with our "automatic closed-form algebraic" exact calibration is probably the most promising, since one can impose exogenously a decent instantaneous-correlation matrix and hope to obtain decent terminal correlations and evolution of the term structure of volatilities.

Our method maps, in a one-to-one correspondence, swaptions volatilities into pieces of instantaneous volatilities of forward rates. This can help also in computing sensitivities with respect to swaptions volatilities, since one knows on which σ's one needs to act in order to influence a single swaption volatility. By a kind of chain rule we can translate sensitivity with respect to the σ's in sensitivity with respect to the swaption volatilities used in the calibration.

We also recall that we have used the swaption matrix as quoted in the market from one broker. However, as we observed in Remark 6.15.1, we must be somehow careful. Further calibration tests are needed with smoothed or adjusted data, before concluding for sure that limitations concern only the model formulations and not a possible "noise" present in the data. One can in fact decide that the model is detecting a misalignment in the market, instead of concluding that the model is not suitable for the joint calibration. Probably the "truth" lies in-between. We have done this for the last piecewise-constant formulation of TABLE 1, and we have actually found that with smoothed data in input the outputs improve.

We finally mention that a synthesis of the results presented in this chapter, concerning the LFM calibration to market data, had already appeared in Brigo, Capitani and Mercurio (2000).

8. Monte Carlo Tests for LFM Analytical Approximations

Hey, Houston, we've got a problem here.
Jack Swigert, April 13 1970, Apollo 13 mission to the moon

In this chapter we test the analytical approximations leading to closed-form formulas for both swaption volatilities and terminal correlations under the LFM, by resorting to Monte Carlo simulation of the LFM dynamics. We first explain what kind of rates we are dealing with, and then move to the volatility part. Section 8.2 gives a plan of the tests on the swaption-volatility approximations and the subsequent section presents results in detail. In particular, we plot, in several cases, the real swap-rate probability density as implied by the LFM dynamics versus a lognormal density characterized by our analytically approximated volatility. We thus measure indirectly the discrepancy between the LFM swap-rate distribution and the lognormal-distribution assumption for the swap rate, as implied instead by the swap market model LSM.

Subsequently, we consider our analytical approximation for terminal correlations, and present our related testing plans in Section 8.4. Again, detailed results follow in the subsequent section. A general section of conclusions closes the chapter.

8.1 The Specification of Rates

In our tests, both on instantaneous volatilities and on terminal correlations, we will consider a family of forward rates whose expiry/maturity pairs are adjacent elements in the array

$$0y \quad 1y \quad 2y \quad \ldots \quad 19y \quad 20y.$$

We thus take as resetting times (expires) of the rates in our family, $T_0 = 1y$, $T_1 = 2y$ up to $T_{18} = 19y$, meaning that we take as initial input the forward rates

$$F_0(0), \ F_1(0), \ F_2(0), \ \ldots, \ F_{19}(0),$$

$F_0(0) = L(0, 1y)$ being the initial spot rate.

The chosen values for these initial rates will always be

$$F_{0\div 9}(0) = [4.69\ 5.01\ 5.60\ 5.84\ 6.00\ 6.13\ 6.28\ 6.27\ 6.29\ 6.23]$$
$$F_{10\div 19}(0) = [6.30\ 6.36\ 6.43\ 6.48\ 6.53\ 6.40\ 6.30\ 6.18\ 6.07\ 5.94],$$

where all rates are expressed as percentages. These values are consistent with the volatility and correlation values to be illustrated later on, in that all such initial inputs reflect a possible calibration of the LFM to caps and swaptions.

At times, we will stress the initial forward rates by uniformly shifting upwards all their values by 2%, thus taking as initial rates

$$F_{0\div 9}(0) = [6.69\ 7.01\ 7.60\ 7.84\ 8.00\ 8.13\ 8.28\ 8.27\ 8.29\ 8.23]$$
$$F_{10\div 19}(0) = [8.30\ 8.36\ 8.43\ 8.48\ 8.53\ 8.40\ 8.30\ 8.18\ 8.07\ 7.94].$$

8.2 The "Testing Plan" for Volatilities

If the art of the detective began and ended in reasoning from an armchair, my brother would be the greatest criminal agent that ever lived.
Sherlock Holmes

We plan to test Rebonato's and Hull and White's formulas, (6.58) and (6.59), against a Monte Carlo evaluation as follows.

We adopt Formulation 7 for instantaneous volatilities, although in some cases we will let it collapse to Formulation 6, or to the formulations of TABLEs 2 or 3 respectively.

We recall the linear-exponential Formulation 7:

$$\sigma_i(t) = \Phi_i \left([a(T_{i-1} - t) + d] e^{-b(T_{i-1}-t)} + c \right),$$

where, as usual, T_{i-1} is the expiry of the relevant forward rate. We will often consider restrictions on the possible parameters values.

Recall the two-factor parametric form in θ for instantaneous correlations,

$$\rho^B_{i,j} = \cos(\theta_i - \theta_j), \qquad \theta = [\theta_1\ \ldots\ \theta_{19}].$$

The values of the parameters a, b, c, d and the values of the θ's, in the general case of Formulation 7, have been built so as to reflect possible joint calibrations of the LFM to caps and swaptions. Such values are reported in point (3.a) below.

A short map of our sets of testing parameters is given in the following.

- (1.a) *Constant instantaneous volatilities, typical rank-two correlations.*
  Formulation 7 with $a = 0$, $b = 0$, $c = 1$, $d = 0$. This amounts actually to Formulation 3, since the ψ part of Formulation 7 is collapsed to one and the Φ's act as the constant instantaneous volatilities s of TABLE 3, which are set to the caplet volatilities taken in input. The correlation angles are $\theta = [\theta_{1\div 8}\ \theta_{9\div 16}\ \theta_{17\div 19}]$, where

8.2 The "Testing Plan" for Volatilities

$$\theta_{1\div 8} = [0.0147\ 0.0643\ 0.1032\ 0.1502\ 0.1969\ 0.2239\ 0.2771\ 0.2950],$$
$$\theta_{9\div 16} = [0.3630\ 0.3810\ 0.4217\ 0.4836\ 0.5204\ 0.5418\ 0.5791\ 0.6496]$$
$$\theta_{17\div 19} = [0.6679\ 0.7126\ 0.7659].$$

This set of angles implies positive and decreasing instantaneous correlations as from Figure 8.1. In the same figure, the values of the Φ's are also shown.

- (1.b) *Constant instantaneous volatilities, perfect correlation.*
  Formulation 7 with a, b, c and d as in (1.a) and $\theta = [0\ 0\ \ldots\ 0\ 0]$, implying that all instantaneous correlations are set to one.

- (1.c) *Constant instantaneous volatilities, some negative rank-two correlations.*
  Formulation 7 with a, b, c and d as in (1.a) and $\theta = [\theta_{1\div 9}\ \theta_{10\div 17}\ \theta_{18,19}]$, where

$$\theta_{1\div 9} = [0\ 0.0000\ 0.0013\ 0.0044\ 0.0096\ 0.0178\ 0.0299\ 0.0474\ 0.0728],$$
$$\theta_{10\div 17} = [0.1100\ 0.1659\ 0.2534\ 0.3989\ 0.6565\ 1.1025\ 1.6605\ 2.0703]$$
$$\theta_{18,19} = [2.2825\ 2.2260].$$

This set of angles implies instantaneous correlations as from Figure 8.6.

- (2.a) *Humped instantaneous volatilities depending only on time to maturity, perfect correlation.*
  Formulation 6 with $a = 0.1908$, $b = 0.9746$, $c = 0.0808$, $d = 0.0134$ and $\theta = [0\ 0\ \ldots\ 0\ 0]$.

- (2.b) *Humped instantaneous volatilities depending only on time to maturity, some negative rank-two correlations.*
  Formulation 6 with a, b, c and d as in (2.a) and θ as in (1.c).

- (3.a) *Humped and maturity-adjusted instantaneous volatilities depending only on time to maturity, typical rank-two correlations.*
  Formulation 7 with $a = 0.1908$, $b = 0.9746$, $c = 0.0808$, $d = 0.0134$ as in (2.a) and θ as in (1.a). This is the "most normal" situation, in that it reflects a qualitatively acceptable evolution of the term structure of volatilities in time and allows for a satisfactory joint calibration to caplets and swaptions volatilities.

- (3.b) *Humped and maturity-adjusted instantaneous volatilities depending only on time to maturity, perfect correlation.*
  Formulation 7 with a, b, c and d as in (3.a) and $\theta = [0\ 0\ \ldots\ 0\ 0]$.

- (3.c) *Humped and maturity-adjusted instantaneous volatilities depending only on time to maturity, some negative rank-two correlations.*
  Formulation 7 with a, b, c and d as in (3.a) and θ as in (1.c).

In each of the above situations, we will display some numerical results and also some plots of probability densities.

The results are based on a comparison of Rebonato's and Hull and White's formulas with the volatilities that plugged into Black's formula lead to the Monte Carlo prices of the corresponding at-the-money swaptions. Indeed, we price a chosen swaption with underlying swap $S_{\alpha,\beta}$ through the Monte Carlo method, and then invert Black's formula by solving the following equation in $v_{\alpha,\beta}^{\text{MC}}$

$$C_{\alpha,\beta}(0)\,\text{Bl}(S_{\alpha,\beta}(0), S_{\alpha,\beta}(0), v_{\alpha,\beta}^{\text{MC}}) = \text{MCprice}_{\alpha,\beta},$$

thus deriving the Black volatility implied by the LFM Monte Carlo price.

Furthermore, we compute the standard error of the method, corresponding to a two-side 98% window around the mean:

$$\text{MCerr}_{\alpha,\beta} = 2.33\,\frac{\text{Std}\{(P(0,T_\alpha)\,C_{\alpha,\beta}^j(T_\alpha)\,(S_{\alpha,\beta}^j(T_\alpha) - K)^+\}}{\sqrt{\text{npath}}},$$

where j denotes the scenario, "npath" is the number of scenarios in the Monte Carlo method, and the standard deviation is taken over the simulated scenarios. We use this quantity to compute the "inf" and "sup" volatilities $v_{\alpha,\beta}^{\text{MCinf}}, v_{\alpha,\beta}^{\text{MCsup}}$ obtained from solving the equations

$$C_{\alpha,\beta}(0)\,\text{Bl}(S_{\alpha,\beta}(0), S_{\alpha,\beta}(0), v_{\alpha,\beta}^{\text{MCinf}}) = \text{MCprice}_{\alpha,\beta} - \text{MCerr}_{\alpha,\beta},$$
$$C_{\alpha,\beta}(0)\,\text{Bl}(S_{\alpha,\beta}(0), S_{\alpha,\beta}(0), v_{\alpha,\beta}^{\text{MCsup}}) = \text{MCprice}_{\alpha,\beta} + \text{MCerr}_{\alpha,\beta},$$

thus deriving the Black implied volatilities corresponding to the extremes of the 99% price window.

Since the differences between Rebonato's and Hull and White's formulas will be shown to be typically negligible, we will just consider the following percentage differences based on Rebonato's volatility:

$$100\,(v_{\alpha,\beta}^{\text{MC}} - v_{\alpha,\beta}^{\text{LFM}})/v_{\alpha,\beta}^{\text{MC}},$$
$$100\,(v_{\alpha,\beta}^{\text{MCinf}} - v_{\alpha,\beta}^{\text{LFM}})/v_{\alpha,\beta}^{\text{MCinf}},$$
$$100\,(v_{\alpha,\beta}^{\text{MCsup}} - v_{\alpha,\beta}^{\text{LFM}})/v_{\alpha,\beta}^{\text{MCsup}}.$$

As far as distributions are concerned, we reason as follows. The Rebonato formula is derived under the assumption that the swap rate is driftless and lognormal under any of the forward measures Q^k, with integrated percentage variance $(v_{\alpha,\beta}^{\text{LFM}})^2$. In particular, the swap rate is assumed to be driftless and lognormal under Q^α, with mean $S_{\alpha,\beta}(0)$ and log-variance $(v_{\alpha,\beta}^{\text{LFM}})^2$. Indeed, as far as the swaption corresponding to $S_{\alpha,\beta}$ is concerned, this is equivalent to assuming $S_{\alpha,\beta}(t) = \overline{S}_{\alpha,\beta}(t)$, the classic driftless geometric Brownian motion:

$$d\overline{S}_{\alpha,\beta}(t) = v(t)\overline{S}_{\alpha,\beta}(t)dW^\alpha, \qquad (8.1)$$

so that we can write

$$Y = \overline{S}_{\alpha,\beta}(T_\alpha) = \overline{S}_{\alpha,\beta}(0)\exp\left\{-\frac{1}{2}(v_{\alpha,\beta}^{\text{LFM}})^2 + v_{\alpha,\beta}^{\text{LFM}}\overline{W}\right\},$$

where v is any function recovering the correct terminal variance, $\int_0^T v(t)^2 dt = (v_{\alpha,\beta}^{\text{LFM}})^2$, and $\overline{W} \sim \mathcal{N}(0,1)$ under Q^α.

To test the quality of this approximation, we compare the sampled distribution under Q^α of the (Monte Carlo) simulated $S_{\alpha,\beta}(T_\alpha)$ with a lognormal distribution corresponding to an initial condition $S_{\alpha,\beta}(0)$ and an integrated percentage variance $(v_{\alpha,\beta}^{\mathrm{LFM}})^2$, consistently with the approximated dynamics (8.1).

The density of this lognormal distribution is given by

$$p_Y(y) = \frac{1}{yv_{\alpha,\beta}^{\mathrm{LFM}}} \frac{1}{\sqrt{2\pi}} \exp\left\{ -\frac{1}{2(v_{\alpha,\beta}^{\mathrm{LFM}})^2} \left[\ln\frac{y}{S_{\alpha,\beta}(0)} + \frac{1}{2}(v_{\alpha,\beta}^{\mathrm{LFM}})^2 \right]^2 \right\}.$$

However, we know that the true S-dynamics is driftless under $Q^{\alpha,\beta}$ but not under Q^α. Therefore, there will be a bias in assuming

$$E_0^\alpha\{S_{\alpha,\beta}(T_\alpha)\} \approx E_0^\alpha\{\overline{S}_{\alpha,\beta}(T_\alpha)\} = S_{\alpha,\beta}(0)$$

since what is true is, instead,

$$E_0^{\alpha,\beta}\{S_{\alpha,\beta}(T_\alpha)\} = S_{\alpha,\beta}(0) \ .$$

To measure this bias, we consider a new lognormal distribution where we replace $S_{\alpha,\beta}(0)$ in the previous one with $\mu_\alpha(0) := E_0^\alpha\{S_{\alpha,\beta}(T_\alpha)\}$. Moreover, we calculate the third central moment and the kurtosis of the swap-rate logarithm, and compare them with those of a standard normal distribution, amounting respectively to 0 and 3.

Further, we plot the density obtained through the Monte Carlo method against the analytical one in both the biased and unbiased cases, so that we can isolate the part of the error in the distribution due to the bias in the drift. We will see that, in almost all cases, the correction in the mean accounts for the whole difference, in that the "true" (Monte Carlo) distribution of S under Q^α is practically indistinguishable from a lognormal distribution with mean $\mu_\alpha(0)$ and log-variance $(v_{\alpha,\beta}^{\mathrm{LFM}})^2$.

8.3 Test Results for Volatilities

I'm running tests, but I don't begin to understand it
Querl Dox (Brainiac 5), Legion Lost 6, 2000, DC Comics

In this section, we present the results obtained when deriving implied volatilities from the swaptions prices calculated through a Monte Carlo method, with four time steps per year and 200000 paths, against Rebonato's and Hull and White's analytical formulas. This will be done for several formulations of instantaneous volatilities and correlations, and for different "expiry/maturity" pairs of forward rates.

In our numerical tests, we will follow the map of cases given in the previous section. For each point of the testing plan, we will also consider the "stressed" results obtained when upwardly shifting the Φ's values by 0.2. Furthermore, going back to unstressed Φ's, we will consider the

322   8. Monte Carlo Tests for LFM Analytical Approximations

"stressed" results obtained when upwardly shifting the initial forward-rate vector $[F(0,0,1), \ldots, F(0,19,20)]$ by 2%.

With reference to the previous section, before showing our numerical results and plots of probability densities, we explain the terms that will appear in the tables.

We denote the first reset date T_α by "res" and take, in all cases, the last payment date to be $T_\beta = 20y$. The volatilities $v_{\alpha,\beta}^{MC}$, $v_{\alpha,\beta}^{MCinf}$, $v_{\alpha,\beta}^{MCsup}$, $v_{\alpha,\beta}^{LFM}$, $\overline{v}_{\alpha,\beta}^{LFM}$, divided by $\sqrt{T_\alpha}$, are denoted, respectively, by "MCVol", "MCinf", "MCsup", "RebVol", "HWVol".

We denote by "differr" and "errperc", respectively, the absolute and percentage differences (errors) between Monte Carlo implied volatilities and Rebonato's volatilities. We also denote by "errinf" and "errsup" the percentage difference between the lower ("inf") and upper ("sup") extremes of the 99% Monte Carlo window and Rebonato's volatility.

We then report the initial swap rate $S_{\alpha,\beta}(0)$, "sw0", the mean $\mu_\alpha(0)$ of the "Monte Carlo" swap rate (generated under the measure Q^α), "mean", and finally the third central moment and the kurtosis of the simulated swap-rate logarithm, "third" and "kurt" (again under the measure Q^α).

8.3.1 Case (1): Constant Instantaneous Volatilities

The first case we analyze is that of constant instantaneous volatilities. We have seen that with such a formulation, as time passes from an expiry date to the next, the volatility term structure is obtained from the previous one by "cutting off" the head, so that its qualitative behaviour can be altered over time. Typically, if the term structure features a hump around two years, this hump will be absent in the term structure occurring in three years.

We can impose structures of this kind via the formulation of TABLE 3. We actually resort to Formulation 7 and take $a = b = d = 0$ and $c = 1$, so that the caplet fitting forces the Φ's to equal the corresponding caplet volatilities, as obtained from the market, $\Phi_i = v_{T_{i-1}-\text{caplet}}$.

(1.a): Typical rank-two correlations

The instantaneous-correlation matrix implied by this choice of parameters comes from a possible joint caps/swaptions calibration and is reported in Figure 8.1. The graph of the parameters Φ's is shown in Figure 8.2.

Numerical results are given in the following:

res	MCVol	MCinf	MCsup	RebVol	HWVol	differr	errperc
9	0.1019	0.1014	0.1023	0.1015	0.1017	0.0003	0.3116

	errinf	errsup	sw0	mean	third	kurt
	-0.1389	0.7581	0.0635	0.0655	0.0000	3.0076

The volatility approximation appears to be excellent. The biased- and unbiased-density plots are shown in Figure 8.3.

8.3 Test Results for Volatilities   323

| Phi | 0.1490 | 0.1590 | 0.1530 | 0.1450 | 0.1360 | 0.1270 | 0.1210 | 0.1180 | 0.1140 | 0.1110 | 0.1080 | 0.1050 | 0.1020 | 0.0989 | 0.0978 | 0.0974 | 0.0969 | 0.0965 | 0.0961 |
| thetas | 0.0147 | 0.0643 | 0.1030 | 0.1500 | 0.1970 | 0.2240 | 0.2770 | 0.2950 | 0.3630 | 0.3810 | 0.4220 | 0.4840 | 0.5200 | 0.5420 | 0.5790 | 0.6500 | 0.6580 | 0.7130 | 0.7660 |
corr	1y	2y	3y	4y	5y	6y	7y	8y	9y	10y	11y	12y	13y	14y	15y	16y	17y	18y	19y
1y	1.0000	0.9988	0.9961	0.9908	0.9834	0.9782	0.9658	0.9610	0.9399	0.9337	0.9183	0.8920	0.8748	0.8642	0.8449	0.8051	0.7941	0.7662	0.7309
2y	0.9988	1.0000	0.9992	0.9963	0.9912	0.9873	0.9774	0.9735	0.9557	0.9503	0.9368	0.9134	0.8978	0.8881	0.8704	0.8335	0.8233	0.7971	0.7638
3y	0.9961	0.9992	1.0000	0.9989	0.9956	0.9927	0.9849	0.9817	0.9664	0.9617	0.9497	0.9285	0.9142	0.9053	0.8889	0.8544	0.8447	0.8200	0.7883
4y	0.9908	0.9963	0.9989	1.0000	0.9989	0.9973	0.9920	0.9895	0.9774	0.9735	0.9634	0.9449	0.9323	0.9243	0.9095	0.8779	0.8690	0.8460	0.8164
5y	0.9834	0.9912	0.9956	0.9989	1.0000	0.9996	0.9968	0.9952	0.9862	0.9831	0.9749	0.9592	0.9481	0.9411	0.9279	0.8993	0.8911	0.8700	0.8425
6y	0.9782	0.9873	0.9927	0.9973	0.9996	1.0000	0.9986	0.9975	0.9903	0.9877	0.9805	0.9665	0.9564	0.9499	0.9376	0.9108	0.9030	0.8830	0.8567
7y	0.9658	0.9774	0.9849	0.9920	0.9968	0.9986	1.0000	0.9998	0.9963	0.9946	0.9896	0.9787	0.9705	0.9652	0.9548	0.9314	0.9246	0.9067	0.8829
8y	0.9610	0.9735	0.9817	0.9895	0.9952	0.9975	0.9998	1.0000	0.9977	0.9963	0.9920	0.9823	0.9747	0.9697	0.9599	0.9378	0.9313	0.9141	0.8912
9y	0.9399	0.9557	0.9664	0.9774	0.9862	0.9903	0.9963	0.9977	1.0000	0.9998	0.9983	0.9927	0.9876	0.9841	0.9768	0.9592	0.9539	0.9395	0.9200
10y	0.9337	0.9503	0.9617	0.9735	0.9831	0.9877	0.9946	0.9963	0.9998	1.0000	0.9992	0.9947	0.9903	0.9871	0.9804	0.9641	0.9591	0.9455	0.9268
11y	0.9183	0.9368	0.9497	0.9634	0.9749	0.9805	0.9896	0.9920	0.9983	0.9992	1.0000	0.9981	0.9951	0.9928	0.9876	0.9741	0.9698	0.9580	0.9413
12y	0.8920	0.9134	0.9285	0.9449	0.9592	0.9665	0.9787	0.9823	0.9927	0.9947	0.9981	1.0000	0.9993	0.9983	0.9954	0.9863	0.9831	0.9739	0.9604
13y	0.8748	0.8978	0.9142	0.9323	0.9481	0.9564	0.9705	0.9747	0.9876	0.9903	0.9951	0.9993	1.0000	0.9998	0.9983	0.9917	0.9891	0.9816	0.9700
14y	0.8642	0.8881	0.9053	0.9243	0.9411	0.9499	0.9652	0.9697	0.9841	0.9871	0.9928	0.9983	0.9998	1.0000	0.9993	0.9942	0.9921	0.9855	0.9750
15y	0.8449	0.8704	0.8889	0.9095	0.9279	0.9376	0.9548	0.9599	0.9768	0.9804	0.9876	0.9954	0.9983	0.9993	1.0000	0.9975	0.9961	0.9911	0.9826
16y	0.8051	0.8335	0.8544	0.8779	0.8993	0.9108	0.9314	0.9378	0.9592	0.9641	0.9741	0.9863	0.9917	0.9942	0.9975	1.0000	0.9998	0.9980	0.9932
17y	0.7941	0.8233	0.8447	0.8690	0.8911	0.9030	0.9246	0.9313	0.9539	0.9591	0.9698	0.9831	0.9891	0.9921	0.9961	0.9998	1.0000	0.9990	0.9952
18y	0.7662	0.7971	0.8200	0.8460	0.8700	0.8830	0.9067	0.9141	0.9395	0.9455	0.9580	0.9739	0.9816	0.9855	0.9911	0.9980	0.9990	1.0000	0.9986
19y	0.7309	0.7638	0.7883	0.8164	0.8425	0.8567	0.8829	0.8912	0.9200	0.9268	0.9413	0.9604	0.9700	0.9750	0.9826	0.9932	0.9952	0.9986	1.0000

Fig. 8.1. Case (1.a): parameters Φ, θ and matrix of instantaneous correlations.

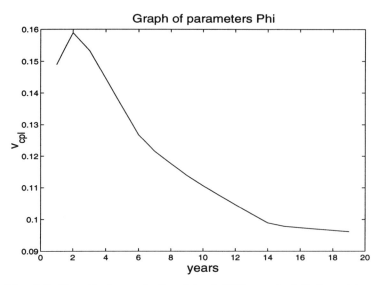

Fig. 8.2. Case (1.a): graph of caplet volatilities against their expiries.

(1.b): Perfect correlations

The set of parameters in this case is as in (1.a) but with perfect instantaneous correlation (matrix of ones, obtained putting $\theta = [0\ldots 0]$). We obtain numerical results similar to those in the previous case.

res	MCVol	MCinf	MCsup	RebVol	HWVol	differr	errperc
9	0.1026	0.1021	0.1030	0.1021	0.1022	0.0004	0.4381
	errinf	errsup	sw0	mean	third	kurt	
	-0.0123	0.8844	0.0635	0.0655	0.0000	2.9986	

Again, the volatility approximations seem to be quite satisfactory. As a "stress test", in this case we also upwardly shifted the parameters Φ by 20%, and, as expected, our results worsened. More specifically, since under this choice of a, b, c and d the Φ's are given by the caplet volatilities, our shift amounts to

$$\Phi_i = v_{T_{i-1}\text{-caplet}} \rightarrow \Phi_i = v_{T_{i-1}\text{-caplet}} + 0.2.$$

This leads to rather large volatilities. In fact, just add 20% to all the original caplet volatilities in input, shown in Figure 8.2, to see that the new ones are more than doubled. Since the "freezing" approximation works better when variability is small, and increasing the Φ's amounts to increasing volatility, it is natural to expect the approximation to worsen, as is happening here.

res	MCVol	MCinf	MCsup	RebVol	HWVol	differr	errperc
2	0.3106	0.3094	0.3119	0.3184	0.3168	-0.0077	-2.4907
9	0.2972	0.2958	0.2987	0.3021	0.3025	-0.0049	-1.6355
	errinf	errsup	sw0	mean	third	kurt	
	-2.8959	-2.0886	0.0623	0.0718	0.0010	2.9995	
	-2.1440	-1.1315	0.0635	0.0894	0.0011	2.8877	

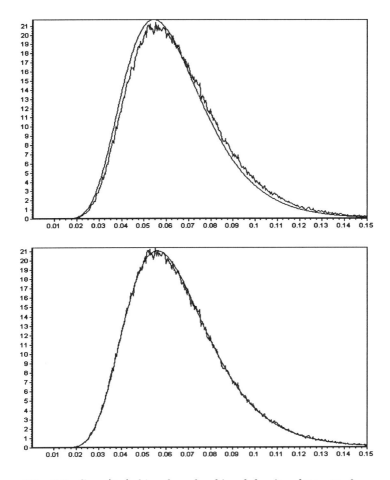

Fig. 8.3. Case (1.a): biased- and unbiased-density plots; res=9y.

Nevertheless, we still believe the approximation to be acceptable, although not as good as in the "unstressed" case.

The biased- and unbiased-density plots, in both the "unstressed" and "stressed" cases, are shown in Figure 8.4 and 8.5, respectively. The last graph shows that a kurtosis far from 3 yields more marked differences even in the unbiased-density plot.

(1.c): Some negative rank-two correlations

Another situation we study is that of a partially negative rank-two correlation matrix, obtained with a new set of θ's. This matrix is reported below, in Figure 8.6.

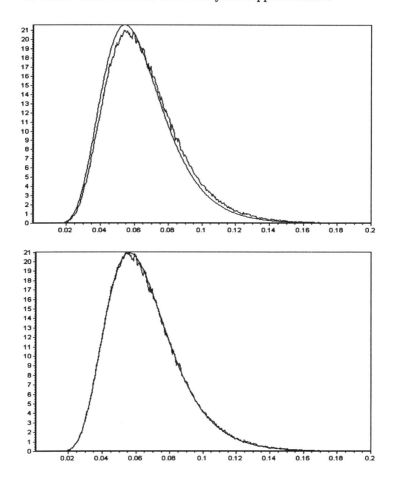

Fig. 8.4. Case (1.b): biased- and unbiased-density plots; $\theta = 0$, res=9y.

We now present the results obtained with this set of parameters and, in the second row, with the Φ's upwardly shifted by 20%:

res	MCVol	MCinf	MCsup	RebVol	HWVol	differr	errperc
15	0.0939	0.0934	0.0944	0.0937	0.0938	0.0002	0.1952
15	0.2862	0.2839	0.2885	0.2870	0.2872	-0.0009	-0.2984
	errinf	errsup	sw0	mean	third	kurt	
	-0.3303	0.7153	0.0619	0.0630	0.0000	2.9933	
	-1.1097	0.5021	0.0619	0.0817	0.0028	3.0334	

Again we obtain small percentage differences for unstressed Φ's, while stressing the Φ's yields relatively small percentage differences for volatilities.

The biased-and unbiased-density plots, in both the "unstressed" and "stressed" cases, are shown in Figures 8.7 and 8.8, respectively.

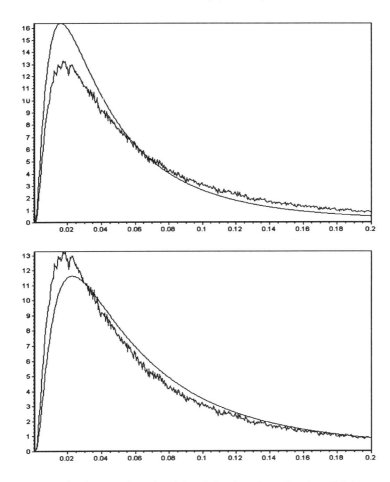

Fig. 8.5. Case (1.b): biased- and unbiased-density plots; $\theta = 0$ and Φ increased of 20%, res=9y.

From Figure 8.8, we see that large differences in both density plots are still observed. Notice also that while in (1.b) we had a log-kurtosis faraway from 3, in this case the kurtosis is close to the correct value, and yet densities appear to be quite different. This might be due to differences in higher order moments. Notice also the large bias in the mean. However, keep in mind that volatilities here were at pathological values, given that we upwardly shifted typical market percentage caplet volatilities by $0.2 = 20\%$.

328   8. Monte Carlo Tests for LFM Analytical Approximations

thetas	0.0000	0.0000	0.0013	0.0044	0.0096	0.0178	0.0299	0.0474	0.0728	0.1100	0.1660	0.2530	0.3990	0.6570	1.1000	1.6600	2.0700	2.2800	2.2300
corr	1y	2y	3y	4y	5y	6y	7y	8y	9y	10y	11y	12y	13y	14y	15y	16y	17y	18y	19y
1y	1.0000	1.0000	1.0000	1.0000	1.0000	0.9998	0.9996	0.9989	0.9974	0.9940	0.9863	0.9681	0.9215	0.7921	0.4514	-0.0896	-0.4790	-0.6531	-0.6093
2y	1.0000	1.0000	1.0000	1.0000	1.0000	0.9998	0.9996	0.9989	0.9974	0.9940	0.9863	0.9681	0.9215	0.7921	0.4514	-0.0896	-0.4790	-0.6531	-0.6093
3y	1.0000	1.0000	1.0000	1.0000	1.0000	0.9999	0.9996	0.9989	0.9974	0.9941	0.9865	0.9684	0.9220	0.7929	0.4525	-0.0883	-0.4778	-0.6521	-0.6083
4y	1.0000	1.0000	1.0000	1.0000	1.0000	0.9999	0.9997	0.9991	0.9977	0.9944	0.9870	0.9692	0.9232	0.7948	0.4553	-0.0852	-0.4751	-0.6498	-0.6058
5y	1.0000	1.0000	1.0000	1.0000	1.0000	1.0000	0.9998	0.9993	0.9980	0.9950	0.9878	0.9704	0.9252	0.7980	0.4599	-0.0800	-0.4705	-0.6458	-0.6017
6y	0.9998	0.9998	0.9999	0.9999	1.0000	1.0000	0.9999	0.9996	0.9985	0.9958	0.9891	0.9724	0.9283	0.8029	0.4672	-0.0718	-0.4633	-0.6395	-0.5951
7y	0.9996	0.9996	0.9996	0.9997	0.9998	0.9999	1.0000	0.9998	0.9991	0.9968	0.9908	0.9751	0.9327	0.8100	0.4778	-0.0598	-0.4525	-0.6302	-0.5853
8y	0.9989	0.9989	0.9989	0.9991	0.9993	0.9996	0.9998	1.0000	0.9997	0.9980	0.9930	0.9789	0.9389	0.8202	0.4931	-0.0423	-0.4369	-0.6165	-0.5711
9y	0.9974	0.9974	0.9974	0.9977	0.9980	0.9985	0.9991	0.9997	1.0000	0.9993	0.9957	0.9837	0.9473	0.8344	0.5151	-0.0169	-0.4139	-0.5963	-0.5500
10y	0.9940	0.9940	0.9941	0.9944	0.9950	0.9958	0.9968	0.9980	0.9993	1.0000	0.9984	0.9897	0.9586	0.8543	0.5466	0.0203	-0.3797	-0.5660	-0.5186
11y	0.9863	0.9863	0.9865	0.9870	0.9878	0.9891	0.9908	0.9930	0.9957	0.9984	1.0000	0.9962	0.9730	0.8821	0.5925	0.0761	-0.3275	-0.5191	-0.4700
12y	0.9681	0.9681	0.9684	0.9692	0.9704	0.9724	0.9751	0.9789	0.9837	0.9897	0.9962	1.0000	0.9894	0.9198	0.6607	0.1630	-0.2436	-0.4424	-0.3911
13y	0.9215	0.9215	0.9220	0.9232	0.9252	0.9283	0.9327	0.9389	0.9473	0.9586	0.9730	0.9894	1.0000	0.9670	0.7625	0.3043	-0.1004	-0.3077	-0.2535
14y	0.7921	0.7921	0.7929	0.7948	0.7980	0.8029	0.8100	0.8202	0.8344	0.8543	0.8821	0.9198	0.9670	1.0000	0.9022	0.5369	0.1564	-0.0552	0.0013
15y	0.4514	0.4514	0.4525	0.4553	0.4599	0.4672	0.4778	0.4931	0.5151	0.5466	0.5925	0.6607	0.7625	0.9022	1.0000	0.8483	0.5671	0.3809	0.4325
16y	-0.0896	-0.0896	-0.0883	-0.0852	-0.0800	-0.0718	-0.0598	-0.0423	-0.0169	0.0203	0.0761	0.1630	0.3043	0.5369	0.8483	1.0000	0.9172	0.8127	0.8443
17y	-0.4790	-0.4790	-0.4778	-0.4751	-0.4705	-0.4633	-0.4525	-0.4369	-0.4139	-0.3797	-0.3275	-0.2436	-0.1004	0.1564	0.5671	0.9172	1.0000	0.9776	0.9879
18y	-0.6531	-0.6531	-0.6521	-0.6498	-0.6458	-0.6395	-0.6302	-0.6165	-0.5963	-0.5660	-0.5191	-0.4424	-0.3077	-0.0552	0.3809	0.8127	0.9776	1.0000	0.9984
19y	-0.6093	-0.6093	-0.6083	-0.6058	-0.6017	-0.5951	-0.5853	-0.5711	-0.5500	-0.5186	-0.4700	-0.3911	-0.2535	0.0013	0.4325	0.8443	0.9879	0.9984	1.0000

Fig. 8.6. Case (1.c): parameters θ and matrix of instantaneous correlation.

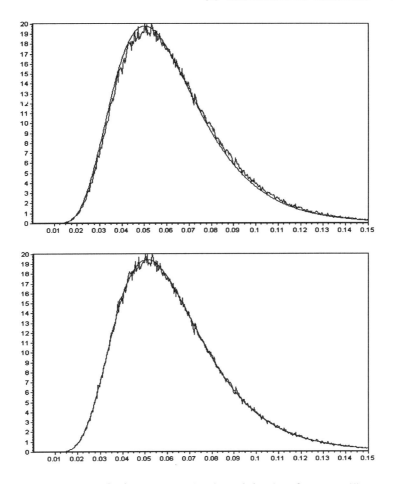

Fig. 8.7. Case (1.c): biased- and unbiased-density plots; res=15y.

8.3.2 Case (2): Volatilities as Functions of Time to Maturity

The second case we analyze is that of instantaneous volatilities depending only on time to maturity.

We obtain this structure when starting from a parametric form allowing for a humped shape in the graph of the instantaneous volatility of the generic forward rate F_i as a function of time to expiry, i.e.

$$T_{i-1} - t \mapsto \sigma_i(t).$$

This is obtained by setting all Φ's equal to one in the previous Formulation 7, so that:

$$\sigma_i(t) = \psi(T_{i-1} - t; a, b, c, d) = [a(T_{i-1} - t) + d]\, e^{-b(T_{i-1}-t)} + c.$$

330   8. Monte Carlo Tests for LFM Analytical Approximations

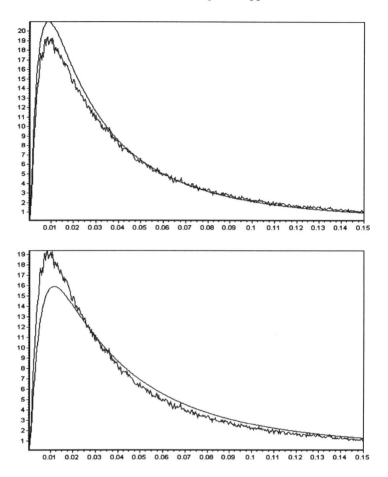

Fig. 8.8. Case (1.c): biased- and unbiased-density plots; Φ increased of 20%; res=15y.

The term structure of volatilities and the function $t \mapsto \psi(t; a, b, c, d)$ for $a = 0.19085664$, $b = 0.97462314$, $c = 0.08089167$, $d = 0.013449479$ are plotted in Figure 8.9. The values of a, b, c, d have been taken from the next case (3).

(2.a): Perfect correlations

We test this instantaneous-volatility structure starting with a perfect instantaneous-correlation matrix, by setting all θ's to be equal to zero. As usual, this implies that all ρ's are equal to one. We obtain the following numerical results:

8.3 Test Results for Volatilities 331

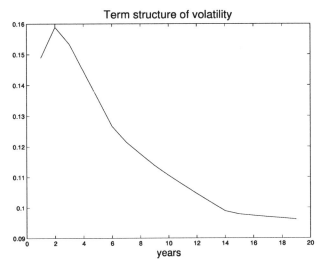

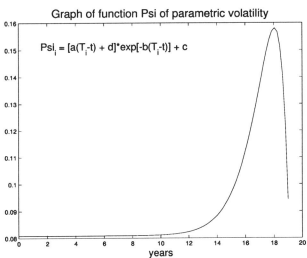

Fig. 8.9. Case (2): term structure of volatility and graph of the function $t \mapsto \psi(t; a, b, c, d)$.

res	MCVol	MCinf	MCsup	RebVol	HWVol	differr	errperc
2	0.0930	0.0927	0.0934	0.0939	0.0935	-0.0009	-0.9584
9	0.0883	0.0879	0.0887	0.0884	0.0885	-0.0001	-0.1112
17	0.0956	0.0951	0.0962	0.0954	0.0954	0.0002	0.2404
	errinf	errsup	sw0	mean	third	kurt	
	-1.3715	-0.5486	0.0623	0.0630	0.0000	3.0289	
	-0.5594	0.3330	0.0635	0.0650	0.0000	3.0098	
	-0.3273	0.8018	0.0607	0.0613	0.0000	3.0024	

The approximation appears to be satisfactory in this case too. Subsequently, as in the previous cases (1), we upwardly shift all Φ's by 20%. How-

332   8. Monte Carlo Tests for LFM Analytical Approximations

ever, we need to keep in mind that this shift has a different impact, due to the different formulation for instantaneous volatilities we are using. We obtain:

res	MCVol	MCinf	MCsup	RebVol	HWVol	differr	errperc
9	0.1064	0.1059	0.1069	0.1061	0.1062	0.0003	0.2649
	errinf	errsup	sw0	mean	third	kurt	
	-0.1899	0.7156	0.0635	0.0657	0.0001	3.0106	

The approximation is still good, contrary to the corresponding case (1.b) with shifted Φ's. We explain the reason for this in the next case (2.b).

Finally, after restoring the original Φ's, we increase of 2% the initial forward rates $[F(0,0,1), \ldots, F(0,19,20)]$, with the following numerical results:

res	MCVol	MCinf	MCsup	RebVol	HWVol	differr	errperc
9	0.0887	0.0883	0.0890	0.0888	0.0889	-0.0001	-0.1473
	errinf	errsup	sw0	mean	third	kurt	
	-0.5763	0.2781	0.0835	0.0860	0.0000	3.0138	

Here the percentage error is close to that of the original unstressed case (2.a).

By looking also at the density plots in Figures 8.10, 8.11, 8.12 and 8.13, we confirm the impression that unlike case (1.b) (volatility-structure constant in time, perfect instantaneous correlation), where the error clearly increases as the Φ's increase, here we do not observe large differences or a pronounced worsening. This may indeed be due to the fact that now the shift in volatilities is *relative*, and that we have only a 20% increase in the Φ values.

(2.b): Negative rank-two correlations

We now repeat the same test with some negative correlations, and precisely with the same instantaneous correlations as in (1.c).

The following table and the density graphs in Figures 8.14, 8.15 and 8.16, show the results obtained respectively with the original parameters (first row) and the two usual shifts, first on the Φ's (second row) and then on $F(0)$'s (third row).

res	MCVol	MCinf	MCsup	RebVol	HWVol	differr	errperc
15	0.0891	0.0886	0.0895	0.0892	0.0893	-0.0001	-0.1644
15	0.1070	0.1064	0.1076	0.1070	0.1071	-0.0000	-0.0382
15	0.1160	0.1150	0.1160	0.1160	0.1160	-0.0006	-0.5020
	errinf	errsup	sw0	mean	third	kurt	
	-0.6872	0.3531	0.0619	0.0629	0.0000	3.0096	
	-0.5869	0.5047	0.0619	0.0634	0.0000	3.0134	
	-0.9160	-0.0903	0.0623	0.0635	0.0001	3.0400	

Again, the approximation appears to be satisfactory in all cases. In particular, shifting the Φ's does not seem to affect sensibly the density plots, contrary to the corresponding case (1.c) of the constant-volatility formulation. As in the previous case (2.a), we recall that now the shift in volatilities is relative. For instance, if we had a caplet volatility of say 16%, now we have a caplet volatility of $1.2 * 16\% = 19.2\%$, and not of $16 + 20 = 36\%$ as before.

8.3 Test Results for Volatilities   333

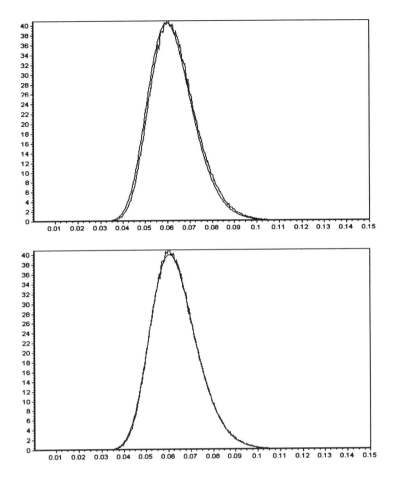

Fig. 8.10. Case (2.a): biased- and unbiased-density plots; $\theta = 0$; res=2y.

We may try and increase volatilities with the current formulation so as to reach again levels such as 36%. This can be done by leaving the Φ's unaffected and by instead increasing both c and d by $0.2 = 20\%$. By doing so, we obtain the following results:

res	MCVol	MCinf	MCsup	RebVol	HWVol	differr	errperc
15	0.2883	0.2859	0.2906	0.2893	0.2895	-0.0010	-0.3475
	errinf	errsup	sw0	mean	third	kurt	
	-1.1630	0.4572	0.0619	0.0835	0.0033	3.0817	

We also show the related biased- and unbiased-density plots in Figure 8.17. Notice that the plots are completely analogous to those of case (1.c), as one could have expected, since the uniform shift of the initial volatility structure amounts to 0.2 in both cases. This seems to suggest that with

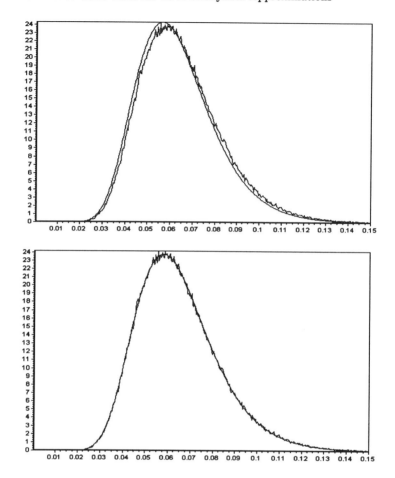

Fig. 8.11. Case (2.a): biased- and unbiased-density plots; $\theta = 0$; res=9y.

huge volatility values, our distributional approximations can become rough, no matter the particular volatility formulation chosen.

8.3.3 Case (3): Humped and Maturity-Adjusted Instantaneous Volatilities Depending only on Time to Maturity, Typical Rank-Two Correlations

The last case we consider is the most general, since it uses the full Formulation 7 for instantaneous volatility:

$$\sigma_i(t) = \Phi_i \left([a(T_{i-1} - t) + d]e^{-b(T_{i-1}-t)} + c \right),$$

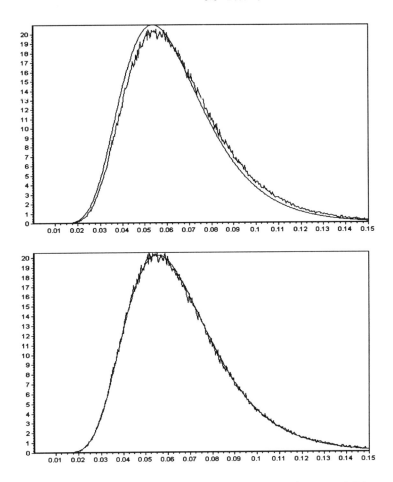

Fig. 8.12. Case (2.a): biased- and unbiased-density plots; $\theta = 0$ and Φ increased of 20%; res=9y.

which reduces to the previous case (2) when all the Φ's are set to one. Recall that this form can be seen as having a parametric core ψ, which is locally altered for each expiry T_{i-1} by the Φ's. As we already noticed in Section 6.5, these local modifications, if small, do not destroy the essential dependence on time to maturity, so as to maintain the desirable behaviour of the term structure of volatilities. In fact, the term structure remains qualitatively the same as time passes, and in particular it can maintain its humped shape if initially humped and if all Φ's are not too different.

We consider the following values of a, b, c and d: $a = 0.19085664$, $b = 0.97462314$, $c = 0.08089167$, $d = 0.013449479$.

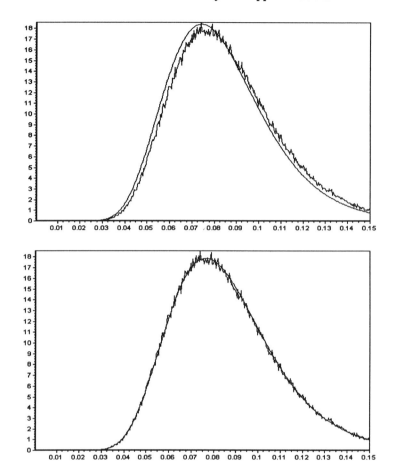

Fig. 8.13. Case (2.a): biased- and unbiased-density plots; $\theta = 0$ and $F(0)$ increased of 2%; res=9y.

Recall that, if the caplet volatilities are given as market input, we may express the Φ's as functions of a, b, c, d through formula (6.27). The computed parameters $\Phi = [\Phi_{1 \div 9}, \Phi_{10 \div 19}]$, matching the caplet volatilities, are in our case:

$$\Phi_{1 \div 8} = [\,1.0500\ 1.0900\ 1.1025\ 1.1025\ 1.0913\ 1.0669\ 1.0624\ 1.0611\,]$$
$$\Phi_{9 \div 17} = [\,1.0544\ 1.0475\ 1.0386\ 1.0270\ 1.0132\ 0.9975\ 0.9979\ 1.0033\ 1.0079\,]$$
$$\Phi_{18 \div 19} = [1.0119\ 1.0152].$$

A plot of the parameters Φ and of the core function $t \mapsto \psi(t; a, b, c, d)$ is shown below in Figure 8.18.

8.3 Test Results for Volatilities   337

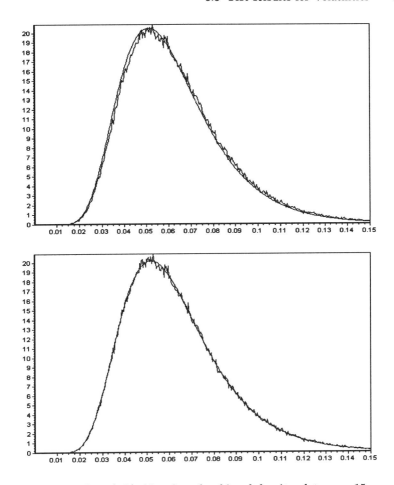

Fig. 8.14. Case (2.b): biased- and unbiased-density plots; res=15y.

(3.a): Typical rank-two correlations

Here we analyze the case where the instantaneous-correlation matrix is the same as in (1.a). We obtain the following results:

res	MCVol	MCinf	MCsup	RebVol	HWVol	differr	errperc
2	0.0969	0.0966	0.0973	0.0977	0.0973	-0.0008	-0.7825
9	0.0895	0.0891	0.0899	0.0897	0.0898	-0.0002	-0.2270
17	0.0968	0.0962	0.0973	0.0965	0.0965	0.0002	0.2401
	errinf	errsup	sw0	mean	third	kurt	
	-1.1977	-0.3707	0.0623	0.0631	0.0001	3.0538	
	-0.6771	0.2191	0.0635	0.0650	0.0000	3.0186	
	-0.3296	0.8034	0.0607	0.0613	0.0000	3.0010	

The approximation appears to be satisfactory. Also with this parametric form of case (3), we consider the usual upward shifts by 20% and 2% respec-

338   8. Monte Carlo Tests for LFM Analytical Approximations

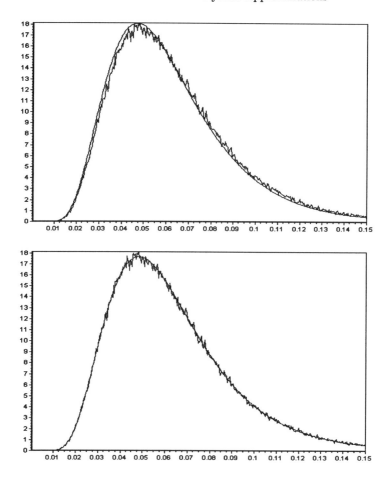

Fig. 8.15. Case (2.b): biased- and unbiased-density plots; Φ increased of 20%; res=15y.

tively for the Φ's and the $F(0)$'s. Notice again that here, too, increasing the Φ's amounts to a relative increase of 20% of the volatility structure.

When we increase the computed Φ's of 20%, we find the following numerical results:

res	MCVol	MCinf	MCsup	RebVol	HWVol	differr	errperc
2	0.1150	0.1150	0.1160	0.1160	0.1160	-0.0012	-0.9980
9	0.1073	0.1068	0.1078	0.1073	0.1074	0.0000	0.0099
17	0.1160	0.1153	0.1167	0.1156	0.1156	0.0004	0.3559
	errinf	errsup	sw0	mean	third	kurt	
	-1.4200	-0.5840	0.0623	0.0634	0.0001	3.0500	
	-0.4483	0.4639	0.0635	0.0658	0.0001	3.0209	
	-0.2543	0.9589	0.0607	0.0617	0.0000	3.0124	

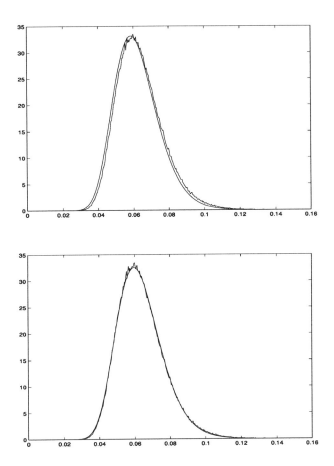

Fig. 8.16. Case (2.b): biased- and unbiased-density plots; $F(0)$ increased of 2%; res=15y.

Results appear to be satisfactory in all cases. We then restore the original Φ's and increase the $F(0)$'s of 2%, obtaining:

res	MCVol	MCinf	MCsup	RebVol	HWVol	differr	errperc
2	0.0987	0.0983	0.0991	0.0995	0.0991	-0.0009	-0.8840
9	0.0898	0.0894	0.0902	0.0901	0.0902	-0.0003	-0.3174
17	0.0967	0.0961	0.0972	0.0965	0.0965	0.0002	0.1565
	errinf	errsup	sw0	mean	third	kurt	
	-1.2900	-0.4850	0.0822	0.0836	0.0001	3.0600	
	-0.7489	0.1104	0.0835	0.0861	0.0000	3.0230	
	-0.4029	0.7099	0.0807	0.0818	0.0000	2.9838	

As we can see from these results, the approximation error remains roughly at the same level in all stressed and unstressed subcases, and is larger for the

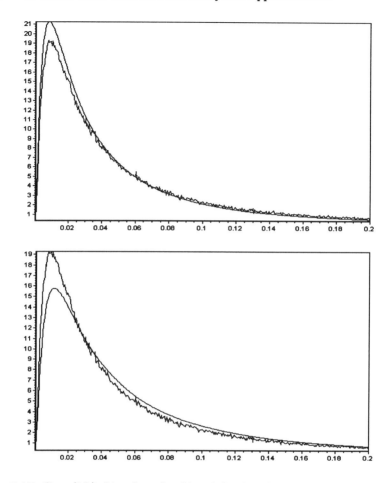

Fig. 8.17. Case (2.b): biased- and unbiased-density plots; c, d increased of 20%; res=15y.

swaption with first date of reset equal to 2 years (first payment at 3 years) and last payment in 19 years. The error appears to be lower for swaptions whose underlying swap is one year long.

The biased- and unbiased-density plots, in the above "unstressed" and "stressed" cases, are shown in Figures 8.19, 8.20, 8.21 and 8.22.

(3.b): Perfect correlations

Now we select the perfect instantaneous-correlation matrix, setting θ's to zero, while maintaining the original a, b, c, d and Φ's. We thus have all ρ's set to one. We obtain the following results:

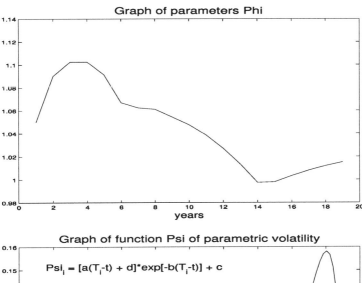
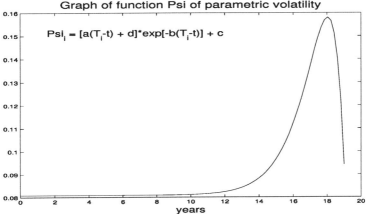

Fig. 8.18. Case (3): parameters Φ and function $t \mapsto \psi(t; a, b, c, d)$.

res	MCVol	MCinf	MCsup	RebVol	HWVol	differr	errperc
2	0.0984	0.0980	0.0988	0.0993	0.0989	-0.0010	-1.0025
9	0.0902	0.0898	0.0906	0.0902	0.0903	0.0000	0.0382
17	0.0964	0.0958	0.0969	0.0965	0.0965	-0.0002	-0.1613
	errinf	errsup	sw0	mean	third	kurt	
	-1.4180	-0.5904	0.0623	0.0631	0.0001	3.0390	
	-0.4096	0.4821	0.0635	0.0650	0.0000	3.0011	
	-0.7323	0.4033	0.0607	0.0613	0.0000	2.9869	

The approximation seems to be working in this case too.

The next step is the usual increase of 20% for the vector Φ. We observe a small worsening in the percentage difference between Monte Carlo's and Rebonato's volatilities. This difference is lower for the swaption resetting at 9 years, as we can see from the following:

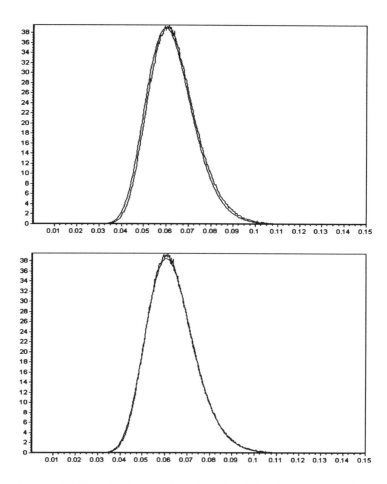

Fig. 8.19. Case (3.a): biased- and unbiased-density plots; res=2y.

res	MCVol	MCinf	MCsup	RebVol	HWVol	differr	errperc
2	0.1168	0.1163	0.1173	0.1181	0.1176	-0.0013	-1.1196
9	0.1078	0.1073	0.1083	0.1079	0.1080	-0.0001	-0.0624
17	0.1160	0.1153	0.1167	0.1156	0.1156	0.0004	0.3353
	errinf	errsup	sw0	mean	third	kurt	
	-1.5374	-0.7052	0.0623	0.0635	0.0001	3.0508	
	-0.5195	0.3906	0.0635	0.0658	0.0001	3.0083	
	-0.2735	0.9370	0.0607	0.0617	0.0000	2.9968	

Next, as usual, we restore the Φ's and increase the $F(0)$'s of 2%. In this case, we notice that the most pronounced (and yet small in absolute terms) worsening with respect to the unstressed situation is obtained for a 9 year reset date.

8.3 Test Results for Volatilities 343

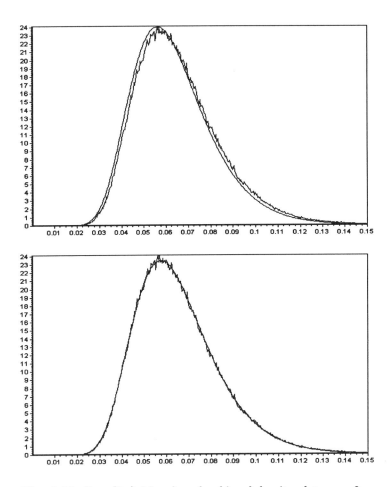

Fig. 8.20. Case (3.a): biased- and unbiased-density plots; res=9y.

res	MCVol	MCinf	MCsup	RebVol	HWVol	differr	errperc
2	0.1150	0.1150	0.1160	0.1160	0.1160	-0.0011	-0.9590
9	0.0904	0.0900	0.0908	0.0906	0.0907	-0.0002	-0.2465
17	0.0965	0.0959	0.0970	0.0965	0.0965	-0.0000	-0.0481
	errinf	errsup	sw0	mean	third	kurt	
	-1.3700	-0.5460	0.0623	0.0634	0.0001	3.0300	
	-0.6778	0.1811	0.0835	0.0861	0.0000	3.0183	
	-0.6102	0.5078	0.0807	0.0818	0.0000	3.0030	

The approximation appears to be working in both stressed cases too. See also the related density plots in Figures 8.23, 8.24 and 8.25.

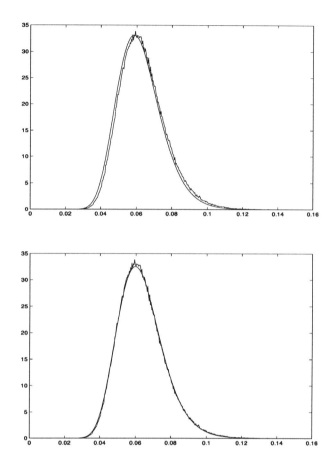

Fig. 8.21. Case (3.a): biased- and unbiased-density plots; Φ increased of 20%; res=2y.

(3.c): Some negative rank-two correlations

Concerning our chosen Formulation 7 of case (3), we finally present the results obtained when the rank-two instantaneous correlation matrix is partially negative, with the same values as in cases (1.c) and (2.b). Again, we first consider the unstressed situation (first row), then we increase Φ's by 20% (second row) and finally, after resetting the Φ's, we increase the $F(0)$'s by 2% (third row). We obtain:

res	MCVol	MCinf	MCsup	RebVol	HWVol	differr	errperc
15	0.0905	0.0900	0.0909	0.0899	0.0899	0.0006	0.6758
15	0.1075	0.1069	0.1081	0.1077	0.1078	-0.0002	-0.2134
15	0.0901	0.0896	0.0905	0.0899	0.0900	0.0002	0.2084

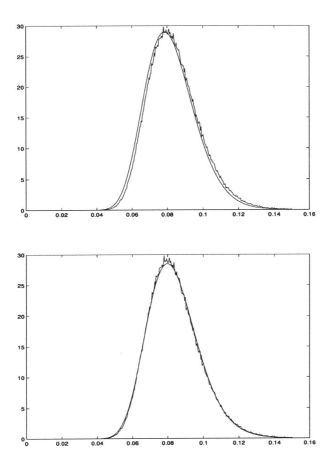

Fig. 8.22. Case (3.a): biased- and unbiased-density plots; $F(0)$ increased of 2%; res=2y.

errinf	errsup	sw0	mean	third	kurt
0.1584	1.1879	0.0619	0.0629	0.0000	2.9957
-0.7640	0.3315	0.0619	0.0634	0.0000	3.0123
-0.3002	0.7120	0.0819	0.0836	0.0000	3.0183

The approximation appears to be working in these last cases too. See also the related density plots in Figures 8.26, 8.27 and 8.28.

8.4 The "Testing Plan" for Terminal Correlations

"I appreciate your confidence in me, Plastic Man. Of course I have a plan."
Batman in JLA 26, 1998, DC Comics.

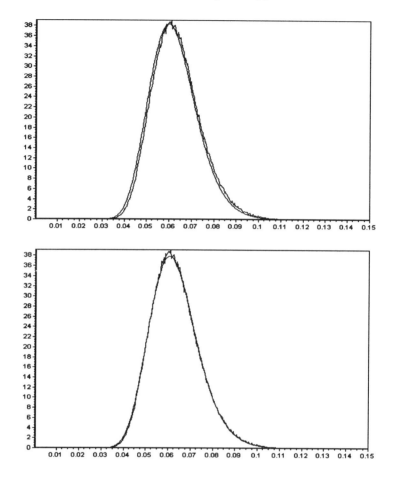

Fig. 8.23. Case (3.b): biased- and unbiased-density plots; $\theta = 0$; res=2y.

We plan to test both our analytical and Rebonato's terminal correlation formulas (6.61) and (6.62) as follows.

Once the initial reset time T_α has been fixed, we will deal with calculations for the terminal correlation between two any forward rates $F_i(T_\alpha) = F(T_\alpha; T_{i-1}, T_i)$ and $F_j(T_\alpha) = F(T_\alpha; T_{j-1}, T_j)$ at time T_α, $\alpha \leq i-1 < j \leq \beta$, under the T_α-forward measure Q^α. We will consider the three cases $\alpha = 1$, $\alpha = 9$, $\alpha = 15$, where the last forward rate has always maturity equal to $T_\beta = 20$ years. We adopt again Formulation 7 for instantaneous volatilities, although in some cases we will let it collapse to the formulation of TABLE 3, here too. Again, we adopt the two-factor parametric form in θ for instantaneous correlations, $\rho^B_{i,j} = \cos(\theta_i - \theta_j)$.

8.4 The "Testing Plan" for Terminal Correlations 347

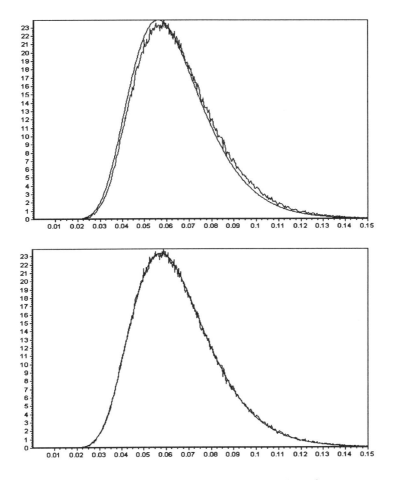

Fig. 8.24. Case (3.b): biased- and unbiased-density plots; $\theta = 0$; res=9y.

In the following, we present a short map of our sets of testing parameters, most of which are retained from our previous volatility testing plan.

- (i) *Humped and maturity-adjusted instantaneous volatilities depending only on time to maturity, typical rank-two correlations.*
  These are the same volatility and correlation parameters as in case (3.a) of the volatility testing plan, and we report them below. We use Formulation 7 with $a = 0.1908$, $b = 0.9746$, $c = 0.0808$, $d = 0.0134$ and $\theta = [\theta_{1 \div 8} \ \theta_{9 \div 17} \ \theta_{18,19}]$, where

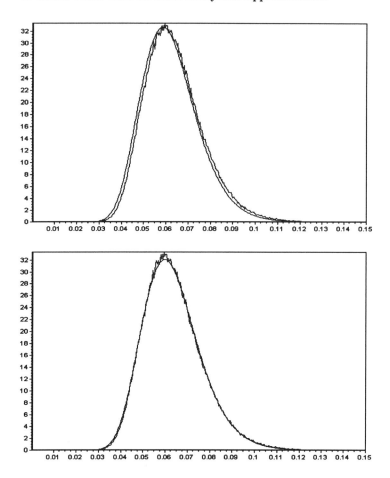

Fig. 8.25. Case (3.b): biased- and unbiased-density plots; Φ increased of 20%; res=2y.

$\theta_{1 \div 8} = [0.0147\ 0.0643\ 0.1032\ 0.1502\ 0.1969\ 0.2239\ 0.2771\ 0.2950]$,

$\theta_{9 \div 17} = [0.3630\ 0.3810\ 0.4217\ 0.4836\ 0.5204\ 0.5418\ 0.5791\ 0.6496\ 0.6679]$,

$\theta_{18,19} = [0.7126\ 0.7659]$.

This set of angles implies positive and decreasing instantaneous correlations as in Figure 8.1. The values of the Φ's in the volatility structure are shown in the same figure.

- (ii) *Constant instantaneous volatilities, typical rank-two correlations.*
  These are the same volatility and correlation parameters as in case (1.a) of the volatility testing plan: Formulation 7 with $a = 0$, $b = 0$, $c = 1$, $d = 0$. This amounts to using the formulation of TABLE 3, since the ψ part of

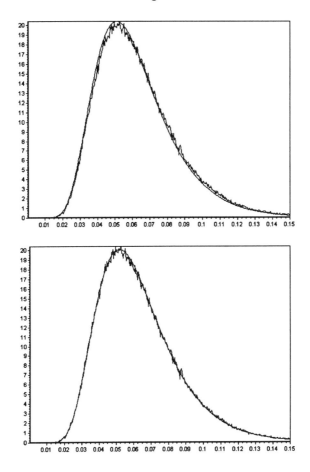

Fig. 8.26. Case (3.c): biased- and unbiased-density plots; res=15y.

Formulation 7 is forced to one and the Φ's now act as the constant volatilities s in TABLE 3. The correlation angles are taken as in the previous case (i).

- (iii) *Humped and maturity-adjusted instantaneous volatilities depending only on time to maturity, some negative rank-two correlations.*

These are the same volatility and correlation parameters as in case (3.c) of the volatility testing plan: Formulation 7 with $a = 0.1908$, $b = 0.9746$, $c = 0.0808$, $d = 0.0134$ as in case (i) and $\theta = [\theta_{1\div 9}\ \theta_{10\div 17}\ \theta_{18,19}]$, where

$$\theta_{1\div 9} = [0\ 0.0000\ 0.0013\ 0.0044\ 0.0096\ 0.0178\ 0.0299\ 0.0474\ 0.0728],$$
$$\theta_{10\div 17} = [0.1100\ 0.1659\ 0.2534\ 0.3989\ 0.6565\ 1.1025\ 1.6605\ 2.0703],$$
$$\theta_{18,19} = [2.2825\ 2.2260].$$

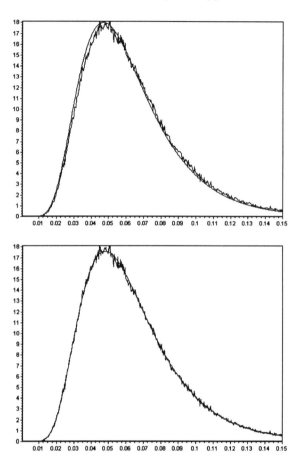

Fig. 8.27. Case (3.c): biased- and unbiased-density plots; Φ increased of 20%; res=15y.

- (iv) *Constant instantaneous volatilities, some negative rank-two correlations.*
  Formulation 7 with a, b, c and d as in case (ii) and θ as in case (iii) (analogous to case (1.c) of the volatility testing plan).
- (v) *Constant instantaneous volatilities, perfect correlation, upwardly shifted Φ's.*
  Formulation 7 with a, b, c and d as in (ii), $\theta = [0\ 0\ \ldots\ 0\ 0]$ implying $\rho^B_{i,j} = 1$ for all i,j, and Φ's upwardly shifted by 0.2.

In each of the above situations we will display some numerical results, consisting of a comparison of both our analytical and Rebonato's formulas with the corresponding terminal correlations of forward rates (as implied by the

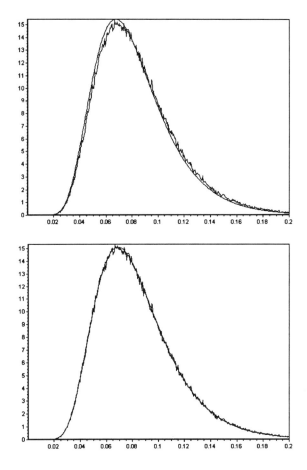

Fig. 8.28. Case (3.c): biased- and unbiased-density plots; $F(0)$ increased of 2%; res=15y.

LFM dynamics) whose computation is based on a Monte Carlo simulation with four time steps per year and 400000 simulated paths. More specifically, we will compare the three formulas (6.60), (6.61), and (6.62).

We thus compute the three correlation matrices:

$$\text{Corr}_{\text{MC}}^\alpha(F_i(T_\alpha), F_j(T_\alpha)), \text{ Corr}^{\text{AN}}(F_i(T_\alpha), F_j(T_\alpha)), \text{ Corr}^{\text{REB}}(F_i(T_\alpha), F_j(T_\alpha)),$$

which we denote, respectively, by "MCcorr", "ancorr", "Rebcorr".

To estimate the Monte Carlo error in the "real" correlation matrix we reason as follows. The i, j entry of the (Monte Carlo) estimated correlation matrix is the mean over scenarios s of the terms

$$\left(\frac{(F_i^s(T_\alpha) - \text{Mean}\{F_i(T_\alpha)\})(F_j^s(T_\alpha) - \text{Mean}\{F_j(T_\alpha)\})}{\text{Std}\{F_i(T_\alpha)\}\text{Std}\{F_j(T_\alpha)\}} \right)_s,$$

where s denotes the scenario index. Actually, this term already contains statistics built upon the scenarios, such as "Mean" and "Std", which denote the mean and standard deviation of the given variables over the different scenarios. We ignore the error in such "inner" statistics leading to the Monte Carlo estimation of the correlation, and consider as an approximation of the standard error of the method, corresponding to a two-side 98% window around the mean, the quantity:

$$\text{MCerr} = \frac{2.33}{\sqrt{\text{npath}}} \text{Std}\left\{ \frac{(F_i^s(T_\alpha) - \text{Mean}\{F_i(T_\alpha)\})(F_j^s(T_\alpha) - \text{Mean}\{F_j(T_\alpha)\})}{\text{Std}\{F_i(T_\alpha)\} \, \text{Std}\{F_j(T_\alpha)\}} \right\}$$

where "npath" denotes again the number of paths in the Monte Carlo method. We are well aware that this definition of Monte Carlo error is far from being ideal, but it is easily computable and gives us a quick estimate of the true error.

We use the quantity "MCerr" to compute the "inf" and "sup" correlation matrices, "corrinf" and "corrsup", corresponding to the extremes of the 99% terminal correlation window, simply by adding and subtracting "MCerr" to the estimated correlation matrix. We will see later on, in several cases, that we can have entries in "corrsup" larger than one. Of course, each of these entries should be replaced by 1 to obtain a tighter upper bound for the correlation matrix. However, we will report, for completeness, the values generated by our definition.

Again, we consider the following percentage differences (errors):

$$100 \, (\text{Corr}_{MC}^\alpha - \text{Corr}^{AN}) / \text{Corr}_{MC}^\alpha,$$
$$100 \, (\text{Corr}_{MC}^\alpha - \text{Corr}^{REB}) / \text{Corr}_{MC}^\alpha,$$
$$100 \, (\text{Corr}^{REB} - \text{Corr}^{AN}) / \text{Corr}^{AN},$$

where each operator acts componentwise. We denote these matrices, respectively, by "percan", "percreb" and "anreb".

As we said before, for each point in the above testing plan we consider three possible values for the initial expiry T_α, the time where the terminal correlation is computed: $T_\alpha = 1y$, $9y$ and $15y$, and the last forward rate is always taken with maturity $T_\beta = 20$.

We display only a few of the obtained results for each choice of α, by taking into consideration the main cases, although we comment all possible situations.

In every case we analyze, we will see that our analytical approximation, formula (6.61), represents a better approximation of the true value (6.60) than Rebonato's formula (6.62), although the difference will be usually rather small. Moreover, we find that the percentage (non-zero) differences between the Monte Carlo method and Rebonato's formula are always negative.

The smallest differences occur in proximity of the diagonal of the correlation matrix, i.e. for (almost) adjacent forward rates, and there the analytical formulas provide an excellent approximation.

The largest differences are observed in case (v): constant instantaneous volatilities, perfect correlation, upwardly shifted Φ's. This was somehow expected, since shifting the volatilities parameters Φ by a large amount, such as 0.2, increases randomness. The increased randomness implies that the freezing procedure, implicit in our analytical formulas, has much heavier consequences on the approximations than in the previous cases (i)-(iv), exactly as we saw for volatility approximations.

8.5 Test Results for Terminal Correlations

8.5.1 Case (i): Humped and Maturity-Adjusted Instantaneous Volatilities Depending only on Time to Maturity, Typical Rank-Two Correlations

We consider Formulation 7 with the volatility and correlation parameters indicated in the previous section for this case (i). By recalling that we may express the parameters Φ's as functions of a, b, c, d through formula (6.27), the computed $\Phi = [\Phi_{1 \div 9}\ \Phi_{10 \div 18}\ \Phi_{19}]$ matching the caplet volatilities are:

$$\Phi_{1 \div 9} = [1.0500\ 1.0900\ 1.1025\ 1.1025\ 1.0913\ 1.0669\ 1.0624\ 1.0611\ 1.0544]$$
$$\Phi_{10 \div 18} = [1.0475\ 1.0386\ 1.0270\ 1.0132\ 0.9975\ 0.9979\ 1.0033\ 1.0079\ 1.0119],$$
$$\Phi_{19} = 1.0152,$$

exactly as in the case (3) of the volatility test plan.

We have now three subcases according to the different time T_α we choose.

(i.a): $T_\alpha = 1$

In this case, the correlation matrices are 18×18 matrices. We only show the correlation matrix obtained with the Monte Carlo method in Tables 8.1 and 8.2, and the first four columns of the percentage difference matrices, "percan", "percreb", which are the most significant ones, in Table 8.3.

As we can see from these results, the percentage errors are rather low, although it is already clear that our analytical formula works slightly better than Rebonato's.

(i.b): $T_\alpha = 9$

The accuracy of our analytical formula for terminal correlations is again pointed out by the results shown in Table 8.4.

354   8. Monte Carlo Tests for LFM Analytical Approximations

MCcorr	2y	3y	4y	5y	6y	7y	8y	9y	10y
2y	1.0000	0.9989	0.9957	0.9900	0.9855	0.9754	0.9712	0.9533	0.9478
3y	0.9989	1.0000	0.9987	0.9948	0.9912	0.9830	0.9795	0.9642	0.9594
4y	0.9957	0.9987	1.0000	0.9987	0.9967	0.9911	0.9885	0.9763	0.9724
5y	0.9900	0.9948	0.9987	1.0000	0.9995	0.9966	0.9949	0.9859	0.9827
6y	0.9855	0.9912	0.9967	0.9995	1.0000	0.9986	0.9974	0.9902	0.9876
7y	0.9754	0.9830	0.9911	0.9966	0.9986	1.0000	0.9998	0.9963	0.9946
8y	0.9712	0.9795	0.9885	0.9949	0.9974	0.9998	1.0000	0.9977	0.9963
9y	0.9533	0.9642	0.9763	0.9859	0.9902	0.9963	0.9977	1.0000	0.9998
10y	0.9478	0.9594	0.9724	0.9827	0.9876	0.9946	0.9963	0.9998	1.0000
11y	0.9343	0.9474	0.9622	0.9744	0.9804	0.9895	0.9920	0.9983	0.9992
12y	0.9108	0.9261	0.9437	0.9587	0.9663	0.9787	0.9822	0.9927	0.9947
13y	0.8952	0.9118	0.9310	0.9477	0.9562	0.9705	0.9747	0.9876	0.9903
14y	0.8855	0.9029	0.9230	0.9406	0.9497	0.9651	0.9697	0.9840	0.9871
15y	0.8677	0.8865	0.9082	0.9274	0.9374	0.9547	0.9599	0.9767	0.9804
16y	0.8309	0.8520	0.8766	0.8988	0.9106	0.9313	0.9377	0.9592	0.9641
17y	0.8206	0.8423	0.8677	0.8906	0.9028	0.9245	0.9312	0.9538	0.9591
18y	0.7945	0.8176	0.8447	0.8694	0.8827	0.9065	0.9140	0.9395	0.9455
19y	0.7612	0.7860	0.8151	0.8419	0.8564	0.8827	0.8911	0.9199	0.9267

Table 8.1. Case (i.a): Monte Carlo correlation matrix, first nine columns.

MCcorr	11y	12y	13y	14y	15y	16y	17y	18y	19y
2y	0.9343	0.9108	0.8952	0.8855	0.8677	0.8309	0.8206	0.7945	0.7612
3y	0.9474	0.9261	0.9118	0.9029	0.8865	0.8520	0.8423	0.8176	0.7860
4y	0.9622	0.9437	0.9310	0.9230	0.9082	0.8766	0.8677	0.8447	0.8151
5y	0.9744	0.9587	0.9477	0.9406	0.9274	0.8988	0.8906	0.8694	0.8419
6y	0.9804	0.9663	0.9562	0.9497	0.9374	0.9106	0.9028	0.8827	0.8564
7y	0.9895	0.9787	0.9705	0.9651	0.9547	0.9313	0.9245	0.9065	0.8827
8y	0.9920	0.9822	0.9747	0.9697	0.9599	0.9377	0.9312	0.9140	0.8911
9y	0.9983	0.9927	0.9876	0.9840	0.9767	0.9592	0.9538	0.9395	0.9199
10y	0.9992	0.9947	0.9903	0.9871	0.9804	0.9641	0.9591	0.9455	0.9267
11y	1.0000	0.9981	0.9951	0.9928	0.9876	0.9741	0.9698	0.9579	0.9413
12y	0.9981	1.0000	0.9993	0.9983	0.9954	0.9862	0.9830	0.9739	0.9604
13y	0.9951	0.9993	1.0000	0.9998	0.9983	0.9917	0.9891	0.9816	0.9700
14y	0.9928	0.9983	0.9998	1.0000	0.9993	0.9942	0.9921	0.9854	0.9750
15y	0.9876	0.9954	0.9983	0.9993	1.0000	0.9975	0.9960	0.9911	0.9826
16y	0.9741	0.9862	0.9917	0.9942	0.9975	1.0000	0.9998	0.9980	0.9932
17y	0.9698	0.9830	0.9891	0.9921	0.9960	0.9998	1.0000	0.9990	0.9952
18y	0.9579	0.9739	0.9816	0.9854	0.9911	0.9980	0.9990	1.0000	0.9986
19y	0.9413	0.9604	0.9700	0.9750	0.9826	0.9932	0.9952	0.9986	1.0000

Table 8.2. Case (i.a): Monte Carlo correlation matrix, second nine columns.

(i.c): $T_\alpha = 15$

In this case, we have 4×4 correlation matrices, so that we can show all the obtained matrices in Table 8.5.

In particular, look at "corrinf" and "corrsup", and notice that their entries are quite close to the corresponding entries of "MCcorr", thus suggesting that, in case (i), the Monte Carlo error is small and that we should not bother about it.

As for the remaining aspects, we can see results analogous to the previous cases.

8.5 Test Results for Terminal Correlations   355

percan	2y	3y	4y	5y
2y	0.0000	0.0025	0.0028	0.0040
3y	0.0025	-0.0000	0.0007	0.0035
4y	0.0028	0.0007	-0.0000	0.0010
5y	0.0040	0.0035	0.0010	0.0000
6y	0.0057	0.0065	0.0029	0.0005
7y	0.0093	0.0101	0.0054	0.0017
8y	0.0106	0.0115	0.0064	0.0023
9y	0.0161	0.0160	0.0099	0.0049
10y	0.0176	0.0171	0.0108	0.0056
11y	0.0216	0.0203	0.0134	0.0078
12y	0.0288	0.0262	0.0184	0.0120
13y	0.0334	0.0300	0.0217	0.0148
14y	0.0362	0.0321	0.0236	0.0166
15y	0.0415	0.0365	0.0274	0.0200
16y	0.0527	0.0461	0.0357	0.0272
17y	0.0555	0.0483	0.0376	0.0291
18y	0.0632	0.0547	0.0433	0.0341
19y	0.0729	0.0628	0.0503	0.0403

percreb	2y	3y	4y	5y
2y	0.0000	-0.0155	-0.0601	-0.1021
3y	-0.0155	-0.0000	-0.0130	-0.0336
4y	-0.0601	-0.0130	-0.0000	-0.0046
5y	-0.1021	-0.0336	-0.0046	0.0000
6y	-0.1320	-0.0503	-0.0117	-0.0016
7y	-0.1478	-0.0595	-0.0162	-0.0034
8y	-0.1547	-0.0636	-0.0183	-0.0045
9y	-0.1659	-0.0709	-0.0224	-0.0066
10y	-0.1711	-0.0745	-0.0245	-0.0078
11y	-0.1793	-0.0802	-0.0279	-0.0097
12y	-0.1914	-0.0886	-0.0330	-0.0127
13y	-0.2010	-0.0955	-0.0371	-0.0152
14y	-0.2088	-0.1010	-0.0405	-0.0172
15y	-0.2151	-0.1057	-0.0435	-0.0192
16y	-0.2270	-0.1144	-0.0494	-0.0233
17y	-0.2300	-0.1170	-0.0513	-0.0247
18y	-0.2389	-0.1238	-0.0561	-0.0282
19y	-0.2510	-0.1332	-0.0628	-0.0333

Table 8.3. Case (i.a): percentage differences, first fourth columns.

8.5.2 Case (ii): Constant Instantaneous Volatilities, Typical Rank-Two Correlations.

Although this would more properly be the formulation in TABLE 3, as usual we equivalently resort to Formulation 7 and take $a = b = d = 0$ and $c = 1$ so that the caplet fitting forces the Φ's to equal the caplet volatilities, which are obtained from the market: $\Phi_i = v_{T_{i-1}-\text{caplet}}$. Such Φ values were plotted in Figure 8.2. The θ's are the same as in the previous case (i).

(ii.a): $T_\alpha = 1$

We show the Monte Carlo terminal correlation matrix in Tables 8.6 and 8.7, the first four columns of the analytical-approximations matrices, "ancorr", "Rebcorr", in Table 8.8 and the related percentage-difference matrices, "percan", "percreb", in Table 8.9.

356   8. Monte Carlo Tests for LFM Analytical Approximations

MCcorr	10y	11y	12y	13y	14y	15y	16y	17y	18y	19y
10y	1.0000	0.9956	0.9828	0.9707	0.9625	0.9534	0.9360	0.9304	0.9167	0.8982
11y	0.9956	1.0000	0.9955	0.9882	0.9827	0.9758	0.9614	0.9567	0.9447	0.9280
12y	0.9828	0.9955	1.0000	0.9983	0.9958	0.9920	0.9823	0.9789	0.9696	0.9560
13y	0.9707	0.9882	0.9983	1.0000	0.9995	0.9976	0.9907	0.9881	0.9804	0.9688
14y	0.9625	0.9827	0.9958	0.9995	1.0000	0.9992	0.9940	0.9918	0.9852	0.9747
15y	0.9534	0.9758	0.9920	0.9976	0.9992	1.0000	0.9975	0.9960	0.9910	0.9825
16y	0.9360	0.9614	0.9823	0.9907	0.9940	0.9975	1.0000	0.9998	0.9980	0.9932
17y	0.9304	0.9567	0.9789	0.9881	0.9918	0.9960	0.9998	1.0000	0.9990	0.9952
18y	0.9167	0.9447	0.9696	0.9804	0.9852	0.9910	0.9980	0.9990	1.0000	0.9986
19y	0.8982	0.9280	0.9560	0.9688	0.9747	0.9825	0.9932	0.9952	0.9986	1.0000

percan	10y	11y	12y	13y	14y	15y	16y	17y	18y	19y
10y	0.0000	-0.0069	-0.0189	-0.0310	-0.0441	-0.0509	-0.0343	-0.0423	-0.0255	0.0107
11y	-0.0069	-0.0000	-0.0015	-0.0051	-0.0110	-0.0122	0.0072	0.0052	0.0256	0.0641
12y	-0.0189	-0.0015	-0.0000	-0.0010	-0.0037	-0.0036	0.0123	0.0136	0.0329	0.0683
13y	-0.0310	-0.0051	-0.0010	-0.0000	-0.0007	-0.0001	0.0133	0.0164	0.0349	0.0680
14y	-0.0441	-0.0110	-0.0037	-0.0007	0.0000	0.0009	0.0125	0.0164	0.0340	0.0655
15y	-0.0509	-0.0122	-0.0036	-0.0001	0.0009	0.0000	0.0062	0.0093	0.0232	0.0495
16y	-0.0343	0.0072	0.0123	0.0133	0.0125	0.0062	-0.0000	0.0002	0.0054	0.0206
17y	-0.0423	0.0052	0.0136	0.0164	0.0164	0.0093	0.0002	-0.0000	0.0030	0.0154
18y	-0.0255	0.0256	0.0329	0.0349	0.0340	0.0232	0.0054	0.0030	-0.0000	0.0047
19y	0.0107	0.0641	0.0683	0.0680	0.0655	0.0495	0.0206	0.0154	0.0047	-0.0000

percreb	10y	11y	12y	13y	14y	15y	16y	17y	18y	19y
10y	0.0000	-0.0497	-0.1539	-0.2444	-0.3107	-0.3571	-0.4053	-0.4324	-0.4665	-0.5010
11y	-0.0497	-0.0000	-0.0285	-0.0725	-0.1098	-0.1374	-0.1667	-0.1838	-0.2042	-0.2235
12y	-0.1539	-0.0285	-0.0000	-0.0101	-0.0265	-0.0400	-0.0548	-0.0640	-0.0743	-0.0829
13y	-0.2444	-0.0725	-0.0101	-0.0000	-0.0038	-0.0096	-0.0171	-0.0220	-0.0271	-0.0303
14y	-0.3107	-0.1098	-0.0265	-0.0038	0.0000	-0.0015	-0.0054	-0.0082	-0.0108	-0.0114
15y	-0.3571	-0.1374	-0.0400	-0.0096	-0.0015	0.0000	-0.0013	-0.0027	-0.0039	-0.0035
16y	-0.4053	-0.1667	-0.0548	-0.0171	-0.0054	-0.0013	-0.0000	-0.0003	-0.0007	-0.0001
17y	-0.4324	-0.1838	-0.0640	-0.0220	-0.0082	-0.0027	-0.0003	-0.0000	-0.0000	0.0007
18y	-0.4665	-0.2042	-0.0743	-0.0271	-0.0108	-0.0039	-0.0007	-0.0000	-0.0000	0.0004
19y	-0.5010	-0.2235	-0.0829	-0.0303	-0.0114	-0.0035	-0.0001	0.0007	0.0004	-0.0000

Table 8.4. Case (i.b): Monte Carlo correlation matrix and percentage differences.

It may be curious to notice that although, in this case, the percentage differences are of the same order of magnitude as in the previous case, the entries of "MCcorr" are centered between the corresponding entries of "ancorr" (slightly smaller) and "Rebcorr" (slightly larger). In this case too, both analytical formulas seem to provide excellent approximations.

(ii.b): $T_\alpha = 9$

The results in this case are similar to previous ones obtained with $T_\alpha = 1$, so that we show only the significant part of the percentage-differences matrix in Table 8.10.

(ii.c): $T_\alpha = 15$

In this long-maturity case, the percentage differences for a constant instantaneous-volatility structure (with typical rank-two correlations) are lower than

8.5 Test Results for Terminal Correlations   357

MCcorr	16y	17y	18y	19y	ancorr	16y	17y	18y	19y
16y	1.0000	0.9964	0.9871	0.9758	16y	1.0000	0.9965	0.9873	0.9762
17y	0.9964	1.0000	0.9968	0.9895	17y	0.9965	1.0000	0.9968	0.9896
18y	0.9871	0.9968	1.0000	0.9977	18y	0.9873	0.9968	1.0000	0.9977
19y	0.9758	0.9895	0.9977	1.0000	19y	0.9762	0.9896	0.9977	1.0000

anreb	16y	17y	18y	19y	Rebcorr	16y	17y	18y	19y
16y	0.0000	-0.0319	-0.1000	-0.1720	16y	1.0000	0.9968	0.9883	0.9779
17y	-0.0319	0.0000	-0.0208	-0.0634	17y	0.9968	1.0000	0.9970	0.9902
18y	-0.1000	-0.0208	0.0000	-0.0124	18y	0.9883	0.9970	1.0000	0.9979
19y	-0.1720	-0.0634	-0.0124	0.0000	19y	0.9779	0.9902	0.9979	1.0000

percan	16y	17y	18y	19y	corrinf	16y	17y	18y	19y
16y	-0.0000	-0.0101	-0.0287	-0.0422	16y	0.9926	0.9889	0.9796	0.9683
17y	-0.0101	0.0000	-0.0035	-0.0059	17y	0.9889	0.9926	0.9894	0.9821
18y	-0.0287	-0.0035	0.0000	0.0005	18y	0.9796	0.9894	0.9926	0.9903
19y	-0.0422	-0.0059	0.0005	-0.0000	19y	0.9683	0.9821	0.9903	0.9926

percreb	16y	17y	18y	19y	corrsup	16y	17y	18y	19y
16y	-0.0000	-0.0421	-0.1287	-0.2143	16y	1.0074	1.0038	0.9945	0.9832
17y	-0.0421	0.0000	-0.0243	-0.0693	17y	1.0038	1.0074	1.0042	0.9970
18y	-0.1287	-0.0243	0.0000	-0.0118	18y	0.9945	1.0042	1.0074	1.0052
19y	-0.2143	-0.0693	-0.0118	-0.0000	19y	0.9832	0.9970	1.0052	1.0074

Table 8.5. Case (i.c): Monte Carlo correlation matrix, correlation matrices from analytical formulas and percentage differences.

MCcorr	2y	3y	4y	5y	6y	7y	8y	9y	10y
2y	1.0000	0.9992	0.9962	0.9910	0.9869	0.9770	0.9729	0.9550	0.9495
3y	0.9992	1.0000	0.9989	0.9955	0.9925	0.9846	0.9812	0.9659	0.9611
4y	0.9962	0.9989	1.0000	0.9989	0.9972	0.9918	0.9893	0.9771	0.9731
5y	0.9910	0.9955	0.9989	1.0000	0.9996	0.9967	0.9951	0.9861	0.9829
6y	0.9869	0.9925	0.9972	0.9996	1.0000	0.9986	0.9974	0.9903	0.9876
7y	0.9770	0.9846	0.9918	0.9967	0.9986	1.0000	0.9998	0.9963	0.9946
8y	0.9729	0.9812	0.9893	0.9951	0.9974	0.9998	1.0000	0.9977	0.9963
9y	0.9550	0.9659	0.9771	0.9861	0.9903	0.9963	0.9977	1.0000	0.9998
10y	0.9495	0.9611	0.9731	0.9829	0.9876	0.9946	0.9963	0.9998	1.0000
11y	0.9359	0.9490	0.9630	0.9746	0.9804	0.9895	0.9920	0.9983	0.9992
12y	0.9124	0.9278	0.9444	0.9589	0.9663	0.9787	0.9822	0.9927	0.9947
13y	0.8968	0.9134	0.9317	0.9478	0.9562	0.9704	0.9747	0.9876	0.9903
14y	0.8870	0.9045	0.9237	0.9407	0.9497	0.9650	0.9696	0.9840	0.9871
15y	0.8693	0.8880	0.9089	0.9275	0.9374	0.9546	0.9599	0.9767	0.9804
16y	0.8325	0.8535	0.8773	0.8989	0.9106	0.9313	0.9377	0.9592	0.9641
17y	0.8222	0.8439	0.8684	0.8907	0.9028	0.9245	0.9312	0.9538	0.9591
18y	0.7961	0.8192	0.8454	0.8696	0.8828	0.9066	0.9140	0.9395	0.9455
19y	0.7628	0.7875	0.8159	0.8421	0.8565	0.8828	0.8912	0.9199	0.9269

Table 8.6. Case (ii.a): Monte Carlo correlation matrix, first nine columns.

the corresponding differences obtained from a humped and maturity-adjusted instantaneous-volatilities structure as in case (i.c). Also in this case, both approximations provide us with excellent results, as we can see from Table 8.11.

358   8. Monte Carlo Tests for LFM Analytical Approximations

MCcorr	11y	12y	13y	14y	15y	16y	17y	18y	19y
2y	0.9359	0.9124	0.8968	0.8870	0.8693	0.8325	0.8222	0.7961	0.7628
3y	0.9490	0.9278	0.9134	0.9045	0.8880	0.8535	0.8439	0.8192	0.7875
4y	0.9630	0.9444	0.9317	0.9237	0.9089	0.8773	0.8684	0.8454	0.8159
5y	0.9746	0.9589	0.9478	0.9407	0.9275	0.8989	0.8907	0.8696	0.8421
6y	0.9804	0.9663	0.9562	0.9497	0.9374	0.9106	0.9028	0.8828	0.8565
7y	0.9895	0.9787	0.9704	0.9650	0.9546	0.9313	0.9245	0.9066	0.8828
8y	0.9920	0.9822	0.9747	0.9696	0.9599	0.9377	0.9312	0.9140	0.8912
9y	0.9983	0.9927	0.9876	0.9840	0.9767	0.9592	0.9538	0.9395	0.9199
10y	0.9992	0.9947	0.9903	0.9871	0.9804	0.9641	0.9591	0.9455	0.9269
11y	1.0000	0.9981	0.9951	0.9928	0.9876	0.9741	0.9698	0.9580	0.9414
12y	0.9981	1.0000	0.9993	0.9983	0.9954	0.9863	0.9831	0.9739	0.9605
13y	0.9951	0.9993	1.0000	0.9998	0.9983	0.9917	0.9891	0.9816	0.9700
14y	0.9928	0.9983	0.9998	1.0000	0.9993	0.9942	0.9921	0.9855	0.9750
15y	0.9876	0.9954	0.9983	0.9993	1.0000	0.9975	0.9961	0.9911	0.9826
16y	0.9741	0.9863	0.9917	0.9942	0.9975	1.0000	0.9998	0.9980	0.9932
17y	0.9698	0.9831	0.9891	0.9921	0.9961	0.9998	1.0000	0.9990	0.9952
18y	0.9580	0.9739	0.9816	0.9855	0.9911	0.9980	0.9990	1.0000	0.9986
19y	0.9414	0.9605	0.9700	0.9750	0.9826	0.9932	0.9952	0.9986	1.0000

Table 8.7. Case (ii.a): Monte Carlo correlation matrix, second nine columns.

ancorr	2y	3y	4y	5y
2y	1.0000	0.9992	0.9962	0.9910
3y	0.9992	1.0000	0.9989	0.9955
4y	0.9962	0.9989	1.0000	0.9989
5y	0.9910	0.9955	0.9989	1.0000
6y	0.9869	0.9925	0.9972	0.9996
7y	0.9769	0.9845	0.9918	0.9967
8y	0.9728	0.9812	0.9893	0.9951
9y	0.9548	0.9658	0.9770	0.9860
10y	0.9493	0.9609	0.9730	0.9828
11y	0.9357	0.9488	0.9628	0.9745
12y	0.9120	0.9274	0.9442	0.9587
13y	0.8963	0.9130	0.9314	0.9475
14y	0.8865	0.9040	0.9233	0.9404
15y	0.8687	0.8875	0.9084	0.9271
16y	0.8317	0.8528	0.8767	0.8984
17y	0.8213	0.8431	0.8677	0.8901
18y	0.7951	0.8182	0.8446	0.8689
19y	0.7617	0.7864	0.8149	0.8413

Rebcorr	2y	3y	4y	5y
2y	1.0000	0.9992	0.9963	0.9912
3y	0.9992	1.0000	0.9989	0.9956
4y	0.9963	0.9989	1.0000	0.9989
5y	0.9912	0.9956	0.9989	1.0000
6y	0.9873	0.9927	0.9973	0.9996
7y	0.9774	0.9849	0.9920	0.9968
8y	0.9735	0.9817	0.9895	0.9952
9y	0.9557	0.9664	0.9774	0.9862
10y	0.9503	0.9617	0.9735	0.9831
11y	0.9368	0.9497	0.9634	0.9749
12y	0.9134	0.9285	0.9449	0.9592
13y	0.8978	0.9142	0.9323	0.9481
14y	0.8881	0.9053	0.9243	0.9411
15y	0.8704	0.8889	0.9095	0.9279
16y	0.8335	0.8544	0.8779	0.8993
17y	0.8233	0.8447	0.8690	0.8911
18y	0.7971	0.8200	0.8460	0.8700
19y	0.7638	0.7883	0.8164	0.8425

Table 8.8. Case (ii.a): correlation matrices from analytical formulas.

8.5 Test Results for Terminal Correlations   359

percan	2y	3y	4y	5y
2y	0.0000	0.0002	0.0010	0.0027
3y	0.0002	0.0000	0.0003	0.0015
4y	0.0010	0.0003	0.0000	0.0004
5y	0.0027	0.0015	0.0004	0.0000
6y	0.0042	0.0026	0.0011	0.0002
7y	0.0083	0.0059	0.0033	0.0014
8y	0.0101	0.0074	0.0045	0.0022
9y	0.0186	0.0149	0.0105	0.0068
10y	0.0215	0.0174	0.0126	0.0084
11y	0.0289	0.0240	0.0183	0.0131
12y	0.0426	0.0364	0.0291	0.0223
13y	0.0523	0.0453	0.0369	0.0291
14y	0.0586	0.0510	0.0420	0.0336
15y	0.0706	0.0621	0.0519	0.0423
16y	0.0969	0.0863	0.0736	0.0617
17y	0.1046	0.0934	0.0800	0.0674
18y	0.1249	0.1121	0.0969	0.0826
19y	0.1520	0.1371	0.1194	0.1028

percreb	2y	3y	4y	5y
2y	0.0000	-0.0015	-0.0084	-0.0204
3y	-0.0015	0.0000	-0.0028	-0.0109
4y	-0.0084	-0.0028	0.0000	-0.0027
5y	-0.0204	-0.0109	-0.0027	0.0000
6y	-0.0347	-0.0222	-0.0094	-0.0022
7y	-0.0488	-0.0336	-0.0171	-0.0062
8y	-0.0578	-0.0412	-0.0227	-0.0098
9y	-0.0727	-0.0537	-0.0318	-0.0159
10y	-0.0811	-0.0610	-0.0376	-0.0200
11y	-0.0916	-0.0701	-0.0446	-0.0251
12y	-0.1037	-0.0805	-0.0527	-0.0308
13y	-0.1128	-0.0885	-0.0591	-0.0356
14y	-0.1202	-0.0952	-0.0646	-0.0399
15y	-0.1242	-0.0987	-0.0671	-0.0415
16y	-0.1274	-0.1011	-0.0684	-0.0417
17y	-0.1284	-0.1020	-0.0689	-0.0419
18y	-0.1289	-0.1022	-0.0686	-0.0411
19y	-0.1279	-0.1012	-0.0672	-0.0392

Table 8.9. Case (ii.a): percentage differences, first fourth columns.

8.5.3 Case (iii): Humped and Maturity-Adjusted Instantaneous Volatilities Depending only on Time to Maturity, Some Negative Rank-Two Correlations.

Here instantaneous volatilities follow again Formulation 7 with $a = 0.1908$, $b = 0.9746$, $c = 0.0808$, $d = 0.0134$, as in case (i), and $\theta = [\theta_{1 \div 9} \; \theta_{10 \div 18} \; \theta_{19}]$, where

$$\theta_{1 \div 9} = [0 \; 0.0000 \; 0.0013 \; 0.0044 \; 0.0096 \; 0.0178 \; 0.0299 \; 0.0474 \; 0.0728],$$
$$\theta_{10 \div 18} = [0.1100 \; 0.1659 \; 0.2534 \; 0.3989 \; 0.6565 \; 1.1025 \; 1.6605 \; 2.0703 \; 2.2825],$$
$$\theta_{19} = 2.2260.$$

The Φ's obtained from caplet volatilities are as in case (i), while the instantaneous-correlation matrix implied by the chosen θ is again the matrix shown earlier in Figure 8.6.

percan	10y	11y	12y	13y
10y	-0.0000	0.0012	0.0079	0.0170
11y	0.0012	0.0000	0.0031	0.0092
12y	0.0079	0.0031	0.0000	0.0016
13y	0.0170	0.0092	0.0016	-0.0000
14y	0.0256	0.0157	0.0049	0.0009
15y	0.0425	0.0292	0.0131	0.0054
16y	0.0886	0.0684	0.0411	0.0263
17y	0.1045	0.0823	0.0519	0.0348
18y	0.1520	0.1243	0.0852	0.0627
19y	0.2243	0.1892	0.1388	0.1092

percreb	10y	11y	12y	13y
10y	-0.0000	-0.0055	-0.0282	-0.0512
11y	-0.0055	0.0000	-0.0088	-0.0230
12y	-0.0282	-0.0088	0.0000	-0.0036
13y	-0.0512	-0.0230	-0.0036	-0.0000
14y	-0.0707	-0.0368	-0.0107	-0.0020
15y	-0.0920	-0.0522	-0.0189	-0.0060
16y	-0.1285	-0.0795	-0.0350	-0.0158
17y	-0.1391	-0.0876	-0.0403	-0.0194
18y	-0.1596	-0.1033	-0.0503	-0.0261
19y	-0.1795	-0.1181	-0.0596	-0.0319

Table 8.10. Case (ii.b): percentage differences, the most significant columns.

MCcorr	16y	17y	18y	19y
16y	1.0000	0.9998	0.9979	0.9928
17y	0.9998	1.0000	0.9989	0.9949
18y	0.9979	0.9989	1.0000	0.9985
19y	0.9928	0.9949	0.9985	1.0000

percan	16y	17y	18y	19y
16y	0.0000	-0.0002	0.0013	0.0095
17y	-0.0002	0.0000	0.0012	0.0084
18y	0.0013	0.0012	0.0000	0.0031
19y	0.0095	0.0084	0.0031	0.0000

anreb	16y	17y	18y	19y
16y	0.0000	-0.0013	-0.0146	-0.0492
17y	-0.0013	0.0000	-0.0072	-0.0346
18y	-0.0146	-0.0072	0.0000	-0.0102
19y	-0.0492	-0.0346	-0.0102	0.0000

percreb	16y	17y	18y	19y
16y	0.0000	-0.0015	-0.0133	-0.0396
17y	-0.0015	0.0000	-0.0060	-0.0261
18y	-0.0133	-0.0060	0.0000	-0.0071
19y	-0.0396	-0.0261	-0.0071	0.0000

Table 8.11. Case (ii.c): Monte Carlo correlation matrix and percentage differences.

We will see that, especially in the cases (iii.a) and (iii.b) the existence of some negative instantaneous correlations can induce larger percentage errors.

(iii.a): $T_\alpha = 1$

We start by showing in Table 8.12 the last columns of the Monte Carlo terminal correlations matrix "MCcorr". Subsequently, we show in Table 8.13 the corresponding matrix "percan" of percentage differences between "MCcorr" and the analytical approximations "ancorr".

We also include in Table 8.14 part of the matrix "anreb" of percentage differences between the analytical approximation "ancorr" and Rebonato's approximation "Rebcorr".

By first looking at the "anreb" matrix we see that, once again, our analytical formula and Rebonato's formula agree well. But how do they relate to the "true" Monte Carlo terminal-correlation matrix? The matrix "percan" shows us that in most cases the approximation is still excellent, with a few

8.5 Test Results for Terminal Correlations

seeming exceptions. Take, for example, the percentage difference relative to the terminal correlations between the 14y-expiry and the 19y-expiry forward rates. Such percentage difference is 155%, and, in "percan", there are also differences of 6%, 2% etc.

However, notice that these large percentage errors are due mostly to the fact that the found correlations are very close to zero. Clearly, if the analytical formula presents us with 0.0013 as an approximation of a Monte Carlo value of −0.0023, we do not worry too much, although the *percentage* difference between this two values is large. Moreover, in our simulations, the analytical values are usually well inside the Monte Carlo window. For example, the Monte Carlo value −0.0023 corresponds to a Monte Carlo window [−0.0077 0.0030].

In order to avoid this kind of misleading percentage errors, we also show in Table 8.15 the matrix of absolute differences, "absan", between the Monte Carlo terminal-correlation matrix and our analytical-approximation matrix. These absolute differences appear to be small, and once again confirm that both our analytical approximation and Rebonato's work indeed well in many situations.

MCcorr	11y	12y	13y	14y	15y	16y	17y	18y	19y
2y	0.9839	0.9657	0.9192	0.7901	0.4499	-0.0891	-0.4745	-0.6457	-0.6027
3y	0.9842	0.9662	0.9199	0.7912	0.4514	-0.0880	-0.4742	-0.6460	-0.6028
4y	0.9859	0.9681	0.9223	0.7942	0.4549	-0.0851	-0.4727	-0.6454	-0.6020
5y	0.9875	0.9701	0.9250	0.7980	0.4600	-0.0801	-0.4690	-0.6426	-0.5989
6y	0.9890	0.9724	0.9284	0.8032	0.4674	-0.0720	-0.4622	-0.6370	-0.5930
7y	0.9908	0.9752	0.9329	0.8105	0.4781	-0.0601	-0.4517	-0.6280	-0.5835
8y	0.9930	0.9789	0.9390	0.8206	0.4934	-0.0427	-0.4362	-0.6145	-0.5695
9y	0.9957	0.9838	0.9475	0.8348	0.5152	-0.0175	-0.4135	-0.5947	-0.5488
10y	0.9984	0.9898	0.9587	0.8547	0.5466	0.0194	-0.3798	-0.5648	-0.5177
11y	1.0000	0.9962	0.9731	0.8823	0.5924	0.0749	-0.3280	-0.5185	-0.4697
12y	0.9962	1.0000	0.9895	0.9200	0.6603	0.1612	-0.2450	-0.4427	-0.3917
13y	0.9731	0.9895	1.0000	0.9670	0.7620	0.3019	-0.1030	-0.3094	-0.2555
14y	0.8823	0.9200	0.9670	1.0000	0.9017	0.5342	0.1525	-0.0586	-0.0023
15y	0.5924	0.6603	0.7620	0.9017	1.0000	0.8469	0.5638	0.3771	0.4288
16y	0.0749	0.1612	0.3019	0.5342	0.8469	1.0000	0.9165	0.8114	0.8431
17y	-0.3280	-0.2450	-0.1030	0.1525	0.5638	0.9165	1.0000	0.9774	0.9878
18y	-0.5185	-0.4427	-0.3094	-0.0586	0.3771	0.8114	0.9774	1.0000	0.9984
19y	-0.4697	-0.3917	-0.2555	-0.0023	0.4288	0.8431	0.9878	0.9984	1.0000

Table 8.12. Case (iii.a): Monte Carlo correlation matrix, last nine columns.

(iii.b): $T_\alpha = 9$

In this case, we find a situation similar to the preceding one. Here, besides showing in Table 8.16 the Monte Carlo terminal correlations matrix, we display directly the absolute differences, "absan", in Table 8.17, ignoring the percentage differences. Once again all approximations work well.

362 8. Monte Carlo Tests for LFM Analytical Approximations

percan	11y	12y	13y	14y	15y	16y	17y	18y	19y
2y	0.0138	0.0270	0.0605	0.1451	0.2957	0.4690	0.2978	0.2175	0.2392
3y	0.0168	0.0298	0.0631	0.1473	0.2971	0.4761	0.2999	0.2192	0.2407
4y	0.0131	0.0259	0.0588	0.1419	0.2881	0.5107	0.3015	0.2186	0.2408
5y	0.0097	0.0222	0.0545	0.1361	0.2771	0.5733	0.3072	0.2211	0.2442
6y	0.0079	0.0199	0.0513	0.1308	0.2647	0.6832	0.3188	0.2278	0.2521
7y	0.0066	0.0179	0.0479	0.1245	0.2493	0.8919	0.3367	0.2387	0.2645
8y	0.0050	0.0152	0.0433	0.1158	0.2286	1.3977	0.3642	0.2551	0.2833
9y	0.0031	0.0117	0.0369	0.1038	0.2008	3.9210	0.4085	0.2808	0.3127
10y	0.0011	0.0074	0.0285	0.0875	0.1644	-4.2307	0.4845	0.3230	0.3613
11y	0.0000	0.0027	0.0179	0.0660	0.1185	-1.3629	0.6310	0.3980	0.4489
12y	0.0027	0.0000	0.0066	0.0396	0.0641	-0.8143	0.9885	0.5510	0.6333
13y	0.0179	0.0066	0.0000	0.0127	0.0098	-0.5617	2.8466	0.9744	1.1906
14y	0.0660	0.0396	0.0127	0.0000	-0.0175	-0.3574	-2.2216	6.2191	155.30
15y	0.1185	0.0641	0.0098	-0.0175	0.0000	-0.1189	-0.4488	-0.8075	-0.6877
16y	-1.3629	-0.8143	-0.5617	-0.3574	-0.1189	0.0000	-0.0510	-0.1003	-0.0873
17y	0.6310	0.9885	2.8466	-2.2216	-0.4488	-0.0510	0.0000	-0.0048	-0.0031
18y	0.3980	0.5510	0.9744	6.2191	-0.8075	-0.1003	-0.0048	0.0000	-0.0001
19y	0.4489	0.6333	1.1906	155.305	-0.6877	-0.0873	-0.0031	-0.0001	0.0000

Table 8.13. Case (iii.a): percentage differences between "MCcorr" and our analytical approximations "ancorr".

anreb	11y	12y	13y	14y	15y	16y	17y	18y	19y
2y	-0.1669	-0.1831	-0.2186	-0.3085	-0.5337	-0.8942	-1.1571	-1.2771	-1.2500
3y	-0.0793	-0.0921	-0.1211	-0.1955	-0.3842	-0.6867	-0.9074	-1.0082	-0.9857
4y	-0.0299	-0.0399	-0.0631	-0.1240	-0.2812	-0.5339	-0.7185	-0.8030	-0.7843
5y	-0.0120	-0.0200	-0.0395	-0.0919	-0.2292	-0.4507	-0.6130	-0.6875	-0.6712
6y	-0.0057	-0.0125	-0.0295	-0.0762	-0.2008	-0.4030	-0.5518	-0.6203	-0.6054
7y	-0.0039	-0.0098	-0.0254	-0.0693	-0.1880	-0.3823	-0.5260	-0.5925	-0.5780
8y	-0.0028	-0.0080	-0.0224	-0.0641	-0.1790	-0.3692	-0.5112	-0.5773	-0.5628
9y	-0.0016	-0.0060	-0.0189	-0.0580	-0.1687	-0.3552	-0.4960	-0.5622	-0.5477
10y	-0.0006	-0.0037	-0.0147	-0.0504	-0.1561	-0.3388	-0.4793	-0.5462	-0.5313
11y	0.0000	-0.0014	-0.0094	-0.0404	-0.1388	-0.3162	-0.4564	-0.5244	-0.5091
12y	-0.0014	0.0000	-0.0036	-0.0271	-0.1141	-0.2830	-0.4225	-0.4920	-0.4760
13y	-0.0094	-0.0036	0.0000	-0.0110	-0.0787	-0.2319	-0.3686	-0.4398	-0.4230
14y	-0.0404	-0.0271	-0.0110	0.0000	-0.0319	-0.1519	-0.2781	-0.3493	-0.3317
15y	-0.1388	-0.1141	-0.0787	-0.0319	0.0000	-0.0498	-0.1427	-0.2049	-0.1885
16y	-0.3162	-0.2830	-0.2319	-0.1519	-0.0498	0.0000	-0.0274	-0.0623	-0.0520
17y	-0.4564	-0.4225	-0.3686	-0.2781	-0.1427	-0.0274	0.0000	-0.0075	-0.0041
18y	-0.5244	-0.4920	-0.4398	-0.3493	-0.2049	-0.0623	-0.0075	0.0000	-0.0005
19y	-0.5091	-0.4760	-0.4230	-0.3317	-0.1885	-0.0520	-0.0041	-0.0005	0.0000

Table 8.14. Case (iii.a): percentage differences between our analytical approximation "ancorr" and Rebonato's approximation "Rebcorr".

(iii.c): $T_\alpha = 15$

Also in this long-maturity subcase, we confirm our previous findings. Moreover, we do not need to resort to absolute differences, since all terminal correlations are far away from zero. From Table 8.18, we can see that percentage differences between the Monte Carlo matrix and our analytical-approximation matrix are at most 0.1348%. Notice that instead Rebonato's approximation leads to percentage differences up to 1%. In this case, our analytical approximation seems to be doing again sensibly better than Rebonato's, although Rebonato's is still good enough for practical purposes.

8.5 Test Results for Terminal Correlations   363

absan	11y	12y	13y	14y	15y	16y	17y	18y	19y
2y	0.0001	0.0003	0.0006	0.0011	0.0013	-0.0004	-0.0014	-0.0014	-0.0014
3y	0.0002	0.0003	0.0006	0.0012	0.0013	-0.0004	-0.0014	-0.0014	-0.0015
4y	0.0001	0.0003	0.0005	0.0011	0.0013	-0.0004	-0.0014	-0.0014	-0.0014
5y	0.0001	0.0002	0.0005	0.0011	0.0013	-0.0005	-0.0014	-0.0014	-0.0015
6y	0.0001	0.0002	0.0005	0.0011	0.0012	-0.0005	-0.0015	-0.0015	-0.0015
7y	0.0001	0.0002	0.0004	0.0010	0.0012	-0.0005	-0.0015	-0.0015	-0.0015
8y	0.0000	0.0001	0.0004	0.0010	0.0011	-0.0006	-0.0016	-0.0016	-0.0016
9y	0.0000	0.0001	0.0003	0.0000	0.0010	0.0007	-0.0017	-0.0017	-0.0017
10y	0.0000	0.0001	0.0003	0.0007	0.0009	-0.0008	-0.0018	-0.0018	-0.0019
11y	0.0000	0.0000	0.0002	0.0006	0.0007	-0.0010	-0.0021	-0.0021	-0.0021
12y	0.0000	0.0000	0.0001	0.0004	0.0004	-0.0013	-0.0024	-0.0024	-0.0025
13y	0.0002	0.0001	0.0000	0.0001	0.0001	-0.0017	-0.0029	-0.0030	-0.0030
14y	0.0006	0.0004	0.0001	0.0000	-0.0002	-0.0019	-0.0034	-0.0036	-0.0036
15y	0.0007	0.0004	0.0001	-0.0002	0.0000	-0.0010	-0.0025	-0.0030	-0.0029
16y	-0.0010	-0.0013	-0.0017	-0.0019	-0.0010	0.0000	-0.0005	-0.0008	-0.0007
17y	-0.0021	-0.0024	-0.0029	-0.0034	-0.0025	-0.0005	0.0000	0.0000	0.0000
18y	-0.0021	-0.0024	-0.0030	-0.0036	-0.0030	-0.0008	0.0000	0.0000	0.0000
19y	-0.0021	-0.0025	-0.0030	-0.0036	-0.0029	-0.0007	0.0000	0.0000	0.0000

Table 8.15. Case (iii.a): absolute differences between the Monte Carlo terminal-correlation matrix and our analytical-approximation matrix.

MCcorr	10y	11y	12y	13y	14y	15y	16y	17y	18y	19y
10y	1.0000	0.9949	0.9778	0.9389	0.8309	0.5272	0.0209	-0.3517	-0.5203	-0.4774
11y	0.9949	1.0000	0.9935	0.9658	0.8714	0.5816	0.0746	-0.3105	-0.4886	-0.4431
12y	0.9778	0.9935	1.0000	0.9882	0.9165	0.6548	0.1593	-0.2355	-0.4237	-0.3753
13y	0.9389	0.9658	0.9882	1.0000	0.9662	0.7587	0.2977	-0.1006	-0.2998	-0.2480
14y	0.8309	0.8714	0.9165	0.9662	1.0000	0.9001	0.5277	0.1465	-0.0606	-0.0057
15y	0.5272	0.5816	0.6548	0.7587	0.9001	1.0000	0.8425	0.5534	0.3655	0.4171
16y	0.0209	0.0746	0.1593	0.2977	0.5277	0.8425	1.0000	0.9132	0.8048	0.8374
17y	-0.3517	-0.3105	-0.2355	-0.1006	0.1465	0.5534	0.9132	1.0000	0.9766	0.9873
18y	-0.5203	-0.4886	-0.4237	-0.2998	-0.0606	0.3655	0.8048	0.9766	1.0000	0.9983
19y	-0.4774	-0.4431	-0.3753	-0.2480	-0.0057	0.4171	0.8374	0.9873	0.9983	1.0000

Table 8.16. Case (iii.b): Monte Carlo terminal-correlation matrix.

absan	10y	11y	12y	13y	14y	15y	16y	17y	18y	19y
10y	0.0000	0.0000	-0.0001	0.0003	0.0016	0.0041	0.0019	-0.0014	-0.0020	-0.0018
11y	0.0000	0.0000	0.0000	0.0003	0.0016	0.0041	0.0018	-0.0017	-0.0024	-0.0023
12y	-0.0001	0.0000	0.0000	0.0002	0.0012	0.0035	0.0013	-0.0024	-0.0032	-0.0030
13y	0.0003	0.0003	0.0002	0.0000	0.0005	0.0023	0.0001	-0.0037	-0.0046	-0.0045
14y	0.0016	0.0016	0.0012	0.0005	0.0000	0.0007	-0.0017	-0.0059	-0.0072	-0.0070
15y	0.0041	0.0041	0.0035	0.0023	0.0007	0.0000	-0.0019	-0.0064	-0.0084	-0.0080
16y	0.0019	0.0018	0.0013	0.0001	-0.0017	-0.0019	0.0000	-0.0017	-0.0033	-0.0030
17y	-0.0014	-0.0017	-0.0024	-0.0037	-0.0059	-0.0064	-0.0017	0.0000	-0.0003	-0.0002
18y	-0.0020	-0.0024	-0.0032	-0.0046	-0.0072	-0.0084	-0.0033	-0.0003	0.0000	0.0000
19y	-0.0018	-0.0023	-0.0030	-0.0045	-0.0070	-0.0080	-0.0030	-0.0002	0.0000	0.0000

Table 8.17. Case (iii.b): absolute differences between the Monte Carlo terminal-correlation matrix and our analytical-approximation matrix.

8.5.4 Case (iv): Constant Instantaneous Volatilities, Some Negative Rank-Two Correlations.

Instantaneous volatilities follow again Formulation 7 with a, b, c, d as in case (ii) and θ as in case (iii).

364   8. Monte Carlo Tests for LFM Analytical Approximations

MCcorr	16y	17y	18y	19y	percan	16y	17y	18y	19y
16y	1.0000	0.9098	0.7962	0.8231	16y	0.0000	0.0457	0.1348	0.0587
17y	0.9098	1.0000	0.9744	0.9818	17y	0.0457	0.0000	0.0170	-0.0092
18y	0.7962	0.9744	1.0000	0.9975	18y	0.1348	0.0170	0.0000	-0.0013
19y	0.8231	0.9818	0.9975	1.0000	19y	0.0587	-0.0092	-0.0013	0.0000

anreb	16y	17y	18y	19y	percreb	16y	17y	18y	19y
16y	0.0000	-0.5534	-1.2219	-1.0526	16y	0.0000	-0.5074	-1.0855	-0.9932
17y	-0.5534	0.0000	-0.1434	-0.1043	17y	-0.5074	0.0000	-0.1263	-0.1135
18y	-1.2219	-0.1434	0.0000	-0.0133	18y	-1.0855	-0.1263	0.0000	-0.0146
19y	-1.0526	-0.1043	-0.0133	0.0000	19y	-0.9932	-0.1135	-0.0146	0.0000

Table 8.18. Case (iii.c): Monte Carlo correlation matrix and percentage differences.

As we observed in the previous case (iii), the presence of negative correlations can lead to some isolated large percentage differences, independently of the values of the instantaneous volatilities. Here, comments are completely analogous to those in the corresponding subcases of case (iii).

(iv.a): $T_\alpha = 1$

Our tests produced a "MCcorr" close to the one obtained in subcase (iii.a), and similar results for "ancorr" and "Rebcorr", so that we show only the matrices "percan" and "absan" in Tables 8.19 and 8.20, respectively. These are shown to point out that the large differences are in the same positions as in case (iii.a) (and mostly concern terminal correlations between the 14y-expiry and 19y-expiry forward rates). Again, all approximations work well, the large percentage differences being due to proximity to zero.

percan	11y	12y	13y	14y	15y	16y	17y	18y	19y
2y	0.0098	0.0224	0.0556	0.1447	0.3414	0.0001	0.2246	0.1755	0.1914
3y	0.0093	0.0219	0.0548	0.1436	0.3390	0.0034	0.2250	0.1754	0.1915
4y	0.0087	0.0210	0.0537	0.1416	0.3344	0.0121	0.2275	0.1768	0.1931
5y	0.0080	0.0200	0.0521	0.1388	0.3274	0.0281	0.2324	0.1798	0.1965
6y	0.0071	0.0187	0.0498	0.1348	0.3169	0.0594	0.2407	0.1851	0.2025
7y	0.0060	0.0169	0.0468	0.1290	0.3021	0.1219	0.2537	0.1932	0.2117
8y	0.0046	0.0145	0.0425	0.1209	0.2813	0.2841	0.2745	0.2061	0.2263
9y	0.0029	0.0112	0.0366	0.1093	0.2527	1.1446	0.3095	0.2273	0.2503
10y	0.0011	0.0071	0.0286	0.0934	0.2145	-1.5707	0.3711	0.2630	0.2911
11y	0.0000	0.0027	0.0183	0.0719	0.1648	-0.6609	0.4929	0.3281	0.3666
12y	0.0027	0.0000	0.0069	0.0445	0.1036	-0.4870	0.7965	0.4645	0.5295
13y	0.0183	0.0069	0.0000	0.0152	0.0378	-0.3983	2.4041	0.8513	1.0331
14y	0.0719	0.0445	0.0152	0.0000	-0.0062	-0.2905	-1.9780	5.7456	163.88
15y	0.1648	0.1036	0.0378	-0.0062	0.0000	-0.1074	-0.4316	-0.7992	-0.6760
16y	-0.6609	-0.4870	-0.3983	-0.2905	-0.1074	0.0000	-0.0542	-0.1118	-0.0961
17y	0.4929	0.7965	2.4041	-1.9780	-0.4316	-0.0542	0.0000	-0.0067	-0.0042
18y	0.3281	0.4645	0.8513	5.7456	-0.7992	-0.1118	-0.0067	0.0000	-0.0003
19y	0.3666	0.5295	1.0331	163.88	-0.6760	-0.0961	-0.0042	-0.0003	0.0000

Table 8.19. Case (iv.a): percentage differences between "MCcorr" and our analytical approximations "ancorr".

8.5 Test Results for Terminal Correlations 365

absan	11y	12y	13y	14y	15y	16y	17y	18y	19y
2y	0.0001	0.0002	0.0005	0.0011	0.0015	0.0000	-0.0011	-0.0011	-0.0012
3y	0.0001	0.0002	0.0005	0.0011	0.0015	0.0000	-0.0011	-0.0011	-0.0012
4y	0.0001	0.0002	0.0005	0.0011	0.0015	0.0000	-0.0011	-0.0011	-0.0012
5y	0.0001	0.0002	0.0005	0.0011	0.0015	0.0000	-0.0011	-0.0012	-0.0012
6y	0.0001	0.0002	0.0005	0.0011	0.0015	0.0000	-0.0011	-0.0012	-0.0012
7y	0.0001	0.0002	0.0004	0.0010	0.0014	-0.0001	-0.0011	-0.0012	-0.0012
8y	0.0000	0.0001	0.0004	0.0010	0.0014	-0.0001	-0.0012	-0.0013	-0.0013
9y	0.0000	0.0001	0.0003	0.0009	0.0013	-0.0002	-0.0013	-0.0013	-0.0014
10y	0.0000	0.0001	0.0003	0.0008	0.0012	-0.0003	-0.0014	-0.0015	-0.0015
11y	0.0000	0.0000	0.0002	0.0006	0.0010	-0.0005	-0.0016	-0.0017	-0.0017
12y	0.0000	0.0000	0.0001	0.0004	0.0007	-0.0008	-0.0019	-0.0020	-0.0021
13y	0.0002	0.0001	0.0000	0.0001	0.0003	-0.0012	-0.0025	-0.0026	-0.0026
14y	0.0006	0.0004	0.0001	0.0000	-0.0001	-0.0016	-0.0030	-0.0033	-0.0033
15y	0.0010	0.0007	0.0003	-0.0001	0.0000	-0.0009	-0.0024	-0.0030	-0.0029
16y	-0.0005	-0.0008	-0.0012	-0.0016	-0.0009	0.0000	-0.0005	-0.0009	-0.0008
17y	-0.0016	-0.0019	-0.0025	-0.0030	-0.0024	-0.0005	0.0000	-0.0001	0.0000
18y	-0.0017	-0.0020	-0.0026	-0.0033	-0.0030	-0.0009	-0.0001	0.0000	0.0000
19y	-0.0017	-0.0021	-0.0026	-0.0033	-0.0029	-0.0008	0.0000	0.0000	0.0000

Table 8.20. Case (iv.a): absolute differences between the Monte Carlo terminal-correlation matrix and our analytical-approximation matrix.

(iv.b): $T_\alpha = 9$

Analogous results and comments as in (iii.b).

(iv.c): $T_\alpha = 15$

Analogous results and comments as in (iii.c).

8.5.5 Case (v): Constant Instantaneous Volatilities, Perfect Correlations, Upwardly Shifted Φ's

We consider again Formulation 7 with a, b, c, d as in (ii) and $\theta = [0 \ldots 0]$, leading to $\rho_{i,j}^B = 1$ for all i, j, but now we upwardly shift the Φ's by $0.2 = 20\%$. Since with this choice of a, b, c and d the Φ's coincide with the caplet volatilities, this amounts to upwardly shifting the caplet volatilities by 0.2%:

$$\Phi_i = v_{T_{i-1}\text{-caplet}} \rightarrow \Phi_i = v_{T_{i-1}\text{-caplet}} + 0.2.$$

This leads to quite large volatilities, since it amounts to more than doubling all the original caplet volatilities taken in input (whose graph is shown in Figure 8.2 and which roughly range from 10% to 16%).

As one can expect, in this case, the approximation results worsen. This is due to the fact that the "freezing" approximation works better when variability is small, and increasing the Φ's amounts to increasing variability. We will also notice that the errors in the Monte Carlo method will be huge, so as to render the comparison difficult to interpret.[1]

---

[1] We have also tried to increase the number of Monte Carlo paths and decrease the time step, but the results did not change substantially.

366   8. Monte Carlo Tests for LFM Analytical Approximations

When all instantaneous correlations are equal to one and instantaneous volatilities are constant as in this case, Rebonato's formula yields immediately a terminal correlation matrix where all entries are equal to one. This does not happen with our analytical formula, which, once again, does better.

(v.a): $T_\alpha = 1$

In this subcase, our results show small percentage errors, lower than 0.1% in the case of Rebonato's formula, and even below 0.013% when using our analytical formula. Being all matrices close to each other, we show only a part of the terminal correlation matrix computed with the Monte Carlo method in Table 8.21.

MCcorr	2y	3y	4y	5y	6y	7y	8y	9y	10y
2y	1.0000	1.0000	1.0000	0.9999	0.9998	0.9997	0.9996	0.9996	0.9995
3y	1.0000	1.0000	1.0000	0.9999	0.9998	0.9998	0.9997	0.9997	0.9996
4y	1.0000	1.0000	1.0000	1.0000	0.9999	0.9999	0.9998	0.9998	0.9997
5y	0.9999	0.9999	1.0000	1.0000	1.0000	1.0000	0.9999	0.9999	0.9999
6y	0.9998	0.9998	0.9999	1.0000	1.0000	1.0000	1.0000	1.0000	0.9999
7y	0.9997	0.9998	0.9999	1.0000	1.0000	1.0000	1.0000	1.0000	1.0000
8y	0.9996	0.9997	0.9998	0.9999	1.0000	1.0000	1.0000	1.0000	1.0000
9y	0.9996	0.9997	0.9998	0.9999	1.0000	1.0000	1.0000	1.0000	1.0000
10y	0.9995	0.9996	0.9997	0.9999	0.9999	1.0000	1.0000	1.0000	1.0000
11y	0.9994	0.9995	0.9997	0.9998	0.9999	1.0000	1.0000	1.0000	1.0000
12y	0.9993	0.9995	0.9996	0.9998	0.9999	0.9999	1.0000	1.0000	1.0000
13y	0.9993	0.9994	0.9996	0.9997	0.9998	0.9999	0.9999	1.0000	1.0000
14y	0.9992	0.9993	0.9995	0.9997	0.9998	0.9999	0.9999	0.9999	1.0000
15y	0.9992	0.9993	0.9995	0.9997	0.9998	0.9998	0.9999	0.9999	1.0000
16y	0.9991	0.9993	0.9995	0.9996	0.9998	0.9998	0.9999	0.9999	0.9999
17y	0.9991	0.9993	0.9994	0.9996	0.9997	0.9998	0.9999	0.9999	0.9999
18y	0.9991	0.9992	0.9994	0.9996	0.9997	0.9998	0.9998	0.9999	0.9999
19y	0.9990	0.9992	0.9994	0.9996	0.9997	0.9998	0.9998	0.9999	0.9999

Table 8.21. Case (v.a): Monte Carlo correlation matrix, first nine columns.

In this case both approximated formulas still work well.

(v.b): $T_\alpha = 9$

In this subcase, we find the worst results of all our correlation tests. This can be due to the fact that terminal correlations are computed at a much later time, so that the freezing procedure has an heavier impact than in the short-time subcase (v.a).

We find that our analytical formula and Rebonato's work about in the same way, so that we show only percentage differences for our analytical matrix. Yet, although both formulas agree, the values of the terminal correlations involving the farthest forward rates and obtained by the Monte Carlo method are very different from the corresponding correlations computed with our analytical method. Differences reach up to 43%, as we can see from Table 8.22.

MCcorr	10y	11y	12y	13y	14y	15y	16y	17y	18y	19y
10y	1.000	0.998	0.989	0.974	0.951	0.917	0.872	0.818	0.758	0.697
11y	0.998	1.000	0.997	0.987	0.970	0.942	0.903	0.854	0.799	0.742
12y	0.989	0.997	1.000	0.997	0.986	0.965	0.932	0.890	0.841	0.788
13y	0.974	0.987	0.997	1.000	0.996	0.983	0.958	0.924	0.881	0.833
14y	0.951	0.970	0.986	0.996	1.000	0.995	0.979	0.953	0.918	0.877
15y	0.917	0.942	0.965	0.983	0.995	1.000	0.994	0.978	0.952	0.919
16y	0.872	0.903	0.932	0.958	0.979	0.994	1.000	0.994	0.979	0.955
17y	0.818	0.854	0.890	0.924	0.953	0.978	0.994	1.000	0.995	0.981
18y	0.758	0.799	0.841	0.881	0.918	0.952	0.979	0.995	1.000	0.995
19y	0.697	0.742	0.788	0.833	0.877	0.919	0.955	0.981	0.995	1.000

percan	10y	11y	12y	13y	14y	15y	16y	17y	18y	19y
10y	0.000	-0.240	-1.057	-2.612	-5.070	-8.995	-14.661	-22.211	-31.762	-43.302
11y	-0.240	0.000	-0.287	-1.251	-3.042	-6.123	-10.752	-17.042	-25.067	-34.775
12y	-1.057	-0.287	0.000	-0.335	-1.433	-3.660	-7.263	-12.337	-18.920	-26.934
13y	-2.612	-1.251	-0.335	0.000	-0.377	-1.739	-4.334	-8.247	-13.489	-19.964
14y	-5.070	-3.042	-1.433	-0.377	0.000	-0.488	-2.097	-4.911	-8.921	-14.019
15y	-8.995	-6.123	-3.660	-1.739	-0.488	0.000	-0.550	-2.229	-5.011	-8.781
16y	-14.661	-10.752	-7.263	-4.334	-2.097	-0.550	0.000	-0.552	-2.165	-4.716
17y	-22.211	-17.042	-12.337	-8.247	-4.911	-2.229	-0.552	0.000	-0.519	-1.974
18y	-31.762	-25.067	-18.920	-13.489	-8.921	-5.011	-2.165	-0.519	0.000	-0.459
19y	-43.302	-34.775	-26.934	-19.964	-14.019	-8.781	-4.716	-1.974	-0.459	0.000

Table 8.22. Case (v.b): Monte Carlo correlation matrix and percentage differences.

We also report in Table 8.23 the two matrices "corrinf" and "corrsup" corresponding to the extremes of the Monte Carlo 99% terminal-correlation window. The two matrices are quite different, showing that the Monte Carlo estimate is still rather uncertain. Caution is therefore in order when comparing the Monte Carlo terminal-correlation matrix to our analytical approximations, since the Monte Carlo window appears to be large.

(v.c): $T_\alpha = 15$

The situation is analogous to the previous subcase. Here, we can report all matrices in Table 8.24, since their size is rather reduced.

8.6 Test Results: Stylized Conclusions

In all our tests, we have found uncertain or even negative results in the cases where shifting the Φ's amounted to shift the caplet volatilities by the corresponding absolute amount of 20%. In such cases, our Monte Carlo windows for terminal correlations become too large, and the volatility approximations seem to worsen with respect to the unstressed cases. Yet, the volatility approximation does not worsen excessively, as one can see from the shifted cases (1.b) and (1.c).

The density plots, instead, appear to be rather affected by the shift in the Φ's, thus confirming that large volatilities render both the "freezing the drift" and the "collapsing all measures" approximations less reliable, bringing the swap-rate distribution far away from the lognormal family of distributions.

368 8. Monte Carlo Tests for LFM Analytical Approximations

corrinf	10y	11y	12y	13y	14y	15y	16y	17y	18y	19y
10y	0.3561	0.3537	0.3455	0.3304	0.3075	0.2731	0.2278	0.1739	0.1146	0.0535
11y	0.3537	0.3561	0.3532	0.3436	0.3264	0.2981	0.2587	0.2102	0.1554	0.0978
12y	0.3455	0.3532	0.3561	0.3527	0.3419	0.3206	0.2882	0.2461	0.1968	0.1437
13y	0.3304	0.3436	0.3527	0.3561	0.3523	0.3389	0.3145	0.2798	0.2372	0.1896
14y	0.3075	0.3264	0.3419	0.3523	0.3561	0.3512	0.3355	0.3093	0.2742	0.2331
15y	0.2731	0.2981	0.3206	0.3389	0.3512	0.3561	0.3506	0.3343	0.3084	0.2754
16y	0.2278	0.2587	0.2882	0.3145	0.3355	0.3506	0.3561	0.3506	0.3349	0.3110
17y	0.1739	0.2102	0.2461	0.2798	0.3093	0.3343	0.3506	0.3561	0.3509	0.3367
18y	0.1146	0.1554	0.1968	0.2372	0.2742	0.3084	0.3349	0.3509	0.3561	0.3515
19y	0.0535	0.0978	0.1437	0.1896	0.2331	0.2754	0.3110	0.3367	0.3515	0.3561

corrsup	10y	11y	12y	13y	14y	15y	16y	17y	18y	19y
10y	1.6439	1.6415	1.6334	1.6182	1.5953	1.5610	1.5156	1.4617	1.4024	1.3413
11y	1.6415	1.6439	1.6410	1.6315	1.6142	1.5860	1.5466	1.4980	1.4432	1.3856
12y	1.6334	1.6410	1.6439	1.6406	1.6297	1.6085	1.5761	1.5339	1.4847	1.4316
13y	1.6182	1.6315	1.6406	1.6439	1.6401	1.6268	1.6023	1.5677	1.5250	1.4774
14y	1.5953	1.6142	1.6297	1.6401	1.6439	1.6391	1.6234	1.5971	1.5620	1.5209
15y	1.5610	1.5860	1.6085	1.6268	1.6391	1.6439	1.6385	1.6221	1.5962	1.5632
16y	1.5156	1.5466	1.5761	1.6023	1.6234	1.6385	1.6439	1.6384	1.6227	1.5989
17y	1.4617	1.4980	1.5339	1.5677	1.5971	1.6221	1.6384	1.6439	1.6388	1.6246
18y	1.4024	1.4432	1.4847	1.5250	1.5620	1.5962	1.6227	1.6388	1.6439	1.6393
19y	1.3413	1.3856	1.4316	1.4774	1.5209	1.5632	1.5989	1.6246	1.6393	1.6439

Table 8.23. Case (v.b): the two matrices corresponding to the extremes of the Monte Carlo 99% terminal-correlation window.

MCcorr	16y	17y	18y	19y	percan	16y	17y	18y	19y
16y	1.0000	0.9851	0.9370	0.8605	16y	0.0000	-1.5084	-6.7230	-16.2117
17y	0.9851	1.0000	0.9828	0.9337	17y	-1.5084	0.0000	-1.7487	-7.1032
18y	0.9370	0.9828	1.0000	0.9835	18y	-6.7230	-1.7487	0.0000	-1.6813
19y	0.8605	0.9337	0.9835	1.0000	19y	-16.2117	-7.1032	-1.6813	0.0000

anreb	16y	17y	18y	19y	percreb	16y	17y	18y	19y
16y	0.0000	-0.0001	-0.0004	-0.0008	16y	0.0000	-1.5085	-6.7234	-16.2126
17y	-0.0001	0.0000	-0.0001	-0.0004	17y	-1.5085	0.0000	-1.7488	-7.1036
18y	-0.0004	-0.0001	0.0000	-0.0001	18y	-6.7234	-1.7488	0.0000	-1.6814
19y	-0.0008	-0.0004	-0.0001	0.0000	19y	-16.2126	-7.1036	-1.6814	0.0000

corrinf	16y	17y	18y	19y	corrsup	16y	17y	18y	19y
16y	0.7043	0.6895	0.6413	0.5648	16y	1.2957	1.2808	1.2327	1.1562
17y	0.6895	0.7043	0.6871	0.6380	17y	1.2808	1.2957	1.2785	1.2293
18y	0.6413	0.6871	0.7043	0.6878	18y	1.2327	1.2785	1.2957	1.2791
19y	0.5648	0.6380	0.6878	0.7043	19y	1.1562	1.2293	1.2791	1.2957

Table 8.24. Case (v.c): Monte Carlo correlation matrix, percentage differences and extreme matrices of the Monte Carlo 99% terminal-correlation window.

However, besides these pathological cases, all suggested analytical approximations seem to work well in "normal" situations.

9. Other Interest-Rate Models

In this chapter we introduce brief sketches of some of the models that are known in the literature and that have not been included in the previous chapters. All models are arbitrage free, and we will not discuss no-arbitrage implications further. Instead, we synthetically explain in what these models differ from the previous models and what are their original features. We also give references for the readers who might wish to deepen their knowledge of a specific approach. Clearly, presenting all the models that have been proposed in the literature is a huge task. We only present a few, without any claim to completeness of the treatment. Indeed, there are certainly several other relevant and worthy models that have not been included here, and we make the excuse that a choice is necessary since it is impossible to do justice to all the models appeared over the years. The reader interested in models that have not appeared in this book can also check other books on interest rate models such as for example James and Webber (2000).

9.1 Brennan and Schwartz's Model

Brennan and Schwartz (1979, 1982) consider a model based on modeling both the short and consol rates. The consol bond is, roughly speaking, a claim paying perpetually a constant dividend rate. If this constant dividend is assumed to be q, the payoff contribution in the infinitesimal time interval $(s, s+ds]$ is the dividend q per unit of time times the time amount ds discounted at time t:

$$D(t,s)q\,ds\ .$$

By adding all these infinitesimal discounted contributions we obtain

$$\int_t^{+\infty} D(t,s)q\,ds$$

whose risk neutral expectation gives the consol bond value at time t:

$$E_t\left[\int_t^{+\infty} D(t,s)q\,ds\right] = q\int_t^{+\infty} P(t,s)\,ds\ .$$

Now notice that if at time t all rates were equal to a single rate $L(t)$ (continuous compounding) the above quantity would reduce to

$$\int_t^{+\infty} qe^{-L(t)(s-t)} ds = \frac{q}{L(t)}.$$

By solving

$$q \int_t^{+\infty} P(t,s) ds = \frac{q}{L(t)}$$

we find the *consol rate* L at time t. The consol rate at time t is that unique rate at time t discounting at which one recovers the correct price at time t of a consol bond.

$L(t)$ is a synthesis of the whole term structure up to infinity, and can be shown to incorporate information on the *steepness* of the yield curve at time t, since it involves discount functions for very large maturities and thus provides an indication of the "height" of the extreme far part of the yield curve at time t.

Brennan and Schwartz model the joint evolution of r and L with two particular diffusion processes and allow for correlation between the two. The dynamics of r and L are given under the objective measure, and market prices of risk are introduced to move to the risk neutral measure and price bonds. It is clear that at a given time t one can interpret the short rate $r(t)$ as the level of the curve at time t. Therefore, since $L(t)$ is related to the steepness of the curve, the Brennan and Schwartz model can be seen as a model incorporating level and steepness in the yield-curve evolution. The advantage of the model is that the two "factors" r and L have a clear and immediate financial interpretation.

Brennan and Schwartz (1979) analyze government bonds of the Canadian market with their proposed model, obtaining satisfactory results. However, besides these attractive features, there are problems concerning analytical tractability and other aspects.

The interested reader is referred also to Chapter 15 of Rebonato (1998).

9.2 Balduzzi, Das, Foresi and Sundaram's Model

Balduzzi, Das, Foresi and Sundaram (1996) propose a general framework to three-factor short-rate models leading to affine term structures. As a fundamental example of their framework they consider the following three-factor model under the objective measure:

$$dr(t) = k(\theta(t) - r(t))dt + \sqrt{v(t)}dZ(t),$$
$$d\theta(t) = \alpha(\beta - \theta(t))dt + \eta dW(t),$$
$$dv(t) = a(b - v(t))dt + \phi\sqrt{v(t)}dV(t),$$

where Z, W and V are Brownian motions such that

$$dZ\, dV = \rho\, dt, \quad dZ\, dW = 0, \quad dZ\, dV = 0,$$

and k, α,β, η, a, b, ϕ and ρ are constants. The short-rate equation features a time-varying "mean reversion" θ that follows a Vasicek-like process, while the instantaneous absolute volatility v follows a CIR-like process.

This model has been often mentioned by our traders in relationship to factor analysis of the term structure of interest rates under the objective measure.

To move under the risk-neutral measure, particular forms of market price of risk have to be chosen for the three processes.

By doing so, one obtains an affine term-structure model, in the sense of the following multidimensional generalization of the formulation given in Section 3.2.4:

$$P(t,T) = A(T-t)\exp[-r(t)B(T-t) - \theta(t)C(T-t) - v(t)D(T-t)].$$

While B and C can be computed analytically, A and D have to be found through numerical solutions.

Balduzzi, Das, Foresi and Sundaram (1996) examine the possible shapes of the zero-coupon interest-rate curves implied by the above three-factor dynamics. That is, for a fixed t they consider the τ-"curve" $\tau \mapsto R(t, t+\tau)$, and they try to see how this whole map changes according to changes in the factors r, v, θ. They thus find that the three proposed factors account for classical features of the evolution of the term-structure "curve": r provides the *level* of the curve, v is strictly related to the *curvature* and centers its influence at medium-term maturities, while finally θ dictates the steepness of the term structure. This work should be checked by researchers interested in the dynamics of the whole yield curve under the objective measure.

9.3 Flesaker and Hughston's Model

After the short-rate setup, Flesaker and Hughston (1996) (FH) were among the first to propose an entirely new approach to interest-rate modeling resulting in concrete models that are not part of the short-rate world. They in fact model quantities related directly to the *state-price density* (known also as *pricing kernel, pricing operator*). In our context the state-price density can also be viewed as the reciprocal $1/U$ of a chosen numeraire U.

The general framework of FH consists of modeling directly the bank-account numeraire as follows. Consider a family $(M(t,T))_{T \geq t}$ of positive martingales such that

$$M(t,T) \geq M(t, T+\tau) \text{ for all } t \geq T, \tau \geq 0.$$

The M's are a specification of

$$M(t,T) = \frac{P(t,T)}{B(t)} = E_t[D(0,T)] \, .$$

For those willing to relate our brief exposition to the Flesaker and Hughston (1996) notation, we specify that our $M(t,T)$ is Flesaker and Hughston's $\Delta_{tT}/\rho_t = \Delta_t P(t,T)/\rho_t$, where ρ is the Radon-Nikodym derivative dQ/dQ^U, Q^U being the particular pricing measure under which we work, and Q being the risk-neutral measure.

You may notice that modeling the M's amounts to modeling the ratio between the two privileged numeraires $P(t,T)$ (T-forward measure) and $B(t)$ (risk-neutral measure).

The model allows for positive interest rates in quite a natural way. Roughly, the fact that the M's are taken decreasing in T makes the resulting zero-coupon-bond prices $P(t,T) = B(t)M(t,T)$ decreasing in T, thus ensuring positive rates.

Flesaker and Hughston develop a nice framework starting from particular specifications of M (or Δ). A particularly interesting model is obtained by them when setting $\Delta_{tT} = A_T + B_T S_t$, where A and B are decreasing and positive deterministic functions and S is a unit-initialized (possibly time-changed) geometric Brownian motion. By this approach, bond prices are obtained as

$$P(t,T) = \frac{A_T + B_T S_t}{A_t + B_t S_t},$$

i.e. as a rational function of a lognormal variable S. The same consideration applies to other classical quantities in the interest-rate world, see for example Rapisarda and Silvotti (2001). This is the reason why this particular version of FH's model is called the *rational lognormal model*.

This model features some appealing properties, which follow easily from the rational bond-price formula above. The model prices caps and swaptions analytically with formulas similar to but different from the corresponding market Black formulas. Moreover, this model interacts well with exchange rates, and is particularly suited to be used in situations involving interest-rate curves of different currencies.

However, as a general framework, the model has to be dealt with carefully. Rogers (1997) shows a possible problem: Not all choices of M as above correspond to interest-rate models. Rogers, in fact, presents the example $M(t,T) = \exp(W_t - t/2 - T)$. This definition of M satisfies the conditions above, but $M(t,t) = \exp(W_t - 3t/2)$ cannot bear the interpretation $P(t,t)/B(t) = 1/B(t)$ as requested, since it is not decreasing in t. We notice that the rational lognormal specification avoids this problem, but in general one has to beware this point. Still, when possible, one can check the models resulting from the FH framework on a case by case basis.

For an empirical investigation on the calibration performances of the FH model to market data, also in comparison with other classical one-factor models seen in Chapter 3, see Rapisarda and Silvotti (2001).

9.4 Rogers's Potential Approach

After pointing out that, notwithstanding all its merits, FH's framework has to be handled with care, Rogers (1997) presents a similar framework, which is more rooted in the classical theory of Markov processes. This approach follows the ideas of Constantinides (1992). The model is basically obtained by specifying two objects: A Markov process X, and a positive function f. Then one defines the state-price density ζ_t in terms of X and f. Recall that the state-price density can be essentially interpreted also as the reciprocal of a chosen numeraire, so that $1/\zeta_t$ can be viewed as our basic numeraire under which the model will work and under which discounting will take place. Rogers sets $\zeta_t = e^{-\alpha t} R_\alpha(\alpha - G)f$ and obtains the short rate as $r_t = [(\alpha - G)f(X_t)]/f(X_t)$, where R_α is the resolvent of the process X and G its generator, see Rogers (1997) for the details. Then different models based on different X and f are derived, most of which are based on taking X as Gaussian. As for several other models, empirical investigations on the calibration to caps or swaptions, on the implied evolution for the term structure of volatilities, on terminal correlations between rates and other features are needed. A final note on this approach is that it is suited to model curves in different currencies in quite an elegant and natural way.

9.5 Markov Functional Models

Hunt, Kennedy and Pelsser (1998b) proposed recently the Markov functional models. Roughly, in this setup it is assumed that the zero-coupon-bond prices P are functionals of some low-dimensional Markov process x, $P(t,T) = \Pi(t,T;x_t)$ for $t \in [0, \mathcal{D}(T)]$, with $\mathcal{D}(T) \leq T$. $\mathcal{D}$ is the "boundary curve for time" and is given as a function of maturities, and is typically $\mathcal{D}(T) = T$ if $T \leq \bar{T}$, or $\mathcal{D}(T) = \bar{T}$ if $T > \bar{T}$, where $\bar{T}$ is some terminal maturity. The process x is assumed to be Markov under a numeraire U, which is in turn a function N of x: $U_t = N(t, x_t)$.

The model is actually specified by assigning the law of x under U, the functional form Π of the discount factors on the boundary, $\Pi(\mathcal{D}(T), T; \cdot)$, and finally the functional form $N(t, \cdot)$.

In other terms, one chooses a fundamental low-dimensional Markov process x, and the model specification requires knowledge of the functional form Π in the chosen process of the zero-coupon-bond price on some boundary $\mathcal{D}$, and the law of a numeraire N. The functional form Π can be chosen so as to calibrate the model to relevant market prices, while the remaining freedom on the choice of the law of the Markov process x is what can allow one to make the model realistic. For example, the freedom in choosing the functional form N can be used to reproduce the marginal distributions of swap rates that are particularly relevant for the calibration considered.

From a methodological point of view, given that the notion of numeraire is close to that of state-price density, and that here the numeraire is taken to depend on a low-dimensional driving Markov process, the reader will notice that this approach looks similar, in some aspects, to the FH and the potential approaches. The main difference is the possibility to choose functional forms so as to match market prices of interest-rate options. Indeed, market distributions of rates at the relevant times are considered and matched as much as possible. Hunt, Kennedy and Pelsser (1998b) note the similarities with the FH and potential models, but observe that both these approaches fail to calibrate well to the distributions of market rates. They point out that the unique feature of their approach is the possibility to recover the correct distributions of the relevant market rates (as in the market models) while keeping low dimensionality at the same time (contrary to the market models). For further details, the interested reader is obviously referred to Hunt, Kennedy and Pelsser (1998b).

Part II

PRICING DERIVATIVES IN PRACTICE

10. Pricing Derivatives on a Single Interest-Rate Curve

> *Increasingly, problems do not rule out practice, but support it. Instead of finding that practice is too difficult, that we have too many problems, we see that the problems themselves are the jewels, and we devote ourselves to be with them in a way we never dreamt of before.*
> Charlotte Joko Beck, "Nothing Special: Living Zen", 1995, HarperCollins.

In this chapter, we present a sample of financial products we believe to be representative of a large portion of the interest-rate market. We will use different models (mostly the LFM and the G2++ model) for different problems, and try to clarify the advantages of each model. All the discounted payoffs will be calculated at time $t = 0$.

Before starting, we remark upon the possible use of an approximated LIBOR market model (LFM) for pricing some of the products we will consider.

It is possible to freeze part of the drift of the LFM dynamics so as to obtain a geometric Brownian motion. This is what was done for example in Section 6.14 to derive approximated formulas for terminal correlations. In that section, we derived such a dynamics under the T_γ-forward-adjusted measure Q^γ:

$$dF_k(t) = \bar{\mu}_{\gamma,k}(t) F_k(t)\, dt + \sigma_k(t) F_k(t)\, dZ_k(t), \qquad (10.1)$$

where

$$\mu_{\gamma,k}(t) := -\sum_{j=k+1}^{\gamma} \frac{\rho_{k,j} \tau_j \sigma_j(t) F_j(0)}{1 + \tau_j F_j(0)}, \quad k < \gamma,$$

$$\mu_{\gamma,\gamma}(t) := 0, \quad k = \gamma,$$

$$\mu_{\gamma,k}(t) := \sum_{j=\gamma+1}^{k} \frac{\rho_{k,j} \tau_j \sigma_j(t) F_j(0)}{1 + \tau_j F_j(0)}, \quad k > \gamma,$$

$$\bar{\mu}_{\gamma,k}(t) := \sigma_k(t) \mu_{\gamma,k}(t).$$

This dynamics gives access, in some cases, to a number of techniques which have been developed for the basic Black and Scholes setup, for example, in equity and FX markets. Moreover, this "freezing-part-of-the-drift" technique can be combined with drift interpolation so as to allow for rates that are not in the fundamental (spanning) family corresponding to the particular LFM being implemented.

We detail this possible "interpolate and freeze (part of the) drift" approach in case of accrual swaps in Section 10.11.1, but the method is rather general and, when used in combination with other possible approximations, can be used for other products.

Finally, even if one keeps on using Monte Carlo evaluation, the frozen-drift approximation leads to a process (geometric Brownian motion) that is much easier to propagate in time and requires no small discretization step in the propagation, allowing instead for "one-shot" simulation also over long periods of time.

In the following, we assume we are given a set of dates $T_\alpha, \ldots, T_i \ldots, T_{\beta+1}$ with associated year fractions $\tau_\alpha, \ldots, \tau_i \ldots, \tau_{\beta+1}$.

10.1 In-Advance Swaps

An in-advance swap is an IRS that resets at dates $T_{\alpha+1}, \ldots, T_\beta$ and pays at the same dates, with unit notional amount and with fixed-leg rate K. More precisely, the discounted payoff of an in-advance swap (of "payer" type) can be expressed via

$$\sum_{i=\alpha+1}^{\beta} D(0, T_i)\tau_{i+1}(F_{i+1}(T_i) - K).$$

The value of such a contract is, therefore,

$$\text{IAS} = E\left[\sum_{i=\alpha+1}^{\beta} D(0, T_i)\tau_{i+1}(F_{i+1}(T_i) - K)\right],$$

where we omit arguments in the "IAS" notation for brevity.

Before calculating the expectations, it is convenient to make some adjustments. We shall use the following identity (obtained easily via iterated conditioning, as seen in Proposition 2.8.1):

$$E[XD(0,T)] = E\left[\frac{XD(0,S)}{P(T,S)}\right] \quad \text{for all } 0 < T < S, \tag{10.2}$$

where X is a T-measurable random variable.

To value the above contract, notice that

$$E\left\{\sum_{i=\alpha+1}^{\beta} D(0,T_i)\tau_{i+1}(F_{i+1}(T_i) - K)\right\}$$

$$= E\left\{\sum_{i=\alpha+1}^{\beta} D(0,T_i)\left[\frac{1}{P(T_i,T_{i+1})} - (1+\tau_{i+1}K)\right]\right\}$$

$$= E\left\{\sum_{i=\alpha+1}^{\beta}\left[\frac{D(0,T_{i+1})}{P(T_i,T_{i+1})^2} - D(0,T_i)(1+\tau_{i+1}K)\right]\right\}$$

$$= \sum_{i=\alpha+1}^{\beta}\left\{P(0,T_{i+1})E^{i+1}\left[\frac{1}{P(T_i,T_{i+1})^2}\right] - P(0,T_i)(1+\tau_{i+1}K)\right\}$$

$$= \sum_{i=\alpha+1}^{\beta}\left\{P(0,T_{i+1})E^{i+1}\left[(1+\tau_{i+1}F_{i+1}(T_i))^2\right] - P(0,T_i)(1+\tau_{i+1}K)\right\}.$$

Computing the expected value is an easy task, since we know that, under Q^{i+1}, F_{i+1} has the driftless (martingale) lognormal dynamics

$$dF_{i+1}(t) = \sigma_{i+1}(t)F_{i+1}(t)dZ_{i+1}(t),$$

so that, remembering the resulting lognormal distribution of $F_{i+1}^2(T_i)$, one has

$$E^{i+1}\left(F_{i+1}^2(T_i)\right) = F_{i+1}^2(0)\exp\left[\int_0^{T_i}\sigma_{i+1}^2(t)dt\right] = F_{i+1}^2(0)\exp(v_{i+1}^2)$$

where the v's have been defined in (6.18) and are deduced from cap prices. We obtain

$$\mathbf{IAS} = \sum_{i=\alpha+1}^{\beta}\{P(0,T_{i+1})\left[1 + 2\tau_{i+1}F_{i+1}(0) + \tau_{i+1}^2 F_{i+1}^2(0)\exp(v_{i+1}^2)\right]$$
$$-(1+\tau_{i+1}K)P(0,T_i)\}. \tag{10.3}$$

Contrary to the plain-vanilla case, this price depends on the volatility of forward rates through the caplet volatilities v. Notice however that correlations between different rates are not involved in this product, as one expects from the additive and "one-rate-per-time" nature of the payoff.

10.2 In-Advance Caps

An in-advance cap is composed by caplets resetting at dates $T_{\alpha+1},\ldots,T_\beta$ and paying at the same dates, with unit notional amount and strike rate K.

More precisely, the discounted payoff of an in-advance cap can be expressed via

$$\sum_{i=\alpha+1}^{\beta} D(0,T_i)\tau_{i+1}(F_{i+1}(T_i) - K)^+.$$

The value of such a contract is, therefore,

$$\text{IAC} = E\left[\sum_{i=\alpha+1}^{\beta} D(0,T_i)\tau_{i+1}(F_{i+1}(T_i) - K)^+\right].$$

The payoff is the same as in the case of in-advance swaps, except for the positive-part operator.

10.2.1 A First Analytical Formula (LFM)

We apply the same reasoning we used for in-advance swaps, obtaining:

$$\begin{aligned}
\text{IAC} &= \sum_{i=\alpha+1}^{\beta} P(0,T_{i+1})E^{i+1}\left[(1 + \tau_{i+1}F_{i+1}(T_i))(F_{i+1}(T_i) - K)^+\right] \\
&= \sum_{i=\alpha+1}^{\beta} P(0,T_{i+1})\left(E^{i+1}[(F_{i+1}(T_i) - K)^+]\right. \\
&\qquad\qquad \left. + \tau_{i+1}E^{i+1}\left[F_{i+1}(T_i)(F_{i+1}(T_i) - K)^+\right]\right) \\
&= \sum_{i=\alpha+1}^{\beta} P(0,T_{i+1})\left[\text{Bl}(K,F_{i+1}(0),v_{i+1}) + \tau_{i+1}g(K,F_{i+1}(0),v_{i+1})\right],
\end{aligned}$$

$$g(K,F,v) := F^2 \exp[v^2]\,\Phi\!\left(\frac{3v}{2} - \frac{1}{v}\ln\frac{K}{F}\right) - FK\,\Phi\!\left(\frac{v}{2} - \frac{1}{v}\ln\frac{K}{F}\right),$$

where "Bl" and v have been defined in (1.26), (6.18) and above. In-advance caps do not depend on the correlation of different rates but just on the caplet volatilities v, as one expects again from the additive and "one-rate-per-time" nature of the payoff.

10.2.2 A Second Analytical Formula (G2++)

The above expectations can also be easily computed under the Gaussian G2++ model, by exploiting the lognormal distribution of bond prices. After lengthy but straightforward calculations we obtain:

$$\mathrm{IAC} = \sum_{i=\alpha+1}^{\beta} P(0,T_i) \left[\frac{P(0,T_i)}{P(0,T_{i+1})} e^{\Sigma(T_i,T_{i+1})^2} \Phi\left(\frac{\ln \frac{P(0,T_i)}{\tilde{K}_i P(0,T_{i+1})} + \frac{3}{2}\Sigma(T_i,T_{i+1})^2}{\Sigma(T_i,T_{i+1})} \right) \right.$$
$$\left. - \tilde{K}_i \Phi\left(\frac{\ln \frac{P(0,T_i)}{\tilde{K}_i P(0,T_{i+1})} + \frac{1}{2}\Sigma(T_i,T_{i+1})^2}{\Sigma(T_i,T_{i+1})} \right) \right],$$

where $\tilde{K}_i = 1 + K\tau_i$ and

$$\Sigma(T,S)^2 = \frac{\sigma^2}{2a^3}\left[1 - e^{-a(S-T)}\right]^2 \left[1 - e^{-2aT}\right] + \frac{\eta^2}{2b^3}\left[1 - e^{-b(S-T)}\right]^2 \left[1 - e^{-2bT}\right]$$
$$+ 2\rho \frac{\sigma\eta}{ab(a+b)}\left[1 - e^{-a(S-T)}\right]\left[1 - e^{-b(S-T)}\right]\left[1 - e^{-(a+b)T}\right].$$

10.3 Autocaps

We adopt the same notation, terminology and conventions as in Section 6.4, and take $\alpha = 0$. An autocap is similar to a cap, but at most $\gamma \leq \beta$ caplets can be exercised, and they *have* to be automatically exercised when in the money. Therefore, the discounted payoff can be written as

$$\sum_{i=1}^{\beta} \tau_i \left[F(T_{i-1};T_{i-1},T_i) - K\right]^+ D(0,T_i)\, 1\{A_i\},$$

$A_i = \{\text{at most } \gamma \text{ among } F_1(T_0),\ldots,F_i(T_{i-1}) \text{ are larger than } K\},$

where $1\{A\}$ denotes the indicator function for the set A.

The pricing of this contract can be obtained by considering the risk-neutral expectation E of its discounted payoff:

$$E\left[\sum_{i=1}^{\beta} \tau_i (F_i(T_{i-1}) - K)^+ D(0,T_i)\, 1\{A_i\}\right]$$
$$= P(0,T_\beta) \sum_{i=1}^{\beta} \tau_i E^\beta \left[\frac{(F_i(T_{i-1}) - K)^+ 1\{A_i\}}{P(T_i,T_\beta)}\right],$$

where we have used (10.2) (equivalently, the remarks of Section 2.8).

Notice that the A_i term depends not only on the forward rate $F_i(T_{i-1})$, but also on $F_1(T_0), ..., F_{i-1}(T_{i-2})$. Therefore, a "path-dependent" feature is introduced in the contract. If we attempt to price this contract by a Monte Carlo method, in order to compute the discounted payoff we need to generate paths under Q^β for the vector (whose dimension decreases over time)

$$F_{\beta(t)}(t),\ldots,F_\beta(t),$$

where we recall that $t \in (T_{\beta(t)-2}, T_{\beta(t)-1}]$. These paths can be deduced from discretizing the dynamics (6.14). In our setting, such dynamics reads $(k = \beta(t), \beta(t) + 1, \ldots, \beta)$

$$dF_k(t) = -\sigma_k(t)F_k(t) \sum_{j=k+1}^{\beta} \frac{\rho_{k,j}\tau_j\sigma_j(t)F_j(t)}{1+\tau_j F_j(t)} dt + \sigma_k(t)F_k(t)dZ_k^{\beta}(t). \quad (10.4)$$

Taking logs and using the Milstein scheme, analogously to what was done for swaptions in Section 6.10, yields the desired simulated paths:

$$\ln F_k^{\Delta t}(t+\Delta t) = \ln F_k^{\Delta t}(t) - \sigma_k(t) \sum_{j=k+1}^{\beta} \frac{\rho_{k,j}\tau_j\sigma_j(t)F_j^{\Delta t}(t)}{1+\tau_j F_j^{\Delta t}(t)} \Delta t - \frac{\sigma_k(t)^2}{2}\Delta t$$
$$+\sigma_k(t)(Z_k^{\beta}(t+\Delta t) - Z_k^{\beta}(t)). \quad (10.5)$$

Actually, here too, one can improve the scheme by resorting to more refined shocks, in the spirit of Remark 6.10.1.

10.4 Caps with Deferred Caplets

These are caps for which all caplets payments occur at the final time T_β. The discounted payoff is, therefore,

$$\sum_{i=1}^{\beta} \tau_i(F_i(T_{i-1}) - K)^+ D(0,T_\beta).$$

The pricing of this "deferred" cap can be obtained by considering the risk-neutral expectation E of its discounted payoff:

$$E\left[\sum_{i=1}^{\beta} \tau_i(F_i(T_{i-1}) - K)^+ D(0,T_\beta)\right] = P(0,T_\beta)\sum_{i=1}^{\beta} \tau_i E^{\beta}\left[(F_i(T_{i-1}) - K)^+\right].$$

The expected value can be computed through a Monte Carlo method based on the discretized Q^β dynamics (10.5).

10.4.1 A First Analytical Formula (LFM)

The above formula requires to compute the expected values

$$E^{\beta}\left[(F_i(T_{i-1}) - K)^+\right].$$

This is a case where the LFM with partially frozen drift can be of help in deriving analytical approximations. Indeed, consider the approximate LFM dynamics (10.1) with $\gamma = \beta$. Then, the above expectation is easily computed as a Black and Scholes call-option price (see Appendix B):

$$\exp\left(\int_0^{T_{i-1}} \bar{\mu}_{\beta,i}(t)dt\right) F_i(0) \Phi\left(\frac{\ln \frac{F_i(0)}{K} + \int_0^{T_{i-1}} \left[\bar{\mu}_{\beta,i}(t) + \frac{\sigma_i^2(t)}{2}\right] dt}{\sqrt{\int_0^{T_{i-1}} v_i^2(t)dt}}\right)$$

$$- K\Phi\left(\frac{\ln \frac{F_i(0)}{K} + \int_0^{T_{i-1}} \left[\bar{\mu}_{\beta,i}(t) - \frac{\sigma_i^2(t)}{2}\right] dt}{\sqrt{\int_0^{T_{i-1}} \sigma_i^2(t)dt}}\right).$$

Replacing the expectation with such expression in the above summation, we obtain an analytical formula for the price of the cap with deferred caplets.

10.4.2 A Second Analytical Formula (G2++)

The expectations

$$E^\beta \left[(F_i(T_{i-1}) - K)^+\right]$$

can also be easily computed under the Gaussian G2++ model, by again exploiting the lognormal distribution of bond prices. After lengthy but straightforward calculations we obtain:

$$\frac{1}{\tau_i}\left[\frac{P(0,T_{i-1})}{P(0,T_i)} e^{\psi(0,T_{i-1},T_i,T_\beta,1)} \Phi\left(\frac{\ln \frac{P(0,T_{i-1})}{\tilde{K}P(0,T_i)} + \psi(0,T_{i-1},T_i,T_\beta,\frac{3}{2})}{\sqrt{\psi(T_{i-1},T_i,T_i,2)}}\right)\right.$$

$$\left. - \tilde{K}\Phi\left(\frac{\ln \frac{P(0,T_{i-1})}{\tilde{K}P(0,T_i)} + \psi(0,T_{i-1},T_i,T_\beta,\frac{1}{2})}{\sqrt{\psi(T_{i-1},T_i,T_i,2)}}\right)\right],$$

where $\tilde{K} := 1 + \tau_i K$ and

$$\psi(T,S,\tau,\lambda)$$
$$:= \frac{\sigma^2}{2a^3}\left[1 - e^{-a(S-T)}\right]\left[1 - e^{-2aT}\right]\left[e^{-a(\tau-T)} - 1 + \lambda - \lambda e^{-a(S-T)}\right]$$
$$+ \frac{\eta^2}{2b^3}\left[1 - e^{-b(S-T)}\right]\left[1 - e^{-2bT}\right]\left[e^{-b(\tau-T)} - 1 + \lambda - \lambda e^{-b(S-T)}\right]$$
$$+ \frac{\rho\sigma\eta}{ab(a+b)}\left[1 - e^{-a(S-T)}\right]\left[1 - e^{-(a+b)T}\right]\left[e^{-b(\tau-T)} - 1 + \lambda - \lambda e^{-b(S-T)}\right]$$
$$+ \frac{\rho\sigma\eta}{ab(a+b)}\left[1 - e^{-b(S-T)}\right]\left[1 - e^{-(a+b)T}\right]\left[e^{-a(\tau-T)} - 1 + \lambda - \lambda e^{-a(S-T)}\right].$$

Notice that, using the previous notation, we can write $\psi(T_{i-1},T_i,T_i,2) = \Sigma(T_{i-1},T_i)^2$.

10.5 Ratchets (One-Way Floaters)

We give a short description of one-way floaters in the following. We assume a unit nominal amount.

- Institution A pays to B (a percentage γ of) a reference floating rate (plus a constant spread S) at dates $\mathcal{T} = \{T_1, \ldots, T_\beta\}$. Formally, at time T_i institution A pays to B

$$(\gamma F_i(T_{i-1}) + S)\tau_i.$$

- Institution B pays to A a coupon that is given by the reference rate plus a spread X at dates $\mathcal{T}$, floored and capped respectively by the previous coupon and by the previous coupon plus an increment Y. Formally, at time T_i with $i > 1$, institution B pays to A the coupon

$$c_i = \begin{cases} (F_i(T_{i-1}) + X)\tau_i & \text{if } c_{i-1} \leq (F_i(T_{i-1}) + X)\tau_i \leq c_{i-1} + Y, \\ c_{i-1} & \text{if } (F_i(T_{i-1}) + X)\tau_i < c_{i-1}, \\ c_{i-1} + Y & \text{if } (F_i(T_{i-1}) + X)\tau_i > c_{i-1} + Y, \end{cases}$$

At the first payment time T_1, institution B pays to A the coupon

$$(F_1(T_0) + X)\tau_1.$$

The discounted payoff as seen from institution A is

$$\sum_{i=1}^{\beta} D(0, T_i) \left[c_i - (\gamma F_i(T_{i-1}) + S)\tau_i \right]$$

and the value to A of the contract is the risk-neutral expectation

$$E \left\{ \sum_{i=1}^{\beta} D(0, T_i) \left[c_i - (\gamma F_i(T_{i-1}) + S)\tau_i \right] \right\}$$

$$= P(0, T_\beta) \sum_{i=1}^{\beta} E^\beta \left[\frac{c_i - (\gamma F_i(T_{i-1}) + S)\tau_i}{P(T_i, T_\beta)} \right].$$

Since the forward-rate dynamics under Q^β of

$$F_{\beta(t)}(t), \ldots, F_\beta(t)$$

is known as from (10.4), a Monte Carlo pricing can be carried out in the usual manner.

10.6 Constant-Maturity Swaps (CMS)

10.6.1 CMS with the LFM

A constant-maturity swap is a financial product structured as follows. We assume a unit nominal amount. Let us denote by $\mathcal{T} = \{T_0, \ldots, T_n\}$ a set of payment dates at which coupons are to be paid. We assume, for simplicity, such dates to be one-year spaced.

- At time T_{i-1} (in some variants at time T_i), $i \geq 1$, institution A pays to B the c-year swap rate resetting at time T_{i-1}. Formally, at time T_{i-1} institution A pays to B

$$S_{i-1,i-1+c}(T_{i-1})\tau_i,$$

where, as usual,

$$S_{i-1,i-1+c}(t) = \frac{P(t, T_{i-1}) - P(t, T_{i-1+c})}{\sum_{k=i}^{i-1+c} \tau_k P(t, T_k)}. \tag{10.6}$$

- Institution B pays to A a fixed rate K.

The net value of the contract to B at time 0 is

$$E\left(\sum_{i=1}^{n} D(0, T_{i-1})(S_{i-1,i-1+c}(T_{i-1}) - K)\tau_i\right)$$

$$= \sum_{i=1}^{n} \tau_i P(0, T_{i-1})\left[E^{i-1}\left(S_{i-1,i-1+c}(T_{i-1})\right) - K\right]$$

$$= \sum_{i=1}^{n} \tau_i \left(P(0, T_n) E^n\left(\frac{S_{i-1,i-1+c}(T_{i-1})}{P(T_{i-1}, T_n)}\right) - KP(0, T_{i-1})\right). \tag{10.7}$$

We need only compute either

$$E^{i-1}\left[S_{i-1,i-1+c}(T_{i-1})\right] \quad \text{or} \quad E^n[S_{i-1,i-1+c}(T_{i-1})/P(T_{i-1}, T_n)]$$

for all i's. At first sight, one might think to discretize equation (6.38) for the dynamics of the forward swap rate and compute the required expectation through a Monte Carlo simulation. However, notice that forward rates appear in the drift m^α of such equation, so that we are forced to evolve forward rates anyway. As a consequence, we can use equation (10.6) jointly with Monte Carlo simulated forward-rate dynamics, and do away with the dynamics (6.38), thus directly recovering the swap rate $S_{i-1,i-1+c}(T_{i-1})$ from the T_{i-1} values of the (Monte Carlo generated) spanning forward rates

$$F_i(T_{i-1}), F_{i+1}(T_{i-1}), \ldots, F_{i-1+c}(T_{i-1}).$$

Analogously to the autocaps case, such forward rates can be generated according to the usual discretized (Milstein) dynamics (10.5) based on Gaussian shocks and under the unique measure Q^n.

10.6.2 CMS with the G2++ Model

It is possible to price a CMS with the G2++ model (4.4). See the related Section 11.2.2 on quanto CMS's in the next chapter. We do not repeat things here, since the CMS pricing procedure can be easily deduced from the procedure for the more general quanto-CMS case.

10.7 The Convexity Adjustment and Applications to CMS

10.7.1 Natural and Unnatural Time Lags

As with so many things, it was simply a matter of time.
The Time Trapper, Zero Hour – End of an Era, LSH 61, 1994, DC Comics

To appropriately introduce the convexity-adjustment technique, we quickly recall the pricing formulas for swaps. We begin by a plain-vanilla swap with natural time lag.

Consider an IRS that resets at dates $T_\alpha, T_{\alpha+1}, \ldots, T_{\beta-1}$ and pays at dates $T_{\alpha+1}, \ldots, T_\beta$, with unit notional amount. The fact that the payment indexed by the LIBOR rate resetting at time T_i for the maturity T_{i+1} occurs precisely at time T_{i+1} is referred to as a "natural time lag". This renders the swap price independent of the volatility of rates. Indeed, let us consider only the variable swap leg. The discounted value of this leg can be expressed either via the swap rate or via forward rates. In effect, the discounted payoff is given by

$$D(0,T_\alpha) S_{\alpha,\beta}(T_\alpha) \sum_{i=\alpha+1}^{\beta} \tau_i P(T_\alpha, T_i),$$

which is equivalent to

$$\sum_{i=\alpha+1}^{\beta} D(0,T_i) \tau_i F_i(T_{i-1}).$$

The value of such a leg is easily computed in both cases as

$$E\left[\sum_{i=\alpha+1}^{\beta} D(0,T_i)\tau_i F_i(T_{i-1})\right] = \sum_{i=\alpha+1}^{\beta} P(0,T_i)\tau_i E^i\left[F_i(T_{i-1})\right]$$

$$= \sum_{i=\alpha+1}^{\beta} P(0,T_i)\tau_i F_i(0) = \sum_{i=\alpha+1}^{\beta} \left[P(0,T_{i-1}) - P(0,T_i)\right]$$

$$= P(0,T_\alpha) - P(0,T_\beta).$$

10.7 The Convexity Adjustment and Applications to CMS

From the last formula notice that, as is well known, neither volatility nor correlation of rates affect this financial product.

Now, let us reconsider in-advance swaps. Consider the variable leg of an IRS that resets at dates $T_{\alpha+1}, \ldots, T_\beta$ and pays *at the same dates*, with unit notional amount. We say this swap has an "unnatural time lag". This term is justified by seeing that the price of such a leg depends on volatility. Indeed, see formula (10.3) with $K = 0$.

Contrary to the plain-vanilla case, the in-advance-swap price depends on the volatility of forward rates through their average volatilities v, which are usually deduced inverting cap prices through Black's formula.

The "natural/unnatural" terminology reflects the above calculations. A *natural time lag* for the variable leg of a swap makes the value of such a leg *independent of the rates volatility*. On the contrary, an *unnatural time lag* makes the value of the variable leg volatility dependent.

As a corollary, we can derive the corresponding formulas for forward-rate agreements. Suppose we are now at time 0, and at time T_2 the contract pays the LIBOR rate resetting at time $T_1 < T_2$ and maturing at T_2. As usual, we denote this rate by $L(T_1, T_2) = F_2(T_1)$ and we denote by τ the year fraction between T_1 and T_2. The contract value is therefore, consistently with the general FRA notation previously established,

$$\begin{aligned}
-\mathbf{FRA}(0, T_1, T_2, 0) &= E[D(0, T_2)\tau F_2(T_1)] \\
&= P(0, T_2)\tau E^2[F_2(T_1)] = P(0, T_2)\tau F_2(0) \\
&= P(0, T_1) - P(0, T_2).
\end{aligned}$$

If we have an in-advance FRA, this time the contract pays at time T_1 the LIBOR rate resetting at the same time $T_1 < T_2$ and maturing at T_2. By reasoning in an analogous way to the case of in-advance swaps, we obtain

$$\begin{aligned}
\mathbf{IAFRA} &= P(0, T_2)\left[1 + 2\tau F_2(0) + \tau^2 F_2^2(0)\exp(v_2^2(T_1))\right] - P(0, T_1) \\
&= P(0, T_1)\left[1 + \frac{\tau F_2(0) + \tau^2 F_2^2(0)\exp(v_2^2(T_1))}{1 + \tau F_2(0)}\right] - P(0, T_1) \\
&= P(0, T_2)\tau F_2(0)\left(1 + \tau F_2(0)\exp(v_2^2(T_1))\right) \\
&\approx P(0, T_2)\tau F_2(0)\left(1 + \tau F_2(0) + \tau v_2^2(T_1)F_2(0)\right) \\
&= P(0, T_1)\tau F_2(0) + P(0, T_2)\tau^2 F_2^2(0)v_2^2(T_1)
\end{aligned} \quad (10.8)$$

10.7.2 The Convexity-Adjustment Technique

The time is out of joint. O cursed spite,
That ever I was born to set it right
Hamlet, I.5

The convexity-adjustment technique can be attempted any time there is an unnatural time lag. We consider its application to a single payment.

Assume a swap rate is involved, and that the payment $\tau_i S_{\alpha,\beta}(T_{i-k})$ is due at time T_i, $i - k \leq i \leq \alpha < \beta$.

Remark 10.7.1. (CMS) This is typical of constant-maturity swaps (CMS) where we have $i = \alpha$ and $k = 1$ or $k = 0$. This is the case where the convexity adjustment works well and is also supported by the output of more sophisticated models like, for instance, the G2++ model. If k is large, the correction can be quite wrong. Therefore, in such cases, the correction discussed here should be considered with care.

We are far from the "usual" IRS case, because the rate being exchanged at each payment instant is a *swap* rate rather than a *LIBOR* rate.

A first adjustment

The forward swap rate $S_{\alpha,\beta}$ is originally defined as related to an IRS that pays at times $\alpha + 1, \ldots, \beta$: $S_{\alpha,\beta}(T)$ at time T, $T \leq T_\alpha$, is the fixed rate such that the fixed leg of the above IRS has value equal to that of the floating leg. In case of reimbursement of the notional amount, such a value at time T is always $P(T, T_\alpha)$ (see Definition 1.5.2 and the subsequent comments), so that we can write ("FL" stands for "floating leg")

$$\mathrm{FL}_{\alpha,\beta}(T) = P(T, T_\alpha) = S_{\alpha,\beta}(T) \sum_{i=\alpha+1}^{\beta} \tau_i P(T, T_i) + P(T, T_\beta) \ .$$

Now rewrite the same expression with the discount factors coming from a flat yield curve fixed at a level y (annually compounded) at time T_α, ("FFL" stands for Flat Floating Leg):

$$\mathrm{FFL}_{\alpha,\beta}(T; y) = S_{\alpha,\beta}(T) \sum_{i=\alpha+1}^{\beta} \tau_i \frac{P(T, T_\alpha)}{(1+y)^{\tau_{\alpha,i}}} + \frac{P(T, T_\alpha)}{(1+y)^{\tau_{\alpha,\beta}}} \ ,$$

where $\tau_{\alpha,i}$ denotes the year fraction between T_α and T_i.

If one allows for the first-order expansion

$$\delta S_{\alpha,\beta}(T) = (1 + S_{\alpha,\beta}(T))^\delta - 1,$$

and takes $T_i = i\delta$, $\tau_{\alpha,i} = (i - \alpha)\delta$ and $\tau_i = \delta$, it is easy to see that the above flat floating leg coincides with $P(T, T_\alpha)$ only for $y = S_{\alpha,\beta}(T)$,

$$\mathrm{FFL}_{\alpha,\beta}(T; S_{\alpha,\beta}(T)) = P(T, T_\alpha).$$

Therefore, the value of y around which the flat-curve approximation is to be considered is the forward swap rate $S_{\alpha,\beta}(T)$, in that it is the flat rate that agrees with the non-flat case as far as the price of the floating leg is concerned:

10.7 The Convexity Adjustment and Applications to CMS

$$\text{FFL}_{\alpha,\beta}(T; S_{\alpha,\beta}(T)) = \text{FL}_{\alpha,\beta}(T).$$

Consider now the expectation

$$E_0^T \left[P(0,T) \text{FFL}_{\alpha,\beta}(T; S_{\alpha,\beta}(T)) - \text{FFL}_{\alpha,\beta}(0; S_{\alpha,\beta}(0)) \right] \quad (10.9)$$

$$= E_0^T \left[P(0,T) \frac{P(T,T_\alpha)}{P(T,T)} - P(0,T_\alpha) \right] = 0.$$

At this point, we proceed by defining the following quantity $\Phi_{\alpha,\beta}(y)$ through a slight approximation of the argument of the above expectation, where we assume $P(0,T)P(T,T_\alpha) \approx P(0,T_\alpha)$:

$$P(0,T)\text{FFL}_{\alpha,\beta}(T,y) \approx S_{\alpha,\beta}(T) \sum_{i=\alpha+1}^{\beta} \tau_i \frac{P(0,T_\alpha)}{(1+y)^{\tau_{\alpha,i}}} + \frac{P(0,T_\alpha)}{(1+y)^{\tau_{\alpha,\beta}}} =: \Phi_{\alpha,\beta}(T,y).$$

We expand Φ through a second-order Taylor expansion around $y = S_{\alpha,\beta}(0)$, we evaluate the resulting expression at $y = S_{\alpha,\beta}(T)$, we then solve for $S_{\alpha,\beta}(T) - S_{\alpha,\beta}(0)$ and introduce a further approximation:

$$S_{\alpha,\beta}(T) - S_{\alpha,\beta}(0) \approx \frac{\Phi_{\alpha,\beta}(T, S_{\alpha,\beta}(T)) - \Phi_{\alpha,\beta}(T, S_{\alpha,\beta}(0))}{\Phi'_{\alpha,\beta}(T, S_{\alpha,\beta}(0))} \quad (10.10)$$

$$- \frac{(S_{\alpha,\beta}(T) - S_{\alpha,\beta}(0))^2}{2} \frac{\Phi''_{\alpha,\beta}(T, S_{\alpha,\beta}(0))}{\Phi'_{\alpha,\beta}(T, S_{\alpha,\beta}(0))}$$

$$\approx \frac{\Phi_{\alpha,\beta}(T, S_{\alpha,\beta}(T)) - \Phi_{\alpha,\beta}(0, S_{\alpha,\beta}(0))}{\Phi'_{\alpha,\beta}(0, S_{\alpha,\beta}(0))}$$

$$- \frac{(S_{\alpha,\beta}(T) - S_{\alpha,\beta}(0))^2}{2} \frac{\Phi''_{\alpha,\beta}(0, S_{\alpha,\beta}(0))}{\Phi'_{\alpha,\beta}(0, S_{\alpha,\beta}(0))},$$

where the superscript $'$ denotes partial derivative with respect to y.

Now take expectation on both sides under the measure Q^T. The first term on the right-hand side has expectation zero, due to equation (10.9).

We further assume that we can approximate the true Q^T-dynamics of $S_{\alpha,\beta}$ by its lognormal $Q^{\alpha,\beta}$-dynamics (6.36), an approximation that has been shown to work well in most situations for the LFM when T is close to T_α (see the tests on the distributions of the swap rate described at the end of Section 8.2 and the related results in Section 8.3). We obtain:

$$E_0^T \left[(S_{\alpha,\beta}(T) - S_{\alpha,\beta}(0))^2 \right] \approx E_0^{\alpha,\beta} \left[(S_{\alpha,\beta}(T) - S_{\alpha,\beta}(0))^2 \right]$$

$$= S_{\alpha,\beta}(0)^2 (e^{v_{\alpha,\beta}^2(T)} - 1) \approx S_{\alpha,\beta}^2(0) \, v_{\alpha,\beta}^2(T),$$

where

$$v_{\alpha,\beta}^2(T) = \int_0^T (\sigma^{(\alpha,\beta)}(t))^2 \, dt$$

is the average variance of the forward swap rate in the interval $[0,T]$ times the interval length. Now, we can evaluate (10.10) by taking expectation on both sides:

$$E_0^T[S_{\alpha,\beta}(T)] \approx S_{\alpha,\beta}(0) - \tfrac{1}{2}S_{\alpha,\beta}^2(0)v_{\alpha,\beta}^2(T)\frac{\Phi_{\alpha,\beta}''(0,S_{\alpha,\beta}(0))}{\Phi_{\alpha,\beta}'(0,S_{\alpha,\beta}(0))}. \quad (10.11)$$

A second adjustment

A second adjustment we can consider is based on neglecting the final reimbursement of the notional amount in the above IRS. We thus define Ψ as Φ without notional reimbursement,

$$\Psi_{\alpha,\beta}(y) := S_{\alpha,\beta}(T) \sum_{i=\alpha+1}^{\beta} \tau_i \frac{P(0,T_\alpha)}{(1+y)^{\tau_{\alpha,i}}}.$$

Assuming that also

$$E_0^T[\Psi_{\alpha,\beta}(S_{\alpha,\beta}(T)) - \Psi_{\alpha,\beta}(S_{\alpha,\beta}(0))] \approx 0,$$

as for Φ when taking expectations on both sides of (10.10), and using again a second-order expansion, it follows that

$$\boxed{E_0^T[S_{\alpha,\beta}(T)] \approx S_{\alpha,\beta}(0) - \tfrac{1}{2}S_{\alpha,\beta}^2(0)v_{\alpha,\beta}^2(T)\frac{\Psi_{\alpha,\beta}''(S_{\alpha,\beta}(0))}{\Psi_{\alpha,\beta}'(S_{\alpha,\beta}(0))}}, \quad (10.12)$$

where the ratio $\Psi_{\alpha,\beta}''(S_{\alpha,\beta}(0))/\Psi_{\alpha,\beta}'(S_{\alpha,\beta}(0))$ is independent of T.

This is the formula that is usually considered in the market for convexity adjustments (especially for CMS), see for example Hull (1997), in particular formula (16.13) and the related Example 16.8. The approximation works well when T is not too far away from T_α, as implied by the "Q^T vs $Q^{\alpha,\beta}$" dynamics approximation for the forward swap rate.

Let us now apply this formula to specific situations.

Floating leg with swap-rate-indexed payments

Suppose we need to compute the present value of our generic payment,

$$E[\tau_i D(0,T_i)S_{\alpha,\beta}(T_{i-k})].$$

Move under the T_i-forward measure, to obtain

$$P(0,T_i)E^i[\tau_i S_{\alpha,\beta}(T_{i-k})].$$

The first rougher approximation is to treat the measure Q^i as if it were the swap measure $Q^{\alpha,\beta}$, under which S can be modeled through the lognormal martingale

10.7 The Convexity Adjustment and Applications to CMS

$$dS_{\alpha,\beta}(t) = \sigma^{(\alpha,\beta)}(t) S_{\alpha,\beta}(t)\, dW_t\,.$$

Under this approximation, we would then have

$$E[\tau_i D(0,T_i) S_{\alpha,\beta}(T_{i-k})] \approx \tau_i P(0,T_i) S_{\alpha,\beta}(0).$$

The convexity adjustment (10.12) leads to the following modification of this last formula:

$$\begin{aligned}&E[\tau_i D(0,T_i) S_{\alpha,\beta}(T_{i-k})]\\ &= \tau_i P(0,T_i) E^i[S_{\alpha,\beta}(T_{i-k})] \approx \tau_i P(0,T_i) E^{i-k}[S_{\alpha,\beta}(T_{i-k})]\\ &\approx \tau_i P(0,T_i) \left[S_{\alpha,\beta}(0) - \tfrac{1}{2} S_{\alpha,\beta}^2(0) v_{\alpha,\beta}^2 (T_{i-k}) \frac{\Psi''_{\alpha,\beta}(S_{\alpha,\beta}(0))}{\Psi'_{\alpha,\beta}(S_{\alpha,\beta}(0))} \right].\end{aligned}$$

As anticipated in Remark 10.7.1, this approximation turns out to work well only for small values of k. Therefore, if k is large, the correction should be considered with due care.

We now check that, in case of an in-advance FRA, this formula is consistent with the value found earlier by exact evaluation. We take $\alpha = i = 1$, $k = 0$, $\beta = 2$ and $\tau_{1,2} = \tau$, so that $S_{\alpha,\beta}(t) = F(t; T_1, T_2)$ and

$$\Psi_{1,2}(y) = \frac{C}{(1+\tau y)},$$

with C a suitable constant, and where we have used simple compounding instead of annual compounding. Notice that

$$\frac{\Psi''_{1,2}(y)}{\Psi'_{1,2}(y)} = \frac{-2\tau}{(1+\tau y)},$$

so that the convexity-adjustment formula (10.12) yields

$$\tau P(0,T_1) \left[F_2(0) + \tau \frac{F_2^2(0) v_2^2(T_1)}{1 + \tau F_2(0)} \right]$$

$$= \tau P(0,T_1) F_2(0) + \tau^2 P(0,T_2) F_2^2(0) v_2^2(T_1),$$

which is the same result found, at first order in v_2^2, by exact evaluation in (10.8).

10.7.3 Deducing a Simple Lognormal Dynamics from the Adjustment

We can easily adjust the approximate driftless dynamics

$$dS_{\alpha,\beta}(t) = \sigma^{(\alpha,\beta)}(t) S_{\alpha,\beta}(t)\, dW_t\,,$$

for which
$$E_0^T[S_{\alpha,\beta}(T_{i-k})] = S_{\alpha,\beta}(0),$$
to a new dynamics
$$dS_{\alpha,\beta}(t) = \mu^{\alpha,\beta} S_{\alpha,\beta}(t)\,dt + \sigma^{(\alpha,\beta)}(t) S_{\alpha,\beta}(t)\,dW_t, \tag{10.13}$$
for which
$$E_0^T[S_{\alpha,\beta}(T_{i-k})] = S_{\alpha,\beta}(0) - \tfrac{1}{2} S_{\alpha,\beta}^2(0) v_{\alpha,\beta}^2(T_{i-k}) \frac{\Psi''_{\alpha,\beta}(S_{\alpha,\beta}(0))}{\Psi'_{\alpha,\beta}(S_{\alpha,\beta}(0))},$$

consistently with the convexity-adjustment evaluation. Since the dynamics (10.13) produces
$$E_0^T[S_{\alpha,\beta}(T_{i-k})] = S_{\alpha,\beta}(0) \exp(\mu^{\alpha,\beta} T_{i-k}) \approx S_{\alpha,\beta}(0)(1 + \mu^{\alpha,\beta} T_{i-k}),$$

at first order in $\mu^{\alpha,\beta} T_{i-k}$, it suffices to set
$$\mu^{\alpha,\beta} = -\tfrac{1}{2} S_{\alpha,\beta}(0) \frac{v_{\alpha,\beta}^2(T_{i-k})}{T_{i-k}} \frac{\Psi''_{\alpha,\beta}(S_{\alpha,\beta}(0))}{\Psi'_{\alpha,\beta}(S_{\alpha,\beta}(0))}.$$

Notice that in case the instantaneous forward-swap-rate volatility $\sigma^{\alpha,\beta}$ is assumed to be constant, we have
$$\frac{v_{\alpha,\beta}^2(T_{i-k})}{T_{i-k}} = (\sigma^{\alpha,\beta})^2.$$

This approximation is however rather rough and should not be used to evaluate nonlinear payoffs, unless a considerable amount of testing has been performed and acceptable errors are found.

10.7.4 Application to CMS

We have seen before that a constant-maturity swap has a floating leg that pays at times $T_{\alpha+1},\ldots,T_\beta$ the swap rates
$$S_{\alpha,\alpha+c}(T_\alpha), S_{\alpha+1,\alpha+1+c}(T_{\alpha+1}),\ldots,S_{\beta-1,\beta-1+c}(T_{\beta-1}).$$

Therefore, at each payment instant $T_{\alpha+k+1}$, such leg pays a certain prespecified c-year swap rate resetting at the previous instant $T_{\alpha+k}$. In some variants, instead, it pays at $T_{\alpha+k+1}$ a certain pre-specified swap rate resetting at the same instant. We will consider here the first version.

The value of the generic CMS payment is given by
$$E\left[D(0,T_{i+1})\tau_{i+1} S_{i,i+c}(T_i)\right] = \tau_{i+1} P(0,T_{i+1}) E^{i+1} S_{i,i+c}(T_i)$$
$$\approx \tau_{i+1} P(0,T_{i+1}) \left[S_{i,i+c}(0) - \tfrac{1}{2} S_{i,i+c}^2(0) v_{i,i+c}^2(T_i) \frac{\Psi''_{i,i+c}(S_{i,i+c}(0))}{\Psi'_{i,i+c}(S_{i,i+c}(0))} \right],$$

10.7 The Convexity Adjustment and Applications to CMS

see also Example 16.8 in Hull (1997). The CMS price is then obtained by adding terms for i ranging from α to $\beta - 1$. Recall that the adjustment used here has been derived under a number of approximations. As such, it can be improved. Indeed, the classical adjustment has been found to be not completely satisfactory by some traders, especially in some market situations involving volatility smiles. For a recent work on CMS adjustments see for example Pugachevsky (2001).

10.7.5 Forward Rate Resetting Unnaturally and Average-Rate Swaps

We consider now the following problem, which can have several applications. Consider two time instants s, u and a payment date T, $s < u < T$. Assume we have a contract that pays at time T the spot LIBOR rate resetting at time s for the maturity u:

$$L(s, u) = F(s; s, u).$$

In case $T = u$ we have a natural time lag. Indeed, the contract value at time 0 is the risk-neutral expectation of the discounted payoff

$$E_0[D(0,T)F(s; s, T)] = P(0,T)E_0^T[F(s; s, T)] = P(0,T)F(0; s, T)$$

and does not depend on volatility specifications.

If $T > u$, the above formula no longer holds. However, we can still evaluate the contract as follows.

Consider the no-arbitrage forward-rate dynamics for $F(t) = F(t; s, u)$ under the T-forward-adjusted measure Q^T:

$$dF(t) = -\sigma_{s,u}\sigma_{u,T}\tau(u,T)F(t)\frac{F(t;u,T)}{1+\tau(u,T)F(t;u,T)}\,dt + \sigma_{s,u}F(t)\,dW_t^T,$$

where $\sigma_{s,u}$ is the instantaneous volatility of $F(t) = F(t; s, u)$ and $\sigma_{u,T}$ is the instantaneous volatility of $F(t; u, T)$, and both are assumed to be constant (otherwise they can be replaced with the square roots of the average variances of $F(t; s, u)$ and $F(t; u, T)$, respectively, over $[0, s]$). The quantity $\tau(a, b)$ denotes in general the time between dates a and b in years.

We assume unit correlation between $F(t; s, u)$ and $F(t; u, T)$, since usually T and u are close. If this is not the case, a ρ parameter can be included in the drift of the above process.

With the usual deterministic-percentage-drift approximation we can write

$$dF(t) = -\sigma_{s,u}\sigma_{u,T}\tau(u,T)F(t)\frac{F(0;u,T)}{1+\tau(u,T)F(0;u,T)}\,dt + \sigma_{s,u}F(t)\,dW_t^T.$$

This new process has lognormal distribution under the T-forward measure and it can be easily seen that its expected value, conditional on the information available at time 0, under the T-forward measure, is

$$E_0^T[F(s;s,u)] = F(0;s,u)\exp\left(-\tau(u,T)\sigma_{s,u}\sigma_{u,T}\, s\, \frac{F(0;u,T)}{1+\tau(u,T)F(0;u,T)}\right).$$

We are now able to price the discounted payoff

$$E_0[D(0,T)L(s,u)] = P(0,T)E_0^T[F(s;s,u)]$$

$$= P(0,T)F(0;s,u)\exp\left(-\tau(u,T)\sigma_{s,u}\sigma_{u,T}\, s\, \frac{F(0;u,T)}{1+\tau(u,T)F(0;u,T)}\right).$$

As an example, consider a contract that pays at a future time T the average value of the 3-month ($3m$) LIBOR rates in the days $t_1 < t_2 < \ldots < t_n$, $t_n < T$, with δ_i denoting the year fraction between t_i and $t_i + 3m$. This is a possible example of a leg of an average-rate swap.

If the notional is N, the contract price is

$$E_0\left[D(0,T)\frac{\sum_{i=1}^n \delta_i NL(t_i,t_i+3m)}{n}\right]$$

$$= \frac{P(0,T)}{n} N \sum_{i=1}^n \delta_i E_0^T[F(t_i;t_i,t_i+3m)]$$

and is given by

$$\frac{P(0,T)}{n} N \sum_{i=1}^n \delta_i F(0;t_i,t_i+3m)$$

$$\cdot \exp\left[-\sigma_{t_i,t_i+3m}\sigma_{t_i+3m,T}\, t_i\, \frac{\tau(t_i+3m,T)F(0;t_i+3m,T)}{1+\tau(t_i+3m,T)F(0;t_i+3m,T)}\right].$$

Notice that the correction to the "brute-force" formula

$$\frac{P(0,T)}{n} N \sum_{i=1}^n \delta_i F(0;t_i,t_i+3m)$$

is multiplicative for each term and is given by the exponentials. The correction effect is to (slightly) reduce the "brute-force" value, since the exponents are negative. The correction might be not negligible for large values of the volatilities.

The difficulty in applying the above formula lies in the fact that the forward rate $F(0;t_i+3m,T)$ can be rather atypical as for expiry or maturity dates. Therefore, apart from few exceptions, its volatility cannot be recovered exactly from market cap prices. However, a synthetic volatility deduced from volatilities of "smaller" forward rates nested in $F(0;t_i+3m,T)$ can be used for this purpose, or arguments similar to those of Section 6.16 can be employed.

At a first stage, the above formula can be used to have a feeling on the order of magnitude of the adjustment due to second-order effects, and to decide whether these should be taken into account or not.

10.8 Captions and Floortions

A caption is an option that gives its holder the right to enter at a future time T_γ a cap whose first caplet resets at date $T_\alpha \geq T_\gamma$ and whose subsequent caplets reset at times $T_{\alpha+1}, \ldots, T_{\beta-1}$ with T_β the last payment date. The strike rate for this cap will be denoted by K. The price the holder of the caption will pay for this future cap is fixed as the caption strike and will be denoted by X. We can therefore express the caption payoff as a call payoff on the underlying cap.

We assume a unit notional amount. The T_γ value of the underlying cap described above is given by the usual Black formula (see for instance Section 6.4.3), computed at time T_γ instead of time 0,

$$\sum_{i=\alpha+1}^{\beta} \tau_i P(T_\gamma, T_i) \, \mathrm{Bl}(K, F_i(T_\gamma), \sqrt{T_{i-1} - T_\gamma} \, V(T_\gamma, T_{i-1})),$$

where the average volatility $V(\cdot, \cdot)$ was defined in Section 6.5.

The caption discounted payoff, expressed as a call payoff, can be written as

$$D(0, T_\gamma) \left\{ \sum_{i=\alpha+1}^{\beta} \tau_i P(T_\gamma, T_i) \, \mathrm{Bl}(K, F_i(T_\gamma), \sqrt{T_{i-1} - T_\gamma} \, V(T_\gamma, T_{i-1})) - X \right\}^+.$$

The caption value is given by the risk-neutral expectation of this payoff, which in turn is given by

$$P(0, T_\gamma) E^\gamma \left\{ \sum_{i=\alpha+1}^{\beta} \tau_i P(T_\gamma, T_i) \, \mathrm{Bl}(K, F_i(T_\gamma), \sqrt{T_{i-1} - T_\gamma} \, V(T_\gamma, T_{i-1})) - X \right\}^+.$$

Once again, the expected value can be computed through a Monte Carlo method, given the simulated values of

$$F_{\gamma+1}(T_\gamma), F_{\gamma+2}(T_\gamma), \ldots, F_\beta(T_\gamma)$$

under Q^γ, obtained through the usual discretized Milstein dynamics (10.5).

10.9 Zero-Coupon Swaptions

In this section we introduce zero-coupon swaptions and explain an approximated analytical method to price them. A payer (receiver) zero-coupon swaption is a contract giving the right to enter a payer (receiver) zero-coupon IRS at a future time. A zero-coupon IRS is an IRS where a single fixed payment is due at the unique (final) payment date T_β for the fixed leg in exchange for a stream of usual floating payments $\tau_i L(T_{i-1}, T_i)$ at times T_i in $T_{\alpha+1}, T_{\alpha+2}, \ldots, T_\beta$ (usual floating leg). In formulas, the discounted payoff of a payer zero-coupon IRS is, at time $t \leq T_\alpha$:

$$D(t, T_\alpha)\left[\sum_{i=\alpha+1}^{\beta} P(T_\alpha, T_i)\tau_i F_i(T_\alpha) - P(T_\alpha, T_\beta)\tau_{\alpha,\beta} K\right],$$

where $\tau_{\alpha,\beta}$ is the year fraction between T_α and T_β. The analogous payoff for a receiver zero-coupon IRS is obviously given by the opposite quantity.

Taking risk-neutral expectation, we obtain easily the contract value as

$$P(t, T_\alpha) - P(t, T_\beta) - \tau_{\alpha,\beta} K P(t, T_\beta),$$

which is the typical value of a floating leg minus the value of a fixed leg with a single final payment.

The value of the strike rate K that renders the contract fair is obtained by equating to zero the above value and solving in K. One obtains $K = F(t; T_\alpha, T_\beta)$. Indeed, we could have reasoned as follows. The value of the swap is independent of the number of payments on the floating leg, since the floating leg always values at par, no matter the number of payments (see Section 1.5.2 and the related remarks). Therefore, we might as well have taken a floating leg paying only in T_β the amount $\tau_{\alpha,\beta} L(T_\alpha, T_\beta)$. This would have given us again a standard swaption, standard in the sense that the two legs of the underlying IRS have the same payment dates (collapsing to T_β) and the unique reset date T_α. In such a one-payment case, the swap rate collapses to a forward rate, so that we should not be surprised to find out that the forward swap rate in this particular case is simply a forward rate.

An option to enter a payer zero-coupon IRS is a payer zero-coupon swaption, and the related payoff is

$$D(t, T_\alpha)\left[\sum_{i=\alpha+1}^{\beta} P(T_\alpha, T_i)\tau_i F_i(T_\alpha) - P(T_\alpha, T_\beta)\tau_{\alpha,\beta} K\right]^+,$$

or, equivalently, by expressing the F's in terms of discount factors,

$$D(t, T_\alpha)\left[1 - P(T_\alpha, T_\beta) - P(T_\alpha, T_\beta)\tau_{\alpha,\beta} K\right]^+,$$

which in turn can be written as

$$D(t, T_\alpha)\tau_{\alpha,\beta} P(T_\alpha, T_\beta)\left[F(T_\alpha; T_\alpha, T_\beta) - K\right]^+.$$

Notice that, from the point of view of the payoff structure, this is merely a caplet. As such, it can be priced easily through Black's formula for caplets. The problem, however, is that such a formula requires the integrated percentage volatility of the forward rate $F(\cdot; T_\alpha, T_\beta)$, which is a forward rate over a

10.9 Zero-Coupon Swaptions

non-standard period. Indeed, $F(\cdot; T_\alpha, T_\beta)$ is not in our usual family of spanning forward rates, unless we are in the trivial case $\beta = \alpha + 1$. Therefore, since the market provides us (through standard caps and swaptions) with volatility data for standard forward rates, we need a formula for deriving the integrated percentage volatility of the forward rate $F(\cdot; T_\alpha, T_\beta)$ from volatility data of the standard forward rates $F_{\alpha+1}, \ldots, F_\beta$. The reasoning is once again based on the "freezing the drift" procedure, leading to an approximately lognormal dynamics for our standard forward rates.

Denote for simplicity $F(t) := F(t; T_\alpha, T_\beta)$ and $\tau := \tau_{\alpha,\beta}$.

We begin by noticing that, through straightforward algebra, we have (write everything in terms of discount factors to check)

$$1 + \tau F(t) = \prod_{j=\alpha+1}^{\beta} (1 + \tau_j F_j(t)).$$

It follows that

$$\ln(1 + \tau F(t)) = \sum_{j=\alpha+1}^{\beta} \ln(1 + \tau_j F_j(t)),$$

so that

$$d\ln(1 + \tau F(t)) = \sum_{j=\alpha+1}^{\beta} d\ln(1 + \tau_j F_j(t)) = \sum_{j=\alpha+1}^{\beta} \frac{\tau_j dF_j(t)}{1 + \tau_j F_j(t)} + (\ldots)dt.$$

Now, since

$$dF(t) = \frac{1 + \tau F(t)}{\tau} d\ln(1 + \tau F(t)) + (\ldots)dt,$$

we obtain from the above expression

$$dF(t) = \frac{1 + \tau F(t)}{\tau} \sum_{j=\alpha+1}^{\beta} \frac{\tau_j dF_j(t)}{1 + \tau_j F_j(t)} + (\ldots)dt.$$

Take variance (conditional on the information up to time t) on both sides:

$$\text{Var}\left(\frac{dF(t)}{F(t)}\right) = \left[\frac{1 + \tau F(t)}{\tau F(t)}\right]^2 \sum_{i,j=\alpha+1}^{\beta} \frac{\tau_i \tau_j \rho_{i,j} \sigma_i(t) \sigma_j(t) F_i(t) F_j(t)}{(1 + \tau_i F_i(t))(1 + \tau_j F_j(t))} dt.$$

Now freeze all t's to zero except for the σ's, and integrate over $[0, T_\alpha]$:

$$(v_{\alpha,\beta}^{zc})^2 := \left[\frac{1 + \tau F(0)}{\tau F(0)}\right]^2 \sum_{i,j=\alpha+1}^{\beta} \frac{\tau_i \tau_j \rho_{i,j} F_i(0) F_j(0)}{(1 + \tau_i F_i(0))(1 + \tau_j F_j(0))} \int_0^{T_\alpha} \sigma_i(t) \sigma_j(t) dt.$$

To price the zero-coupon swaption it is then enough to put this quantity into the related Black's formula:

$$\mathbf{ZCPS} = \tau P(0, T_\beta)\mathrm{Bl}(K, F(0), v^{zc}_{\alpha,\beta}).$$

We can check the accuracy of this formula against the usual Monte Carlo pricing based on the exact dynamics of the forward rates. In the tests all swaptions are at-the-money. We have done this under the data of case (3.a) of the volatility tests of Section 8.2, and in other situations. Under the data of Section 8.2, we considered first the case $T_\alpha = 2y$, $T_\beta = 19y$. We obtained the implied volatility $v^{zcMC}_{\alpha,\beta}/\sqrt{T_\alpha}$ by inverting the Monte Carlo price through Black's formula:

$$\mathbf{MCZCPS} = \tau P(0, T_\beta)\mathrm{Bl}(F(0), F(0), v^{zcMC}_{\alpha,\beta}).$$

We found, in this case:

$$\frac{v^{zcMC}_{\alpha,\beta}}{\sqrt{T_\alpha}} = 0.1410, \quad \frac{v^{zc}_{\alpha,\beta}}{\sqrt{T_\alpha}} = 0.1455.$$

A two-side 98% window for the Monte Carlo volatility defined as in Section 8.2 is in this case [0.1404 0.1416]. Our algebraic approximation falls out of the 98% window, but of a small amount if compared with the distance from the volatility of the corresponding plain-vanilla European swaption. In fact, the standard at-the-money plain-vanilla swaption with the same initial reset date and final payment date, whose algebraic approximation has been found to be accurate in Section 8.2, has volatility

$$\frac{v^{\mathrm{LFM}}_{\alpha,\beta}}{\sqrt{T_\alpha}} = 0.0997.$$

We have also considered the case $T_\alpha = 10y$, $T_\beta = 19y$. We obtained

$$\frac{v^{zcMC}_{\alpha,\beta}}{\sqrt{T_\alpha}} = 0.1081, \quad \frac{v^{zc}_{\alpha,\beta}}{\sqrt{T_\alpha}} = 0.1114.$$

Now a two-side 98% window for the Monte Carlo volatility defined as in Section 8.2 is [0.1076 0.1086]. Again, our algebraic approximation falls out of the 98% window of a small amount when compared with the discrepancy with respect to the corresponding standard swaption, resulting in a volatility

$$\frac{v^{\mathrm{LFM}}_{\alpha,\beta}}{\sqrt{T_\alpha}} = 0.0897.$$

In the two examples above we notice that the at-the-money standard swaption has always a lower volatility (and hence price) than the corresponding at-the-money zero-coupon swaption. We may wonder whether this is a general feature. Indeed, we have the following.

Remark 10.9.1. (**Comparison between zero-coupon swaptions and corresponding standard swaptions**). A first remark is due for a comparison between the zero-coupon swaption volatility $v^{zc}_{\alpha,\beta}$ and the corresponding European-swaption approximation $v^{\mathrm{LFM}}_{\alpha,\beta}$. If we rewrite the latter as

$$(v^{\text{LFM}}_{\alpha,\beta})^2 = \sum_{i,j=\alpha+1}^{\beta} \rho_{i,j}\lambda_i\lambda_j \int_0^{T_\alpha} \sigma_i(t)\sigma_j(t)dt, \quad \lambda_i = \frac{w_i(0)F_i(0)}{S_{\alpha,\beta}(0)},$$

it is easy to check that

$$(v^{\text{ZC}}_{\alpha,\beta})^2 = \sum_{i,j=\alpha+1}^{\beta} \rho_{i,j}\mu_i\mu_j \int_0^{T_\alpha} \sigma_i(t)\sigma_j(t)dt,$$

where

$$\mu_i = \frac{P(0,T_\alpha)}{P(0,T_i)}\lambda_i \geq \lambda_i,$$

the discrepancy increasing with the payment index i. It follows that, for positive correlations, the zero-coupon swaption volatility is always larger than the corresponding plain vanilla swaption volatility, the difference increasing with the tenor $T_\beta - T_\alpha$, for each given T_α.

A final remark concerns the possibility to price zero-coupon swaptions with other models.

Remark 10.9.2. **(Pricing zero-coupon swaptions with other models).** Zero-coupon swaptions can be priced analytically under all short-rate models admitting explicit formulas for European options on zero-coupon bonds and, accordingly, for caplets. For instance, under the CIR++ model (3.76) we can use formula (3.79), whereas under the G2++ model (4.4) we can resort to formula (4.28).

10.10 Eurodollar Futures

A Eurodollar-futures contract gives its owner the payoff

$$X(1 - L(S_1, S_2))$$

at the future time $S_1 < S_2$, where X is a notional amount, and the year fraction between S_1 and S_2 is denoted by τ. The fair price of this contract at time t is

$$V_t = E_t[X(1 - L(S_1, S_2))] = X(1 - E_t[L(S_1, S_2)]) \tag{10.14}$$

$$= X\left(1 + \frac{1}{\tau} - \frac{1}{\tau}E_t\left[\frac{1}{P(S_1, S_2)}\right]\right),$$

and takes into account continuous rebalancing (see for example Sandmann and Sondermann (1997) and their reference to the related work of Cox Ingersoll and Ross).

The problem is computing the expectation

$$E_t\left[\frac{1}{P(S_1,S_2)}\right] = E_t\left[\frac{P(S_1,S_1)}{P(S_1,S_2)}\right].$$

If we were under the S_2-forward-adjusted measure this would be simply

$$\frac{P(0,S_1)}{P(0,S_2)} = 1 + \tau F(0;S_1,S_2),$$

and the price would reduce to

$$X(1 - F(0;S_1,S_2)).$$

Instead, we need the expectation under the risk-neutral measure. Since we need to compute

$$E_t[L(S_1,S_2)] = E_t[F(S_1;S_1,S_2)],$$

the result will depend on the interest-rate model we are using.

10.10.1 The Shifted Two-Factor Vasicek G2++ Model

We can use the two-additive-factor Gaussian model described in Chapter 4. Consistently with the notation adopted there, recall that

$$P(t,T) = \frac{P^M(0,T)}{P^M(0,t)} \exp\{\mathcal{A}(t,T)\},$$

$$\mathcal{A}(t,T) = \frac{1}{2}[V(t,T) - V(0,T) + V(0,t)] - \frac{1-e^{-a(T-t)}}{a}x(t)$$
$$- \frac{1-e^{-b(T-t)}}{b}y(t),$$

with V defined as in (4.10), so that

$$E_t\left[\frac{1}{P(T_1,T_2)}\right] = \frac{P^M(0,T_1)}{P^M(0,T_2)} E_t\left[\exp\{-\mathcal{A}(T_1,T_2)\}\right]$$
$$= \frac{P^M(0,T_1)}{P^M(0,T_2)} \exp\left\{-\frac{1}{2}[V(T_1,T_2) - V(0,T_2) + V(0,T_1)]\right.$$
$$+ \frac{1-e^{-a(T_2-T_1)}}{a}x(t)e^{-a(T_1-t)}$$
$$\left.+ \frac{1-e^{-b(T_2-T_1)}}{b}y(t)e^{-b(T_1-t)}\right.$$

10.10 Eurodollar Futures

$$+ \left(\frac{1-e^{-a(T_2-T_1)}}{a}\right)^2 \frac{\sigma^2}{4a}\left[1-e^{-2a(T_1-t)}\right]$$

$$+ \left(\frac{1-e^{-b(T_2-T_1)}}{b}\right)^2 \frac{\eta^2}{4b}\left[1-e^{-2b(T_1-t)}\right]$$

$$+ \frac{(1-e^{-a(T_2-T_1)})(1-e^{-b(T_2-T_1)})}{ab}$$

$$\left. \cdot \rho\frac{\sigma\eta}{a+b}\left[1-e^{-(a+b)(T_1-t)}\right]\right\}.$$

By substituting this algebraic formula in (10.14) one has the value of the Eurodollar-futures contract. Notice that if this is evaluated at time 0, since $x_0 = y_0 = t = 0$ the above formula simplifies a little. Typically $X = 100$ and $\tau = 0.25$.

We have then considered a set of parameters coming from a typical calibration of the G2++ model to swaptions volatilities and to the zero-coupon curve of the Euro market. The values of these parameters are: $a = 0.0234$; $b = 0.0015$; $\sigma = 0.0081429$; $\eta = 0.0020949$; $\rho = -0.2536$.

We have finally computed prices for increasing maturities T_1 (from three months to ten years), while keeping $T_2 = T_1 + 0.25$, and we considered the differences

$$\text{Spread}(T_1) := E_0[L(T_1, T_1 + 0.25)] - F(0; T_1, T_1 + 0.25)$$

as T_1 increases. Such differences, in basis points (hundredths of a percentage point), are shown in Figure 10.1 below.

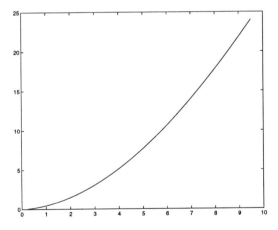

Fig. 10.1. Spread(T_1) in basis points plotted against T_1

The (upward concave) qualitative behaviour of the correction agrees with what is usually experienced in the market.

10.10.2 Eurodollar Futures with the LFM

Since we need to compute

$$E_t[L(S_1, S_2)] = E_t[F(S_1; S_1, S_2)],$$

we need the dynamics of the forward rate $F(\cdot; S_1, S_2)$ under the risk-neutral measure. This can be obtained starting from the martingale dynamics under the numeraire $P(\cdot, S_2)$ and moving to the bank-account numeraire via the "change of numeraire toolkit". This new dynamics involves the bond-price dynamics of $P(\cdot, S_2)$, which in turn can be expressed in terms of spanning forward rates. Therefore, in the relevant dynamics, correlation and volatilities of all spanning forward rates are involved. Subsequently, the forward-rates dynamics need be discretized and a Monte Carlo method can be applied to compute the relevant expectation under the risk-neutral measure.

In detail, assume we have a set of expiry/maturity dates $\{T_0, T_1, \ldots, T_M\}$ for a family of spanning forward rates, with $T_{M-1} = S_1$ and $T_M = S_2$. As we explained in Section 6.3, the forward-rate dynamics under the risk-neutral measure is given by (6.16).

Here, we would need to model the instantaneous forward rate f in the initial interval $(t, T_{\beta(t)-1}]$ to close the equations, but if we discretize these equations (for the logarithm of F's) with a Milstein scheme *exactly at the time instants* $\{T_0, T_1, \ldots, T_M\}$, we are in no need to model f. One sees easily that this is the same as discretizing the LFM dynamics (6.17) under the spot LIBOR measure whose numeraire is the discretely-rebalanced bank account. As usual, a Monte Carlo method, based on the jointly Gaussian distributions of the shocks for different components, can be applied to propagate all F's up to time $T_{M-1} = S_1$ in order to evaluate the final expectation

$$E_0[F_M(T_{M-1})] = E_0[F(S_1; S_1, S_2)].$$

Again, we can freeze part of the drift in the Spot-LIBOR-measure dynamics (6.17) thus obtaining

$$dF_k(t) = \sigma_k(t) F_k(t) \sum_{j=\beta(t)}^{k} \frac{\tau_j \rho_{j,k} \sigma_j(t) F_j(0)}{1 + \tau_j F_j(0)} dt + \sigma_k(t) F_k(t) dZ_k^d(t),$$

which is a geometric Brownian motion. Under this dynamics, the above expected value is easily computed in terms of the now deterministic percentage drift.

10.11 LFM Pricing with "In-Between" Spot Rates

Let us assume the current time to be $t = 0$, and let us denote by $\mathcal{T} = \{T_0, \ldots, T_n\}$ a set of payment dates, at which coupons of a certain financial

10.11 LFM Pricing with "In-Between" Spot Rates

instrument are to be paid. Such dates are assumed to be equally δ-spaced for simplicity. We also denote by $\mathcal{G} = \{g_1, \ldots, g_m\}$ the set of future dates at which a reference rate (typically the six-month LIBOR rate) is quoted in the market up to time T_n.

We denote by $L(t)$ the relevant reference rate at time t with maturity $t + \delta$. Forward-rate dynamics will be considered under the forward-adjusted measure Q^n corresponding to the final payment time T_n.

Consider a financial product whose payoff depends on the (for example daily) evolution of the reference rate L in each reset/payment interval.

In order to Monte Carlo price this product based on the forward-rate dynamics of the LFM, we need to recover at any time t the reference rate, which we assume to be the δ spot rate $L(t) = F(t; t, t+\delta)$, from the family of spanning forward rates at our disposal at times $T_{\beta(t)-2}$ and $T_{\beta(t)-1}$, i.e. at the dates in $\mathcal{T}$ that are closest to the current time t: $T_{\beta(t)-2} < t \leq T_{\beta(t)-1}$. In particular, we have both

$$L(T_{\beta(t)-2}) = F_{\beta(t)-1}(T_{\beta(t)-2})$$

and

$$L(T_{\beta(t)-1}) = F_{\beta(t)}(T_{\beta(t)-1}) \ .$$

How do we obtain $L(t)$ from $L(T_{\beta(t)-2})$ and $L(T_{\beta(t)-1})$? We have faced this problem earlier in Sections 6.17.1 and 6.17.2, proposing both a "drift interpolation" and a "bridging" technique.

We now present some particular products depending on "in-between" rates, and we will tacitly assume that "in-between" rates have been obtained through one of these methods.

10.11.1 Accrual Swaps

We give a short description of accrual swaps in the following. We assume a unit nominal amount.

- Institution A pays to B (a percentage γ of) the reference rate L (plus a spread S) at dates $\mathcal{T}$. Formally, at time T_i institution A pays to B

$$(\gamma L(T_{i-1}) + S)\tau_i \ ,$$

where τ_i is the year fraction between the payment dates T_{i-1} and T_i.

- Institution B pays to A, at time T_i, a percentage α of the reference rate plus a spread Q, times the relative number of days between T_{i-1} and T_i where the reference rate L was in the corridor $L_1 \leq L \leq L_2$. Formally, at time T_i, institution B pays to A the coupon

$$c(T_i) = (\alpha L(T_{i-1}) + Q) \tau_i \frac{\sum_{g \in \mathcal{G} \cap [T_{i-1}, T_i)} 1\{L_1 \leq L(g) \leq L_2\}}{\#\{\mathcal{G} \cap [T_{i-1}, T_i)\}} \ , \quad (10.15)$$

where, as usual, # denotes the number of elements of a set (cardinality).

When the simulated paths for L are available, we are able to evaluate the accrual swap. The discounted payoff as seen from institution A is

$$\sum_{i=1}^{n} D(0,T_i)\tau_i \left[(\alpha L(T_{i-1}) + Q) \frac{\sum_{g \in \mathcal{G} \cap [T_{i-1},T_i)} 1\{L_1 \leq L(g) \leq L_2\}}{\#\{\mathcal{G} \cap [T_{i-1},T_i)\}} \right.$$

$$\left. - (\gamma L(T_{i-1}) + S) \right],$$

so that the value to A of the accrual swap is the risk-neutral expectation

$$E\left\{ \sum_{i=1}^{n} D(0,T_i)\tau_i \left[(\alpha\, L(T_{i-1}) + Q) \frac{\sum_{g \in \mathcal{G} \cap [T_{i-1},T_i)} 1\{L_1 \leq L(g) \leq L_2\}}{\#\{\mathcal{G} \cap [T_{i-1},T_i)\}} \right.\right.$$

$$\left.\left. -(\gamma L(T_{i-1}) + S) \right] \right\}$$

$$= P(0,T_n) \sum_{i=1}^{n} \tau_i E^n\left\{ \frac{1}{P(T_i,T_n)} \right.$$

$$\left. \cdot \left[(\alpha\, L(T_{i-1}) + Q) \frac{\sum_{g \in \mathcal{G} \cap [T_{i-1},T_i)} 1\{L_1 \leq L(g) \leq L_2\}}{\#\{\mathcal{G} \cap [T_{i-1},T_i)\}} - (\gamma L(T_{i-1}) + S) \right] \right\}$$

Both the forward-rate dynamics and the related approximated L dynamics under Q^n are known and a Monte Carlo pricing can be carried out.

Analytical Formula for Accrual Swaps

Alternatively, we may study an analytical formula based on a drift approximation in the LIBOR market model, similar to the one used in deriving approximated swaptions volatilities and terminal correlations in Chapter 6. We proceed as follows.

We concentrate on the non-trivial leg, paid by B to A. Let us focus on the single discounted payment occurring at time T_i. It will suffice to add up all contributions after each one has been priced. We have seen above the payment at time T_i to be given by (10.15). Instead of expressing every coupon under the terminal measure, let us write

$$E_0[D(0,T_i)c(T_i)] = P(0,T_i)E_0^i[c(T_i)].$$

Our task is then reduced to computing the expected value $E_0^i[c(T_i)]$, which, in turn, amounts to computing, by additive decomposition, expected values such as

$$E_0^i[L(T_{i-1})\,1\{L_1 \leq L(u) \leq L_2\}], \quad E_0^i[1\{L_1 \leq L(u) \leq L_2\}]$$

10.11 LFM Pricing with "In-Between" Spot Rates   405

under the T_i-forward-adjusted measure Q^i and for $T_{i-1} \le u < T_i$. These may be rewritten in terms of forward rates as

$$E_0^i[F_i(T_{i-1})\mathbf{1}\{L_1 \le F(u; u, u+\delta) \le L_2\}], \quad Q^i\{L_1 \le F(u; u, u+\delta) \le L_2\},$$

where we have expressed the expected value of an indicator function directly as a probability.

Now, in order to handle such expressions we consider approximated forward-rate dynamics. Actually, no approximation is needed for F_i under Q^i, since its drift is zero and we have a nice geometric Brownian motion. Instead, we act on the dynamics of $F(t; u, u+\delta)$. Since $F(\cdot; u, u+\delta)$ is not in our fundamental family of forward rates, we use the drift-interpolation technique seen in Section 6.17.1. If we set $F_u(t) = F(t; u, u+\delta)$ for brevity, by applying formula (6.69), with partially frozen coefficients and a few rearrangements, we obtain (notice that $T_i \le u + \delta < T_{i+1}$)

$$dF_u(t) = \mu(t)F_u(t)\,dt + \sigma(t; u, u+\delta)F_u(t)\,dZ^i(t),$$
$$\mu(t) := \frac{(u+\delta-T_i)F_{i+1}(0)}{1+\tau_{i+1}F_{i+1}(0)}\sigma(t; u, u+\delta)\rho_{i,i+1}\sigma_{i+1}(t) \quad (10.16)$$

where $\sigma(t; u, u+\delta)$ is the instantaneous volatility of the related forward rate, and is usually obtained by some kind of interpolation from the "standard rates" volatilities σ_k's.

Our approximation has produced a fundamental effect. The process (10.16) is now a geometric Brownian motion, and we can apply a standard "Black and Scholes technology" to our pricing problem.

Let us recall the following Black and Scholes fundamental setup. Assume we are given two asset prices following correlated geometric Brownian motions under the relevant measure,

$$dS_t = \mu_1(t)S_t\,dt + v_1(t)S_t\,dZ_1(t),$$
$$dA_t = \mu_2(t)A_t\,dt + v_2(t)A_t\,dZ_2(t), \quad dZ_1\,dZ_2 = \rho\,dt$$

(all coefficients being deterministic). We can easily calculate, through laborious but straightforward computations, for $T < u$,

$$E_0[S(T)\mathbf{1}\{L_1 \le A(u) \le L_2\}] = S(0)\exp\left(\int_0^T \mu_1(s)ds\right)$$
$$\cdot\left[\Phi\left(\frac{\ln(L_2/A(0)) - \int_0^u \left(\mu_2(s) - \frac{v_2(s)^2}{2}\right)ds - \rho\int_0^T v_1(s)v_2(s)ds}{\sqrt{\int_0^u v_2(s)^2 ds}}\right)\right.$$
$$\left. - \Phi\left(\frac{\ln(L_1/A(0)) - \int_0^u \left(\mu_2(s) - \frac{v_2(s)^2}{2}\right)ds - \rho\int_0^T v_1(s)v_2(s)ds}{\sqrt{\int_0^u v_2(s)^2 ds}}\right)\right].$$

We can apply this formula to our case, by setting $T = T_{i-1}$, $S(t) = F_i(t)$, $A(t) = F_u(t)$, $\mu_1(t) = 0$, $\mu_2(t) = \mu(t)$, $v_1(t) = \sigma_{i+1}(t)$, $v_2(t) = \sigma(t; u, u+\delta)$, $\rho = \rho_{i,i+1}$. This provides us with the terms

$$E_0^i[F_i(T_{i-1})\,1\{L_1 \leq F(u; u, u+\delta) \leq L_2\}],$$

where the impact of correlation is evident. On the other hand, we may compute

$$E_0[1\{L_1 \leq A(u) \leq L_2\}] = \left[\Phi\left(\frac{\ln(L_2/A(0)) - \int_0^u \left(\mu_2(s) - \frac{v_2(s)^2}{2}\right)ds}{\sqrt{\int_0^u v_2(s)^2 ds}}\right) - \Phi\left(\frac{\ln(L_1/A(0)) - \int_0^u \left(\mu_2(s) - \frac{v_2(s)^2}{2}\right)ds}{\sqrt{\int_0^u v_2(s)^2 ds}}\right)\right],$$

from which terms

$$Q^i\{L_1 \leq F(u; u, u+\delta) \leq L_2\}$$

are readily computed. Now, putting all the pieces together, we obtain the accrual-swap price. The "frozen drift" approximation guarantees us that for short maturities this formula should work well. However, we have seen that the drift "freezing approximation" above usually does not take us far away from the lognormal distribution even for large maturities, as we have observed in the density plots given in Chapter 8. Problems might only occur with pathological or very large volatilities.

Finally, we would like to point out that the "freezing part of the drift" method can usually be used to transform the distributionally unknown LFM dynamics into the geometric-Brownian-motion dynamics of the basic Black and Scholes lognormal setup. As a consequence, this method can be of help in all cases where forward rates play the role of underlying assets under the relevant measure and where the basic Black and Scholes setup leads to analytical formulas. Before adopting the thus derived approximated formulas, however, one should test them against a Monte Carlo pricing, carried out through the true LFM dynamics, in a sufficiently large number of market situations.

10.11.2 Trigger Swaps

A trigger swap is an interest-rate swap periodically paying a certain *reference* rate against a fixed payment. This swap "comes to life" or "terminates" when a certain *index* rate hits a prespecified level. It is somehow similar to barrier options in the FX or equity markets. Usually, the two rates coincide, but the index rate is observed at a higher frequency than the payment frequency. For example, the index rate and the reference rate can both coincide with the

10.11 LFM Pricing with "In-Between" Spot Rates

six-month LIBOR rate, which can be observed daily for the indexing and every six months for the payments.

There are four standard basic types of trigger swaps: Down and Out (DO), Up and Out (UO), Down and In (DI), Up and In (UI). Let the prespecified level be H.

- DO: The initial index rate is above H. The swap terminates its payments ("goes OUT") as soon as the index rate hits the level H (from above, i.e. going "DOWN").
- UO: The initial index rate is below H. The swap terminates its payments ("goes OUT") as soon as the index rate hits the level H (from below, i.e. going "UP").
- DI: The initial index rate is above H. The swap starts its payments ("goes IN") as soon as the index rate hits the level H (from above, i.e. going "DOWN").
- UI: The initial index rate is below H. The swap starts its payments ("goes IN") as soon as the index rate hits the level H (from below, i.e. going "UP").

The payoff from a DO trigger swap can be expressed formally as follows. As for accrual swaps, we assume the current time to be $t = 0$, and we denote by $\mathcal{T} = \{T_0, \ldots, T_n\}$ a set of payment dates, at which payments occur. Such dates are assumed to be equally δ-spaced. We also denote by $\mathcal{G} = \{g_1, \ldots, g_m\}$ the set of future dates at which the reference rate (typically the six-month LIBOR rate) is quoted in the market up to time T_n.

We assume the index rate and reference rate to coincide. We denote by $L(t)$ the reference rate at the generic time instant t with maturity $t + \delta$. Forward-rate dynamics will be considered under the forward-adjusted measure Q^n corresponding to the final payment time T_n.

We assume unit nominal amount. If the swap is still alive at time $t = T_{i-1}$, then at time T_i the following will occur:

- Institution A pays to B the fixed rate K at time T_i *if* at all previous instants in the interval $(T_{i-1}, T_i]$ the index rate L is above the triggering barrier H. Formally, if the swap is still alive at time T_{i-1}, at time T_i institution A pays to B

$$K\tau_i \prod_{g \in \mathcal{G} \cap (T_{i-1}, T_i]} 1\{L(g) > H\}$$
$$= K\tau_i 1\{ \min\{L(g),\ g \in \mathcal{G} \cap (T_{i-1}, T_i]\} > H \} ,$$

where τ_i is the year fraction between the payment dates T_{i-1} and T_i.

- Institution B pays to A (a percentage α of) the reference rate L at the last reset date T_{i-1} (plus a spread Q) *if* at all previous instants of the interval $(T_{i-1}, T_i]$ the index rate L is above the triggering barrier H. Formally, at time T_i institution B pays to A

$$(\alpha L(T_{i-1}) + Q)\,\tau_i\,\mathbf{1}\{\min\{L(g),\ g \in \mathcal{G} \cap (T_{i-1}, T_i]\} > H\}$$

The complete discounted payoff as seen from institution A can be expressed as

$$\sum_{i=1}^{n} D(0,T_i)\,(\alpha L(T_{i-1}) + Q - K)\,\tau_i\,\mathbf{1}\{\min\{L(g),\ g \in \mathcal{G} \cap (T_0, T_i]\} > H\}$$

and the contract value to institution A is

$$E\left[\sum_{i=1}^{n} D(0,T_i)\,(\alpha L(T_{i-1}) + Q - K)\,\tau_i\,\mathbf{1}\{\min\{L(g),\ g \in \mathcal{G} \cap (T_0, T_i]\} > H\}\right]$$

$$= P(0,T_n)\sum_{i=1}^{n} \tau_i\,E^n\left[\frac{(\alpha L(T_{i-1}) + Q - K)\,\mathbf{1}\{\min\{L(g),\ g \in \ldots\} > H\}}{P(T_i, T_n)}\right].$$

Once again, it is enough to recover spot rates $L(T_i) = F_{i+1}(T_i)$ and discount factors $P(T_i, T_n)$ by generating for all i's spanning forward rates

$$F_{i+1}(T_i), F_{i+2}(T_i), \ldots, F_n(T_i)$$

under Q^n according to the usual discretized (Milstein) dynamics (analogously to the autocaps case (10.5)), and apply either the "drift interpolation" or the "bridging" technique of Sections 6.17.1 and 6.17.2 to recover in-between rates $L(g)$.

10.12 LFM Pricing with Early Exercise and Possible Path Dependence

Here we shortly present Longstaff and Schwartz's (2000) method for pricing early-exercise (and possibly path-dependent) products through Monte Carlo simulation in the LFM. Indeed, what we will present here can be intended as a solution of the following two different and yet related problems.

1. How can we use Monte Carlo for early-exercise (non path-dependent) products? This can be necessary when in presence of non-Markovian dynamics or of large dimensionality of the underlying process, as we shall see in a moment.
2. How can we price derivatives that show at the same time path dependence and early-exercise features, even in the favorable cases of low dimensionality and Markovian dynamics?

In Chapter 3 we observed that the pricing of early-exercise products can be carried out through binomial/trinomial trees, and that Monte Carlo is instead suited to treat path-dependent products. Here, before proposing a

10.12 LFM Pricing with Early Exercise and Possible Path Dependence

recent promising extension of the Monte Carlo method, we shortly recall what was already remarked there, in the beginning of Sections 3.11.2 and 3.11.3.

Trees can be used for early-exercise products when the fundamental underlying variable is low-dimensional (say one or two-dimensional), as happens typically with short-rate models. In such cases, the tree is the ideal instrument, given its "backward-in-time" nature. We know the value of the payoff in each final node, and move backward in time, by updating the value of continuation through discounting. At each node of the tree we can compare the backwardly propagated value of continuation with the payoff evaluated at that node, and decide whether exercise is to be considered or not at that point. After the exercise decision has been taken, the backward induction restarts and we continue to propagate backwards the updated value. When we reach the initial node of the tree, at time 0, we have (an approximation of) the price of our early-exercise product. Thus trees are ideally suited to "travel backward in time".

The other family of products that is usually considered is the family of "path-dependent" payoffs. Such products can be exercised only at a final date, but their final payoffs depend on the history of the underlying variable up to the final time, and not only on the value of the underlying variable at maturity. For path-dependent products, the Monte Carlo method is ideally suited, since it works through forward propagation in time of the underlying variable, by simulating its transition density between dates where the underlying-variable history matters to the final payoff. Monte Carlo is thus ideally suited to "travel forward in time".

In principle, trees have problems mainly in two situations. The first case concerns high dimensionality. If the underlying variable follows a high-dimensional process (in practice with dimension larger than two or three, as in case of the LFM, for example), the tree is practically impossible to consider, since the computational time grows roughly exponentially with the dimension. Moreover, there are also difficulties in handling correlations and other aspects, so that trees become extremely difficult to use.

The second case where trees have major problems is with path-dependent products. When we try and propagate backwards the contract value from the final nodes we are immediately in trouble, since to value the payoff at a given node (and at any final node in particular) we need to know the past history of the underlying variable. But this past history is not determined yet, since we move backward in time. This method, therefore, is not applicable in a standard way.

Actually, there are ad-hoc procedures to render trees able to price particular path-dependent products in the basic Black and Scholes setting, for example barrier and lookback options. However, in general there is no consolidated and realistic recipe on how using a tree for path-dependent payoffs, and moreover, when dealing with interest-rate derivatives models, we are usually outside the Black and Scholes framework.

As for the Monte Carlo method, it does better with respect to high dimensionality, in that computational time grows roughly *linearly* with the dimension, and it is also suited to parallel computing. However, there are problems with early exercise. Since we propagate trajectories forward in time, we have no means to know whether it is optimal to continue or to exercise at a certain time. Therefore, Monte Carlo cannot be used, in its original formulation, for the large range of products involving early exercise features.

However, Longstaff and Schwartz (2000) have proposed an approximated method to make Monte Carlo techniques work also in presence of early-exercise features. The resulting method is very promising, since it allows for the pricing of instruments with high-dimensional underlying variables, path dependence and early exercise at the same time.

Clearly, the method needs further testing beyond what shown in Longstaff and Schwartz (2000), especially on practical cases concerning interest-rate models. Still, test results in Longstaff and Schwartz (2000) look rather encouraging so as to justify a general exposition of the method and of its possible developments even before extensive testing has been carried out. This generality, together with the potential of the method of not exceeding the true value of the early-exercise contract, could result in a supremacy of Monte Carlo over trees and finite-difference methods in general, especially if numerically efficient Monte Carlo methods are brought into play.

Recently, research on improvements of the basic Monte Carlo setup, based on weighted paths and other techniques have received considerable attention in the literature, thus further strengthening the interest in the Longstaff and Schwartz (2000) method.

We now review this method for a generic product whose final payoff depends on a (possibly multi-dimensional) underlying variable X.

Assume we have a product that can be exercised at times $t_1, \ldots, t_N$, whose immediate-exercise value at each time t_k depends on part of the history of an underlying process $X(t)$ up to time t_k itself. Typically, the value can depend on $X(s_1), \ldots, X(s_{j_k})$, where the times $s_1 < s_2 < \ldots < s_{j_k} \leq t_k$ are the ones contributing to the immediate-exercise payoff at time t_k. In detail, we assume that, if exercised at time t_k, the product pays immediately the Cash flow from Exercise (CE) given by

$$\text{CE}(t_k) := \text{CE}(t_k; X(s_1), \ldots, X(s_{j_k})).$$

This value has to be compared with the backwardly Cumulated discounted cash flows from Continuation (CC) at the same time, namely the value of the contract at t_k when this has not been exercised before or at t_k itself,

$$\text{CC}(t_k) := \text{CC}(t_k; X(s_1), \ldots, X(s_{j_k})).$$

We assume we are computing prices under a generic numeraire asset $U(t)$. In their paper, Longstaff and Schwartz (2000) take the bank account as fundamental numeraire, and work under the risk-neutral measure.

10.12 LFM Pricing with Early Exercise and Possible Path Dependence

The method can be summarized through the following scheme:

1. Choose a number of paths, np.
2. (Choice of the basis functions). For each time t_k, choose i_k basis functions

$$\phi_1(t_k, x_1, \ldots, x_{j_k}), \ldots, \phi_{i_k}(t_k, x_1, \ldots, x_{j_k})$$

   that will be used in approximating the continuation value as a function of the past and present values $X(s_1), X(s_2), \ldots, X(s_{j_k})$ of the underlying variable up to time t_k (see step 7 below).

3. (Simulating the underlying variables). Simulate np paths for both the underlying variable X and the numeraire U from time t_1 to time t_n. Make sure of including the reset times $s_1, \ldots, s_{j_n}$ among the dates at which X and U are simulated. Typically, this simulation is "exact" if the transition distributions of X and U are known, like, for example, in the case of geometric Brownian motion or linear-Gaussian processes as in Hull and White's models. Alternatively, a numerical discretization scheme for SDEs such as the Euler or Milstein schemes can be employed if this transition density is not known. In any case, denote by

$$X^j(t_k), U^j(t_k)$$

   the simulated values of X and U respectively under the j-th scenario at time t_k. More generally, the superscript on a stochastic quantity will denote the quantity itself under the scenario given by the superscript index.

4. (Computing the payoff at final time). Set

$$CC^j(t_n) := CE(t_n; X^j(s_1), \ldots, X^j(s_{j_n})).$$

   (The backwardly Cumulated discounted cash flow from Continuation at final time is simply the exercise value at that time).

5. (Positioning the initial step at final time). Set $k = n$. We position ourselves at the final exercise time. Now the iterative part of the scheme begins.

6. (Consider only scenarios where the immediate-exercise value of the contract is positive). Set

$$I_{k-1} := \{j \in \{1, 2, \ldots, np\} : CE(t_{k-1}; X^j(s_1), \ldots, X^j(s_{j_{k-1}})) > 0\}.$$

   We thus focus only on scenarios where the exercise value is strictly positive at the current evaluation time t_{k-1};

7. (Regressing the discounted continuation value on the chosen basis functions). In this step, we aim at approximating the discounted continuation value at current time t_{k-1} as a linear combination of the basis functions

$$\phi_1(t_{k-1}, x_1, \ldots, x_{j_{k-1}}), \ldots, \phi_{i_{k-1}}(t_{k-1}, x_1, \ldots, x_{j_{k-1}})$$

through a regression, so as to estimate the combinators λ in

$$\frac{U^j(t_{k-1})}{U^j(t_k)} CC^j(t_k) = \sum_{h=1}^{i_{k-1}} \lambda_h(t_{k-1}) \, \phi_h(t_{k-1}, X^j(s_1), \ldots, X^j(s_{j_{k-1}})),$$

where $j \in I_{k-1}$.

On the left-hand side of the above equation, we have the continuation value an instant later discounted back at current time t_{k-1} through the chosen numeraire U. Notice that if the numeraire is the bank account $B(t) = \exp(rt)$, with deterministic constant r, as in Longstaff and Schwartz (2000), then the U's ratio reduces to $\exp(-r(t_k - t_{k-1}))$.

On the right-hand side of the same equation, we have a linear combination of the chosen basis functions, corresponding ideally to a truncated L^2 expansion. The step could be made exact with an infinite expansion ($i_{k-1} = \infty$), when the conditional expectation defining the actual continuation value above behaves nicely in an L^2 sense. See Longstaff and Schwartz (2000) for further details.

8. Store the exercise flag (EF) over scenarios at time t_{k-1}:

$$\mathrm{EF}(j, t_{k-1}) := 1\Big\{ \mathrm{CE}(t_{k-1}; X^j(s_1), \ldots, X^j(s_{j_{k-1}}))$$

$$> \sum_{h=1}^{i_{k-1}} \lambda_h(t_{k-1}) \, \phi_h(t_{k-1}, X^j(s_1), \ldots, X^j(s_{j_{k-1}})) \Big\}.$$

This flag is set to one when exercise is the convenient choice, and to zero when continuation is in order. Again, $1\{\cdots\}$ denotes the indicator function of the set between curly brackets.

When $\mathrm{EF}(j, t_{k-1})$ is one, set all its subsequent values to zero, $\mathrm{EF}(j, t_h) := 0$ for all $h > k - 1$.

9. Set

$$CC^j(t_{k-1}) := \frac{U^j(t_{k-1})}{U^j(t_k)} CC^j(t_k) \quad \text{if } \mathrm{EF}(j, t_{k-1}) = 0 \quad \text{(continuation)},$$

and set

$$CC^j(t_{k-1}) := \mathrm{CE}(t_{k-1}; X^j(s_1), \ldots, X^j(s_{j_{k-1}})) \quad \text{if } \mathrm{EF}(j, t_{k-1}) = 1 \quad \text{(exercise)}$$

10. If $k = 0$ stop, otherwise replace k with $k - 1$ and restart from point 6.

10.13 LFM: Pricing Bermudan Swaptions

Bermudan swaptions are options to enter an IRS not only at its first reset date, but also at subsequent reset dates of the underlying IRS, at least in some of the simplest formulations.

10.13 LFM: Pricing Bermudan Swaptions

Let again $\mathcal{T} = \{T_1, \ldots, T_n\}$ be a set of reset and payment dates. Recall that we denote by $\mathbf{PS}(T_i, T_k, \{T_k, \ldots, T_n\}, K)$ the price at time T_i of a (payer) swaption maturing at time T_k, which gives its holder the right to enter at time T_k an interest-rate swap with first reset date T_k and payment dates $T_{k+1}, \ldots, T_n$ at the fixed strike rate K. We will abbreviate this price by $\mathbf{PS}_{k,n}(T_i)$. This price is known as a function of the present value for basis point $C_{k,n}(T_i)$ and of the forward swap rate $S_{k,n}(T_i)$ through Black's formula for swaptions.

Definition 10.13.1. (Bermudan Swaption). *A (payer) Bermudan swaption is a swaption characterized by three dates $T_k < T_h < T_n$, giving its holder the right to enter at any time T_l in-between T_k and T_h (included) into an interest-rate swap with first reset in T_l, last payment in T_n and fixed rate K. Thus, the swap start and length depend on the instant T_l when the option is exercised. We denote by $\mathbf{PBS}_{k,h,n}(T_i)$ the value of such a Bermudan swaption at time T_i, with $T_i \leq T_k$.*[1]

Pricing Bermudan swaptions with the LFM has to be handled through tailor-made methods, since the model is not ideally suited for the implementation of recombining lattices. A possible alternative to the tailor-made techniques is the general Longstaff Schwartz "Monte Carlo Regression" (LSMC) approach reviewed in Section (10.12), which is indeed quite general and usually results in good approximations. However, the method itself has to be tailored (choice of the basis functions, ...) when applied to Bermudan swaptions in the LFM.

10.13.1 Longstaff and Schwartz's Approach

As we just noticed, the LSMC method can be used to price Bermudan swaptions in the LFM. Longstaff and Schwartz (2000), however, tested the LSMC method (in the section "valuing swaptions in a string model" of their paper) by actually considering a version of the so called string model. In practice, when working in a finite set of expiries/maturities, string models are often equivalent to the LIBOR market model (LFM). For more details on string models, see for example Santa Clara and Sornette (2001) or Longstaff, Santa Clara and Schwartz (2001). In the specific application of string models we are considering here, Longstaff and Schwartz (2000) used directly bond-prices dynamics and bond-prices volatilities instead of forward-rates volatilities. For completeness, we here illustrate their procedure.

A Bermudan swaption is considered, where the underlying swap starts at the initial time with given reset and payment dates. The swaption's holder has the right to exercise the option at some fixed dates and enter the swap, whose life span decreases as time moves forward.

---
[1] There are other types of Bermudan swaptions, but for our purposes the type described here suffices.

The underlying swap, with a ten-year maturity, resets semi-annually, and exercise can occur at any reset date after one year, one year included and ten years excluded. There are therefore nineteen exercise dates. In propagating the zero-coupon-bond prices,

$$P(\cdot,\ 0.5y), P(\cdot, 1y), P(\cdot,\ 1.5y), \ldots, P(\cdot, 10y),$$

the LSMC method starts from the twenty-dimensional vector above, and the dimension decreases by one each six months. The bond-price dynamics has as percentage risk-neutral drift the risk-free rate, which is approximated with the corresponding six-month continuously-compounded rate

$$r(t) \approx -2 \ln P(t, t + 0.5y),$$

thus closing the set of equations in P for the discretized approximate dynamics once the volatility has been assigned. Indeed, the approximate dynamics reads now

$$dP(t, T_i) = -2 \ln P(t, t + 0.5y) P(t, T_i) dt + \sigma_{P_i}(t) P(t, T_i) dZ_i(t), \quad T_i = 0.5i,$$

$i = 1, \ldots, 20$. When these equations are discretized at times T_i, we obtain a closed set of equations, since the drift rate now involves a bond price in the family.

The Z_i's are correlated Brownian motions under the risk-neutral measure. In the simulation it is assumed that

$$dZ_i dZ_j = \exp(-k|i - j|)dt,$$

where k is a positive constant, and the Z vector is kept twenty-dimensional.

At the exercise time T_i, the basis functions of the algorithm are selected as:

$$1, P(\cdot, T_i), \ldots, P(\cdot, T_{20}), \frac{1 - P(T_i, T_{20})}{\sum_{j=i+1}^{20} 0.5 P(T_i, T_j)}, \left[\frac{1 - P(T_i, T_{20})}{\sum_{j=i+1}^{20} 0.5 P(T_i, T_j)} \right]^2,$$

$$\left[\frac{1 - P(T_i, T_{20})}{\sum_{j=i+1}^{20} 0.5 P(T_i, T_j)} \right]^3,$$

where the last three terms are simply the underlying swap rate $S_{i,20}(T_i)$ and its second and third powers.

At the first exercise time ($i = 3$), there are 22 basis functions, their number decreasing as time goes by. Longstaff and Schwartz state that adding further functions does not change the option value, so that one can infer the valuation to be correct, given that the approximated value never exceeds the real value. See also the related discussion in Longstaff and Schwartz (2000).

Notice that Longstaff and Schwartz have assumed a deterministic bond-price percentage volatility. This is not really consistent with the LFM distribution for lognormal forward rates. Therefore, as already mentioned above, the model analyzed in this section is not a LFM, from a theoretical point of view.

10.13.2 Carr and Yang's Approach

Carr and Yang (1997) use simulations to develop a Markov-chain approximation for the valuation of Bermudan swaptions in the LFM. Their method stems from the observation that, given the tenor structure

$$T_1, \ldots, T_n,$$

one can represent the whole yield curve along the structure by just knowing the evolution of a chosen numeraire. Take for example the numeraire $P(\cdot, T_n)$, associated with the terminal measure Q^n. At a time T_i in the tenor structure, the whole (Zero-bond) curve

$$P(T_i, T_{i+1}), P(T_i, T_{i+2}), \ldots, P(T_i, T_n)$$

can be obtained as follows. Recall that by definition of numeraire we have

$$\frac{P(T_i, T_j)}{P(T_i, T_n)} = E^n_{T_i}\left[\frac{P(T_j, T_j)}{P(T_j, T_n)}\right],$$

or

$$P(T_i, T_j) = P(T_i, T_n) E^n_{T_i}\left[\frac{1}{P(T_j, T_n)}\right], \tag{10.17}$$

so that we can compute each $P(T_i, T_j)$ by knowing the current value of the numeraire $P(\cdot, T_n)$ and its distribution under its own measure Q^n. The exercise decision, at any instant, can thus be reduced to knowledge of the distributional properties of the single process $P(\cdot, T_n)$.

Based on this observation, Carr and Young (1997) found a way to construct a Markov chain approximating the migration of $P(\cdot, T_n)$ in between areas of a selected partition of $[0, 1]$. Partitioning $[0, 1]$ in $I_1(t), I_2(t), \ldots, I_{l(t)}(t)$, so that $[0, 1]$ is given by the disjoint union of the sets I, the Markov chain is constructed as follows.

First, simulate spanning forward-rate dynamics $F_{i+1}(t)^j, \ldots, F_n(t)^j$ under several scenarios, each scenario denoted by a superscript j, up to a generic time $t = T_i$. Second, obtain the numeraire bond price $P(t, T_n)^j$ from these simulations under each scenario j. Third, define the transition matrix between "state" h at time $t = T_i$ and "state" k at time $t + \Delta = T_{i+1}$ as

$$p_{h,k}(t) := \frac{\#\{j: \ P(t, T_n)^j \in I_h(t) \ \text{and} \ P(t+\Delta, T_n)^j \in I_k(t)\}}{\#\{j: \ P(t, T_n)^j \in I_h(t)\}}.$$

Then one defines $\bar{P}_h(t, T_n)$ as the average of the $P(t, T_n)^j$'s in $I_h(t)$,

$$\bar{P}_h(t, T_n) := \frac{\sum_{j: \ P(t, T_n)^j \in I_h(t)} P(t, T_n)^j}{\#\{j: \ P(t, T_n)^j \in I_h(t)\}}.$$

Consider the chain $X(t)$ with states $\{1, 2, \ldots, l(t)\}$ and probability $p_{h,k}(t)$ of going from $X(t) = h$ to $X(t + \Delta) = k$. Our $\bar{P}_h(t, T_n)$ can be considered as a discrete-space approximation of the numeraire $P(t, T_n)$ when $X(t) = h$.

The chain X summarizes the true dynamics of $P(t, T_n)$ into a Markov process that can be used for approximately simulating $P(t, T_n)$. We can therefore simulate the whole yield curve in the spirit of the relationship (10.17). Then backward induction becomes possible by using the Markov chain instead of the original paths for the numeraire.

We move backwards in time by means of the process $\bar{P}_{X(t)}(t, T_n)$ in place of the process $P(t, T_n)$, with $\bar{P}_{X(t)}(t, T_n)$ that assumes only a finite set of possible values at each instant. The transition probabilities allow us to roll back the relevant expectations and the Bermudan swaption can be easily priced through backward induction, see Carr and Yang (1997) for the details and for numerical tests.

A similar approach has been suggested in Clewlow and Strickland (1998) for a Gaussian multi-factor Heath-Jarrow-Morton model (and not the LFM), where again the early-exercise opportunity is evaluated in terms of a single variable. This variable is taken to be the fixed leg of the underlying interest-rate swap. Since the floating leg is always valued on par at reset dates, this choice amounts roughly to considering the value of the underlying interest-rate swap as fundamental single process at the reset dates.

The approximate specification of the early-exercise region as a function of the underlying variable is found by using a single-factor extended Vasicek (Hull and White) approximation of the multi-factor model. With the one-factor model one obtains the approximate early-exercise region via a recombining tree for the short rate, by determining the critical values of the underlying interest-rate swap at the early-exercise dates through backward induction on the tree. Choosing only one factor allows for a richer discretization in time and this yields an accurate exercise region.

Once the exercise decision has been estimated as a function of the underlying swap through the tree, one runs a Monte Carlo simulation for the original multi-factor model, where each early-exercise opportunity, when encountered, is evaluated as the (known) approximate function of the underlying swap.

This method seems to be robust. It provides one with a lower bound for the Bermudan swaption price, due to the sub-optimal exercise region, as in the LSMC method. A similar method for the LFM has been proposed by Andersen (1999), and we review it in the following.

10.13.3 Andersen's Approach

Andersen (1999) proposed a method similar to that of Clewlow and Strickland (1998). Again, the early-exercise region is extracted by a low-dimensional parameterization, consisting of a small number of key variables (these including the underlying interest-rate swap as in Clewlow and Strickland), but the approximated early-exercise region, as a function of these variables, is not determined through a *one-factor* model. Rather, an optimization on a separate simulation *for the whole multi-factor model* is considered in order to deter-

10.13 LFM: Pricing Bermudan Swaptions

mine this function. The method can be summarized as follows. We adopt the notation introduced earlier in Definition 10.13.1.

We now provide a scheme summarizing a possible formulation of Andersen's method for approximately computing $\mathbf{PBS}_{k,h,n}(T_k)$.

1) Choose a function f approximating for each T_l the optimal exercise flag $\mathcal{I}(T_l)$, depending for example on the nested European swaptions and on a function $H = H(T_l)$ to be determined,

$$\mathcal{I}(T_l) \approx f(\mathbf{PS}_{l,n}(T_l), \mathbf{PS}_{l+1,n}(T_l), \ldots, \mathbf{PS}_{h,n}(T_l), H(T_l)).$$

The optimal exercise flag $\mathcal{I}(T_l)(\omega)$ at time T_l, under the path ω, is defined to be one when exercise is optimal at T_l along the trajectory ω and 0 when the continuation value at T_l is larger than the exercise value along ω. As usual, ω is omitted in the notation.

2) Simulate, through the LFM dynamics for the forward LIBOR rates, in a set of scenarios indexed by j, all the variables

$$\mathbf{PS}^j_{l,n}(T_l),\ \mathbf{PS}^j_{l+1,n}(T_l), \ldots,\ \mathbf{PS}^j_{h,n}(T_l),\ B^j_d(T_l)$$

entering in f's expression above, for all $l = k, k+1, \ldots, h$. The last quantity is the discrete-bank-account numeraire that is used for discounting, i.e.

$$B_d(T_l) = \prod_{m=1}^{l} [1 + \tau_m F_m(T_{m-1})],$$

which is determined by the simulated forward-rate dynamics of the LFM, with T_0 denoting the initial time. Notice that the first variable $\mathbf{PS}^j_{l,n}(T_l)$ involves the interest-rate swap whose swap rate is $S_{l,n}(T_l)$, which was the (unique) "early-exercise flag" variable in the Clewlow and Strickland method.

3) Compute by backward induction all values of $H(T_l)$ from $T_l = T_h$ to $T_l = T_k$ as follows:

- 3.a) The final $H(T_h)$ has to be known from the requirement

$$f(\mathbf{PS}_{h,n}(T_h), H(T_h)) = 1\{\mathbf{PS}_{h,n}(T_h) > 0\}.$$

This is to say that at the last possible exercise date we simply exercise if the underlying European swaption has strictly positive value, as should be. Set $m = h$.

- 3.b) Find $H(T_{m-1})$ as follows. For each simulated path j, solve the optimization problem

$$H^j(T_{m-1}) = \arg\sup_{H} \Big\{ f(\mathbf{PS}^j_{m-1,n}(T_{m-1}), \mathbf{PS}^j_{m,n}(T_{m-1}), \ldots$$

$$\ldots, \mathbf{PS}^j_{h,n}(T_{m-1}), H)\ \mathbf{PS}^j_{m-1,n}(T_{m-1})$$

$$+ \frac{B^j_d(T_{m-1})}{B^j_d(T_m)} (1 - f)\ \mathbf{PBS}^j_{m,h,n}(T_m) \Big\},$$

where we omit f's arguments in the second half of the expression for brevity, and where the expression between curly brackets basically reads as:

if (exercise(H)) then (current underlying European swaption)
else (present value of one-period ahead Bermudan swaption).

We thus look for the value of H in the exercise strategy that maximizes the option value in each scenario.

Notice also that we can write

$$\frac{B_d(T_{m-1})}{B_d(T_m)} = \frac{1}{1 + \tau_m F_m(T_{m-1})}.$$

Let $\mathbf{PBS}_{m-1,h,n}^{j}(T_{m-1})$ be the supremum corresponding to the above $H^j(T_{m-1})$.

Average over all scenarios j and find $H(T_{m-1})$ from the $H^j(T_{m-1})$'s.
- 3.c) If $m-1$ equals k then move to point 4), otherwise decrease m by one and restart from point 3.b).
4) Now that H is known at all times, compute the Bermudan-swaption price $\mathbf{PBS}_{k,h,n}(T_k)$ through a new simulation with a larger number of paths and with the approximated exercise function given by f.

Andersen (1999) proposed as possible examples of approximate early-exercise function f two possibilities. First, one can set

$$\mathcal{I}(T_l) = 1\{\mathbf{PS}_{l,n}(T_l) > H(T_l)\}.$$

With this choice we say that early exercise will depend on the longest nested European swaption exceeding a level H. A second possibility is setting

$$\mathcal{I}(T_l) = 1\{\mathbf{PS}_{l,n}(T_l) > H(T_l) \text{ and } \max_{p=l+1,\ldots,h} \mathbf{PS}_{p,n}(T_l) \leq \mathbf{PS}_{l,n}(T_l)\}.$$

This choice is more refined than the previous one and amounts to adding the requirement that all the other nested future European swaptions, when valued at T_l, have a lower value than the current longest one. This intuitively amounts to saying that, in the context of European swaptions evaluated now, the most convenient is the current longest one. Then, as before, the option is to be exercised if this longest swaption exceeds a level $H(T_l)$.

The second choice is more refined but also more computationally demanding. Indeed, with the first choice, f depends only on the present value per basis point C and on the underlying swap rate (both defining the relevant European swaption), so that backward induction concerns only these two variables and memory requirements are not a problem.

Andersen (1999) also made several considerations on the possible computational efficiency of the method and on low memory requirements. The first Monte Carlo simulation involved in steps 1)-3) usually requires a low number of paths, whereas the evaluation in step 4) requires usually a higher

10.13 LFM: Pricing Bermudan Swaptions 419

number of scenarios. For other considerations and numerical results, see Andersen (1999). We also mention that Pedersen (1999), among several other issues, considers a comparison of the Andersen method with the Longstaff and Schwartz Monte Carlo method summarized in Section 10.13.1.

11. Pricing Derivatives on Two Interest-Rate Curves

So curiosity is in a sense the heart of practice.
Charlotte Joko Beck, "Nothing Special: Living Zen", 1995, HarperCollins.

In this chapter, we explain how one can model both a first (domestic) and a second (foreign) interest-rate curve, each by a two-factor additive Gaussian short-rate model, in order to Monte Carlo price a quanto constant-maturity swap and similar contracts, which we will present in the following sections.

We need to compute the payoff of a contract of this kind: party A pays party B a certain rate associated with a first currency "1", say Euro to fix ideas. Party B pays A the currency "1" amount expressed by a certain rate associated with a second currency "2", say British Pounds (GBP). To compute the expectations involved in pricing this kind of contracts, a good strategy can be to use a short-rate model with analytical formulas for zero-coupon bond prices in terms of the factors concurring to the short-rate. In such cases, we carry out the different rates involved in the contract in terms of the discount factors, so that the above expectations are easily computed once the distribution of the short-rate factors is known.

At the same time, the chosen model has to allow for a realistic volatility structure for each curve, and also correlation between different rates has to be modeled realistically, since usually swap rates are involved. A good choice, therefore, can be given by selecting a two-factor Gaussian short-rate model for each curve.

11.1 The Attractive Features of G2++ for Multi-Curve Payoffs

11.1.1 The Model

The basic model we adopt for each curve is described in detail in Chapter 4. Here, we briefly reintroduce it, and slightly adapt the notation to the present context.

We assume that the dynamics of the instantaneous short-rate process for curve "1", under the associated risk-adjusted measure Q_1, is given by

$$\text{Curve ``1'' (Euro): } r_1(t) = x_1(t) + y_1(t) + \varphi_1(t), \quad r_1(0) = r_0^1, \quad (11.1)$$

11. Pricing Derivatives on Two Interest-Rate Curves

where the processes $\{x_1(t) : t \geq 0\}$ and $\{y_1(t) : t \geq 0\}$ satisfy

$$dx_1(t) = -a_1 x_1(t)dt + \sigma_1 dW_1^x(t), \quad x_1(0) = 0, \qquad (11.2)$$
$$dy_1(t) = -b_1 y_1(t)dt + \eta_1 dW_1^y(t), \quad y_1(0) = 0,$$

where W_1^x and W_1^y are two correlated Brownian motions, $dW_1^x(t)dW_1^y(t) = \rho_1 dt$, and r_0^1, a_1, b_1, σ_1, η_1 and ρ_1 are suitable constants. The function φ_1 is deterministic and well defined in the time interval $[0, T_n]$, with T_n a given time horizon, for example 10, 30 or 50 (years). In particular, $\varphi_1(0) = r_0^1$. We denote by $\mathcal{F}_t^1$ the sigma-field generated by the pair (x_1, y_1) up to time t.

Simple integration of equations (11.2) implies that for each $s < t$

$$r_1(t) = x_1(s)e^{-a_1(t-s)} + y_1(s)e^{-b_1(t-s)} + \sigma_1 \int_s^t e^{-a_1(t-u)} dW_1^x(u)$$
$$+ \eta_1 \int_s^t e^{-b_1(t-u)} dW_1^y(u) + \varphi_1(t),$$

meaning that $r_1(t)$ conditional on $\mathcal{F}_s^1$ is normally distributed with mean and variance given respectively by

$$E\{r_1(t)|\mathcal{F}_s^1\} = x_1(s)e^{-a_1(t-s)} + y_1(s)e^{-b_1(t-s)} + \varphi_1(t),$$
$$\text{Var}\{r_1(t)|\mathcal{F}_s^1\} = \frac{\sigma_1^2}{2a_1}\left[1 - e^{-2a_1(t-s)}\right] + \frac{\eta_1^2}{2b_1}\left[1 - e^{-2b_1(t-s)}\right] \qquad (11.3)$$
$$+ 2\rho_1 \frac{\sigma_1 \eta_1}{a_1 + b_1}\left[1 - e^{-(a_1+b_1)(t-s)}\right].$$

This was already noticed in (4.6) of Chapter 4.

The second curve is modeled analogously. We in fact assume that the dynamics of the instantaneous short-rate process for curve "2", under the associated risk-adjusted measure Q_2, is given by

Curve "2" (GBP): $r_2(t) = x_2(t) + y_2(t) + \varphi_2(t), \quad r_2(0) = r_0^2, \qquad (11.4)$

where the processes $\{x_2(t) : t \geq 0\}$ and $\{y_2(t) : t \geq 0\}$ satisfy

$$dx_2(t) = -a_2 x_2(t)dt + \sigma_2 dW_2^x(t), \quad x_2(0) = 0, \qquad (11.5)$$
$$dy_2(t) = -b_2 y_2(t)dt + \eta_2 dW_2^y(t), \quad y_2(0) = 0,$$

where W_2^x and W_2^y are two correlated Brownian motions, $dW_2^x(t)dW_2^y(t) = \rho_2 dt$, and r_0^2, a_2, b_2, σ_2, η_2 and ρ_2 are suitable constants. The function φ_2 is deterministic and well defined in the time interval $[0, T_n]$. In particular, $\varphi_2(0) = r_0^2$. We denote by $\mathcal{F}_t^2$ the sigma-field generated by the pair (x_2, y_2) up to time t. The explicit expression and transition densities of r_2 are completely analogous to those for curve "1". It is indeed sufficient to replace subscripts and superscripts "1" with "2".

11.1 The Attractive Features of G2++ for Multi-Curve Payoffs

Now consider the market instantaneous forward rates for the two curves "1" and "2", respectively, at the initial time 0:

$$f_1^M(0,T) = -\frac{\partial \ln P_1^M(0,T)}{\partial T},$$

$$f_2^M(0,T) = -\frac{\partial \ln P_2^M(0,T)}{\partial T},$$

where the superscript "M" denotes financial quantities as observed in the market.

It is shown in Section 4.2.2 that by choosing

$$\varphi_1(T) = f_1^M(0,T) + \frac{\sigma_1^2}{2a_1^2}\left(1-e^{-a_1 T}\right)^2 + \frac{\eta_1^2}{2b_1^2}\left(1-e^{-b_1 T}\right)^2$$
$$+ \rho_1 \frac{\sigma_1 \eta_1}{a_1 b_1}\left(1-e^{-a_1 T}\right)\left(1-e^{-b_1 T}\right),$$

$$\varphi_2(T) = f_2^M(0,T) + \frac{\sigma_2^2}{2a_2^2}\left(1-e^{-a_2 T}\right)^2 + \frac{\eta_2^2}{2b_2^2}\left(1-e^{-b_2 T}\right)^2$$
$$+ \rho_2 \frac{\sigma_2 \eta_2}{a_2 b_2}\left(1-e^{-a_2 T}\right)\left(1-e^{-b_2 T}\right),$$

as in (4.12) of Chapter 4, the initial term structures of discount factors $T \mapsto P_1^M(0,T)$ and $T \mapsto P_2^M(0,T)$ for the curves "1" and "2" are perfectly reproduced by the above models for r_1 and r_2, respectively.

An equivalent condition is (we just write it for curve "1")

$$\exp\left\{-\int_t^T \varphi_1(u)du\right\} = \frac{P_1^M(0,T)}{P_1^M(0,t)}\exp\left\{-\frac{1}{2}[V_1(0,T) - V_1(0,t)]\right\}, \quad (11.6)$$

where in general

$$V_1(t,T) = \frac{\sigma_1^2}{a_1^2}\left[T-t+\frac{2}{a_1}e^{-a_1(T-t)} - \frac{1}{2a_1}e^{-2a_1(T-t)} - \frac{3}{2a_1}\right]$$
$$+ \frac{\eta_1^2}{b_1^2}\left[T-t+\frac{2}{b_1}e^{-b_1(T-t)} - \frac{1}{2b_1}e^{-2b_1(T-t)} - \frac{3}{2b_1}\right]$$
$$+ 2\rho_1 \frac{\sigma_1 \eta_1}{a_1 b_1}\left[T-t+\frac{e^{-a_1(T-t)}-1}{a_1} + \frac{e^{-b_1(T-t)}-1}{b_1}\right. \quad (11.7)$$
$$\left. - \frac{e^{-(a_1+b_1)(T-t)}-1}{a_1+b_1}\right],$$

as in (4.10) of Chapter 4.

The corresponding condition and definition for curve "2" are analogous, with the subscript "2" replacing the subscript "1".

11.1.2 Interaction Between Models of the Two Curves "1" and "2"

So far we have worked on the models for the curves "1" (Euro) and "2" (GBP) separately. We now consider quantities describing the interaction of the two curves. We assume the following instantaneous correlations between the factors of the two curves:

$$d \begin{bmatrix} W_1^x \\ W_1^y \\ W_2^x \\ W_2^y \end{bmatrix} d \begin{bmatrix} W_1^x & W_1^y & W_2^x & W_2^y \end{bmatrix} = \begin{bmatrix} 1 & \rho_1 & \gamma_{x1,x2} & \gamma_{x1,y2} \\ \cdot & 1 & \gamma_{y1,x2} & \gamma_{y1,y2} \\ \cdot & \cdot & 1 & \rho_2 \\ \cdot & \cdot & \cdot & 1 \end{bmatrix} dt$$

where the entries that are not specified are determined by symmetry.

We are therefore assuming that:

- The instantaneous correlation between shocks in the first factor of the first curve and the second factor of the first curve is the previously introduced ρ_1.
- The instantaneous correlation between shocks in the first factor of the first curve and the first factor of the second curve is the new parameter $\gamma_{x1,x2}$.
- The instantaneous correlation between shocks in the first factor of the first curve and the second factor of the second curve is the new parameter $\gamma_{x1,y2}$.
- The instantaneous correlation between shocks in the second factor of the first curve and the second factor of the second curve is the new parameter $\gamma_{y1,y2}$.
- The instantaneous correlation between shocks in the second factor of the first curve and the first factor of the second curve is the new parameter $\gamma_{y1,x2}$.
- The instantaneous correlation between shocks in the first factor of the second curve and the second factor of the second curve is the previously introduced ρ_2.

However, a trader may find it difficult to express views on correlations between single factors. Indeed, it would be preferable to express views on the instantaneous correlations between the two rates r_1 and r_2 themselves. This can be obtained by observing that

$$\text{Corr}\{dr_1, dr_2\} = \frac{\sigma_1 \sigma_2 \gamma_{x1,x2} + \eta_1 \sigma_2 \gamma_{y1,x2} + \sigma_1 \eta_2 \gamma_{x1,y2} + \eta_1 \eta_2 \gamma_{y1,y2}}{\sqrt{\sigma_1^2 + \eta_1^2 + 2\rho_1 \sigma_1 \eta_1} \sqrt{\sigma_2^2 + \eta_2^2 + 2\rho_2 \sigma_2 \eta_2}}.$$
(11.8)

Now notice that the parameters σ_1, η_1, ρ_1 can be determined by calibration of r_1 to the cap or swaption markets related to curve "1" (Euro), whereas σ_2, η_2, ρ_2 can be determined by calibration of r_2 to the cap or swaption markets related to curve "2" (GBP). At this point, after these two separate calibrations, the trader may be willing to express a view on the instantaneous correlation $\text{Corr}\{dr_1, dr_2\}$ between the two curves. However, we still

11.1 The Attractive Features of G2++ for Multi-Curve Payoffs

have four unknowns $\gamma_{x,y}$ and just one equation. A simplifying assumption that can be made is that

$$\gamma_{x1,x2} = \gamma_{x1,y2} = \gamma_{y1,x2} = \gamma_{y1,y2} =: \gamma.$$

Now, when the trader expresses her view on $\text{Corr}\{dr_1, dr_2\}$, equation (11.8) can be solved in the unique unknown γ:

$$\gamma = \text{Corr}\{dr_1, dr_2\} \frac{\sqrt{\sigma_1^2 + \eta_1^2 + 2\rho_1\sigma_1\eta_1} \sqrt{\sigma_2^2 + \eta_2^2 + 2\rho_2\sigma_2\eta_2}}{(\sigma_1 + \eta_1)(\sigma_2 + \eta_2)}. \quad (11.9)$$

It is immediate to see that γ is always a number whose absolute value is smaller than one, and therefore a "legal" correlation parameter:

$$-1 \le -|\text{Corr}\{dr_1, dr_2\}| \le \gamma \le |\text{Corr}\{dr_1, dr_2\}| \le 1.$$

Notice that extreme cases are allowed: $\gamma = 0$ translates the view of no instantaneous correlation between curves "1" and "2", $\text{Corr}\{dr_1, dr_2\} = 0$, whereas perfect positive or negative correlations, $\text{Corr}\{dr_1, dr_2\} = \pm 1$, can be attained respectively with

$$\gamma = \pm \frac{\sqrt{\sigma_1^2 + \eta_1^2 + 2\rho_1\sigma_1\eta_1} \sqrt{\sigma_2^2 + \eta_2^2 + 2\rho_2\sigma_2\eta_2}}{(\sigma_1 + \eta_1)(\sigma_2 + \eta_2)}.$$

We remind that the above dynamics for x and y are given under the respective risk-neutral measures: x_1 and y_1 are modeled under the risk-neutral measure Q_1 for curve "1", whereas x_2 and y_2 are modeled under the risk-neutral measure Q_2 for curve "2".

11.1.3 The Two-Models Dynamics under a Unique Convenient Forward Measure

We need to express the equations for the factors x_2 and y_2 describing curve "2" (GBP) under the risk-neutral measure Q_1, related to curve "1" (Euro). This in turn requires modeling the exchange rate between markets "1" and "2" (GBP). Let $X(t)$ denote the amount of currency "2" (GBP) needed to buy one unit of currency "1" (Euro). We assume the following no-arbitrage dynamics for X under Q_2:

$$dX(t) = ((r_2(t) - r_1(t))X(t)\,dt + \nu X(t)\,dW^X(t).$$

We need also to model the instantaneous correlations between the exchange rate X and the two factors of curve "2" (GBP), x_2 and y_2. We assume

$$d \begin{bmatrix} W^X \\ W_2^x \\ W_2^y \end{bmatrix} d\begin{bmatrix} W^X & W_2^x & W_2^y \end{bmatrix} = \begin{bmatrix} 1 & c_{x2,X} & c_{y2,X} \\ . & 1 & \rho_2 \\ . & . & 1 \end{bmatrix} dt$$

where the entries that are not specified are determined by symmetry. We are therefore assuming that:

- The instantaneous correlation between shocks in the first factor of the second curve "2" and the exchange rate between markets "1" and "2" is the new parameter $c_{x2,X}$.
- The instantaneous correlation between shocks in the second factor of the second curve "2" and the exchange rate between markets "1" and "2" is the new parameter $c_{y2,X}$.
- The instantaneous correlation between shocks in the first factor of the second curve and the second factor of the second curve is the previously introduced ρ_2.

The trader may find it difficult to express views or to estimate correlation between single factors of curve "2" and the exchange rate. What is reasonable to expect, instead, is some view on the correlation between the instantaneous rate of curve "2" as a whole and the exchange rate. Notice that such a correlation is given by

$$\mathrm{Corr}\{dX, dr_2\} = \frac{\sigma_2 c_{x2,X} + \eta_2 c_{y2,X}}{\sqrt{\sigma_2^2 + \eta_2^2 + 2\rho_2 \sigma_2 \eta_2}}, \qquad (11.10)$$

so that by inverting this formula one can translate the trader's view into model parameters. However, equation (11.10) cannot be inverted as it stands, since there are two unknowns. A simplifying assumption that can be made at this point is

$$c_{x2,X} = c_{y2,X} =: c_X .$$

Following this assumption (11.10) can now be inverted so as to yield

$$c_X = \mathrm{Corr}\{dX, dr_2\} \frac{\sqrt{\sigma_2^2 + \eta_2^2 + 2\rho_2 \sigma_2 \eta_2}}{\sigma_2 + \eta_2}, \qquad (11.11)$$

so that, by means of this formula, one can translate the trader's view into the model parameter c_X. Notice that the fraction in the above formula is a positive number smaller or equal than one. As a consequence,

$$-1 \leq -|\mathrm{Corr}\{dX, dr_2\}| \leq c_X \leq |\mathrm{Corr}\{dX, dr_2\}| \leq 1,$$

i.e. c_X is always a viable correlation. Notice also that extreme cases are allowed: $c_X = 0$ translates the view of no instantaneous correlation between curve "2" and the exchange rate, $\mathrm{Corr}\{dX, dr_2\} = 0$, whereas perfect positive or negative correlation, $\mathrm{Corr}\{dX, dr_2\} = \pm 1$, can be attained respectively with

$$c_X = \pm\sqrt{\sigma_2^2 + \eta_2^2 + 2\rho_2 \sigma_2 \eta_2}/(\sigma_2 + \eta_2).$$

We can now express the equations for the factors x_2 and y_2, describing curve "2" (GBP) under the risk-neutral measure Q_1 associated with market "1". We use the change-of-numeraire technique and move from the bank-account numeraire for market "2" to the numeraire

$$X(t) \times (\text{bank account of market ``1''}).$$

This is the right change of numeraire for moving from measure Q_2 to measure Q_1 as explained in Section 2.9 on numeraire changes between domestic and foreign markets. Beware that, in Section 2.9, our exchange rate X was $1/Q$, while X was an asset price. Moreover B^f denoted the bank account of market "2".

We can apply the change-of-numeraire toolkit (as described in Section 2.3) obtaining

$$dx_2(t) = [-a_2 x_2(t) + \sigma_2 \nu c_{x2,X}] dt + \sigma_2 dU_2^x(t),$$
$$dy_2(t) = [-b_2 y_2(t) + \eta_2 \nu c_{y2,X}] dt + \eta_2 dU_2^y(t), \qquad (11.12)$$

where U_2^x and U_2^y are Brownian motions under Q_1, whose correlation structure is the same as that of W_2^x and W_2^y.

11.2 Quanto Constant-Maturity Swaps

A quanto CMS is a financial product involving both multi-currency issues and unnatural time lags for rate payments. Therefore, we will develop this example to a certain degree, up to a scheme for implementing the G2++ model for quanto CMS.

11.2.1 Quanto CMS: The Contract

Consider the following contract. Party A pays to party B an amount in a first currency "1" (Euro) expressed by the c-year swap rate associated with a second currency "2" (GBP). Payments occur once every δ years (typically $\delta = 0.5$ years $= 6$ months). Notice that here we slightly generalize the CMS case of Section 10.6 to payment intervals different from one year. On the same payment dates, party B pays to party A the δ-year simply-compounded LIBOR rate associated with the currency "1". Formally, the payoff can be expressed as follows.

We assume unit nominal amount. Let us assume the current time to be $t = 0$, and let us denote by $\mathcal{T} = \{T_1, \ldots, T_n\}$ the set of payment dates at which the flows of the two legs are to be exchanged. We denote by τ_i the year fraction between T_{i-1} and T_i, which will be close in general to the single value δ. We take $T_0 = 0$.

- At time T_i, $i \geq 1$, party A pays to B, in currency "1", the amount expressed by the c-year swap rate $S_{i-1,i-1+c/\delta}(T_{i-1})$ for the curve "2" as reset in time T_{i-1}:

$$S_{i-1,i-1+c/\delta}(T_{i-1})\tau_i \text{ units of currency ``1''}.$$

In general, the (forward) swap rate at time $t \leq T_{j-1}$ for a swap whose payments occur at times $T_j, T_{j+1}, \ldots, T_{m-1}$ and with first reset date at T_{j-1} is defined as

$$S_{j-1,m-1}(t) = \frac{P_2(t, T_{j-1}) - P_2(t, T_{m-1})}{\sum_{k=j}^{m-1} \tau_k P_2(t, T_k)} \qquad (11.13)$$

(where indices are shifted by one for later notation convenience). In general, $P_2(t, T)$ denotes the discount factor for the curve "2" at time t for maturity T. An analogous definition is given for curve "1".

- At time T_i, $i \geq 1$, Institution B pays to A the LIBOR rate for curve "1":

$$\tau_i L_1(T_{i-1}, T_i) \text{ units of currency "1"}.$$

The quantity

$$L_1(T_{i-1}, T_i) = \frac{1}{\tau_i} \left(\frac{1}{P_1(T_{i-1}, T_i)} - 1 \right)$$

denotes the LIBOR rate at time T_{i-1} for maturity T_i for curve "1".

The net value of the contract as seen by B at time 0 is

$$E_1 \left\{ \sum_{i=1}^{n} \exp\left(-\int_0^{T_i} r_1(s)\, ds\right) [S_{i-1,i-1+c/\delta}(T_{i-1}) - L_1(T_{i-1}, T_i)]\tau_i \right\}$$

$$= P_1(0, T_n) \sum_{i=1}^{n} \tau_i E_1^n \left\{ \frac{S_{i-1,i-1+c/\delta}(T_{i-1})}{P_1(T_i, T_n)} \right\} - (1 - P_1(0, T_n)), \qquad (11.14)$$

where E_1 denotes the risk-neutral expectation for curve "1", and E_1^i is in general expectation with respect to the T_i-forward measure $Q_1^{T_i}$ for curve "1".

We need to compute

$$E_1^n[S_{i-1,i-1+c/\delta}(T_{i-1})/P_1(T_i, T_n)] \qquad (11.15)$$

for all i.

To this purpose, we remember the considerations made in the introduction of this chapter. We can use a short-rate model with analytical formulas for zero-coupon bond prices in terms of the factors concurring to the short rate. In such a case, (11.13) provides an analytical formula for swap rates in terms of such factors, and it is enough to simulate paths for the factors of the short rates up to T_n under the measure Q_1^n. If such a dynamics is linear, the transition densities are Gaussian. The transition of the factors of the short rate between two instants T_{i-1}, T_i can thus be simulated one-shot from the such Gaussian densities, without resorting to a time-discretization of the dynamics between T_{i-1} and T_i.

At the same time, the chosen model has to allow for a realistic volatility structure for each curve, and moreover correlation between different rates has to be modeled in curve "2", since swap rates are involved. A good choice can therefore be given by selecting a two-factor Gaussian short rate model for each curve.

11.2.2 Quanto CMS: The G2++ Model

Starting from the models described in Section 11.1, we deduce the analytical formulas needed for the quanto-CMS payoff.

The model allows for the following bond-price formula for curve "1":

$$P_1(t,T;x_1(t),y_1(t)) = \frac{P_1^M(0,T)}{P_1^M(0,t)} \exp\left\{\frac{1}{2}[V_1(t,T) - V_1(0,T) + V_1(0,t)] - \frac{1-e^{-a_1(T-t)}}{a_1}x_1(t) - \frac{1-e^{-b_1(T-t)}}{b_1}y_1(t)\right\},$$

where $V_1(t,T)$ is defined in (11.7) (see also (4.14) in Chapter 4).

The corresponding bond-price formula for curve "2" is analogous, with the subscript "2" replacing the subscript "1".

Swap rates can be computed analytically in terms of x and y from the above formula for discount factors. For the (forward) swap rate of curve "2" we can rewrite (11.13) in terms of model quantities:

$$S_{j-1,m-1}(t;x_2(t),y_2(t)) = \frac{P_2(t,T_{j-1};x_2(t),y_2(t)) - P_2(t,T_{m-1};x_2(t),y_2(t))}{\sum_{k=j}^{m-1}\tau_k P_2(t,T_k;x_2(t),y_2(t))}.$$
(11.16)

In order to compute (11.15), we need the dynamics of x_1, y_1, x_2 and y_2 under the T_n-forward measure Q_1^n for curve "1" (Euro).

To this purpose, recall that in Section 11.1.3 we derived the equations for the factors x_2 and y_2 of curve "2" (GBP) under the risk-neutral measure Q_1 associated with market "1" (Euro). After the exchange rate between markets "1" and "2", we modeled the instantaneous correlations between the exchange rate "X" and the two factors, x_2 and y_2, of curve "2" (GBP), thus obtaining the dynamics (11.12).

Consider now the dynamics for x_1, y_1, x_2 and y_2, as from (11.2) and (11.12). A second change of numeraire for market "1" can now be performed, moving from the (market-"1") bank-account to the (market-"1") T_n-maturity bond price. We set $T_n = T$. By applying again the change-of-numeraire toolkit, and taking into account the equivalence shown in Section 2.9 or at the end of Section 11.1.3, we obtain the following dynamics:

$$dx_1(t) = \left[-a_1 x_1(t) - \frac{\sigma_1^2}{a_1}(1 - e^{-a_1(T-t)}) - \rho_1 \frac{\sigma_1 \eta_1}{b_1}(1 - e^{-b_1(T-t)})\right] dt$$
$$+ \sigma_1 dZ_1^x(t),$$

11. Pricing Derivatives on Two Interest-Rate Curves

$$dy_1(t) = \left[-b_1 y_1(t) - \frac{\eta_1^2}{b_1}(1 - e^{-b_1(T-t)}) - \rho_1 \frac{\sigma_1 \eta_1}{a_1}(1 - e^{-a_1(T-t)}) \right] dt$$
$$+ \eta_1 dZ_1^y(t),$$

$$dx_2(t) = \left[-a_2 x_2(t) - \sigma_2 \left(\gamma_{x1,x2} \frac{\sigma_1}{a_1}(1 - e^{-a_1(T-t)}) \right. \right.$$
$$\left. \left. + \gamma_{x2,y1} \frac{\eta_1}{b_1}(1 - e^{-b_1(T-t)}) - \nu c_{x2,X} \right) \right] dt + \sigma_2 dZ_2^x(t),$$

$$dy_2(t) = \left[-b_2 y_2(t) - \eta_2 \left(\gamma_{y1,y2} \frac{\eta_1}{b_1}(1 - e^{-b_1(T-t)}) \right. \right.$$
$$\left. \left. + \gamma_{x1,y2} \frac{\sigma_1}{a_1}(1 - e^{-a_1(T-t)}) - \nu c_{y2,X} \right) \right] dt + \eta_2 dZ_2^y(t),$$

where the Z's are Brownian motions under Q_1^n with the same correlation structure as the W's.

By integrating the four-dimensional Gaussian process above and using a four-dimensional version of Ito's isometry, one obtains the exact transition density as follows:

$$\begin{aligned} x_1(t) &= e^{-a_1(t-s)} x_1(s) - M_{x1}^T(s,t) + N_1(t-s), \\ y_1(t) &= e^{-b_1(t-s)} y_1(s) - M_{y1}^T(s,t) + N_2(t-s), \\ x_2(t) &= e^{-a_2(t-s)} x_2(s) - M_{x2}^T(s,t) + N_3(t-s), \\ y_2(t) &= e^{-b_2(t-s)} y_2(s) - M_{y2}^T(s,t) + N_4(t-s), \end{aligned} \tag{11.17}$$

for $s \leq t \leq T$, where

$$M_{x1}^T(s,t) = \left(\frac{\sigma_1^2}{a_1^2} + \rho_1 \frac{\sigma_1 \eta_1}{a_1 b_1} \right) \left[1 - e^{-a_1(t-s)} \right]$$
$$- \frac{\sigma_1^2}{2 a_1^2} \left[e^{-a_1(T-t)} - e^{-a_1(T+t-2s)} \right]$$
$$- \frac{\rho_1 \sigma_1 \eta_1}{b_1(a_1 + b_1)} \left[e^{-b_1(T-t)} - e^{-b_1 T - a_1 t + (a_1 + b_1)s} \right],$$

$$M_{y1}^T(s,t) = \left(\frac{\eta_1^2}{b_1^2} + \rho_1 \frac{\sigma_1 \eta_1}{a_1 b_1} \right) \left[1 - e^{-b_1(t-s)} \right]$$
$$- \frac{\eta_1^2}{2 b_1^2} \left[e^{-b_1(T-t)} - e^{-b_1(T+t-2s)} \right]$$
$$- \frac{\rho_1 \sigma_1 \eta_1}{a_1(a_1 + b_1)} \left[e^{-a_1(T-t)} - e^{-a_1 T - b_1 t + (a_1 + b_1)s} \right],$$

and

11.2 Quanto Constant-Maturity Swaps

$$M_{x2}^T(s,t) = \frac{\sigma_1\sigma_2\gamma_{x1,x2}}{a_1}\left(\frac{1-e^{-a_2(t-s)}}{a_2} - \frac{e^{-a_1(T-t)}-e^{-a_1(T-s)-a_2(t-s)}}{a_1+a_2}\right)$$

$$+\frac{\eta_1\sigma_2\gamma_{x2,y1}}{b_1}\left(\frac{1-e^{-a_2(t-s)}}{a_2} - \frac{e^{-b_1(T-t)}-e^{-b_1(T-s)-a_2(t-s)}}{b_1+a_2}\right)$$

$$-\sigma_2\nu c_{X,x2}\frac{1-e^{-a_2(t-s)}}{a_2}$$

$$M_{y2}^T(s,t) = \frac{\eta_1\eta_2\gamma_{y1,y2}}{b_1}\left(\frac{1-e^{-b_2(t-s)}}{b_2} - \frac{e^{-b_1(T-t)}-e^{-b_1(T-s)-b_2(t-s)}}{b_1+b_2}\right)$$

$$+\frac{\sigma_1\eta_2\gamma_{y2,x1}}{a_1}\left(\frac{1-e^{-b_2(t-s)}}{b_2} - \frac{e^{-a_1(T-t)}-e^{-a_1(T-s)-b_2(t-s)}}{a_1+b_2}\right)$$

$$-\eta_2\nu c_{X,y2}\frac{1-e^{-b_2(t-s)}}{b_2}$$

and where, finally, $N(t-s)$ is a four-dimensional Gaussian random vector with zero mean and covariance matrix $C(t-s)$ given by

$$\begin{bmatrix} \sigma_1^2\frac{1-e^{-2a_1(t-s)}}{2a_1} & & & \\ \sigma_1\eta_1\frac{1-e^{-(a_1+b_1)(t-s)}}{a_1+b_1}\rho_1 & \eta_1^2\frac{1-e^{-2b_1(t-s)}}{2b_1} & & \\ \sigma_1\sigma_2\frac{1-e^{-(a_1+a_2)(t-s)}}{a_1+a_2}\gamma_{x1,x2} & \eta_1\sigma_2\frac{1-e^{-(b_1+a_2)(t-s)}}{b_1+a_2}\gamma_{y1,x2} & (***) & \\ \sigma_1\eta_2\frac{1-e^{-(a_1+b_2)(t-s)}}{a_1+b_2}\gamma_{x1,y2} & \eta_1\eta_2\frac{1-e^{-(b_1+b_2)(t-s)}}{b_1+b_2}\gamma_{y1,y2} & (**) & (*) \end{bmatrix}$$

$$(***) = \sigma_2^2\frac{1-e^{-2a_2(t-s)}}{2a_2},$$

$$(**) = \eta_2\sigma_2\frac{1-e^{-(b_2+a_2)(t-s)}}{b_2+a_2}\rho_2, \quad (*) = \eta_2^2\frac{1-e^{-2b_2(t-s)}}{2b_2}.$$

Quanto CMS: Monte Carlo Pricing and Examples. We now explain how one can price the contract described in Subsection 11.2.1. Recall that the net value of the contract as seen by B at time 0 is

$$E_1\left\{\sum_{i=1}^n \exp\left(-\int_0^{T_i} r_1(s)\,ds\right)[S_{i-1,i-1+c/\delta}(T_{i-1}) - L_1(T_{i-1},T_i)]\tau_i\right\}$$

$$= P_1^M(0,T_n)\sum_{i=1}^n \tau_i\, E_1^n\left[\frac{S_{i-1,i-1+c/\delta}(T_{i-1};x_2(T_{i-1}),y_2(T_{i-1}))}{P_1(T_i,T_n;x_1(T_i),y_1(T_i))}\right]$$

$$-(1-P_1^M(0,T_n)). \tag{11.18}$$

We need therefore to compute

$$E_1^n\left[\frac{S_{i-1,i-1+c/\delta}(T_{i-1};x_2(T_{i-1}),y_2(T_{i-1}))}{P_1(T_i,T_n;x_1(T_i),y_1(T_i))}\right] \tag{11.19}$$

for all i's and then substitute back in the above formula. Proceed as follows.

11. Pricing Derivatives on Two Interest-Rate Curves

Inputs We first describe the inputs necessary for the evaluation of the contract.

1. The number np of scenarios for the Monte Carlo evaluation.
2. The initial curves "1" (Euro) and "2" (GBP) at time 0 (as curves of discount factors, possibly interpolated): $T \mapsto P_1^M(0,T)$ and $T \mapsto P_2^M(0,T)$.
3. The parameters $a_1, b_1, \sigma_1, \eta_1$ and ρ_1, obtained by calibrating model (11.1) to the caps or swaptions markets "1" (Euro).
4. The parameters $a_2, b_2, \sigma_2, \eta_2$ and ρ_2, obtained by calibrating model (11.4) to the caps or swaptions markets "2" (GBP).
5. The instantaneous correlation between the two curves, $\text{Corr}\{dr_1, dr_2\}$, from which the corresponding model parameter γ can be computed via (11.9).
6. The instantaneous correlation between curve "2" (GBP) and the exchange rate Euro/GBP, $\text{Corr}\{dX, dr_2\}$, from which the corresponding model parameter c_X can be computed via (11.11).
7. The percentage annualized volatility ν of the exchange rate Euro/GBP.

Scheme We can now give a schematic description of the use of the G2++ model for Monte Carlo pricing a quanto CMS, i.e. for computing (11.18).

1. Set the current time t to $t = T_0 = 0$. Accordingly, set $x_1(T_0) = y_1(T_0) = x_2(T_0) = y_2(T_0) = 0$. Set $i = 1$.
2. If $i \geq 2$, compute the following (curve-"2") discount factors at time $t = T_{i-1}$, in each scenario "p", by using formula (11.16) for curve "2":

$$P_2(T_{i-1}, T_i; x_2^p(T_{i-1}), y_2^p(T_{i-1})),$$
$$P_2(T_{i-1}, T_{i+1}; x_2^p(T_{i-1}), y_2^p(T_{i-1})),$$
$$\ldots, P_2(T_{i-1}, T_{i-1+c/\delta}; x_2^p(T_{i-1}), y_2^p(T_{i-1})).$$

Else, if $i = 1$, the above discount factors at time $T_{i-1} = 0$ are known as an input.

3. If $i \geq 2$, compute the model swap rate at time $t = T_{i-1}$ in each scenario "p", $S_{i-1,i-1+c/\delta}^p = S_{i-1,i-1+c/\delta}(T_{i-1}; x_2^p(T_{i-1}), y_2^p(T_{i-1}))$, via formula (11.16) with $j = i$, $m = i + c/\delta$ and $t = T_{i-1}$. Again, if $i = 1$, such swap rates at time $T_{i-1} = 0$ are known from the input.
4. Use formula (11.17) to generate np realizations

$$x_1^p(T_i), \ y_1^p(T_i), \ x_2^p(T_i), \ y_2^p(T_i), \quad p = 1, 2, \ldots, \text{np}$$

of $x_1(T_i), y_1(T_i), x_2(T_i), y_2(T_i)$ starting from the previously generated $x_1(T_{i-1}), y_1(T_{i-1}), x_2(T_{i-1}), y_2(T_{i-1})$. Formula (11.17) is to be applied with $s = T_{i-1}$, $t = T_i$, and with the np new realizations of $N_1, \ldots, N_4$ generated from a four-dimensional Gaussian variable with mean zero and covariance matrix $C(T_i - T_{i-1})$.

11.2 Quanto Constant-Maturity Swaps

5. Compute the T_n-discount factor for curve "1" at time $t = T_i$ in each scenario "p" by using formula (11.16):

$$P_1(T_i, T_n; x_1^p(T_i), y_1^p(T_i)).$$

6. Compute in each scenario "p" the ratio

$$\frac{S^p_{i-1,i-1+c/\delta}(T_{i-1})}{P_1(T_i, T_n; x_1^p(T_i), y_1^p(T_i))}$$

7. Average the above quantities over all scenarios. We obtain the Monte Carlo evaluation of (11.19). Store this value.
8. Increase i by one.
9. If $i \leq n$ then go back to step 2, otherwise
10. Compute the final price by adding all the terms computed and stored at step 7, according to formula (11.18).

The above scheme can be easily generalized to cases where optional features are added to either leg of the contract.

Analysis of a specific contract We now present a specific example that is based on a concrete problem we solved via the above method. Precisely, we consider the following contract, which is specified in two parts.

First Part. First, it is given a fixed leg switching, at a given future instant, to a "floored" CMS fraction, which is exchanged for a canonical floating leg. Both legs are based on rates from market "1". Formally, maintaining notation of Subsection 11.2.1,

- party A pays to B a fixed-rate payment

$$\tau_i R_F \text{ units of currency "1"}, \quad i = 1, 2, \ldots, f,$$

at instants T_i for $i \leq f$, and pays the "floored" fraction of the swap rate

$$\max[Y S^1_{i-1,i-1+c/\delta}(T_{i-1}), R_c]\tau_i \text{ units of currency "1"},$$

at later instants $T_{f+1}, \ldots, T_n$, where Y is a fraction-parameter to be determined. Notice that here S^1 is the swap rate for curve "1".
- At time T_i, $i \geq 1$, Institution B pays to A the LIBOR rate for curve "1":

$$\tau_i L_1(T_{i-1}, T_i) \text{ units of currency "1"}.$$

One needs to find the value of Y such that the present value of this first CMS-like contract is zero. The net value of the contract as seen by B at time 0 is

$$E_1\left(\sum_{i=1}^{f}\exp\left(-\int_0^{T_i}r_1(s)\,ds\right)(R_F-L_1(T_{i-1},T_i))\tau_i\right)$$

$$+E_1\left[\sum_{i=f+1}^{n}\exp\left(-\int_0^{T_i}r_1(s)\,ds\right)(\max[YS^1_{i-1,i-1+c/\delta}(T_{i-1}),R_c]\right.$$

$$\left.-L_1(T_{i-1},T_i))\tau_i\right]$$

$$=\sum_{i=1}^{f}R_F\tau_i P_1(0,T_i)+P_1(0,T_n)\sum_{i=f+1}^{n}\tau_i E_1^n\left[\frac{\max[YS^1_{i-1,i-1+c/\delta}(T_{i-1}),R_c]}{P_1(T_i,T_n)}\right]$$

$$-(1-P_1(0,T_n)).$$

Therefore, if such a value has to be zero, we need to solve the following nonlinear equation in Y:

$$\sum_{i=1}^{f}R_F\tau_i P_1(0,T_i)+P_1(0,T_n)\sum_{i=f+1}^{n}\tau_i E_1^n\left[\frac{\max[YS^1_{i-1,i-1+c/\delta}(T_{i-1}),R_c]}{P_1(T_i,T_n)}\right]$$

$$-(1-P_1(0,T_n))=0.$$

The difficulty is in computing the expected values. These can be evaluated through a Monte Carlo method, which will be iterated in order to solve for the unknown Y. Fortunately, the Gaussian G2++ model allows for a quick Monte Carlo evaluation, so as to render the task feasible.

Second Part. Now that the "fair" swap-fraction Y has been determined, we consider the following quanto "floored" CMS.

We are given the currency-"1" amount paid by a fixed leg switching, at a given future instant, to a "floored" CMS fraction for curve "2", which is exchanged for a canonical floating leg from curve "1". Formally, maintaining notation of Subsection 11.2.1,

- party A pays a fixed rate payment

$$\tau_i R_F \text{ units of currency "1"}, \quad i=1,2,\ldots,f,$$

at instants T_i for $i\leq f$, and pays the "floored" fraction of the swap rate

$$\max[YS_{i-1,i-1+c/\delta}(T_{i-1}),R_c]\tau_i \text{ units of currency "1"},$$

at later instants $T_{f+1},\ldots,T_n$, where Y is the fraction parameter determined before. Notice that here S is the swap rate for curve "2" (thus introducing the quanto feature of the contract).
- At time T_i, $i\geq 1$, Institution B pays to A the LIBOR rate for curve "1":

$$\tau_i L_1(T_{i-1},T_i) \text{ units of currency "1"}.$$

11.2 Quanto Constant-Maturity Swaps

The net value of the contract as seen by B at time 0 is

$$E_1\left(\sum_{i=1}^{f}\exp\left(-\int_0^{T_i}r_1(s)ds\right)(R_F - L_1(T_{i-1},T_i))\tau_i\right) \qquad (11.20)$$

$$+E_1\left[\sum_{i=f+1}^{n}\exp\left(-\int_0^{T_i}r_1(s)ds\right)(\max[YS_{i-1,i-1+c/\delta}(T_{i-1}),R_c]\right.$$

$$\left.-L_1(T_{i-1},T_i))\tau_i\right]$$

$$=\sum_{i=1}^{f}R_F\tau_iP_1(0,T_i) + P_1(0,T_n)\sum_{i=f+1}^{n}\tau_iE_1^n\left[\frac{\max[YS_{i-1,i-1+c/\delta}(T_{i-1}),R_c]}{P_1(T_i,T_n)}\right]$$

$$-(1-P_1(0,T_n)).$$

Again, the only difficulty is in evaluating the expected values. However, a Monte Carlo method and related scheme can be applied in a completely analogous way to the one presented earlier for standard quanto CMS.

11.2.3 Quanto CMS: Quanto Adjustment

Suppose we are again pricing a payoff that involves swap rates *in a foreign currency*. Denote quantities related to the foreign currency by the superscript or subscript "(2)". We need to compute

$$E[\tau_i D(0,T_i) S_{\alpha,\beta}^{(2)}(T_{i-k})] ,$$

where E denotes expectation under the domestic risk-neutral measure.

As in the original derivation of Black's formula for caps, we now assume that, in both markets, discounting occurs with deterministic rates, thus implicitly assuming equivalence between every T_i-forward measure and the risk-neutral measure for the same market.

Consider the foreign-market convexity-adjusted dynamics

$$dS_{\alpha,\beta}^{(2)}(t) = \mu_{(2)}^{\alpha,\beta}S_{\alpha,\beta}^{(2)}(t)\,dt + \sigma_{(2)}^{(\alpha,\beta)}(t)S_{\alpha,\beta}^{(2)}(t)\,dW_S^{(2)}(t).$$

This is basically the dynamics (10.13) rewritten for the foreign market, i.e. under Q_2. However, to evaluate the above expectation, we need the dynamics of the foreign swap rate under the *domestic* risk-neutral measure. This in turn requires modeling the exchange rate between the domestic and foreign markets.

Let $X(t)$ denote, as before, the amount of foreign currency needed to buy one unit of domestic currency. Assume the following no-arbitrage dynamics for X under Q_2:

436   11. Pricing Derivatives on Two Interest-Rate Curves

$$dX(t) = ((r_2(t) - r(t))X(t)dt + \nu X(t)dW_X^{(2)}(t).$$

We need also to model the instantaneous correlations between the exchange rate X and the foreign swap rate:

$$dW_S^{(2)}(t)dW_X^{(2)}(t) = cdt.$$

We use the change-of-numeraire technique, described in Section 2.3, and move from the bank-account numeraire for the foreign market (2) to the numeraire

$$X(t) \times \text{(bank account of domestic market)}.$$

This is the right change of numeraire for moving from measure Q_2 to the domestic measure Q, as explained in Section 2.9. We obtain the following dynamics under Q:

$$dS_{\alpha,\beta}^{(2)}(t) = \left[\mu_{(2)}^{\alpha,\beta} + \sigma_{(2)}^{(\alpha,\beta)}(t)\nu c\right] S_{\alpha,\beta}^{(2)}(t)dt + \sigma_{(2)}^{(\alpha,\beta)}(t) S_{\alpha,\beta}^{(2)}(t)dW_S(t).$$

With this last dynamics it is immediate to check that

$$E\left[\tau_i D(0,T_i) S_{\alpha,\beta}^{(2)}(T_{i-k})\right] = \tau_i P(0,T_i) E^i\left[S_{\alpha,\beta}^{(2)}(T_{i-k})\right]$$

$$\approx \exp\left[\nu c \int_0^{T_{i-k}} \sigma_{(2)}^{(\alpha,\beta)}(t)dt\right] \tau_i P(0,T_i)$$

$$\cdot \left[S_{\alpha,\beta}^{(2)}(0) - \tfrac{1}{2}(S_{\alpha,\beta}^{(2)}(0))^2 (v_{\alpha,\beta}^{(2)}(T_{i-k}))^2 \frac{\Psi''_{\alpha,\beta}(S_{\alpha,\beta}^{(2)}(0))}{\Psi'_{\alpha,\beta}(S_{\alpha,\beta}^{(2)}(0))}\right].$$

If the instantaneous volatility of the foreign swap rate is assumed to be constant,

$$\sigma_{(2)}^{(\alpha,\beta)}(t) = \sigma_{(2)}^{(\alpha,\beta)},$$

then the above formula reduces to

$$\approx \exp\left[\sigma_{(2)}^{(\alpha,\beta)} T_{i-k}\nu c\right] \tau_i P(0,T_i)$$

$$\cdot \left[S_{\alpha,\beta}^{(2)}(0) - \tfrac{1}{2}(S_{\alpha,\beta}^{(2)}(0))^2 (\sigma_{(2)}^{(\alpha,\beta)})^2 T_{i-k} \frac{\Psi''_{\alpha,\beta}(S_{\alpha,\beta}^{(2)}(0))}{\Psi'_{\alpha,\beta}(S_{\alpha,\beta}^{(2)}(0))}\right].$$

Notice that the quantity inside square brackets is the convexity-adjusted expectation for the foreign market alone. One can compute it as shown in Chapter 10 (for the domestic market), and subsequently multiply by the domestic-market discount factor $\tau_i P(0,T_i)$. The further correction due to the quanto feature is given by the factor

$$\exp\left[\sigma_{(2)}^{(\alpha,\beta)} T_{i-k}\nu c\right].$$

If either the correlation c or the volatilities are small, the effect of this correction is negligible.

11.3 Differential Swaps

We now introduce differential swaps, which are also called quanto swaps. In order to maintain this section as self-contained as possible, we will reintroduce some notation used in earlier sections. The main difference with quanto CMS, of which differential swaps are a particular case, is that a closed form formula is available within the G2++ model. A market-like closed-form formula is also available for differential swaps, and we will derive it in the later Section 11.4, in the context of general quanto derivatives.

11.3.1 The Contract

Consider the following contract. Party A pays to party B in a first (domestic) currency "1" (Euro) an amount expressed by the δ-simply-compounded LIBOR rate associated with a second (foreign) currency "2" (GBP). Payments occur once every δ years (typically $\delta = 0.5$ years = 6 months). On the same payment dates, party B pays to party A the δ-simply-compounded LIBOR rate associated with currency "1". Formally, the payoff can be expressed as follows.

We assume unit nominal amount. Let us assume the current time to be $t = 0$, and let us denote by $\mathcal{T} = \{T_1, \ldots, T_n\}$ a set of payment dates at which the flows of the two legs are to be exchanged. We denote by τ_i the year fraction between T_{i-1} and T_i, which will be close in general to the single value δ. We take $T_0 = 0$.

- At time T_i, $i \geq 1$, party A pays to B, in currency "1", the amount expressed by the δ-simply-compounded LIBOR rate for curve "2":

$$L_2(T_{i-1}, T_i)\tau_i \text{ units of currency "1"}.$$

where the quantity

$$L_2(T_{i-1}, T_i) = \frac{1}{\tau_i}\left(\frac{1}{P_2(T_{i-1}, T_i)} - 1\right) \qquad (11.21)$$

denotes the LIBOR rate at time T_{i-1} for maturity T_i for curve "2", and where in general $P_2(t, T)$ denotes the discount factor, for curve "2", at time t for maturity T. Analogous definitions of LIBOR rates and discount factors are given for curve "1".

- At time T_i, $i \geq 1$, institution B pays to A the LIBOR rate for curve "1" plus a spread K:

$$\tau_i(L_1(T_{i-1}, T_i) + K) \text{ units of currency "1"}.$$

The net value of the contract as seen by B at time 0 is

438   11. Pricing Derivatives on Two Interest-Rate Curves

$$E_1\left(\sum_{i=1}^{n}\exp\left(-\int_0^{T_i}r_1(s)ds\right)(L_2(T_{i-1},T_i)-L_1(T_{i-1},T_i)-K)\tau_i\right)$$

$$=\sum_{i=1}^{n}\tau_i P_1(0,T_i)[E_1^i(L_2(T_{i-1},T_i))-K]-(1-P_1(0,T_n)), \quad (11.22)$$

where E_1^i denotes the T_i-forward-measure expectation for curve "1". The related measure is denoted by Q_1^i.

We need to compute

$$E_1^i[L_2(T_{i-1},T_i)] \quad (11.23)$$

for all i's.

To this purpose, as in the case of a quanto CMS, a good strategy can be to use a short-rate model with analytical formulas for zero-coupon rates in terms of the factors concurring to the short rate. In such a case, the above expectations are easily computed once the distribution of the short-rate factors is known.

At the same time, the chosen model has to allow for a realistic volatility structure for each curve. A good choice can again be given by selecting the two-factor Gaussian G2++ short-rate model for each curve.

11.3.2 Differential Swaps with the G2++ Model

Starting from the models described in Subsection 11.1.1, we deduce the analytical formulas needed for the differential-swap payoff.

The model allows for the following bond-price formula for curve "2":

$$P_2(t,T;x_2(t),y_2(t))=\frac{P_2^M(0,T)}{P_2^M(0,t)}\exp\left\{\frac{1}{2}[V_2(t,T)-V_2(0,T)+V_2(0,t)]\right.$$
$$\left.-\frac{1-e^{-a_2(T-t)}}{a_2}x_2(t)-\frac{1-e^{-b_2(T-t)}}{b_2}y_2(t)\right\}(11.24)$$

where $V_2(t,T)$ is defined analogously to $V_1(t,T)$ given in (11.7).

We now determine the dynamics of x_2 and y_2 under Q_1^i.

As shown in Subsection 11.2.2, using the change-of-numeraire technique we can express the equations for the factors x_2 and y_2 describing curve "2" (GBP) under the measure Q_1^i associated with market "1", obtaining the transition equation between $s=0$ and any instant t, with $T=T_i$. In particular, we obtain

$$x_2(t)=-M_{x2}^T(0,t)+N_x(t), \quad (11.25)$$
$$y_2(t)=-M_{y2}^T(0,t)+N_y(t),$$

where $(N_x(t),N_y(t))$ is a two-dimensional Gaussian random vector with zero mean and covariance matrix $C(t)$ given by

11.3 Differential Swaps

$$\begin{bmatrix} \sigma_2^2 \frac{1-e^{-2a_2 t}}{2a_2} & \eta_2\sigma_2 \frac{1-e^{-(b_2+a_2)t}}{b_2+a_2}\rho_2 \\ \cdot & \eta_2^2 \frac{1-e^{-2b_2 t}}{2b_2} \end{bmatrix}.$$

It is now possible to compute the following expected value through formula (11.24) combined with the just derived distribution of $x_2(t)$ and $y_2(t)$, starting from $x_2(0)$ and $y_2(0)$, under Q_1^i:

$$E_1^i\left[\frac{1}{P_2(t,T;x_2(t);y_2(t))}\right] = \frac{P_2^M(0,t)}{P_2^M(0,T)}\exp\{-\tfrac{1}{2}[V_2(t,T)-V_2(0,T)+V_2(0,t)]\}$$
$$\cdot E_1^i\left\{\exp\left[\frac{1-e^{-a_2(T-t)}}{a_2}x_2(t)+\frac{1-e^{-b_2(T-t)}}{b_2}y_2(t)\right]\right\}.$$

Now use the joint distribution of $x_2(t)$ and $y_2(t)$ under Q_1^i as from (11.25) to obtain

$$\psi_2(t,T) := E_1^T\left[\frac{1}{P_2(t,T;x_2(t);y_2(t))}\right]$$
$$= \frac{P_2^M(0,t)}{P_2^M(0,T)}\exp\left\{-\frac{1}{2}[V_2(t,T)-V_2(0,T)+V_2(0,t)]\right.$$
$$-\frac{1-e^{-a_2(T-t)}}{a_2}M_{x2}^T(0,t)-\frac{1-e^{-b_2(T-t)}}{b_2}M_{y2}^T(0,t)$$
$$+\frac{1}{2}\left[\sigma_2^2\left(\frac{1-e^{-a_2(T-t)}}{a_2}\right)^2\frac{1-e^{-2a_2 t}}{2a_2}\right.$$
$$+2\eta_2\sigma_2\rho_2\frac{(1-e^{-a_2(T-t)})(1-e^{-b_2(T-t)})(1-e^{-(a_2+b_2)(T-t)})}{a_2b_2(a_2+b_2)}$$
$$\left.\left.+\eta_2^2\left(\frac{1-e^{-b_2(T-t)}}{b_2}\right)^2\frac{1-e^{-2b_2 t}}{2b_2}\right]\right\}$$

Therefore, by taking into account definition (11.21), we have that

$$E_1^T[L_2(t,T)] = \frac{1}{\tau}(\psi_2(t,T)-1),$$

where τ denotes the year fraction between t and T. We can then compute the contract value (11.22) as follows:

$$\sum_{i=1}^n \tau_i P_1(0,T_i)E_1^i[L_2(T_{i-1},T_i)] - (1-P_1(0,T_n))$$
$$= \sum_{i=1}^n P_1(0,T_i)(\psi_2(T_{i-1},T_i)-1-\tau_i K) - (1-P_1(0,T_n))$$

There is no need for a numerical scheme here, since we have found a closed-form formula. This formula requires the following inputs:

1. The initial bond prices $P_1(0, T_1), \ldots, P_1(0, T_n)$ for the zero curve "1" (Euro) and the analogous quantities for the initial zero curve "2" (GBP).
2. The parameters $a_2, b_2, \sigma_2, \eta_2$ and ρ_2, obtained by calibration of model (11.4, 11.5) to caps or swaptions in market "2" (GBP).
3. The instantaneous correlation between curve "1" (Euro) and curve "2" (GBP), $\text{Corr}\{dr_1, dr_2\}$, from which the corresponding model parameter γ can be computed via (11.9).
4. The instantaneous correlation between curve "2" (GBP) and the exchange rate Euro/GBP, $\text{Corr}\{dX, dr_2\}$, from which the corresponding model parameter c_X can be computed via (11.11).
5. The percentage annualized volatility ν of the exchange rate Euro/GBP.

11.3.3 A Market-Like Formula

In Subsection 11.4.3 we will present a market-like closed formula for differential swaps, as an easy consequence of the general formulas for quanto caps and floors.

11.4 Market Formulas for Basic Quanto Derivatives

This section is self-contained and can be read independently of the previous sections in this chapter. Here we derive the arbitrage-free price of some fundamental derivatives with multi-currency features. We start with the pricing of a quanto caplet by modeling the relevant quantities through lognormal martingales under a given forward measure. Such assumptions immediately lead to a Black-like pricing formula with coefficients that can be all expressed in terms of relevant financial quantities. Analogously, we also derive the arbitrage-free price of a quanto floorlet and, by extension, of a quanto cap, of a quanto floor and of a quanto swap.

11.4.1 The Pricing of Quanto Caplets/Floorlets

Given a domestic market and a foreign market, let us assume that the term structure of discount factors that are observed in the domestic and foreign markets at time t are respectively given by $T \mapsto P(t,T)$ and $T \mapsto P^f(t,T)$ for $T \geq t$. Let us denote by $\mathcal{X}(t)$ the exchange rate at time t between the currencies in the two markets, in that 1 unit of the foreign currency equals $\mathcal{X}(t)$ unit of the domestic currency. Notice that $\mathcal{X}(t)$ is the reciprocal of the exchange rate $X(t)$ defined in Subsection 11.1.3.

Given the future times T_1 and T_2, a quanto caplet pays off at time T_2

$$\left[F^f(T_1; T_1, T_2) - K\right]^+ \tau_{1,2} N \quad \text{in domestic currency,} \tag{11.26}$$

11.4 Market Formulas for Basic Quanto Derivatives

where N is the nominal value, K is the caplet rate (strike), $\tau_{1,2}$ is the year fraction between times T_1 and T_2 and $F^f(t;T_1,T_2)$ is the forward rate in the foreign market at time t for the interval $[T_1, T_2]$, i.e.,

$$F^f(t;T_1,T_2) = \frac{P^f(t,T_1) - P^f(t,T_2)}{\tau_{1,2} P^f(t,T_2)},$$

where the year fraction is assumed to be the same in both markets. The no-arbitrage value at time t of the payoff (11.26) is then given by

$$\mathbf{QCpl}(t,T_1,T_2,N,K) = \tau_{1,2} N P(t,T_2) E^2 \left\{ [F^f(T_1;T_1,T_2) - K]^+ | \mathcal{F}_t \right\},$$

where E^2 denotes the expectation under the domestic forward measure Q^2 induced by the numeraire $P(t,T_2)$.

In order to compute this expectation we must know the distribution of $F^f(T_1;T_1,T_2)$ under the measure Q^2. Notice that, under the foreign forward measure associated with the numeraire $P^f(t,T_2)$, $F^f(t;T_1,T_2)$ is indeed a martingale. However, under the domestic forward measure Q^2, $F^f(t;T_1,T_2)$ displays a drift that we shall derive as follows.

Since $P^f(t,T_1) - P^f(t,T_2)$ and $P^f(t,T_2)$ are prices of tradable assets in the foreign market, then $\mathcal{X}(t)[P^f(t,T_1) - P^f(t,T_2)]$ and $\mathcal{X}(t)P^f(t,T_2)$ are both prices of tradable assets in the domestic market. Therefore the processes Y and Z defined by

$$Y(t) = \mathcal{X}(t) \frac{P^f(t,T_1) - P^f(t,T_2)}{P(t,T_2)},$$

$$Z(t) = \mathcal{X}(t) \frac{\tau_{1,2} P^f(t,T_2)}{P(t,T_2)} = \tau_{1,2} F_{\mathcal{X}}(t,T_2)$$

are both martingales under Q^2, with $F_{\mathcal{X}}(t,T_2)$ denoting the forward exchange rate at time t maturing at time T_2. More precisely, under lognormal assumptions, there exist real numbers σ_Y, σ_Z and ρ_{YZ} such that, under Q^2,

$$dY(t) = \sigma_Y Y(t) dW_Y(t),$$
$$dZ(t) = \sigma_Z Z(t) dW_Z(t),$$

where W_Y and W_Z are two Brownian motions with $dW_Y dW_Z = \rho_{YZ} dt$.

We have that, under Q^2,

$$dF^f(t;T_1,T_2) = d\left(\frac{Y(t)}{Z(t)}\right)$$
$$= Y(t)\left[-\frac{1}{Z^2(t)}\sigma_Z Z(t) dW_Z(t) + \frac{1}{2}\frac{2}{Z^3(t)}\sigma_Z^2 Z^2(t) dt\right]$$
$$+ \frac{1}{Z(t)}\sigma_Y Y(t) dW_Y(t) - \frac{Y(t)}{Z(t)}\rho_{YZ}\sigma_Y\sigma_Z dt,$$

that is
$$dF^f(t;T_1,T_2) = F^f(t;T_1,T_2)[(\sigma_Z^2 - \rho_{YZ}\sigma_Y\sigma_Z)dt + \sigma_Y dW_Y(t) - \sigma_Z dW_Z(t)],$$
$$= F^f(t;T_1,T_2)[\mu dt + \sigma dW(t)],$$

where
$$\mu = \sigma_Z^2 - \rho_{YZ}\sigma_Y\sigma_Z,$$
$$\sigma = \sqrt{\sigma_Y^2 + \sigma_Z^2 - 2\rho_{YZ}\sigma_Y\sigma_Z}$$

and W is a Brownian motion under Q^2. Therefore, under Q^2, $F^f(T_1;T_1,T_2)$ is lognormally distributed with

$$E^2\left\{\ln\frac{F^f(T_1;T_1,T_2)}{F^f(t;T_1,T_2)}\bigg|\mathcal{F}_t\right\} = \left(\mu - \frac{1}{2}\sigma^2\right)(T_1 - t),$$

$$\text{Var}^2\left\{\ln\frac{F^f(T_1;T_1,T_2)}{F^f(t;T_1,T_2)}\bigg|\mathcal{F}_t\right\} = \sigma^2(T_1 - t),$$

which implies that, see also (B.2) in Appendix B,

$$\mathbf{QCpl}(t,T_1,T_2,N,K) = \tau_{1,2}NP(t,T_2)\left[F^f(t;T_1,T_2)e^{\mu(T_1-t)}\Phi(d_1) - K\Phi(d_2)\right]$$

$$d_1 = \frac{\ln\frac{F^f(t;T_1,T_2)}{K} + \left(\mu + \frac{1}{2}\sigma^2\right)(T_1 - t)}{\sigma\sqrt{T_1 - t}}$$

$$d_2 = d_1 - \sigma\sqrt{T_1 - t}.$$

In order to compute the price $\mathbf{QCpl}(t,T_1,T_2,N,K)$, we need to know the values of the parameters μ and σ in terms of observable market quantities. The value of σ can be implied from market data since it is the (proportional) volatility of the foreign forward rate $F^f(t;T_1,T_2)$. The value of μ, instead, can be derived either by applying the toolkit formula (2.12) with $U = P(\cdot,T_2)$, $S = P^f(\cdot,T_2)$, and $m^S = 0$, or by the following equivalent direct argument. Let us define $\widetilde{Y} = \ln(Y)$ and $\widetilde{Z} = \ln(Z)$, so that

$$\mu dt = d\widetilde{Z}d\widetilde{Z} - d\widetilde{Y}d\widetilde{Z}$$
$$= d\widetilde{Z}(d\widetilde{Z} - d\widetilde{Y})$$
$$= -d\widetilde{Z}\, d\ln F^f(t;T_1,T_2)$$
$$= -\text{Cov}(d\ln(Z), d\ln F^f(t;T_1,T_2))$$
$$= -\text{Cov}(d\ln(\tau_{1,2}F_{\mathcal{X}}(t,T_2)), d\ln F^f(t;T_1,T_2))$$
$$= -\text{Cov}(d\ln(F_{\mathcal{X}}(t,T_2)), d\ln F^f(t;T_1,T_2)).$$

Therefore, denoting by $\sigma_{F_{\mathcal{X}}}$ the (proportional) volatility of the forward exchange rate $F_{\mathcal{X}}(t,T_2)$ and by ρ the instantaneous correlation between $F_{\mathcal{X}}(t,T_2)$ and $F^f(t;T_1,T_2)$, we have that

$$\mu = -\rho\sigma_{FX}\sigma,$$

with ρ and σ_{FX} obtainable from market data.

Analogously, the arbitrage-free price of a quanto floorlet that pays off at time T_2

$$\left[K - F^f(T_1; T_1, T_2)\right]^+ \tau_{1,2} N \quad \text{in domestic currency,}$$

is

$$\text{QFll}(t, T_1, T_2, N, K) = \tau_{1,2} N P(t, T_2)$$
$$\cdot \left[-F^f(t; T_1, T_2) e^{\mu(T_1-t)} \Phi(-d_1) + K \Phi(-d_2) \right].$$

11.4.2 The Pricing of Quanto Caps/Floors

As to the pricing of caps and floors, we denote by $D = \{d_1, d_2, \ldots, d_n\}$ the set of the cap/floor payment dates and by $\mathcal{T} = \{t_0, t_1, \ldots, t_n\}$ the set of the corresponding times, meaning that t_i is the difference in years between d_i and the settlement date t, and where t_0 is the first reset time. Moreover, we denote by τ_i the year fraction for the time interval $(t_{i-1}, t_i]$, $i = 1, \ldots, n$, by σ_i the (proportional) volatility of $F^f(t; t_{i-1}, t_i)$, by ρ_i the instantaneous correlation between $F_{\mathcal{X}}(t, t_i)$ and $F^f(t; t_{i-1}, t_i)$, by σ_{FX}^i the (proportional) volatility of the forward exchange rate $F_{\mathcal{X}}(t, t_i)$. We set $\tau := \{\tau_1, \ldots, \tau_n\}$ and

$$\mu_i = -\rho_i \sigma_{FX}^i \sigma_i.$$

Since the price of a cap (floor) is the sum of the prices of the underlying caplets (floorlets), the price at time t of a cap with cap rate (strike) K, nominal value N and set of times $\mathcal{T}$ is then given by

$$\text{QCap}(t, \mathcal{T}, \tau, N, K)$$
$$= \sum_{i=1}^{n} \tau_i N P(t, t_i) \left[F^f(t; t_{i-1}, t_i) e^{\mu_i(t_{i-1}-t)} \right.$$
$$\left. \cdot \Phi\left(\frac{\ln \frac{F^f(t; t_{i-1}, t_i)}{K} + \left(\mu_i + \frac{1}{2}\sigma_i^2\right)(t_{i-1} - t)}{\sigma_i \sqrt{t_{i-1} - t}} \right) \right. \quad (11.27)$$
$$\left. - K \Phi\left(\frac{\ln \frac{F^f(t; t_{i-1}, t_i)}{K} + \left(\mu_i - \frac{1}{2}\sigma_i^2\right)(t_{i-1} - t)}{\sigma_i \sqrt{t_{i-1} - t}} \right) \right],$$

and the price of the corresponding floor is

$\mathbf{QFlr}(t, \mathcal{T}, \tau, N, K)$

$$= \sum_{i=1}^{n} \tau_i NP(t, t_i) \Bigg[- F^f(t; t_{i-1}, t_i) e^{\mu_i(t_{i-1}-t)} $$

$$\cdot \Phi\left(\frac{\ln \frac{K}{F^f(t; t_{i-1}, t_i)} - (\mu_i + \frac{1}{2}\sigma_i^2)(t_{i-1}-t)}{\sigma_i \sqrt{t_{i-1}-t}} \right) \quad (11.28)$$

$$+ K\Phi\left(\frac{\ln \frac{K}{F^f(t; t_{i-1}, t_i)} - (\mu_i - \frac{1}{2}\sigma_i^2)(t_{i-1}-t)}{\sigma_i \sqrt{t_{i-1}-t}} \right) \Bigg].$$

11.4.3 The Pricing of Differential Swaps

A differential swap, or quanto swap, has been defined in the previous Section 11.3. Suppose for a moment that we take away the LIBOR payment in the leg paid by institution B to institution A. Then the price at time t of the corresponding quanto swap where we receive the floating rate and we pay the fixed rate K is simply the difference

$$\mathbf{QS}(t, \mathcal{T}, \tau, N, K) = \mathbf{QCap}(t, \mathcal{T}, \tau, N, K) - \mathbf{QFlr}(t, \mathcal{T}, \tau, N, K),$$

since $(F - K)^+ - (K - F)^+ = F - K$. Hence,

$$\mathbf{QS}(t, \mathcal{T}, \tau, N, K) = \sum_{i=1}^{n} \tau_i NP(t, t_i) \left[F^f(t; t_{i-1}, t_i) e^{\mu_i(t_{i-1}-t)} - K \right].$$

If we put back the domestic LIBOR rate together with the fixed-rate payment K in the leg from B to A, we have to subtract the corresponding value to the above formula, thus obtaining

$$\sum_{i=1}^{n} \tau_i NP(t, t_i) \left[F^f(t; t_{i-1}, t_i) e^{\mu_i(t_{i-1}-t)} - K \right] - N(1 - P(t, T_n)).$$

11.4.4 The Pricing of Quanto Swaptions

We now consider a further multi-currency derivative, namely a European-style quanto swaption, and show how to price it by means of a general formula for spread options. To this end, we first price an option on the spread between two given assets, by assuming that the assets evolve according to two geometric Brownian motions with nonzero instantaneous correlation.

The financial market constituted by these assets and a deterministic bond is complete, meaning that arbitrage-free prices are given by risk-neutral valuation. However, no closed formula is obtainable for the price of the spread option. We will have instead to resort to numerical integration of a function involving both the cumulative distribution and probability density functions of a standard normal random variable.

11.4 Market Formulas for Basic Quanto Derivatives

The Pricing of a Spread Option. Consider two assets whose prices S_1 and S_2 evolve, under the real-world measure, according to

$$dS_1(t) = S_1(t)[m_1 dt + \sigma_1 dW_1(t)], \quad S_1(0) = s_1,$$
$$dS_2(t) = S_2(t)[m_2 dt + \sigma_2 dW_2(t)], \quad S_2(0) = s_2, \quad (11.29)$$
$$dW_1(t)dW_2(t) = \rho\, dt,$$

where m_1, m_2, σ_1, σ_2, s_1 and s_2 are positive real numbers and W_1 and W_2 are Brownian motions with instantaneous correlation ρ.

Assume that the assets pay continuous dividend yields q_1 and q_2, with q_1 and q_2 positive real numbers, and that interest rates are constant for all maturities and equal to the positive real number r.

The dynamics (11.29) implies there exists a unique equivalent martingale measure Q under which S_1 and S_2 evolve according to

$$dS_1(t) = S_1(t)[(r-q_1)dt + \sigma_1 dW_1^Q(t)], \quad S_1(0) = s_1,$$
$$dS_2(t) = S_2(t)[(r-q_2)dt + \sigma_2 dW_2^Q(t)], \quad S_2(0) = s_2, \quad (11.30)$$

where W_1^Q and W_2^Q are Brownian motions under Q with instantaneous correlation ρ.

Consider now an option on the spread between the two assets. Precisely, fix a maturity T, a positive real number a, a negative real number b, a strike price K. The spread-option payoff at time T is then defined by

$$H = (awS_1(T) + bwS_2(T) - wK)^+, \quad (11.31)$$

where $w = 1$ for a call and $w = -1$ for a put.

The existence of a unique equivalent martingale measure implies that there exists a unique arbitrage-free price for the spread option at any time $t \in [0,T]$. Such a price is given by, see also (2.2),

$$\pi_t = e^{-r(T-t)} E^Q\left\{(awS_1(T) + bwS_2(T) - wK)^+ \big| \mathcal{F}_t\right\}, \quad (11.32)$$

where E^Q denotes expectation under Q and $\mathcal{F}_t$ is the sigma-field generated by (S_1, S_2) up to time t.

For a nonzero strike price, the expectation in (11.32) cannot be computed in an explicit fashion. However, a pseudo-analytical formula can be derived in terms of improper integrals. This is explained in the following.

Proposition 11.4.1. *The unique arbitrage-free price of the payoff (11.31) at maturity T is*

$$\pi_t = \int_{-\infty}^{+\infty} \frac{1}{\sqrt{2\pi}} e^{-\frac{1}{2}v^2} f(v) dv, \quad (11.33)$$

where

$$f(v) = awS_1(t)\exp\left[-q_1\tau - \frac{1}{2}\rho^2\sigma_1^2\tau + \rho\sigma_1\sqrt{\tau}v\right] \cdot$$

$$\cdot \Phi\left(w\frac{\ln\frac{aS_1(t)}{h(v)} + [\mu_1 + (\frac{1}{2} - \rho^2)\sigma_1^2]\tau + \rho\sigma_1\sqrt{\tau}v}{\sigma_1\sqrt{\tau}\sqrt{1-\rho^2}}\right)$$

$$- wh(v)e^{-r\tau}\Phi\left(w\frac{\ln\frac{aS_1(t)}{h(v)} + (\mu_1 - \frac{1}{2}\sigma_1^2)\tau + \rho\sigma_1\sqrt{\tau}v}{\sigma_1\sqrt{\tau}\sqrt{1-\rho^2}}\right)$$

and

$$h(v) = K - bS_2(t)e^{(\mu_2 - \frac{1}{2}\sigma_2^2)\tau + \sigma_2\sqrt{\tau}v}$$
$$\mu_1 = r - q_1$$
$$\mu_2 = r - q_2$$
$$\tau = T - t$$

with $\Phi(\cdot)$ denoting the standard normal cumulative distribution function.

Proof. Defining

$$X := \ln\frac{S_1(T)}{S_1(t)},$$
$$Y := \ln\frac{S_2(T)}{S_2(t)},$$

the joint density function $f_{X,Y}$ of (X,Y) under the measure Q is bivariate normal with mean vector

$$M_{X,Y} = \begin{bmatrix}\mu_x \\ \mu_y\end{bmatrix} := \begin{bmatrix}(\mu_1 - \frac{1}{2}\sigma_1^2)\tau \\ (\mu_2 - \frac{1}{2}\sigma_2^2)\tau\end{bmatrix}$$

and covariance matrix

$$V_{X,Y} = \begin{bmatrix}\sigma_x^2 & \rho\sigma_x\sigma_y \\ \rho\sigma_x\sigma_y & \sigma_y^2\end{bmatrix} := \begin{bmatrix}\sigma_1^2\tau & \rho\sigma_1\sigma_2\tau \\ \rho\sigma_1\sigma_2\tau & \sigma_2^2\tau\end{bmatrix},$$

that is

$$f_{X,Y}(x,y) = \frac{1}{2\pi\sigma_x\sigma_y\sqrt{1-\rho^2}}\exp\left[-\frac{\left(\frac{x-\mu_x}{\sigma_x}\right)^2 - 2\rho\frac{x-\mu_x}{\sigma_x}\frac{y-\mu_y}{\sigma_y} + \left(\frac{y-\mu_y}{\sigma_y}\right)^2}{2(1-\rho^2)}\right].$$

It is well known that

$$f_{X,Y}(x,y) = f_{X|Y}(x,y)f_Y(y),$$

where

11.4 Market Formulas for Basic Quanto Derivatives

$$f_{X|Y}(x,y) = \frac{1}{\sigma_x\sqrt{2\pi}\sqrt{1-\rho^2}} \exp\left[-\frac{\left(\frac{x-\mu_x}{\sigma_x} - \rho\frac{y-\mu_y}{\sigma_y}\right)^2}{2(1-\rho^2)}\right] \quad (11.34)$$

$$f_Y(y) = \frac{1}{\sigma_y\sqrt{2\pi}} \exp\left[-\frac{1}{2}\left(\frac{y-\mu_y}{\sigma_y}\right)^2\right].$$

This implies that

$$\pi_t = e^{-r\tau} \int_{-\infty}^{+\infty} \int_{-\infty}^{+\infty} (awS_1(t)e^x + bwS_2(t)e^y - wK)^+ f_{X,Y}(x,y)dxdy$$

$$= e^{-r\tau} \int_{-\infty}^{+\infty} \int_{-\infty}^{+\infty} (awS_1(t)e^x + bwS_2(t)e^y - wK)^+ f_{X|Y}(x,y)f_Y(y)dxdy$$

$$= e^{-r\tau} \int_{-\infty}^{+\infty} \left[\int_{-\infty}^{+\infty} (awS_1(t)e^x + bwS_2(t)e^y - wK)^+ f_{X|Y}(x,y)dx\right] f_Y(y)dy. \quad (11.35)$$

The inner integral can be explicitly computed, since the term $bwS_2(t)e^y$ can be viewed as a constant when integrating with respect to x, and turns out to be of the Black-Scholes type since $f_{X|Y}(x,y)$, for a fixed y, is itself the density function of a Gaussian random variable. More precisely, one applies the following formula, see also (B.1) in Appendix B, where $w \in \{-1,1\}$ and $A > 0$, $K > 0$, M and $V > 0$ are real numbers,

$$\int_{-\infty}^{+\infty} \frac{1}{\sqrt{2\pi}V}(wAe^z - wK)^+ e^{-\frac{1}{2}\frac{(z-M)^2}{V^2}} dz$$

$$= wAe^{M+\frac{1}{2}V^2} \Phi\left(w\frac{M - \ln\frac{K}{A} + V^2}{V}\right) - wK\Phi\left(w\frac{M - \ln\frac{K}{A}}{V}\right),$$

to get that

$$\int_{-\infty}^{+\infty} (awS_1(t)e^x + bwS_2(t)e^y - wK)^+ f_{X|Y}(x,y)dx$$

$$= awS_1(t)\exp\left[\mu_x + \rho\sigma_x\frac{y-\mu_y}{\sigma_y} + \frac{1}{2}\sigma_x^2(1-\rho^2)\right]$$

$$\cdot \Phi\left(w\frac{\mu_x + \rho\sigma_x\frac{y-\mu_y}{\sigma_y} - \ln\frac{K-bS_2(t)e^y}{aS_1(t)} + \sigma_x^2(1-\rho^2)}{\sigma_x\sqrt{1-\rho^2}}\right)$$

$$- w(K - bS_2(t)e^y)\Phi\left(w\frac{\mu_x + \rho\sigma_x\frac{y-\mu_y}{\sigma_y} - \ln\frac{K-bS_2(t)e^y}{aS_1(t)}}{\sigma_x\sqrt{1-\rho^2}}\right). \quad (11.36)$$

The definition of μ_x, μ_y, σ_x and σ_y and the variable change $v = \frac{y-\mu_y}{\sigma_y}$ then lead to (11.33). □

The Pricing of a Quanto Swaption. Let $\mathcal{T} = \{T = t_0, t_1, t_2, \ldots, t_n\}$ be a set of $n+1$ times with $0 < T < t_1 < t_2 < \cdots < t_n$. Assume we are given a domestic and a foreign financial markets, where discount factors and forward swap rates corresponding to $\mathcal{T}$ are defined. As before, we denote respectively by $P(t, t_i)$ and $P^f(t, t_i)$ the domestic and foreign discount factors at time t for maturity t_i, $i = 0, 1, \ldots, n$. The exchange rate between the two market currencies is again assumed to evolve according to the above process $\mathcal{X}$, meaning that, at any time t, one unit of foreign currency is worth $\mathcal{X}(t)$ units of domestic currency.

The domestic forward swap rate at time $0 < t < T$ corresponding to the swap starting at time T and with payment times t_i, $i = 1, \ldots, n$, is given by

$$S(t) = \frac{P(t,T) - P(t,t_n)}{C(t)}, \tag{11.37}$$

where

$$C(t) = \sum_{i=1}^{n} \tau_i P(t, t_i), \tag{11.38}$$

and τ_i is again the year fraction over the period from time t_{i-1} to t_i, and $\tau := \{\tau_1, \ldots, \tau_n\}$. Analogously, the foreign forward swap-rate at time $0 < t < T$ corresponding to the swap starting at time T and with payment times t_i, $i = 1, \ldots, n$, is given by

$$S^f(t) = \frac{P^f(t,T) - P^f(t,t_n)}{C^f(t)}, \tag{11.39}$$

where

$$C^f(t) = \sum_{i=1}^{n} \tau_i P^f(t, t_i), \tag{11.40}$$

and, for simplicity, it is assumed that the year fractions are the same in both markets.

A (European-style) quanto swaption is, roughly speaking, an option on the difference between these two swap rates, with this difference being denominated in units of domestic currency. In formulas, given the swap rates (11.37) and (11.39), the strike $K > 0$ and assuming that the option maturity is T, the quanto-swaption payoff at time T is

$$C(T)[wS^f(T) - wS(T) - wK]^+, \tag{11.41}$$

where either $w = 1$ or $w = -1$. Choosing $C(t)$ as numeraire, the arbitrage-free price of (11.41) at time t can be written as

$$\mathbf{QES}(t, \mathcal{T}, \tau, K) = C(t) E^C \left\{ [wS^f(T) - wS(T) - wK]^+ | \mathcal{F}_t \right\}, \tag{11.42}$$

where E^C denotes expectation under the domestic forward-swap measure Q^C induced by $C(t)$ and $\mathcal{F}_t$ is the sigma-field generated by (S^f, S) up to time t.

11.4 Market Formulas for Basic Quanto Derivatives

By definition of domestic forward-swap measure, the forward-swap-rate process S is a martingale under such a measure. Assuming lognormal dynamics, the domestic forward-swap process under Q^C is then described by

$$dS(t) = \sigma S(t) dW(t), \tag{11.43}$$

where σ is a positive real number and W is a Q^C-Brownian motion.

The forward-swap-rate process S^f is a martingale under the foreign forward-swap measure, but under Q^C displays the drift that has been calculated by Hunt and Pelsser (1998). Precisely, they assumed that under Q^C the two following martingales

$$M_1(t) = \frac{\mathcal{X}(t)(P^f(t,T) - P^f(t,t_n))}{C(t)}$$

$$M_2(t) = \frac{\mathcal{X}(t) C^f(t)}{C(t)}$$

are lognormally distributed, so that also $S^f(t) = \frac{M_1(t)}{M_2(t)}$ is lognormally distributed with dynamics given by

$$dS^f(t) = S^f(t)[\mu^f dt + \sigma^f dW^f(t)],$$

where μ^f and σ^f are real constants and W^f is a Brownian motion under Q^C. They also showed that the drift rate $\mu^f dt$ is equal to minus the instantaneous covariance of $\ln M_2(t)$ and $\ln S^f(t)$ and stated that, for practical purposes,[1] $M_2(t)$ can be replaced by the time-T forward exchange rate

$$F_{\mathcal{X}}(t,T) = \frac{\mathcal{X}(t) P^f(t,T)}{P(t,T)}.$$

By doing so, we can get the approximated equality

$$\mu^f = -\rho_{F,f} \sigma^f \sigma_F,$$

where σ_F is the (assumed constant) proportional volatility of the process $F_{\mathcal{X}}$ and $\rho_{F,f}$ is the instantaneous correlation between $F_{\mathcal{X}}$ and S^f, i.e. between the forward exchange rate and the foreign swap rate.

Knowing the distribution of S and S^f under the measure Q^C, we can now calculate the expectation in (11.42) by means of the results of the previous section. Notice, in fact, that under Q^C the joint distribution of

$$\left(\ln \frac{S^f(T)}{S^f(t)}, \ln \frac{S(T)}{S(t)} \right) \text{ conditional on } \mathcal{F}_t$$

is bivariate normal with with mean vector

---
[1] That is, to be able to explicitly calculate this covariance from market data.

450   11. Pricing Derivatives on Two Interest-Rate Curves

$$\begin{bmatrix} \mu_x \\ \mu_y \end{bmatrix} = \begin{bmatrix} (\mu^f - \frac{1}{2}(\sigma^f)^2) \\ -\frac{1}{2}\sigma^2 \end{bmatrix} \tau$$

and covariance matrix

$$\begin{bmatrix} \sigma_x^2 & \rho\sigma_x\sigma_y \\ \rho\sigma_x\sigma_y & \sigma_y^2 \end{bmatrix} = \begin{bmatrix} (\sigma^f)^2 & \rho\sigma^f\sigma \\ \rho\sigma^f\sigma & \sigma^2 \end{bmatrix} \tau$$

where $\tau = T - t$ and ρ is the instantaneous correlation between the two forward swap-rates S^f and S, i.e., $dW(t)dW^f(t) = \rho\,dt$. Therefore, we can apply the same decomposition as in (11.35) to obtain

$$\mathbf{QES}(t) = C(t) \int_{-\infty}^{+\infty} \left[\int_{-\infty}^{+\infty} \left(wS^f(t)e^x - wS(t)e^y - wK \right)^+ f_{X|Y}(x,y)dx \right]$$
$$\cdot f_Y(y)dy, \qquad (11.44)$$

with $f_{X|Y}$ and f_Y defined in (11.34), which, by formula (11.36), becomes

$$C(t) \int_{-\infty}^{+\infty} \left\{ wS^f(t) \exp\left[\mu_x + \rho\sigma_x \frac{y - \mu_y}{\sigma_y} + \frac{1}{2}\sigma_x^2(1-\rho^2) \right] \right.$$

$$\cdot \Phi\left(w \frac{\mu_x + \rho\sigma_x \frac{y-\mu_y}{\sigma_y} - \ln\frac{K+S(t)e^y}{S^f(t)} + \sigma_x^2(1-\rho^2)}{\sigma_x\sqrt{1-\rho^2}} \right)$$

$$\left. - w(K + S(t)e^y) \Phi\left(w \frac{\mu_x + \rho\sigma_x \frac{y-\mu_y}{\sigma_y} - \ln\frac{K+S(t)e^y}{S^f(t)}}{\sigma_x\sqrt{1-\rho^2}} \right) \right\} f_Y(y)dy.$$
$$(11.45)$$

We are then ready to state the following.

Proposition 11.4.2. *The unique arbitrage-free price of the quanto swaption described by the payoff (11.41) is*

$$\mathbf{QES}(t) = C(t) \int_{-\infty}^{+\infty} \frac{1}{\sqrt{2\pi}} e^{-\frac{1}{2}v^2} f(v)dv, \qquad (11.46)$$

where

$$f(v) = wS^f(t) \exp\left[(\mu^f - \frac{1}{2}\rho^2(\sigma^f)^2)\tau + \rho\sigma^f\sqrt{\tau}v \right] \cdot$$

$$\cdot \Phi\left(w \frac{\ln\frac{S^f(t)}{h(v)} + [\mu^f + (\frac{1}{2} - \rho^2)(\sigma^f)^2]\tau + \rho\sigma^f\sqrt{\tau}v}{\sigma^f\sqrt{\tau}\sqrt{1-\rho^2}} \right)$$

$$- wh(v)\Phi\left(w \frac{\ln\frac{S^f(t)}{h(v)} + [\mu^f - \frac{1}{2}(\sigma^f)^2]\tau + \rho\sigma^f\sqrt{\tau}v}{\sigma^f\sqrt{\tau}\sqrt{1-\rho^2}} \right)$$

11.4 Market Formulas for Basic Quanto Derivatives

and
$$h(v) = K + S(t)e^{-\frac{1}{2}\sigma^2 \tau + \sigma\sqrt{\tau}v}$$

Proof. Formula (11.46) immediately follows from (11.45) by remembering the definition of μ_x, μ_y, σ_x and σ_y and performing the variable change $v = \frac{y - \mu_y}{\sigma_y}$. □

The formula (11.46) can be computed through an easy numerical integration. As to the parameters being involved in it, besides the obvious inputs $S^f(t)$, $S(t)$, T and K, we have to remark that σ^f, σ and σ_F can be immediately obtained from market data of implied volatilities, whereas ρ and $\rho_{F,f}$ can be estimated historically.

12. Pricing Equity Derivatives under Stochastic Rates

All possible definitions of probability fall short of the actual practice.
William Feller, "An Introduction to Probability Theory and Its Applications", Vol. 1, Wiley.

The well consolidated theory for pricing equity derivatives under the Black and Scholes (1973) model is based on the assumption of deterministic interest rates. Such an assumption is harmless in most situations since the interest-rates variability is usually negligible if compared to the variability observed in equity markets. When pricing a long-maturity option, however, the stochastic feature of interest rates has a stronger impact on the option price. In such cases it is therefore advisable to relax the assumption of deterministic rates.

The general interest rate theory developed in the first part of this book can be applied to the pricing of equity derivatives under the assumption of stochastic interest rates. We then do so and consider a continuous-time economy where interest rates are stochastic and asset prices evolve according to a geometric Brownian motion. Explicit formulas for European options on a given asset are provided when the instantaneous spot rate follows the Hull and White (1994a) process (3.33). Our pricing procedure is based on the derivation of the asset and rate dynamics under the forward measure whose numeraire is the bond price with the same maturity as the option's. The option prices are then obtained by computing the expectation (2.21).

If the instantaneous spot rate follows a lognormal process like that of Black and Karasinski (1991), as in (3.54), it is not possible to derive closed form formulas for the above European option prices. We can anyway construct an approximating tree, which can also be used for pricing more complex derivatives.

12.1 The Short Rate and Asset-Price Dynamics

We consider a continuous-time economy where interest rates are stochastic and the price of a given tradable asset evolves according to a geometric Brownian motion. We assume that under the risk adjusted measure Q, the dynamics of the instantaneous short rate is given by the Hull and White (1994a) process, see (3.33),

454   12. Pricing Equity Derivatives under Stochastic Rates

$$dr(t) = [\theta(t) - ar(t)]dt + \sigma dW(t), \quad r(0) = r_0, \qquad (12.1)$$

and that the asset price evolves according to

$$dS(t) = S(t)[(r(t) - y)dt + \eta dZ(t)], \quad S(0) = S_0, \qquad (12.2)$$

where W and Z are two correlated Brownian motions with

$$dZ(t)\,dW(t) = \rho\,dt,$$

ρ being the instantaneous-correlation parameter between the asset price and the short interest rate, and where $y > 0$ is the asset dividend yield, r_0, a, σ, η and S_0 are positive constants and θ is a deterministic function that is well defined in the time interval $[0, T^*]$, with T^* a given time horizon. We denote by $\mathcal{F}_t$ the sigma-field generated by (r, S) up to time t.

We know from Section 3.3 that we can write

$$r(t) = x(t) + \varphi(t), \quad r(0) = r_0, \qquad (12.3)$$

where the process x satisfies

$$dx(t) = -ax(t)dt + \sigma dW(t), \quad x(0) = 0, \qquad (12.4)$$

and the function φ is deterministic and well defined in the time interval $[0, T^*]$. In particular, $\varphi(0) = r_0$.

Let us now assume that the term structure of discount factors that is currently observed in the market is given by the sufficiently smooth function $T \mapsto P^M(0, T)$.

We remember from Section 3.3 that, denoting by $f^M(0, T)$ the instantaneous forward rate at time 0 for a maturity T implied by the term structure $T \mapsto P^M(0, T)$, i.e.,

$$f^M(0, T) = -\frac{\partial \ln P^M(0, T)}{\partial T},$$

in order to exactly fit the observed term structure, we must have that, for each T, see (3.36),

$$\varphi(T) = f^M(0, T) + \frac{\sigma^2}{2a^2}\left(1 - e^{-aT}\right)^2. \qquad (12.5)$$

Simple integration of the previous SDE's implies that, for each $s < t$,

$$r(t) = r(s)e^{-a(t-s)} + \int_s^t e^{-a(t-u)}\theta(u)du + \sigma\int_s^t e^{-a(t-u)}dW(u)$$

$$= x(s)e^{-a(t-s)} + \sigma\int_s^t e^{-a(t-u)}dW(u) + \varphi(t)$$

and

12.1 The Short Rate and Asset-Price Dynamics

$$S(t) = S(s) \exp\left\{\int_s^t r(u)du - y(t-s) - \tfrac{1}{2}\eta^2(t-s) + \eta(Z(t) - Z(s))\right\}.$$

This means that, conditional on $\mathcal{F}_s$, $r(t)$ is normally distributed with mean and variance given respectively by

$$E\{r(t)|\mathcal{F}_s\} = x(s)e^{\,a(t-s)} + \varphi(t),$$

$$\text{Var}\{r(t)|\mathcal{F}_s\} = \frac{\sigma^2}{2a}\left[1 - e^{-2a(t-s)}\right]. \tag{12.6}$$

(see (3.37)). Moreover, also $\ln S(t)$ conditional on $\mathcal{F}_s$ is normally distributed. Its mean and variance are calculated as follows.

By using the following equality, see also Appendix A at the end of Chapter 4,

$$\int_t^T x(u)du = \frac{1 - e^{-a(T-t)}}{a}x(t) + \frac{\sigma}{a}\int_t^T \left[1 - e^{-a(T-u)}\right]dW(u), \tag{12.7}$$

we have that

$$\ln \frac{S(t)}{S(s)} = \int_s^t r(u)du - y(t-s) - \tfrac{1}{2}\eta^2(t-s) + \eta(Z(t) - Z(s))$$

$$= \frac{1 - e^{-a(t-s)}}{a}x(s) + \frac{\sigma}{a}\int_s^t (1 - e^{-a(t-u)})dW(u) + \int_s^t \varphi(u)du$$

$$- y(t-s) - \tfrac{1}{2}\eta^2(t-s) + \eta\int_s^t dZ(u),$$

so that

$$E\left\{\ln \frac{S(t)}{S(s)}\Big|\mathcal{F}_s\right\} = \frac{1 - e^{-a(t-s)}}{a}\left[r(s) - f^M(0,s) - \frac{\sigma^2}{2a^2}(1 - e^{-as})^2\right]$$

$$- y(t-s) - \tfrac{1}{2}\eta^2(t-s) + \ln \frac{P^M(0,s)}{P^M(0,t)}$$

$$+ \frac{\sigma^2}{2a^2}\left[t - s + \frac{2}{a}(e^{-at} - e^{-as}) - \frac{1}{2a}(e^{-2at} - e^{-2as})\right]$$

and

$$\text{Var}\left\{\ln \frac{S(t)}{S(s)}\Big|\mathcal{F}_s\right\} = \frac{\sigma^2}{a^2}\int_s^t (1 - e^{-a(t-u)})^2 du + \eta^2(t-s)$$

$$+ 2\rho\frac{\sigma\eta}{a}\int_s^t (1 - e^{-a(t-u)})du$$

$$= \frac{\sigma^2}{a^2}\left[t - s - \frac{2}{a}\left(1 - e^{-a(t-s)}\right) + \frac{1}{2a}\left(1 - e^{-2a(t-s)}\right)\right]$$

$$+ \eta^2(t-s) + 2\rho\frac{\sigma\eta}{a}\left[t - s - \frac{1}{a}\left(1 - e^{-a(t-s)}\right)\right].$$

12.1.1 The Dynamics under the Forward Measure

The dynamics of the processes r and S can be also expressed in terms of two independent Brownian motions $\widetilde{W}$ and $\widetilde{Z}$ as follows (Cholesky decomposition):

$$dr(t) = [\theta(t) - ar(t)]dt + \sigma d\widetilde{W}(t),$$
$$dS(t) = S(t)[(r(t) - y)dt + \eta\rho d\widetilde{W}(t) + \eta\sqrt{1-\rho^2}d\widetilde{Z}(t)], \quad (12.8)$$

where

$$dW(t) = d\widetilde{W}(t),$$
$$dZ(t) = \rho d\widetilde{W}(t) + \sqrt{1-\rho^2}d\widetilde{Z}(t).$$

This decomposition makes it easier to perform a measure transformation. In fact, for any fixed maturity T, let us denote by Q^T the T-forward (risk-adjusted) measure, i.e., the probability measure that is defined by the Radon-Nikodym derivative, see Chapter 2,

$$\frac{dQ^T}{dQ} = \frac{\exp\left\{-\int_0^T r(u)du\right\}}{P(0,T)}$$

$$= \frac{\exp\left\{-\int_0^T x(u)du - \int_0^T \varphi(u)du\right\}}{P(0,T)}$$

$$= \frac{\exp\left\{-\frac{\sigma}{a}\int_0^T [1 - e^{-a(T-u)}]d\widetilde{W}(u) - \int_0^T f^M(0,u)du - \int_0^T \frac{\sigma^2}{2a^2}(1-e^{-au})^2 du\right\}}{P(0,T)}$$

$$= \exp\left\{-\frac{\sigma}{a}\int_0^T \left[1 - e^{-a(T-u)}\right] d\widetilde{W}(u) - \int_0^T \frac{\sigma^2}{2a^2}\left[1 - e^{-a(T-u)}\right]^2 du\right\}, \quad (12.9)$$

where equalities (12.5) and (12.7) have been taken into account.

The Girsanov theorem then implies that the two processes $\widetilde{W}^T$ and $\widetilde{Z}^T$ defined by

$$d\widetilde{W}^T(t) = d\widetilde{W}(t) + \frac{\sigma}{a}\left[1 - e^{-a(T-t)}\right]dt$$
$$d\widetilde{Z}^T(t) = d\widetilde{Z}(t) \quad (12.10)$$

are two independent Brownian motions under the measure Q^T. Therefore the (joint) dynamics of r and S under Q^T are given by

$$dr(t) = \left[\theta(t) - \frac{\sigma^2}{a}\left(1 - e^{-a(T-t)}\right) - ar(t)\right]dt + \sigma d\widetilde{W}^T(t),$$
$$dS(t) = S(t)\left[(r(t) - y) - \rho\frac{\sigma\eta}{a}\left(1 - e^{-a(T-t)}\right)\right]dt \quad (12.11)$$
$$+ S(t)\left[\eta\rho d\widetilde{W}^T(t) + \eta\sqrt{1-\rho^2}d\widetilde{Z}^T(t)\right].$$

12.1 The Short Rate and Asset-Price Dynamics

Integrating equation (12.11) yields, for each $t < T$,

$$r(t) = r(s)e^{-a(t-s)} + \int_s^t e^{-a(t-u)}\theta(u)du - \frac{\sigma^2}{a}\int_s^t e^{-a(t-u)}\left[1 - e^{-a(T-u)}\right]du$$

$$+ \sigma \int_s^t e^{-a(t-u)}d\widetilde{W}^T(u)$$

$$= x(s)\,e^{-a(t-s)} - \frac{\sigma^2}{a}\int_s^t e^{-a(t-u)}\left[1 - e^{-a(T-u)}\right]du$$

$$+ \sigma \int_s^t e^{-a(t-u)}d\widetilde{W}^T(u) + \varphi(t)$$

and

$$S(T) = S(t)\exp\left\{\int_t^T r(u)du - y(T-t) - \rho\frac{\sigma\eta}{a}\int_t^T\left(1 - e^{-a(T-u)}\right)du\right.$$

$$\left. - \tfrac{1}{2}\eta^2(T-t) + \eta\rho(\widetilde{W}^T(T) - \widetilde{W}^T(t)) + \eta\sqrt{1-\rho^2}(\widetilde{Z}^T(T) - \widetilde{Z}^T(t))\right\}$$

$$= S(t)\exp\left\{\frac{1 - e^{-a(T-t)}}{a}x(t) + \frac{\sigma}{a}\int_t^T\left[1 - e^{-a(T-u)}\right]d\widetilde{W}^T(u)\right.$$

$$- \frac{\sigma^2}{a}\int_t^T\int_t^u e^{-a(u-s)}\left[1 - e^{-a(T-s)}\right]ds\,du + \int_t^T f^M(0,u)du$$

$$+ \frac{\sigma^2}{2a^2}\int_t^T\left(1 - e^{-au}\right)^2du - y(T-t) - \rho\frac{\sigma\eta}{a}\int_t^T\left(1 - e^{-a(T-u)}\right)du$$

$$\left. - \tfrac{1}{2}\eta^2(T-t) + \eta\rho(\widetilde{W}^T(T) - \widetilde{W}^T(t)) + \eta\sqrt{1-\rho^2}(\widetilde{Z}^T(T) - \widetilde{Z}^T(t))\right\}.$$

Straightforward calculations lead to

$$E^T\left\{\ln\frac{S(T)}{S(t)}\Big|\mathcal{F}_t\right\}$$

$$= \frac{1 - e^{-a(T-t)}}{a}x(t) - \frac{\sigma^2}{a^2}\left[T - t + \frac{2}{a}e^{-a(T-t)} - \frac{1}{2a}e^{-2a(T-t)} - \frac{3}{2a}\right]$$

$$+ \ln\frac{P^M(0,t)}{P^M(0,T)} + \frac{\sigma^2}{2a^2}\left[T - t + \frac{2}{a}(e^{-aT} - e^{-at}) - \frac{1}{2a}(e^{-2aT} - e^{-2at})\right]$$

$$- y(T-t) - \rho\frac{\sigma\eta}{a}\left[T - t - \frac{1}{a}\left(1 - e^{-a(T-t)}\right)\right] - \tfrac{1}{2}\eta^2(T-t)$$

$$= \frac{1 - e^{-a(T-t)}}{a}x(t) - V(t,T) + \ln\frac{P^M(0,t)}{P^M(0,T)} + \tfrac{1}{2}[V(0,T) - V(0,t)]$$

$$- y(T-t) - \rho\frac{\sigma\eta}{a}\left[T - t - \frac{1}{a}\left(1 - e^{-a(T-t)}\right)\right] - \tfrac{1}{2}\eta^2(T-t)$$

$$= -\ln(P(t,T)) - \rho\frac{\sigma\eta}{a}\left[T - t - \frac{1}{a}\left(1 - e^{-a(T-t)}\right)\right]$$

$$- (y + \tfrac{1}{2}\eta^2)(T-t) - \tfrac{1}{2}V(t,T),$$

where

$$V(t,T) := \frac{\sigma^2}{a^2}\left[T - t + \frac{2}{a}e^{-a(T-t)} - \frac{1}{2a}e^{-2a(T-t)} - \frac{3}{2a}\right],$$

and

$$\mathrm{Var}^T\left\{\ln\frac{S(T)}{S(t)}\Big|\mathcal{F}_t\right\} = V(t,T) + \eta^2(T-t)$$
$$+ 2\rho\frac{\sigma\eta}{a}\left[T - t - \frac{1}{a}\left(1 - e^{-a(T-t)}\right)\right].$$

12.2 The Pricing of a European Option on the Given Asset

The results of the previous section allow us to explicitly calculate the price of a European option when the asset price is modeled by a geometric Brownian motion and interest rates are stochastic and evolve according to the Hull and White (1994a) process.

From the general results and assumptions of Chapter 2, we know that the arbitrage-free option price, for a strike K and a maturity T, is

$$\mathcal{O}(t,T,K) = P(t,T)E^{Q^T}\left\{[\psi(S(T) - K)]^+\big|\mathcal{F}_t\right\},$$

with $\psi \in \{-1, 1\}$, and can be calculated by means of formula (B.2). Indeed, we have just to replace M and V^2 in such a formula respectively with $m(t,T)$ and $v^2(t,T)$:

$$m(t,T) := \ln\frac{S(t)}{P(t,T)} - \rho\frac{\sigma\eta}{a}\left[T - t - \frac{1}{a}\left(1 - e^{-a(T-t)}\right)\right]$$
$$- (y + \tfrac{1}{2}\eta^2)(T-t) - \tfrac{1}{2}V(t,T),$$

$$v^2(t,T) := V(t,T) + \eta^2(T-t) + 2\rho\frac{\sigma\eta}{a}\left[T - t - \frac{1}{a}\left(1 - e^{-a(T-t)}\right)\right].$$

We therefore have the following.

Proposition 12.2.1. *The price at time t of a European option with maturity T, strike K, and written on the asset S is given by*

$$\mathcal{O}(t,T,K) = \psi S(t)e^{-y(T-t)}\Phi\left(\psi\frac{\ln\frac{S(t)}{KP(t,T)} - y(T-t) + \tfrac{1}{2}v^2(t,T)}{v(t,T)}\right)$$
$$- \psi KP(t,T)\Phi\left(\psi\frac{\ln\frac{S(t)}{KP(t,T)} - y(T-t) - \tfrac{1}{2}v^2(t,T)}{v(t,T)}\right).$$
(12.12)

where $\psi = 1$ for a call and $\psi = -1$ for a put.

Formula (12.12) is in agreement with the result of Merton (1973), who derived a closed-form solution for the price of a European call option on a risky asset when interest rates are stochastic. He assumed the existence of a zero-coupon-bond price following a diffusion process where the variance rate of the instantaneous bond's return is deterministic and vanishes at the bond's maturity. Merton's model gives the following formula for the price of the above European call option ($\psi = 1$)

$$S(t)e^{-y(T-t)}\Phi\left(\frac{\ln\frac{S(t)}{KP(t,T)} - y(T-t) + \frac{1}{2}u_t^2}{u_t}\right)$$

$$-KP(t,T)\Phi\left(\frac{\ln\frac{S(t)}{KP(t,T)} - y(T-t) - \frac{1}{2}u_t^2}{u_t}\right),$$

where u_t^2 satisfies

$$u_t^2 = \int_t^T \text{Var}\left\{d\ln\frac{S(\tau)}{P(\tau,T)}\right\},$$

meaning that u_t^2 is the integrated variance of the instantaneous return of $S(t)/P(t,T)$. This formula holds in general for Gaussian short-rate models, not necessarily one-factor. In particular, one can use this expression under a multi-factor Gaussian model such as G2++.

Remark 12.2.1. The difference $v^2(t,T) - \eta^2(T-t)$ is an increasing function of the time to maturity $T - t$. Therefore, as expected, the larger the option maturity, the larger the impact of the stochastic behaviour of interest rates on the option price.

12.3 A More General Model

We now assume that the dynamics of the instantaneous short rate under the risk-adjusted measure Q is given by the more general process

$$\begin{cases} dx(t) = -ax(t)dt + \sigma dW(t), \quad x(0) = 0, \\ r(t) = f(x(t) + \alpha(t)), \end{cases} \quad (12.13)$$

where f is a deterministic real function with inverse g (i.e., $g(f(x)) = x$) and α is a deterministic function that is properly chosen so as to exactly fit the current term structure of spot rates. The asset price is still assumed to evolve according to (12.2), i.e.,

$$dS(t) = S(t)[(r(t) - y)dt + \eta dZ(t)], \quad S(0) = S_0,$$

460   12. Pricing Equity Derivatives under Stochastic Rates

where W and Z are again two Brownian motions with $dZ(t)dW(t) = \rho dt$. Putting $\bar{S}(t) = \ln(S(t)/S_0)$, we denote by $\mathcal{F}_t$ the sigma-field generated by $(x, \bar{S})$ up to time t.

Assuming enough regularity of the deterministic functions above, by Ito's lemma, the dynamics of r is given by

$$dr(t) = \left[f'(g(r(t)))\alpha'(t) - af'(g(r(t)))(g(r(t)) - \alpha(t)) + \tfrac{1}{2}f''(g(r(t)))\sigma^2\right] dt \\ + \sigma f'(g(r(t)))dW(t),$$

(12.14)

with $'$ denoting a derivative, so that r's absolute volatility is $\sigma f'(g(r(t)))$ and the instantaneous correlation between r and S is ρ.

For example, we can retrieve the Black and Karasinski (1991) model, see (3.54), by setting $f(x) = \exp(x)$, so that $g(x) = \ln(x)$. In this case, (12.14) becomes

$$dr(t) = \left[r(t)(\alpha'(t) + a\alpha(t)) - ar(t)\ln(r(t)) + \tfrac{1}{2}\sigma^2 r(t)\right] dt + \sigma r(t)dW(t).$$

The model (12.13) is not analytically tractable in general. We have, therefore, to resort to numerical procedures even to price the simplest derivatives. Indeed, even the fundamental quantities of interest-rate modeling, i.e. bond prices, do not have a closed-form expression with this model.

To this end, we first construct two trinomial trees, one for the process r and the other for the process S, and then we merge the two trees accounting for the proper correlation. The construction of the trinomial trees for r and S is reviewed in the following.

12.3.1 The Construction of an Approximating Tree for r

We first construct a trinomial tree for the process x based on a generalization of the Hull and White (1993) procedure so as to allow for a variable time step. See also Appendix C.

Let us assume that we are given a sequence of times $t_0, t_1, \ldots, t_m$. We set $\Delta t_i = t_{i+1} - t_i$. We denote the value of x on the j-th node at time t_i by $x_{i,j}$ and we set $x_{i,j} := j\Delta x_i$, where the vertical step Δx_i is constant at each time t_i. The mean and standard deviation of $x(t_{i+1})$ conditional on $x = x_{i,j}$ are $x_{i,j} + M_i x_{i,j}$ and $\sqrt{V_i}$, respectively, where

$$M_i := e^{-a\Delta t_i} - 1,$$

$$V_i := \frac{\sigma^2}{2a}\left(1 - e^{-2a\Delta t_i}\right).$$

The branching procedure is established as follows. From $x_{i,j}$, the variable x can move to $x_{i+1,k-1}$, to $x_{i+1,k}$ or to $x_{i+1,k+1}$, where k is chosen so that $x_{i+1,k}$ is as close as possible to $x_{i,j} + M_i x_{i,j}$. The associated probabilities are calculated by matching the mean and variance of the increments of the original continuous-time process. We obtain

$$\begin{cases} p_u(i,j) = \dfrac{\sigma^2 \Delta t_i}{2\Delta x_{i+1}^2} + \dfrac{\eta^2(i,j)}{2\Delta x_{i+1}^2} + \dfrac{\eta(i,j)}{2\Delta x_{i+1}}, \\ p_m(i,j) = 1 - \dfrac{\sigma^2 \Delta t_i}{\Delta x_{i+1}^2} - \dfrac{\eta^2(i,j)}{\Delta x_{i+1}^2}, \\ p_d(i,j) = \dfrac{\sigma^2 \Delta t_i}{2\Delta x_{i+1}^2} + \dfrac{\eta^2(i,j)}{2\Delta x_{i+1}^2} - \dfrac{\eta(i,j)}{2\Delta x_{i+1}}, \end{cases}$$

where $p_u(i,j), p_m(i,j)$ and $p_d(i,j)$ denote the up, middle and down branching probabilities on the j-th node at time t_i, respectively, and where

$$\eta(i,j) = x_{i,j} M_i + j\Delta x_i - k\Delta x_{i+1}, \qquad (12.15)$$

with k to be determined at each node in the tree as previously explained. A portion of this trinomial tree for x is displayed in Figure 12.1. The branching

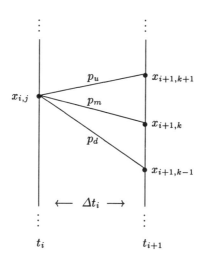

Fig. 12.1. Evolution of the process x at node (i,j).

probabilities should always lie in the interval $[0,1]$. This can be ensured by choosing Δx_{i+1} in such a way that, at each node in the tree, $x_{i+1,k-1}$ and $x_{i+1,k+1}$ bracket the expected value of x in the next time interval conditional to $x_{i,j}$. Hull and White (1993) suggested $\Delta x_{i+1} = \sqrt{3V_i}$ as an appropriate choice. We denote by $\underline{j}_i$ and $\overline{j}_i$ the minimum and the maximum levels at each time step i.

Proceeding in the tree construction, the following step is the displacement of the tree nodes according to the function α. To this end, we denote by $Q(i,j)$ the present value of an instrument paying 1 if node (i,j) is reached and zero otherwise. The values of the displacement $\alpha_i := \alpha(t_i)$ at period i and of $Q(i,j)$ are calculated recursively from $\alpha(0) = g(r_0)$. Precisely, as soon as the value of α_i has been determined, the values $Q(i+1,j)$, $j = \underline{j}_i, \ldots, \overline{j}_i$, are calculated through

$$Q(i+1,j) = \sum_k Q(i,k) p(k,j) \exp(-f(\alpha_i + k\Delta x_i)\Delta t_i),$$

where $p(k,j)$ is the probability of moving from node (i,k) to node $(i+1,j)$ and the sum is over all values of k for which such probability is non-zero.

After deriving the value of $Q(i,j)$, for each j, the value of α_i is calculated by numerically solving

$$\sum_{j=-n_i}^{n_i} Q(i,j) \exp\left[-f(\alpha_i + j\Delta x_i)\Delta t_i\right] - P^M(0,t_{i+1}) = 0$$

where $P^M(0, t_{i+1})$ denotes (as before) the market price at the initial time of a discount bond with maturity t_{i+1} and n_i is the number of nodes on each side of the central node at the i-th time step in the tree.

To obtain the approximating tree for r we finally have to apply the function f to each node value.

12.3.2 The Approximating Tree for S

We first build an approximating tree for $\bar{S}$ assuming that interest rates are constant and equal to $\bar{r}$.

Setting $\Delta \bar{S}_{i+1} := \eta\sqrt{3\Delta t_i}$, for each i, we denote by $q_u(i,l)$, $q_m(i,l)$ and $q_d(i,l)$ the probabilities at period i of moving from $\bar{S}_{i,l} := l\Delta \bar{S}_i$ to $\bar{S}_{i+1,h+1} = \bar{S}_{i+1,h} + \Delta \bar{S}_{i+1}$, $\bar{S}_{i+1,h}$ and $\bar{S}_{i+1,h-1} = \bar{S}_{i+1,h} - \Delta \bar{S}_{i+1}$, respectively. Applying the same "first-two-moments-matching" procedure as before, we obtain

$$\begin{cases} q_u(i,l) = \dfrac{1}{6} + \dfrac{\xi_{l,h}^2}{6\eta^2 \Delta t_i} + \dfrac{\xi_{l,h}}{2\sqrt{3}\eta\sqrt{\Delta t_i}}, \\[2mm] q_m(i,l) = \dfrac{2}{3} - \dfrac{\xi_{l,h}^2}{3\eta^2 \Delta t_i}, \\[2mm] q_d(i,l) = \dfrac{1}{6} + \dfrac{\xi_{l,h}^2}{6\eta^2 \Delta t_i} - \dfrac{\xi_{l,h}}{2\sqrt{3}\eta\sqrt{\Delta t_i}}, \end{cases} \quad (12.16)$$

where

$$h = \text{round}\left(\frac{\bar{S}_{i,l} + (\bar{r} - y - \tfrac{1}{2}\eta^2)\Delta t_i}{\Delta \bar{S}_{i+1}}\right),$$

$$\xi_{l,h} = \bar{S}_{i,l} + (\bar{r} - y - \tfrac{1}{2}\eta^2)\Delta t_i - \bar{S}_{i+1,h}.$$

12.3 A More General Model

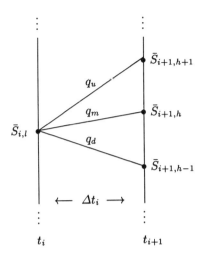

Fig. 12.2. Evolution of the process $\bar{S}$ starting from the value $\bar{S}_{i,l}$ at time t_i.

The resulting tree geometry is displayed in Figure 12.2. To finally obtain the approximating tree for S we have simply to transform each node value by applying the relation $S(t) = S_0 \exp(\bar{S}(t))$.

12.3.3 The Two-Dimensional Tree

In the construction of the joint two-dimensional tree, we have to remember that the marginal probabilities (12.16) explicitly depend on the spot rate r, so that they vary as soon as r varies in the joint tree.

Along the procedure suggested by Hull and White (1994b), we build a preliminary tree by assuming zero correlation. See again Appendix C.

Denoting by (i, j, l) the tree node at time t_i where $x = x_{i,j} = j\Delta x_i$ and $\bar{S} = \bar{S}_{i,l} = l\Delta \bar{S}_i$, if k and h are chosen as before, we denote by

- π_{uu} the probability of moving from (i, j, l) to $(i+1, k+1, h+1)$;
- π_{um} the probability of moving from (i, j, l) to $(i+1, k+1, h)$;
- π_{ud} the probability of moving from (i, j, l) to $(i+1, k+1, h-1)$;
- π_{mu} the probability of moving from (i, j, l) to $(i+1, k, h+1)$;
- π_{mm} the probability of moving from (i, j, l) to $(i+1, k, h)$;
- π_{md} the probability of moving from (i, j, l) to $(i+1, k, h-1)$;
- π_{du} the probability of moving from (i, j, l) to $(i+1, k-1, h+1)$;
- π_{dm} the probability of moving from (i, j, l) to $(i+1, k-1, h)$;
- π_{dd} the probability of moving from (i, j, l) to $(i+1, k-1, h-1)$;

Under zero correlation, these probabilities are simply the product of the corresponding marginal probabilities, where in (12.16) we replace $\bar{r}$ with

$f(x_{i,j} + \alpha_i)$. For example,

$$\pi_{ud} = \left[\frac{\sigma^2 \Delta t_i}{2\Delta x_{i+1}^2} + \frac{\eta^2(i,j)}{2\Delta x_{i+1}^2} + \frac{\eta(i,j)}{2\Delta x_{i+1}}\right]\left[\frac{1}{6} + \frac{\xi_{l,h}^2}{6\eta^2 \Delta t_i} - \frac{\xi_{l,h}}{2\sqrt{3}\eta\sqrt{\Delta t_i}}\right].$$

Let us now define Π_0 as the matrix of these probabilities, i.e.,

$$\Pi_0 := \begin{pmatrix} \pi_{ud} & \pi_{um} & \pi_{uu} \\ \pi_{md} & \pi_{mm} & \pi_{mu} \\ \pi_{dd} & \pi_{dm} & \pi_{du} \end{pmatrix}.$$

To account for the proper correlation ρ we have to shift each probability in Π_0 in such a way that the sum of the shifts in each row and each column is zero (so that the marginal distributions are maintained) and the following is verified:

$$\Delta x_{i+1} \Delta \bar{S}_{i+1} (-\pi_{ud} - \varepsilon_{ud} + \pi_{uu} + \varepsilon_{uu} + \pi_{dd} + \varepsilon_{dd} - \pi_{du} - \varepsilon_{du})$$
$$= \text{Cov}\{x(t_{i+1}), \bar{S}(t_{i+1}) | \mathcal{F}_{t_i}\}$$
$$= \sigma \eta \rho \Delta t_i,$$

at first order in Δt_i. Clearly, no middle "m" term appears, since such terms correspond to at least one of the two components being displaced by 0, thus giving no contribution to the above covariance. Now, recalling that

$$\Delta x_{i+1} = \sqrt{3V_i} = \sqrt{3\frac{\sigma^2}{2a}(1 - e^{-2a\Delta t_i})} \approx \sigma\sqrt{3\Delta t_i}$$

and that

$$\Delta \bar{S}_{i+1} = \eta\sqrt{3\Delta t_i},$$

we obtain that the above equation becomes, in the limit,

$$-\varepsilon_{ud} + \varepsilon_{uu} + \varepsilon_{dd} - \varepsilon_{du} = \frac{\rho}{3},$$

where ε_{ud} denotes the shift for π_{ud} (the definition of the other shifts is analogous).

Solving for the previous constraints, a possible solution, which is similar to that proposed by Hull and White (1994b), is

$$\Pi_\rho = \Pi_0 + \rho(\Pi_1^l - \Pi_0^l), \text{ if } \rho > 0$$
$$\Pi_\rho = \Pi_0 - \rho(\Pi_{-1}^l - \Pi_0^l), \text{ if } \rho < 0$$

where Π_ρ is the probability matrix that accounts for the right correlation ρ and Π_1^l, Π_0^l and Π_{-1}^l are the probability matrices in the limit ($\Delta t_i = 0$) respectively for $\rho = 1$, $\rho = 0$ and $\rho = -1$. More explicitly,

12.3 A More General Model

$$\Pi_\rho = \Pi_0 + \frac{\rho}{36} \begin{pmatrix} -1 & -4 & 5 \\ -4 & 8 & -4 \\ 5 & -4 & -1 \end{pmatrix} \quad \text{if } \rho > 0$$

and

$$\Pi_\rho = \Pi_0 - \frac{\rho}{36} \begin{pmatrix} 5 & -4 & -1 \\ -4 & 8 & -4 \\ -1 & -4 & 5 \end{pmatrix} \quad \text{if } \rho < 0.$$

However, it may happen that one or more entries in Π_ρ are negative. To overcome this drawback, we are compelled to modify the correlation on node (i, j, l), for example substituting Π_ρ with Π_0. Assuming a different correlation on such node, and on all the tree nodes where some probabilities are negative, has a negligible impact on a derivative price as long as the mesh of $\{t_0, t_1, \ldots, t_m\}$ is sufficiently small. Notice, in fact, that in the limit for such a mesh going to zero, the elements of Π_ρ are all positive, being Π_ρ a convex linear combination of two matrices with positive entries.

Part III

APPENDICES

A. A Crash Introduction to Stochastic Differential Equations

> *The principle of generating small amounts of finite improbability by simply hooking the logic circuits of a Bambleweeny 57 Sub-Meson Brain to an atomic vector plotter suspended in a strong Brownian motion producer (say a nice hot cup of tea) were of course well understood [...]*
> Douglas Adams, "The Hitch-Hiker's Guide to the Galaxy"

This book uses continuous time stochastic calculus as a mathematical tool for financial modeling. In this appendix we plan to give a quick (informal) introduction to stochastic differential equations (SDEs) for the reader who is not familiar with this field. These notes are far from being complete or fully rigorous, in that we privilege the intuitive aspect, but we give references for the reader who is willing to deepen her knowledge on such matters.

We note that the understanding, and subsequent implementation, of most of the essential and important issues in interest rate modeling do not require excessively-exotic tools of stochastic calculus. The basic paradigms, risk neutral valuation and change of numeraire, in fact, essentially involve Ito's formula and the Girsanov theorem. We therefore introduce quickly and intuitively such results, which are used especially in Chapter 2. The fact that we do not insist upon more advanced tools of stochastic calculus is mainly due to our belief that the fundamental questions to address in practice can be very often solved with the basic tools above.

A.1 From Deterministic to Stochastic Differential Equations

Here we present a quick and informal introduction to SDEs. We consider the scalar case to simplify exposition.

We consider a probability space $(\Omega, \mathcal{F}, (\mathcal{F}_t)_t, P)$. The usual interpretation of this space as an experiment can help intuition. The generic experiment result is denoted by $\omega \in \Omega$; Ω represents the set of all possible outcomes of the random experiment, and the σ-field $\mathcal{F}$ represents the set of events $A \subset \Omega$ with which we shall work. The σ-field $\mathcal{F}_t$ represents the information available up to time t. We have $\mathcal{F}_t \subseteq \mathcal{F}_u \subseteq \mathcal{F}$ for all $t \leq u$, meaning that "the information increases in time", never exceeding the whole set of events $\mathcal{F}$. The family of σ-fields $(\mathcal{F}_t)_{t \geq 0}$ is called filtration.

If the experiment result is ω and $\omega \in A \in \mathcal{F}$, we say that the event A occurred. If $\omega \in A \in \mathcal{F}_t$, we say that the event A occurred at a time smaller or equal to t.

We use the symbol E to denote expectation, and $E[\cdot|\mathcal{F}]$ denotes expectation conditional on the information contained in $\mathcal{F}$.

We begin by a simple example. Consider a population growth model. Let $x(t) = x_t \in \mathbb{R}$, $x_t \geq 0$, be the population at time $t \geq 0$. The simplest model for the population growth is obtained by assuming that the growth rate dx_t/dt is proportional to the current population. This can be translated into the differential equation:

$$dx_t = Kx_t dt, \quad x_0,$$

where K is a real constant. Now suppose that, due to some complications, it is no longer realistic to assume the initial condition x_0 to be a deterministic constant. Then we may decide to let x_0 be a random variable $X_0(\omega)$, and to model the population growth by the differential equation:

$$dX_t(\omega) = KX_t(\omega)dt, \quad X_0(\omega).$$

The solution of this last equation is $X_t(\omega) = X_0(\omega)\exp[Kt]$. Note that $X_t(\omega)$ is a random variable, but all its randomness comes from the initial condition $X_0(\omega)$. For each experiment result ω, the map $t \mapsto X_t(\omega)$ is called the path of X associated to ω.

As a further step, suppose that not even K is known for certain, but that also our knowledge of K is perturbed by some randomness, which we model as the "increment" of a stochastic process $\{W_t(\omega), t \geq 0\}$, so that

$$dX_t(\omega) = (Kdt + dW_t(\omega))X_t(\omega), \quad X_0(\omega), \quad K \geq 0. \quad (A.1)$$

Here, $dW_t(\omega)$ represents a noise process that adds randomness to K.

Equation (A.1) is an example of stochastic differential equation (SDE). More generally, a SDE is written as

$$dX_t(\omega) = f_t(X_t(\omega))dt + \sigma_t(X_t(\omega))dW_t(\omega), \quad X_0(\omega). \quad (A.2)$$

The function f, corresponding to the deterministic part of the SDE, is called the *drift*. The function σ_t (or sometimes its square $a_t := \sigma_t^2$) is called the *diffusion coefficient*. Note that the randomness enters the differential equation from two sources: The "noise term" $\sigma_t(\cdot)dW_t(\omega)$ and the initial condition $X_0(\omega)$.

Usually, the solution X of the SDE is called also a *diffusion process*, because of the fact that some particular SDE's can be used to arrive at a model of physical diffusion. In general the paths $t \mapsto X_t(\omega)$ of a diffusion process X are continuous.

A.1 From Deterministic to Stochastic Differential Equations

Brownian Motion

The process whose "increments" $dW_t(\omega)$ are candidate for representing the noise process in (A.2) is the Brownian motion. This process has important properties: It has stationary and independent Gaussian increments "$dW_t(\omega)$", which means intuitively that, for example, $W_{T+\Delta}(\omega) - W_T(\omega)$ is independent of the history of W up to time T. Therefore, $W_{T+\Delta}(\omega) - W_T(\omega)$ can assume *any* value independently of $\{W_t(\omega), t \leq T\}$.

The definition of Brownian motion requires also the paths $t \mapsto W_t(\omega)$ to be continuous.

It turns out that the properties listed above imply that although the path be continuous, they are (almost surely) nowhere differentiable. In fact, the paths have unbounded variation, and hence $\dot{W}_t(\omega) = dW_t(\omega)/dt$ does not exist.

Stochastic Integrals

Since $\dot{W}_t(\omega)$ does not exist, what meaning can we give to equation (A.2)? The answer relies on rewriting (A.2) in integral form:

$$X_t(\omega) = X_0(\omega) + \int_0^t f_s(X_s(\omega))\, ds + \int_0^t \sigma_s(X_s(\omega))\, dW_s(\omega), \qquad (A.3)$$

so that, from now on, all differential equations involving terms like dW are meant as integral equations, in the same way as (A.2) will be an abbreviation for (A.3).

However, we are not done yet, since we have to deal with the new problem of defining an integral like $\int_0^t \sigma_s(X_s(\omega))dW_s(\omega)$. A priori it is not possible to define it as a Stieltjes integral on the paths, since they have unbounded variation. Nonetheless, under some "reasonable" assumptions that we do not mention (see for example Chapter 3 of Øksendal (1992)), it is still possible to define such integrals a la Stieltjes. The price to be paid is that the resulting integral will depend on the chosen points of the sub-partitions (whose mesh tends to zero) used in the limit that defines the integral. More specifically, consider the following definition. Take an interval $[0,T]$ and consider the following dyadic partition of $[0,T]$ depending on an integer n,

$$T_i^n = \min\left(T, \frac{i}{2^n}\right), \quad i = 0, 1, \ldots, \infty.$$

Notice that from a certain i on all terms collapse to T, i.e. $T_i^n = T$ for all $i > 2^n T$. For each n we have such a partition, and when n increases the partition contains more elements, giving a better discrete approximation of the continuous interval $[0,T]$. Then define the integral as

$$\int_0^T \phi_s(\omega) dW_s(\omega) = \lim_{n \to \infty} \sum_{i=0}^{\infty} \phi_{t_i^n}(\omega)[W_{T_{i+1}^n}(\omega) - W_{T_i^n}(\omega)]$$

where t_i^n is any point in the interval $[T_i^n, T_{i+1}^n)$. Now, by choosing $t_i^n := T_i^n$ (initial point of the subinterval) we have the definition of the Ito integral, whereas by taking $t_i^n := (T_i^n + T_{i+1}^n)/2$ (middle point) we obtain a different result, the Stratonovich integral.

The Ito integral has interesting probabilistic properties (for example, it is a *martingale*, an important type of stochastic process that will be briefly defined below), but leads to a calculus where the standard chain rule is not preserved since there is a non-zero contribution of the second order terms. On the contrary, although probabilistically less interesting, the Stratonovich integral does preserve the ordinary chain rule, and is preferable from the viewpoint of properties of the paths.

To better understand the difference between these two definitions, we can resort to the following classical example of stochastic integral computed both with the Ito calculus and the Stratonovich calculus:

$$\text{Ito} \Rightarrow \int_0^t W_s(\omega) dW_s(\omega) = \frac{W_t(\omega)^2}{2} - \frac{1}{2}t,$$
$$\text{Stratonovich} \Rightarrow \int_0^t W_s(\omega) dW_s(\omega) = \frac{W_t(\omega)^2}{2}.$$

To distinguish between the two definition, a symbol "∘" is often introduced to denote the Stratonovich version as follows:

$$\int_0^t W_s(\omega) \circ dW_s(\omega).$$

In differential notation, one then has

$$\text{Ito} \Rightarrow dW_t(\omega)^2 = dt + 2W_t(\omega)dW_t(\omega), \tag{A.4}$$
$$\text{Stratonovich} \Rightarrow dW_t(\omega)^2 = 2W_t(\omega) \circ dW_t(\omega), \tag{A.5}$$

In the Ito version, the "dt" term originates from second order effects, which are not negligible like in ordinary calculus. Note that the first integral is a martingale (so that, for example, it has constant expected value equal to zero, which is an important probabilistic property), but does not satisfy formal rules of calculus, as instead does the second one (which is not a martingale).[1]

In general, stochastic integrals are defined a la Lebesgue rather than a la Riemann-Stieltjes. One defines the stochastic integral for increasingly more sophisticated integrands (indicators, simple functions...), and then takes the limit in some sense. The reader interested in the deeper mathematical aspects of stochastic integration in connection with SDEs may consult books such as, for example, Øksendal (1992) or, for a more advanced treatment, Rogers and Williams (1987).

---

[1] We must point out, anyway, that it is possible to transform an equation written in the Ito form into an equation in the Stratonovich form with the same solution by altering the drift, and vice-versa. This is referred to as the Ito-Stratonovich transformation, see Øksendal (1992), Chapter 3.

Martingales, Driftless SDEs and Semimartingales

In our discussion above, we have mentioned the concept of martingale. To give a quick idea, consider a process X satisfying the following measurability and integrability conditions.

Measurability: $\mathcal{F}_t$ includes all the information on X up to time t, usually expressed in the literature by saying that $(X_t)_t$ is *adapted* to $(\mathcal{F}_t)_t$;

Integrability: the relevant expected values exist.

A martingale is a process satisfying these two conditions and such that the following property holds for each $t \leq T$:

$$E[X_T|\mathcal{F}_t] = X_t.$$

This definition states that, if we consider t as the present time, the expected value at a future time T given the current information is equal to the current value. This is, among other things, a picture of a "fair game", where it is not possible to gain or lose on average. It turns out that the martingale property is also suited to model the absence of arbitrage in mathematical finance. To avoid arbitrage, one requires that certain fundamental processes of the economy be martingales, so that there are no "safe" ways to make money from nothing out of them.

Consider an SDE admitting a unique solution (resulting in a diffusion process, as stated above). This solution is a martingale when the equation has zero drift. In other terms, the solution of the SDE (A.2) is a martingale when $f_t(\cdot) = 0$ for all t:

$$dX_t(\omega) = \sigma_t(X_t(\omega))dW_t(\omega), \quad X_0(\omega).$$

Therefore, in diffusion-processes language, martingale means driftless diffusion process.

A submartingale is a similar process X satisfying instead

$$E[X_T|\mathcal{F}_t] \geq X_t.$$

This means that the expected value of the process grows in time, and that averages of future values of the process given the current information always exceed (or at least are equal to) the current value.

Similarly, a supermartingale satisfies

$$E[X_T|\mathcal{F}_t] \leq X_t,$$

and the expected value of the process decreases in time, so that averages of future values of the process given the current information are always smaller than (or at most are equal to) the current value.

A process X that is either a supermartingale or a submartingale is usually termed semimartingale.

Quadratic Variation

The quadratic variation of a stochastic process Y_t with continuous paths $t \mapsto Y_t(\omega)$ is defined as follows:

$$\langle Y \rangle_T = \lim_{n \to \infty} \sum_{i=1}^{\infty} (Y_{T_i^n}(\omega) - Y_{T_{i-1}^n}(\omega))^2.$$

Intuitively this could be written as a "second order" integral:

$$\langle Y \rangle_T = \text{``} \int_0^T (dY_s(\omega))^2 \text{''},$$

or, even more intuitively, in the differential form

$$\text{``} d\langle Y \rangle_t = dY_t(\omega) dY_t(\omega) \text{''}.$$

It is easy to check that a process Y whose paths $t \mapsto Y_t(\omega)$ are differentiable for almost all ω satisfies $\langle Y \rangle_t = 0$. In case Y is a Brownian motion, it can be proved, instead, that

$$\langle W \rangle_T = T, \quad \text{for each } T,$$

which can be written in a more informal way as

$$\boxed{dW_t(\omega)\, dW_t(\omega) = dt}.$$

Again, this comes from the fact that the Brownian motion moves so quickly that second order effects are not negligible. Instead, a process whose trajectories are differentiable cannot move so quickly, and therefore its second order effects do not contribute.

In case the process Y is equal to the deterministic process $t \mapsto t$, so that $dY_t = dt$, we immediately retrieve the classical result from (deterministic) calculus:

$$\boxed{dt\, dt = 0}.$$

Quadratic Covariation

One can also define the quadratic covariation of two processes Y and Z with continuous paths as

$$\langle Y, Z \rangle_T = \lim_{n \to \infty} \sum_{i=1}^{\infty} (Y_{T_i^n}(\omega) - Y_{T_{i-1}^n}(\omega))(Z_{T_i^n}(\omega) - Z_{T_{i-1}^n}(\omega)).$$

Intuitively this could be written as a "second order" integral:

A.1 From Deterministic to Stochastic Differential Equations

$$\langle Y, Z \rangle_T = \text{``} \int_0^T dY_s(\omega) dZ_s(\omega) \text{''},$$

or, in differential form,

$$\text{``} d\langle Y, Z \rangle_t = dY_t(\omega) dZ_t(\omega) \text{''}$$

It is then easy to check that, denoting again by t the deterministic process $t \mapsto t$,

$$\langle W, t \rangle_T = 0, \quad \text{for each } T,$$

which can be informally written as

$$\boxed{dW_t(\omega) \, dt = 0}.$$

Solution to a General SDE

Let us go back to our general SDE, and let us take time-homogeneous coefficients for simplicity:

$$dX_t(\omega) = f(X_t(\omega))dt + \sigma(X_t(\omega))dW_t(\omega), \quad X_0(\omega). \quad (A.6)$$

Under which conditions does it admit a unique solution in the Ito sense? Standard theory tells us that it is enough to have both the f and σ coefficients satisfying Lipschitz continuity (and linear growth, which does not follow automatically in the time-inhomogeneous case or with *local* Lipschitz continuity only). These *sufficient* conditions are valid for deterministic differential equations as well, and can be weakened, especially in dimension one. Typical examples showing how, without Lipschitz continuity or linear growth, existence and uniqueness of solutions can fail are the following:

$$dx_t = x_t^2 \, dt, \quad x_0 = 1 \quad \Rightarrow \quad x_t = \frac{1}{1-t}, \quad t \in [0, 1),$$

(explosion in finite time) and

$$dx_t = 3x_t^{2/3} \, dt, \quad x_0 = 0 \quad \Rightarrow \quad x_t = 1_{(a,\infty)}(t) \, (t-a)^3, \quad t \in [0, +\infty)$$

for any positive a (no uniqueness).

The proof of the fact that the existence and uniqueness of a solution to a SDE is guaranteed by Lipschitz continuity and linear growth of the coefficients, is similar in spirit to the proof for deterministic equations. See again Øksendal (1992) for the details.

Interpretation of the Coefficients of the SDE

We conclude by presenting an interpretation of the drift and diffusion coefficient for a SDE. For deterministic differential equations such as

$$dx_t = f(x_t)dt,$$

with a smooth function f, one clearly has

$$\lim_{h \to 0} \frac{x_{t+h} - x_t}{h}\bigg|_{x_t = y} = f(y),$$

$$\lim_{h \to 0} \frac{(x_{t+h} - x_t)^2}{h}\bigg|_{x_t = y} = 0.$$

The "analogous" relations for the SDE

$$dX_t(\omega) = f(X_t(\omega))dt + \sigma(X_t(\omega))dW_t(\omega),$$

are the following:

$$\lim_{h \to 0} E\left\{ \frac{X_{t+h}(\omega) - X_t(\omega)}{h} \bigg| X_t = y \right\} = f(y)$$

$$\lim_{h \to 0} E\left\{ \frac{[X_{t+h}(\omega) - X_t(\omega)]^2}{h} \bigg| X_t = y \right\} = \sigma^2(y).$$

The second limit is non-zero because of the "infinite velocity" of Brownian motion, while the first limit is the analogous of the deterministic case.

A.2 Ito's Formula

Now we are ready to introduce the famous Ito formula, which gives the "chain rule" for differentials in a stochastic context.

For deterministic differential equations such as

$$dx_t = f(x_t)dt,$$

given a smooth transformation $\phi(t, x)$, one can write the evolution of $\phi(t, x_t)$ via the chain rule:

$$d\phi(t, x_t) = \frac{\partial \phi}{\partial t}(t, x_t)dt + \frac{\partial \phi}{\partial x}(t, x_t)dx_t. \tag{A.7}$$

We already observed in (A.4) that whenever a Brownian motion is involved, such a fundamental rule of calculus needs to be modified. The general formulation of the chain rule for stochastic differential equations is the following.

A.2 Ito's Formula

Let $\phi(t,x)$ be a smooth function and let $X_t(\omega)$ be the unique solution of the stochastic differential equation (A.6). Then, Ito's formula reads as

$$d\phi(t, X_t(\omega)) = \frac{\partial \phi}{\partial t}(t, X_t(\omega))dt + \frac{\partial \phi}{\partial x}(t, X_t(\omega))dX_t(\omega)$$
$$+ \frac{1}{2}\frac{\partial^2 \phi}{\partial x^2}(t, X_t(\omega))dX_t(\omega)dX_t(\omega), \tag{A.8}$$

or, in a more compact notation,

$$\boxed{d\phi(t, X_t) = \frac{\partial \phi}{\partial t}(t, X_t)dt + \frac{\partial \phi}{\partial x}(t, X_t)dX_t + \frac{1}{2}\frac{\partial^2 \phi}{\partial x^2}(t, X_t)d\langle X \rangle_t}.$$

Comparing equation (A.8) with its "deterministic" counterpart (A.7), we notice that the extra term

$$\frac{1}{2}\frac{\partial^2 \phi}{\partial x^2}(t, X_t(\omega))dX_t(\omega)dX_t(\omega)$$

appears in our stochastic context, and this is the term due to the Ito integral.[2]

The term $dX_t(\omega)dX_t(\omega)$ can be developed algebraically by taking into account the rules on the quadratic variation and covariation seen above:

$$dW_t(\omega)\,dW_t(\omega) = dt, \quad dW_t(\omega)\,dt = 0, \quad dt\,dt = 0.$$

We thus obtain

$$d\phi(t, X_t(\omega)) = \frac{\partial \phi}{\partial t}(t, X_t(\omega))dt + \frac{\partial \phi}{\partial x}(t, X_t(\omega))f(X_t(\omega))dt$$
$$+ \frac{1}{2}\frac{\partial^2 \phi}{\partial x^2}(t, X_t(\omega))\sigma^2(X_t)dt + \frac{\partial \phi}{\partial x}(t, X_t(\omega))\sigma(X_t(\omega))dW_t(\omega).$$

Stochastic Leibnitz Rule

Also the classical Leibnitz rule for differentiation of a product of functions is modified, analogously to the chain rule. The related formula can be derived as a corollary of Ito's formula in two dimensions, and is reported below.

For deterministic and differentiable functions x and y we have the deterministic Leibnitz rule

$$d(x_t\,y_t) = x_t\,dy_t + y_t\,dx_t.$$

---

[2] Notice that, when writing instead the equation for $d\phi(t, X_t(\omega))$ in the Stratonovich sense, we would re-obtain

$$d\phi(t, X_t(\omega)) = \frac{\partial \phi}{\partial t}(t, X_t(\omega))dt + \frac{\partial \phi}{\partial x}(t, X_t(\omega)) \circ dX_t(\omega),$$

since formal rules of calculus are preserved.

For two diffusion processes (and more generally semimartingales) $X_t(\omega)$ and $Y_t(\omega)$ we have instead[3]

$$d(X_t(\omega) Y_t(\omega)) = X_t(\omega) \, dY_t(\omega) + Y_t(\omega) \, dX_t(\omega) + dX_t(\omega) \, dY_t(\omega),$$

or, in more compact notation,

$$\boxed{d(X_t Y_t) = X_t \, dY_t + Y_t \, dX_t + d\langle X, Y\rangle_t}.$$

A.3 Discretizing SDEs for Monte Carlo: Euler and Milstein Schemes

When one cannot solve an SDE explicitly, it is possible to simulate its trajectories through a discretization scheme. Here we briefly review the Euler scheme. For a more detailed treatment the reader is referred to Klöden and Platen (1995).

Consider again the SDE

$$dX_t(\omega) = f(X_t(\omega))dt + \sigma(X_t(\omega))dW_t(\omega), \quad x_0$$

where for simplicity we took a deterministic initial condition. Let us integrate this equation between s and $s + \Delta s$:

$$X_{s+\Delta s}(\omega) = X_s(\omega) + \int_s^{s+\Delta s} f(X_t(\omega))dt + \int_s^{s+\Delta s} \sigma(X_t(\omega))dW_t(\omega).$$

The Euler scheme consists of approximating this integral equation by

$$\bar{X}_{s+\Delta s}(\omega) = \bar{X}_s(\omega) + f(\bar{X}_s(\omega))\Delta s + \sigma(\bar{X}_s(\omega))(W_{s+\Delta s}(\omega) - W_s(\omega)) \quad \text{(A.9)}$$

with $\bar{X}_0(\omega) = x_0$. If we apply this formula iteratively for a given set of s's, say

$$s = s_1, s_2, \ldots, s_m, \quad s_1 = 0, \quad s_m = T,$$

we obtain a discretized approximation $\bar{X}$ of the solution X of the above SDE. For a definition of the order of convergence of this and other schemes see Klöden and Platen (1995).

A stronger convergence is attained with a more refined scheme, called Milstein scheme. We do not review the Milstein scheme here. We only hint at the fact that when the diffusion coefficient is deterministic, say $\sigma(X_t(\omega)) = \sigma(t)$, with $t \mapsto \sigma(t)$ a deterministic function of time, the Euler and Milstein schemes coincide. Therefore it is preferable, when possible, to apply the Euler scheme to SDEs with deterministic diffusion coefficients, since this ensures the same stronger convergence of the Milstein scheme.

---

[3] Clearly, using the Stratonovich calculus, we would still have $d(X_t Y_t) = X_t \circ dY_t + Y_t \circ dX_t$.

A.3 Discretizing SDEs for Monte Carlo: Euler and Milstein Schemes 479

These discretization schemes can be useful for Monte Carlo simulation. Indeed, suppose we need to compute the expected value of a functional of the solution X of the above SDE, say for simplicity

$$E_0[\phi(X_{s_1}(\omega),\ldots,X_{s_m}(\omega))], \quad s_1 = 0, \quad s_m = T$$

(this is a typical pricing problem for path-dependent payoffs in mathematical finance). Assume also that the times s are close to each other.

We may decide to compute an approximation of this expectation as follows.

1. Select the number N of scenarios for the Monte Carlo method.
2. Set the initial value to $\bar{X}_0^j = x_0$ for all scenarios $j = 1,\ldots,N$.
3. Set $k = 1$.
4. Set $s = s_k$ and $\Delta s = s_{k+1} - s_k$ (so that $s + \Delta s = s_{k+1}$).
5. Generate N new realizations ΔW^j ($j = 1,\ldots,N$) of a standard Gaussian distribution $\mathcal{N}(0,1)$ multiplied by $\sqrt{\Delta s}$, thus simulating the distribution of

$$(W_{s+\Delta s}(\omega) - W_s(\omega)).$$

6. Apply formula (A.9) for each scenario $j = 1,\ldots,N$ with the generated shocks:

$$\bar{X}_{s+\Delta s}^j = \bar{X}_s^j + f(\bar{X}_s^j)\Delta s + \sigma(\bar{X}_s^j)\Delta W^j.$$

7. Store $\bar{X}_{s+\Delta s}^j$ for all j.
8. If $s + \Delta s = s_m$ then stop, otherwise increase k by one and start again from point 4.
9. Approximate the expected value by

$$\frac{\sum_{j=1}^{N} \phi(\bar{X}_{s_1}^j, \ldots, \bar{X}_{s_m}^j)}{N}.$$

Notice that the increments can be generated as new independent draws from a Gaussian distribution at each iteration because the Brownian motion has independent (stationary Gaussian) increments.

Notice also that in case the s's are not close to each other, one needs to infra-discretize the equation between two of the s's, and apply several times the basic scheme (A.9) at the infra-instants. For simplicity, we assumed above that the times s's in the expectation are close enough for the scheme to be applied directly without further discretization.

Discretization schemes are useful and powerful because they allow us to replace, in small intervals, the unknown exact distribution for the transition $X_s \to X_{s+\Delta s}$ with the known (and easy to simulate) Gaussian distribution in the shocks of the transition $\bar{X}_s \to \bar{X}_{s+\Delta s}$.

A.4 Examples

We now present some relevant examples of SDEs that are often encountered throughout the book.

Linear SDEs with Deterministic Diffusion Coefficient

A SDE is said to be linear if both its drift and diffusion coefficients are first order polynomials (or affine functions) in the state variable. We here consider the particular case:

$$dX_t(\omega) = (\beta_t X_t(\omega) + \alpha_t)dt + v_t dW_t(\omega), \quad X_0(\omega) = x_0, \quad (A.10)$$

where α, β, v are deterministic functions of time that are regular enough to ensure existence and uniqueness of a solution.

It can be shown that a stochastic integral of a deterministic function is the same both in the Stratonovich and in the Ito sense. As a consequence, by writing (A.10) in integral form we see that the same equation holds in the Stratonovich sense:

$$dX_t(\omega) = (\beta_t X_t(\omega) + \alpha_t)dt + v_t \circ dW_t(\omega), \quad X_0(\omega) = x_0,$$

so that we can solve it by ordinary calculus for linear differential equations: We obtain

$$X_t(\omega) = e^{\int_0^t \beta_s ds} \left[x_0 + \int_0^t e^{-\int_0^s \beta_u du} \alpha_s ds + \int_0^t e^{-\int_0^s \beta_u du} v_s dW_s(\omega) \right]$$

$$= x_0 e^{\int_0^t \beta_s ds} + \int_0^t e^{\int_s^t \beta_u du} \alpha_s ds + \int_0^t e^{\int_s^t \beta_u du} v_s dW_s(\omega).$$

A remarkable fact is that the distribution of the solution X_t is normal at each time t. Intuitively, this holds since the last stochastic integral is a limit of a sum of independent normal random variables. Indeed, we have

$$X_t \sim \mathcal{N}\left(x_0 e^{\int_0^t \beta_s ds} + \int_0^t e^{\int_s^t \beta_u du} \alpha_s ds, \int_0^t e^{2\int_s^t \beta_u du} v_s^2 ds \right).$$

The major examples of models based on a SDE like (A.10) are that of Vasicek (1978) and that of Hull and White (1990). See also equations (3.5) and (3.32).

Lognormal Linear SDEs

Another interesting example of linear SDE is that where the diffusion coefficient is a first order homogeneous polynomial in the underlying variable. This SDE can be obtained as an exponential of a linear equations with deterministic diffusion coefficient. Indeed, let us take $Y_t = \exp(X_t)$, where X_t evolves according to (A.10), and write by Ito's formula

$$dY_t(\omega) = de^{X_t(\omega)} = e^{X_t(\omega)}dX_t(\omega) + \tfrac{1}{2}e^{X_t(\omega)}dX_t(\omega)\,dX_t(\omega)$$
$$= Y_t(\omega)[\alpha_t + \beta_t \ln Y_t(\omega) + \tfrac{1}{2}v_t^2]dt + v_t Y_t(\omega)dW_t(\omega).$$

As a consequence, the process Y has a lognormal marginal density. A major example of model based on such a SDE is the Black and Karasinski (1991) model. See also (3.53).

Geometric Brownian Motion

The geometric Brownian motion is a particular case of a process satisfying a lognormal linear SDE. Its evolution is defined according to

$$dX_t(\omega) = \mu X_t(\omega)dt + \sigma X_t(\omega)dW_t(\omega),$$

where μ and σ are positive constants. To check that X is indeed a lognormal process, one can compute $d\ln(X_t)$ via Ito's formula and obtain

$$X_t(\omega) = X_0 \exp\left\{\left(\mu - \tfrac{1}{2}\sigma^2\right)t + \sigma W_t(\omega)\right\}.$$

From the seminal work of Black and Scholes (1973) on, processes of this type are frequently used in option pricing theory to model general asset price dynamics. Notice that this process is a submartingale, in that clearly

$$E[X_T|\mathcal{F}_t] = e^{\mu(T-t)}X_t \geq X_t.$$

Finally, notice also that by setting $Y_t(\omega) = e^{-\mu t}X_t(\omega)$, we obtain

$$dY_t(\omega) = \sigma Y_t(\omega)dW_t(\omega).$$

Therefore, since the drift of this last SDE is zero, $e^{-\mu t}X_t$ is a martingale.

Square-Root Processes

An interesting case of non-linear SDE is given by

$$dX_t(\omega) = (\beta_t X_t(\omega) + \alpha_t)dt + v_t\sqrt{X_t(\omega)}\,dW_t(\omega), \quad X_0(\omega) = x_0. \quad (A.11)$$

A process following such dynamics is commonly referred to as square-root process. Major examples of models based on this dynamics are the Cox, Ingerssoll and Ross (1985) instantaneous interest rate model, see also (3.21), and a particular case of the constant-elasticity of variance (CEV) model for stock prices:

$$dX_t(\omega) = \mu X_t(\omega)dt + \sigma\sqrt{X_t(\omega)}\,dW_t(\omega).$$

In general, square-root processes are naturally linked to non-central χ-square distributions.[4] In particular, there are simplified versions of (A.11) for which the resulting process X is strictly positive and analytically tractable, like in the case of the Cox, Ingerssoll and Ross (1985) model.

---

[4] They may also display exit boundaries, for example the CEV process features absorption in the origin.

A.5 Two Important Theorems

We now introduce (informally) two important results. See Øksendal (1992) for more details.

The Feynman-Kač Theorem

The Feynman-Kač theorem, under certain assumptions, allows us to express the solution of a given partial differential equation (PDE) as the expected value of a function of a suitable diffusion process whose drift and diffusion coefficient are defined in terms of the PDE coefficients.

Theorem A.5.1. [**The Feynman-Kač Theorem**] *Given Lipschitz continuous $f(x)$ and $\sigma(x)$ and a smooth ϕ, the solution of the PDE*

$$\frac{\partial V}{\partial t}(t,x) + \frac{\partial V}{\partial x}(t,x)f(x) + \frac{1}{2}\frac{\partial^2 V}{\partial x^2}(t,x)\sigma^2(x) = rV(t,x) \qquad (A.12)$$

with terminal boundary condition

$$V(T,x) = \phi(x) \qquad (A.13)$$

can be expressed as the following expected value

$$V(t,x) = e^{-r(T-t)}\,\widetilde{E}\{\phi(X_T)|X_t = x\}, \qquad (A.14)$$

where the diffusion process X has dynamics, starting from x at time t, given by

$$dX_s(\omega) = f(X_s(\omega))ds + \sigma(X_s(\omega))d\widetilde{W}_s(\omega), \quad s \geq t,\ X_t(\omega) = x \qquad (A.15)$$

under the probability measure $\widetilde{P}$ under which the expectation $\widetilde{E}\{\cdot\}$ is taken. The process $\widetilde{W}$ is a standard Brownian motion under $\widetilde{P}$.

Notice that the terminal condition determines the function ϕ of the diffusion process whose expectation is relevant, whereas the PDE coefficients determine the dynamics of the diffusion process.

This theorem is important because it establishes a link between the PDE's of traditional analysis and physics and diffusion processes in stochastic calculus. Solutions of PDE's can be interpreted as expectations of suitable transformations of solutions of stochastic differential equations and vice versa.

The Girsanov Theorem

The Girsanov theorem shows how a SDE changes due to changes in the underlying probability measure. It is based on the fact that the SDE drift depends on the particular probability measure P in our probability space

$(\Omega, \mathcal{F}, (\mathcal{F}_t)_t, P)$, and that, if we change the probability measure in a "regular" way, the drift of the equation changes while the diffusion coefficient remains the same. The Girsanov theorem can be thus useful when we want to modify the drift coefficient of a SDE. Indeed, suppose that we are given two measures P^* and P on the space $(\Omega, \mathcal{F}, (\mathcal{F}_t)_t)$. Two such measures are said to be equivalent, written $P^* \sim P$, if they share the same sets of null probability (or of probability one, which is equivalent). Therefore two measures are equivalent when they agree on which events of $\mathcal{F}$ hold almost surely. Accordingly, a proposition holds almost surely under P if and only if it holds almost surely under P^*. Similar definitions apply also for the measures restriction to $\mathcal{F}_t$, thus expressing equivalence of the two measures up to time t.

When two measures are equivalent, it is possible to express the first in terms of the second through the Radon-Nikodym derivative. Indeed, there exists a martingale ρ_t on $(\Omega, \mathcal{F}, (\mathcal{F}_t)_t, P)$ such that

$$P^*(A) = \int_A \rho_t(\omega) dP(\omega), \quad A \in \mathcal{F}_t,$$

which can be written in a more concise form as

$$\left. \frac{dP^*}{dP} \right|_{\mathcal{F}_t} = \rho_t.$$

The process ρ_t is called the Radon-Nikodym derivative of P^* with respect to P restricted to $\mathcal{F}_t$.

When in need of computing the expected value of an integrable random variable X, it may be useful to switch from one measure to another equivalent one. Indeed, it is possible to prove that the following equivalence holds:

$$E^*[X] = \int_\Omega X(\omega) dP^*(\omega) = \int_\Omega X(\omega) \frac{dP^*}{dP}(\omega) dP(\omega) = E\left[X \frac{dP^*}{dP}\right],$$

where E^* and E denote expected values with respect to the probability measures P^* and P, respectively. More generally, when dealing with conditional expectations, we can prove that

$$E^*[X|\mathcal{F}_t] = \frac{E\left[X \frac{dP^*}{dP} | \mathcal{F}_t\right]}{\rho_t}.$$

Theorem A.5.2. [**The Girsanov theorem**] *Consider again the stochastic differential equation, with Lipschitz coefficients,*

$$dX_t(\omega) = f(X_t(\omega))dt + \sigma(X_t(\omega))dW_t(\omega), \quad x_0,$$

under P. Let be given a new drift $f^*(x)$ and assume $(f^*(x) - f(x))/\sigma(x)$ to be bounded. Define the measure P^* by

484 A. A Crash Introduction to Stochastic Differential Equations

$$\frac{dP^*}{dP}(\omega)\bigg|_{\mathcal{F}_t} = \exp\left\{-\frac{1}{2}\int_0^t \left(\frac{f^*(X_s(\omega)) - f(X_s(\omega))}{\sigma(X_s(\omega))}\right)^2 ds \right.$$
$$\left. + \int_0^t \frac{f^*(X_s(\omega)) - f(X_s(\omega))}{\sigma(X_s(\omega))} dW_s(\omega)\right\}.$$

Then P^* is equivalent to P. Moreover, the process W^* defined by

$$dW_t^*(\omega) = -\left[\frac{f^*(X_t(\omega)) - f(X_t(\omega))}{\sigma(X_t(\omega))}\right] dt + dW_t(\omega)$$

is a Brownian motion under P^*, and

$$dX_t(\omega) = f^*(X_t(\omega))dt + \sigma(X_t(\omega))dW_t^*(\omega), \quad x_0.$$

As already noticed, this theorem is fundamental when we wish to change the drift of a SDE. It now clear that we can do this by defining a new probability measure P^*, via a suitable Radon-Nikodym derivative, in terms of the difference "desired drift - given drift".

In mathematical finance, a classical example of application of the Girsanov theorem is when one moves from the "real-world" asset price dynamics

$$dS_t(\omega) = \mu S_t(\omega)dt + \sigma S_t(\omega)dW_t(\omega)$$

to the risk-neutral ones

$$dS_t(\omega) = rS_t(\omega)dt + \sigma S_t(\omega)dW_t^*(\omega).$$

This is accomplished by setting

$$\frac{dP^*}{dP}(\omega)\bigg|_{\mathcal{F}_t} = \exp\left\{-\frac{1}{2}\left(\frac{\mu-r}{\sigma}\right)^2 t - \frac{\mu-r}{\sigma}W_t(\omega)\right\}. \quad (A.16)$$

We finally stress that above we assumed boundedness for simplicity, but less stringent assumptions are possible for the theorem to hold. See for example Øksendal (1992).

B. A Useful Calculation

Lemma. *Let M, V and K be real numbers with V and K positive. Then, for $\omega \in \{-1, 1\}$,*

$$\int_{-\infty}^{+\infty} \frac{1}{\sqrt{2\pi}V} [\omega(e^y - K)]^+ e^{-\frac{1}{2}\frac{(y-M)^2}{V^2}} dy$$

$$= \omega e^{M + \frac{1}{2}V^2} \Phi\left(\omega \frac{M - \ln(K) + V^2}{V}\right) - \omega K \Phi\left(\omega \frac{M - \ln(K)}{V}\right).$$

(B.1)

Proof. The above integral

$$\int_{-\infty}^{+\infty} \frac{1}{\sqrt{2\pi}V} [\omega(e^y - K)]^+ e^{-\frac{1}{2}\frac{(y-M)^2}{V^2}} dy$$

is calculated as follows:

$$= \int_{\ln(K)}^{+\infty \cdot \omega} \frac{1}{\sqrt{2\pi}V} (e^y - K) e^{-\frac{1}{2}\frac{(y-M)^2}{V^2}} dy$$

$$= \int_{\frac{\ln(K)-M}{V}}^{+\infty \cdot \omega} \frac{1}{\sqrt{2\pi}} (e^{M+Vz} - K) e^{-\frac{1}{2}z^2} dz$$

$$= e^{M + \frac{1}{2}V^2} \int_{\frac{\ln(K)-M}{V}}^{+\infty \cdot \omega} \frac{1}{\sqrt{2\pi}} e^{-\frac{1}{2}(z-V)^2} dz - K \int_{\frac{\ln(K)-M}{V}}^{+\infty \cdot \omega} \frac{1}{\sqrt{2\pi}} e^{-\frac{1}{2}z^2} dz$$

$$= e^{M + \frac{1}{2}V^2} \left[\Phi(+\infty \cdot \omega) - \Phi\left(\frac{\ln(K) - M - V^2}{V}\right)\right]$$

$$- K \left[\Phi(+\infty \cdot \omega) - \Phi\left(\frac{\ln(K) - M}{V}\right)\right]$$

$$= e^{M + \frac{1}{2}V^2} \omega \Phi\left(-\omega \frac{\ln(K) - M - V^2}{V}\right) - K\omega \Phi\left(-\omega \frac{\ln(K) - M}{V}\right).$$

□

Proposition. *Let X be random variable that is lognormally distributed, and denote by M and V the mean and standard deviation of $Y := \ln(X)$. Then*

B. A Useful Calculation

$$E\left\{[\omega(X-K)]^+\right\} = \omega e^{M+\frac{1}{2}V^2} \Phi\left(\omega \frac{M-\ln(K)+V^2}{V}\right) \quad (B.2)$$

$$-\omega K \Phi\left(\omega \frac{M-\ln(K)}{V}\right),$$

for each $K > 0$, $\omega \in \{-1,1\}$, where E denotes expectation with respect to X's distribution and Φ denotes the cumulative standard normal distribution function.

Proof. We just have to notice that

$$E\left\{[\omega(X-K)]^+\right\} = \int_{-\infty}^{+\infty} \frac{1}{\sqrt{2\pi}V} [\omega(e^y-K)]^+ e^{-\frac{1}{2}\frac{(y-M)^2}{V^2}} dy,$$

and apply the previous lemma. □

C. Approximating Diffusions with Trees

The Holy One directed his steps to that blessed Bodhi-tree beneath whose shade he was to accomplish his search.
Paul Carus, "The Gospel of Buddha", 1894.

In this appendix, we show how to approximate a diffusion process with a tree. The general procedure we outline is used throughout the book in the tree construction for both one-factor and two-factor short-rate models. In the one-factor case, the tree is constructed by imposing that the conditional local mean and variances at each node are equal to those of the basic continuous-time process. The geometry of the tree is then designed so as to ensure the positivity of all branching probabilities. In the two-factor case, instead, we first construct the trees for the two factors along the procedure that applies to one-factor diffusions. We then construct a two-dimensional tree by imposing that the tree marginal distributions match those of the two factors' trees and by imposing the correct local correlation structure so as to preserve the positivity of all branching probabilities as well.

The tree construction procedure we propose in this appendix is mostly based on heuristic arguments. Indeed, the proof of formal convergence results is beyond the scope of this section. When deriving an approximating tree, we simply have in mind that, at first order in the amplitude of the time step, the diffusion transition density is normal and hence completely specified by its mean and variance.

Approximating a one-factor diffusion

Let us consider the diffusion process X that evolves according to

$$dX_t = \mu(t, X_t)dt + \sigma(t, X_t)dW_t, \qquad (C.1)$$

where μ and σ are smooth scalar real functions and W is a scalar standard Brownian motion.

We want to discretize the dynamics (C.1) both in time and in space. Precisely, we want to construct a trinomial tree that suitably approximates the evolution of the process X.

To this end, we fix a finite set of times $0 = t_0 < t_1 < \cdots < t_n = T$ and we set $\Delta t_i = t_{i+1} - t_i$. At each time t_i, we have a finite number of equispaced

488   C. Approximating Diffusions with Trees

states, with constant vertical step Δx_i to be suitably determined. We set $x_{i,j} = j\Delta x_i$.

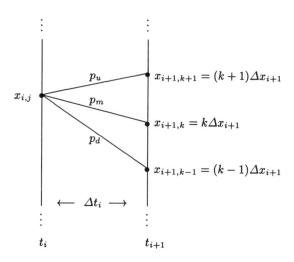

Fig. C.1. Evolution of the process x starting from $x_{i,j}$ at time t_i and moving to $x_{i+1,k+1}$, $x_{i+1,k}$ or $x_{i+1,k-1}$ at time t_{i+1} with probabilities p_u, p_m and p_d, respectively.

The tree geometry is displayed in Figure C.1. Assuming that at time t_i we are on the j-th node with associated value $x_{i,j}$, the process can move to $x_{i+1,k+1}$, $x_{i+1,k}$ or $x_{i+1,k-1}$ at time t_{i+1} with probabilities p_u, p_m and p_d, respectively. The central node is therefore the k-th node at time t_{i+1}, where also the level k is to be suitably determined.

Denoting by $M_{i,j}$ and $V_{i,j}^2$ the mean and the variance of X at time t_{i+1} conditional on $X(t_i) = x_{i,j}$, i.e.,

$$E\{X(t_{i+1})|X(t_i) = x_{i,j}\} = M_{i,j}$$
$$\text{Var}\{X(t_{i+1})|X(t_i) = x_{i,j}\} = V_{i,j}^2,$$

we want to find p_u, p_m and p_d such that these conditional mean and variance match those in the tree. Precisely, noting that $x_{i+1,k+1} = x_{i+1,k} + \Delta x_{i+1}$ and $x_{i+1,k-1} = x_{i+1,k} - \Delta x_{i+1}$, we look for positive constants p_u, p_m and p_d summing up to one and satisfying

$$\begin{cases} p_u(x_{i+1,k} + \Delta x_{i+1}) + p_m x_{i+1,k} + p_d(x_{i+1,k} - \Delta x_{i+1}) = M_{i,j} \\ p_u(x_{i+1,k} + \Delta x_{i+1})^2 + p_m x_{i+1,k}^2 + p_d(x_{i+1,k} - \Delta x_{i+1})^2 = V_{i,j}^2 + M_{i,j}^2. \end{cases}$$

Simple algebra leads to

C. Approximating Diffusions with Trees

$$\begin{cases} x_{i+1,k} + (p_u - p_d)\Delta x_{i+1} = M_{i,j} \\ x_{i+1,k}^2 + 2x_{i+1,k}\Delta x_{i+1}(p_u - p_d) + \Delta x_{i+1}^2(p_u + p_d) = V_{i,j}^2 + M_{i,j}^2. \end{cases}$$

Setting $\eta_{j,k} = M_{i,j} - x_{i+1,k}$,[1] we finally obtain

$$\begin{cases} (p_u - p_d)\Delta x_{i+1} = \eta_{j,k} \\ (p_u + p_d)\Delta x_{i+1}^2 = V_{i,j}^2 + \eta_{j,k}^2, \end{cases}$$

so that, remembering that $p_m = 1 - p_u - p_d$, the candidate probabilities are

$$\begin{cases} p_u = \frac{V_{i,j}^2}{2\Delta x_{i+1}^2} + \frac{\eta_{j,k}^2}{2\Delta x_{i+1}^2} + \frac{\eta_{j,k}}{2\Delta x_{i+1}}, \\ p_m = 1 - \frac{V_{i,j}^2}{\Delta x_{i+1}^2} - \frac{\eta_{j,k}^2}{\Delta x_{i+1}^2}, \\ p_d = \frac{V_{i,j}^2}{2\Delta x_{i+1}^2} + \frac{\eta_{j,k}^2}{2\Delta x_{i+1}^2} - \frac{\eta_{j,k}}{2\Delta x_{i+1}}. \end{cases}$$

In general, there is no guarantee that p_u, p_m and p_d are actual probabilities, because the expressions defining them could be negative. We then have to exploit the available degrees of freedom in order to obtain quantities that are always positive. To this end, we make the assumption that $V_{i,j}$ is independent of j,[2] so that from now on we simply write V_i instead of $V_{i,j}$. We then set $\Delta x_{i+1} = V_i\sqrt{3}$,[3] and we choose the level k, and hence $\eta_{j,k}$, in such a way that $x_{i+1,k}$ is as close as possible to $M_{i,j}$. As a consequence,

$$k = \text{round}\left(\frac{M_{i,j}}{\Delta x_{i+1}}\right), \tag{C.2}$$

where round(x) is the closest integer to the real number x. Moreover,

$$\begin{cases} p_u = \frac{1}{6} + \frac{\eta_{j,k}^2}{6V_i^2} + \frac{\eta_{j,k}}{2\sqrt{3}V_i}, \\ p_m = \frac{2}{3} - \frac{\eta_{j,k}^2}{3V_i^2}, \\ p_d = \frac{1}{6} + \frac{\eta_{j,k}^2}{6V_i^2} - \frac{\eta_{j,k}}{2\sqrt{3}V_i}. \end{cases} \tag{C.3}$$

It is easily seen that both p_u and p_d are positive for every value of $\eta_{j,k}$, whereas p_m is positive if and only if $|\eta_{j,k}| \leq V_i\sqrt{2}$. However, defining k as in (C.2) implies that $|\eta_{j,k}| \leq V_i\sqrt{3}/2$, hence the condition for the positivity of p_m is satisfied, too.

As a conclusion, (C.3), under the condition (C.2), are actual probabilities such that the corresponding trinomial tree has conditional (local) mean and variance that match those of the continuous-time process X.

---

[1] We omit to express the dependence on the index i to lighten the notation.
[2] This assumption indeed holds true in all the specific examples we consider in this book. In general, however, one has to look for a suitable transformation leading to a deterministic-volatility process, as in Sections 3.9.1 and 5.3.
[3] This choice, motivated by convergence purposes, is a standard one. See for instance Hull and White (1993, 1994).

Approximating a two-factor diffusion

Consider a process Z that is the sum of two diffusion processes X and Y, i.e.,
$$Z_t = X_t + Y_t,$$
where
$$dX_t = \mu^X(t, X_t)dt + \sigma^X(t, X_t)dW_t^X,$$
$$dY_t = \mu^Y(t, Y_t)dt + \sigma^Y(t, Y_t)dW_t^Y,$$
with μ^X, μ^Y, σ^X and σ^Y smooth real functions and W^X and W^Y two correlated standard Brownian motions with $dW_t^X dW_t^Y = \rho dt$, $\rho \in [-1, 1]$.

To construct an approximating tree for Z, we first construct the two trees approximating the processes X and Y along the procedure being previously illustrated. We then build a two-dimensional tree by locally imposing the right marginal distributions and the right correlation. To this end, we follow a similar procedure to that suggested by Hull and White (1994c), thus building a preliminary tree based on the assumption of zero correlation.

We denote by $M_{i,j}^X$ and $(V_i^X)^2$ the mean and the variance of X at time t_{i+1} conditional on $X(t_i) = x_{i,j}$, and by $M_{i,l}^Y$ and $(V_i^Y)^2$ the mean and the variance of Y at time t_{i+1} conditional on $Y(t_i) = y_{i,l}$, i.e.,
$$E\{X(t_{i+1})|X(t_i) = x_{i,j}\} = M_{i,j}^X,$$
$$\mathrm{Var}\{X(t_{i+1})|X(t_i) = x_{i,j}\} = (V_i^X)^2,$$
$$E\{Y(t_{i+1})|Y(t_i) = y_{i,l}\} = M_{i,l}^Y,$$
$$\mathrm{Var}\{Y(t_{i+1})|Y(t_i) = y_{i,l}\} = (V_i^Y)^2,$$

where, analogously to the one-factor case, we assume that the two variances are independent of j and l, respectively. We then denote by (i,j,l) the tree node at time t_i where X equals $x_{i,j} = j\Delta x_i$ and $Y = y_{i,l} = l\Delta y_i$, and by

- π_{uu} the probability of moving from (i,j,l) to $(i+1, k+1, h+1)$,
- π_{um} the probability of moving from (i,j,l) to $(i+1, k+1, h)$,
- π_{ud} the probability of moving from (i,j,l) to $(i+1, k+1, h-1)$,
- π_{mu} the probability of moving from (i,j,l) to $(i+1, k, h+1)$,
- π_{mm} the probability of moving from (i,j,l) to $(i+1, k, h)$,
- π_{md} the probability of moving from (i,j,l) to $(i+1, k, h-1)$,
- π_{du} the probability of moving from (i,j,l) to $(i+1, k-1, h+1)$,
- π_{dm} the probability of moving from (i,j,l) to $(i+1, k-1, h)$,
- π_{dd} the probability of moving from (i,j,l) to $(i+1, k-1, h-1)$,

where k and h are chosen so that $x_{i+1,k} = k\Delta x_{i+1}$ is as close as possible to $M_{i,j}^X$ and $y_{i+1,h} = h\Delta y_{i+1}$ is as close as possible to $M_{i,l}^Y$.

Under zero correlation, these probabilities are simply the product of the corresponding marginal probabilities. For instance,

C. Approximating Diffusions with Trees

$$\pi_{ud} = \left(\frac{1}{6} + \frac{(\eta_{j,k}^X)^2}{6(V_i^X)^2} + \frac{\eta_{j,k}^X}{2\sqrt{3}V_i^X}\right)\left(\frac{1}{6} + \frac{(\eta_{l,h}^Y)^2}{6(V_i^Y)^2} - \frac{\eta_{l,h}^Y}{2\sqrt{3}V_i^Y}\right),$$

where $\eta_{j,k}^X = M_{i,j}^X - x_{i+1,k}$ and $\eta_{l,h}^Y = M_{i,l}^Y - y_{i+1,h}$.

Let us now define Π_0 as the matrix of the "zero-correlation" probabilities, i.e.,

$$\Pi_0 = \begin{pmatrix} \pi_{ud} & \pi_{um} & \pi_{uu} \\ \pi_{md} & \pi_{mm} & \pi_{mu} \\ \pi_{dd} & \pi_{dm} & \pi_{du} \end{pmatrix}$$

To account for the proper correlation ρ we have to shift each probability in Π_0 in such a way that the sum of the shifts in each row and each column is zero (so that the marginal distributions are maintained) and the following is verified:

$$\frac{\text{Cov}\{X(t_{i+1}), Y(t_{i+1})|X(t_i), Y(t_i)\}}{V_i^X V_i^Y}$$

$$= \frac{\Delta x_{i+1} \Delta y_{i+1} \left(-\pi_{ud} - \varepsilon_{ud} + \pi_{uu} + \varepsilon_{uu} + \pi_{dd} + \varepsilon_{dd} - \pi_{du} - \varepsilon_{du}\right)}{V_i^X V_i^Y} = \rho,$$

at first order in Δt_i, where ε_{ud} denotes the shift for π_{ud} (the definition of the other shifts is analogous). Since $\Delta x_{i+1} = V_i^X \sqrt{3}$ and $\Delta y_{i+1} = V_i^Y \sqrt{3}$, this condition can be written as

$$-\pi_{ud} - \varepsilon_{ud} + \pi_{uu} + \varepsilon_{uu} + \pi_{dd} + \varepsilon_{dd} - \pi_{du} - \varepsilon_{du} = \frac{\rho}{3},$$

becoming, in the limit for $\Delta t_i \to 0$,

$$-\varepsilon_{ud} + \varepsilon_{uu} + \varepsilon_{dd} - \varepsilon_{du} = \frac{\rho}{3}.$$

Solving for the previous constraints, a possible solution, that is similar to that proposed by Hull and White (1994c), is

$$\Pi_\rho = \Pi_0 + \rho(\Pi_1^l - \Pi_0^l), \text{ if } \rho > 0$$
$$\Pi_\rho = \Pi_0 - \rho(\Pi_{-1}^l - \Pi_0^l), \text{ if } \rho < 0$$

where Π_ρ is the probability matrix that accounts for the right correlation ρ and Π_1^l, Π_0^l and Π_{-1}^l are the probability matrices in the limit, respectively for $\rho = 1$, $\rho = 0$ and $\rho = -1$. More explicitly,

$$\Pi_\rho = \Pi_0 + \frac{\rho}{36}\begin{pmatrix} -1 & -4 & 5 \\ -4 & 8 & -4 \\ 5 & -4 & -1 \end{pmatrix} \text{ if } \rho > 0$$

and

$$\Pi_\rho = \Pi_0 - \frac{\rho}{36}\begin{pmatrix} 5 & -4 & -1 \\ -4 & 8 & -4 \\ -1 & -4 & 5 \end{pmatrix} \text{ if } \rho < 0.$$

492   C. Approximating Diffusions with Trees

However, it may happen that one or more entries in Π_ρ are negative. To overcome this drawback, we are compelled to modify the correlation on node (i,j,l), for example substituting Π_ρ with Π_0. Assuming a different correlation on such node, and on all the tree nodes where some probabilities are negative, has a negligible impact on a derivative price as long as the mesh of $\{t_0, t_1, \ldots, t_m\}$ is sufficiently small. Notice, in fact, that in the limit for such mesh going to zero, the elements of Π_ρ are all positive, being Π_ρ a convex linear combination of two matrices with positive entries.

D. Talking to the Traders

'A logical division of Labor.
Interviews require a certain facility for interpersonal interaction,
sometimes referred to as "people skills" '
J'onn J'onnz in reply to Batman in "Martian Manhunter annual" 2,
1999, DC Comics.

In this appendix, we would like to reproduce an hypothetical conversation between a trader in interest rate derivatives and a quantitative analyst eager to get acquainted with some specific market practice. This virtual interview reflects our personal experience of interaction with traders, considering some of the traders' opinions we have collected over the years.

We would also like to warn the reader that he/she must not be deceived by the smooth flowing of this interview. Traders and quants tend to speak different languages and sentences or reasoning must be often rephrased for a reciprocal understanding. However, after an unavoidable initial effort, and many imprecations that (hopefully) fade away on the tip of both tongues, traders and quants get to know each other, becoming (more and more) aware of how important can be maintaining a reciprocal cooperation. Thus, in the interview below we have tried to reproduce this atmosphere at times. We took the symbol ":-)" from the internet jargon to mean that a humorous remark needs not to be taken as offence or hostility.

Short rate models, HJM and market models: Where now?

1. Quant: *Among all the short rate models we implemented for you in the software unit you call "big toy", namely* HW, CIR++, BK, EEV, G2++, *which ones do you use most frequently and why?*
   Trader: It depends. For our purposes, we mostly use BK (or alternatively EEV) and G2++ (equivalent to two factor Hull and White model, HW2).
2. Q: *Why this choice?*
   T: Well, first off, these are among the most popular models in the short rate world, and as such we tend to trust them more. Next, they involve kind of opposite assumptions on the short rate distribution. One model takes it to be Gaussian, the other lognormal.

3. Q: *So what's your problem with CIR++ for example?*
   T: The problem is that this noncentral chi square distribution stays in-between the other two cases as far as tails are concerned, and so we prefer ranging a wider interval by considering the lognormal and normal models. Not to talk about the numerical problems one is likely to face when dealing with chi square distribution functions.
4. Q: *Why considering two so different models?*
   T: Their difference is exactly the reason why we take them both. We expect, and have actually experienced, that when calibrated on the same data and used for managing products that are *at the money* (ATM), the two models yield roughly the same answers. Instead, as soon as moneyness moves, the tails of the rates distributions become more relevant, so that one model tends to amplify prices while the other underestimates them. Therefore, when dealing with *in the money* (ITM) or *out of the money* (OTM) options, we take as final solution a (somehow weighted) average of the answers provided by the two models.
5. Q: *Can you give me some further insight? I mean, why is moneyness so important?*
   T: Ok, it's very simple. In the market, we usually observe what is called the "smile effect". This basically means that implied volatilities are not all traded at a same level. It usually happens that OTM volatilities are higher than the corresponding ITM ones, and this can have a serious impact in the pricing of derivatives.
6. Q: *Can you be more explicit?*
   T: Yes, just to fix ideas, let's say you must price an OTM receiver swaption of Bermudan type. If you calibrate your favorite interest rate model to ATM volatilities (they are the only ones to be actively quoted in the market), you miss the fact that an OTM receiver European swaption is priced with a higher volatility than the corresponding ATM one. Assuming a normal distribution of rates can then be helpful, since it assigns a higher probability to low rates, thus rendering the OTM swaption more valuable. At the same time, however, the market implicitly assumes that rates are lognormally distributed. This is why the correct price is somehow in between the Gaussian and lognormal prices.
7. Q: *Do you think short rate models still have a future?*
   T: Who knows. The HJM developments did not threaten seriously the short-rate world but rather incorporated it. However, the recent growth of market models certainly threatens the short-rate model survival.
8. Q: *What can be impressing is the calibrating capability: The LIBOR market model can calibrate a lot of swaptions and caps at the same time, at least in principle...*
   T: Sure, but that is not necessarily a uniform advantage. Remember what I call the "uncertainty principle of modeling", the more a model fits, the less it explains.

9. Q: *What do you mean exactly?*
   T: I mean, if a model recovers a huge number of financial observables by construction, it cannot be able to explain well what is happening, exactly because it "eats" everything you feed it on. On the contrary, a model that cannot calibrate a huge number of prices will signal problems. Then, I can watch the data and decide whether the problem is with the model limitations or there are some pathologies with the market structures given as input. So the "poor" model warns me, while a too rich model is always happy and does not help me that much to sense danger.
10. Q: *I think I read something similar on Rebonato's book on interest rate models*
    T: Yes, I agree completely with Rebonato on this issue, as well as on so many other things.
11. Q: *This looks like a point in favor of short rate models.*
    T: Yes. Let me ask you, are there reasons to prefer short rate models as far as implementation issues are concerned?
12. Q: *Well, a lot of features are simplified with short rate models. You can build trees and price early exercise products easily enough, but only in case of low dimension, say two or at most three. Notice also that, in case of correlated factors, the tree construction can be tricky already in the two-factor case. With high dimensional trees, moreover, you easily have problems concerning the speed of convergence and the computer memory required to run your pricing routines. However, the tree construction for the short-rate models we have implemented is rather straightforward.*
    T: Which is not the case with the market model. What about Monte Carlo?
13. Q: *Well, with tractable models such as HW2 or CIR++, from the short rate factors all kind of rates (forward and swap rates for any expiry/maturity/payment dates) can be recovered easily. Instead, in the market model you need to interpolate when leaving the preassigned dates, since you simulate finite rates with preassigned expiries and maturities. Also, with HW2 and CIR++, Monte Carlo simulation is easy, since the factors dynamics under the forward measure are known explicitly as well as their transition densities. In fact, "one-shot" simulations are possible and no discretizations of the factors equations are required, contrary to the market models. These can be attractive features especially for risk management purposes.*
    T: So you see some further advantages of short rate models.

Some hedging issues

1. Q: *Well, I was also wondering about hedging. What about hedging?*
   T: Hedging is a different matter. Hedging can be done naively on the basis of shifting the required market observable, recalibrate, and compute the

difference in prices divided by the shift amount. This is good to compute "global" sensitivities, i.e. sensitivities with respect to uniform shifts of input market structures, but...

2. Q: *Yes?*

   T: But we have problems with breakdowns of sensitivities. When we need sensitivities to single inputs, your big toy©doesn't help that much, you see... [gets excited, moves his hands frantically]... these @##!@%%&@!! volatility surfaces move like this [traces surfaces contours and tangent planes in the air]... How the @##!@%%&@!! can I compute sensitivities say to single swaption prices? And I need them...

3. Q: *Ahem, ever the diplomat :-) ...let me rephrase it... are you saying that the influence of a local shift in a market observable is distributed globally on the parameters by the calibration, so that the local effects of the variations are lost? I see... Hedging seems possible when shifting uniformly market curves and surfaces in input, but when shifting single points, the effect is probably lost and possibly confused with other possible causes.*

   T: Right, that's what I said, but you took all the colour out of it :-). A short rate model with only one time-dependent function (used for exact calibration to the zero-coupon curve) has too few parameters to appreciate the influence of local changes in the input volatility structures. Shifting two rather different points can cause the same change in the parameters, due to the flattening of the information implied by the low number of parameters. A market model, instead, can often appreciate such a change by distinguishing the two cases. In the book you even described a volatility parameterization in the LIBOR market model (LFM) where you have a one to one correspondence between forward rates volatilities and swaption prices. This can be incredibly helpful for hedging.

4. Q: *Can you elaborate more on this?*

   T: What I wanted to say is that this can allow for computation of sensitivities with respect to a single swaption price (volatility) used in the calibration. This is important for Vega breakdown analysis and sensitivity to volatility in general. Indeed, it helps us understand what portion of the (implied) volatility surface has a significant effect on the price of the considered derivative. We can then construct a hedging strategy based on the European swaptions corresponding to that volatility surface portion.

5. Q: *I understand. But don't you think you can achieve similar results by introducing further time-dependent coefficients in the short rate dynamics? This is actually the original approach followed by Hull and White, Black, Derman and Toy or Black and Karasinski.*

   T: You're right. However, too many time-dependent parameters can be dangerous. And not only due to the overparameterization problems I mentioned before. Indeed, the price to pay for a better, possibly exact fitting of market data, is an unplausible evolution of the future term structure of volatility, in that the implied future volatility surfaces can have unrealistic shapes.

6. Q: *Uhm, [an amused look appears on the quant's face...] how comes that traders give so much importance to aesthetics? :-) ...*
   T: Heh, of course it's deeper than that! See, an unplausible future volatility structure can badly affect the pricing of instruments that implicitly depend on such a volatility structure, like a Bermudan swaption for instance. To fix ideas, suppose you have to price a Bermudan swaption, and you decide to calibrate to the underlying European swaptions. Using the general formulation of the Hull and White one-factor model, you can find infinitely many specifications of the parameters leading to an exact fitting of these European swaptions. This is not harmless. You'll indeed find that the corresponding prices wanders freely in such a wide range that you can't give any meaningful answer.

7. Q: *So there are both advantages and disadvantages with short rate models. Which do you think are dominant?*
   T: Disadvantages. Probably, in due time, market models will finally replace short rate models completely. We will see. But let me give you a warning. When the HJM theory came out, a lot of people thought that the interest rate theory was dead and everything had been done. Yet, we have seen how many things happened since then. So we should be careful in thinking market models are the final and complete solution to all the problems in interest rate models... and who knows, maybe short rate models will come back one day, after they have vanished.

Are market models completely developed?

1. Q: *What are - in your opinion - the areas where market models are not yet completely effective?*
   T: Let me think... I guess that multi-currency products are still an issue. In the book you pointed out that products such as quanto CMS with optional features are easily priced with two correlated two-factor extended Vasicek models. You were also able to translate the relevant model correlations into correlations I would be able to express a view upon. However, pricing similar products with correlated LIBOR market models for each curve can be more complicated and computationally demanding.

2. Q: *Actually, I have recently seen some interesting work on the multi-currency extensions of the LIBOR market model, so it seems some research is being done in that direction, too. What else?*
   T: Well, market models are relatively young and there is still so much to learn about them. You know, models such as HW or BK have been on the market for years, have been also implemented by several commercial softwares, and their limits are well known. I feel market models are yet

at a preliminary stage in that respect, although more and more studies on their practical implementation are being done and about to become public domain.

3. *Q: So you feel that LIBOR market models are yet to be fully tested?*
   T: I mean, just think of how many commercial software companies implemented a comprehensive version of the LIBOR market model, compared to older short rate models. I think there are not those many. At the same time, however, I must say that the big market players have their own knowledge and expertise. Just think that the top financial institutions were using some kind of market models well before they appeared in the literature.

4. *Q: I heard several commercial software companies are working on the LIBOR market model, and probably two or three of them already have a version of it. However, I think it is difficult to check whether the proposed implementations are satisfactory in the sense we outlined in the chapter we devoted to market models.*
   T: In some cases the possibility for a full investigation would allow the client to deduce key features of the model implementation, of the covariance structure, of the method for computing sensitivities, and so on. And we know that the added value of a model often lies in an efficient implementation rather than on nice "mathematical features". A company may not be willing to give that away, not even to a good client.

5. *Q: Does that mean that you don't care that much about mathematical rigor?*
   T: Wait, don't put words in my mouth. But consider an example: Many financial institutions have been pricing products with BK's model for years. This is a model where the average bank account is infinite after an infinitesimal time...

6. *Q: And yet BK was and, to some extent, still is quite a successful model...*
   T: Indeed. Now let me ask you: If you put your money in a bank account, and your model tells you that the expected value of the money you'll have in say one second is infinite, would you consider such a model satisfactory?

7. *Q: Mmmmhhh... as a mathematician I would probably have perplexities about this infinite expected future bank account.*
   T: Right. However, when you implement this model you usually resort to a tree, and with trees the danger of infinite expectations is avoided. The tree has a finite number of states, and thus avoids this undesirable feature. So, as a practitioner, I would not worry too much about this explosion issue, given that, in change for this, we get a lognormal distribution for the short rate and usually a good fitting to a low number of selected swaption prices. To sum up: Mathematical rigor is important, but it is not everything, and at times it is a good idea to put the accent on different aspects.

Modeling the smile in market models

1. Q: *All right. Back to market models: Other open issues?*
   T: Of course, smile modeling. Andersen and Andreasen had the idea of applying the CEV structure to the LIBOR market model and have developed interesting approximations. Yet, the CEV structure does not have enough flexibility for many practical purposes.

2. Q: *What do you think of our "shifted lognormal mixture dynamics"?*
   T: I would like to see some approximations leading to a quick pricing of swaptions, something similar to the tricks leading to Rebonato's or Hull and White's formulas for swaption volatilities in the LIBOR market model (LFM). Then with enough parameters in your mixture dynamics you might try a calibration to part of the swaptions smile. And... well...

3. Q: *Yes?*
   T: It would be interesting to see something on the transition densities implied by your dynamics, not merely on the marginal distribution. This could give information on the volatility structures implied by your dynamics at future times.

4. Q: *You know, I almost feel we have exchanged places, with you discussing such technical matters. I mean, you are asking quite technical questions. Ok, I'll put that on schedule. Not that we hadn't thought about it...*
   T: Don't worry. This is a kind of aspect that has been ignored in several models I have seen around. However, you should not be too surprised by my questions. Quantitative traders often have quite some experience, and can thus appreciate the models implications and limitations in connections with their market applications.

5. Q: *I have no doubts about it. It's just that in other banks we have had more difficulties in interacting with traders. Thanks for your time. Even if this "interview" has been slightly unfocused and informal, I think our readers might appreciate it.*
   T: My pleasure. It's not so many times that a trader can give an opinion or some teachings in a book about the doctrine. I am glad I have been given this possibility, and let me wish good luck to all your readers in their efforts either in the market or in quantitative research.

   Q: *Good idea:* Good luck!

'It looks like our troubles are over for the time being, Batman'
'Our troubles are just beginning, Superman'
Day of Judgment 5, 1999, DC Comics

References

1. Aït-Sahalia, Y. (1996) Testing Continuous-Time Models of the Spot Rate. *The Review of Financial Studies* 9, 385-426.
2. Amin, K., and Morton, A. (1994) Implied Volatility Functions in Arbitrage Free Term Structure Models. *Journal of Financial Economics* 35, 141-180.
3. Andersen, L. (1999). A Simple Approach to the Pricing of Bermudan Swaptions in the Multi-Factor LIBOR Market Model, preprint.
4. Andersen, L., and Andreasen, J. (2000). Volatility Skews and Extensions of the LIBOR Market Model. *Applied Mathematical Finance* 7, 1-32.
5. Avellaneda, M., Newman, J. (1998) Positive Interest Rates and Nonlinear Term Structure Models. Preprint, Courant Institute of Mathematical Sciences, New York University.
6. Balduzzi, P., Das, S.R., Foresi, S., Sundaram, R. (1996) A Simple Approach to Three-Factor Term Structure Models. *The Journal of Fixed Income* 6, 43-53.
7. Baxter, M.W. (1997) General Interest-Rate Models and the Universality of HJM. *Mathematics of Derivative Securities*, M.A.H. Dempster, S.R. Pliska, eds. Cambridge University Press (1997), Cambridge, pp. 315-335.
8. Björk, T. (1997) Interest rate Theory. In *Financial Mathematics, Bressanone 1996*, W. Runggaldier, ed. *Lecture Notes in Math.* 1656, Springer, Berlin Heidelberg New York, pp. 53-122.
9. Björk, T. (1998) Arbitrage Theory in Continuous Time. Oxford University Press.
10. Black, F., Derman, E., Toy, W. (1990) A One-Factor Model of Interest Rates and its Application to Treasury Bond Options. *Financial Analysts Journal* 46, 33-39.
11. Black, F., Karasinski, P. (1991) Bond and Option Pricing when Short Rates are Lognormal. *Financial Analysts Journal* 47, 52-59.
12. Black, F., Scholes, M. (1973) The Pricing of Options and Corporate Liabilities. *Journal of Political Economy* 81, 637-654.
13. Bliss, R., Ritchken, P. (1996) Empirical Tests of Two State-Variable Heath, Jarrow and Morton Models. *Journal of Money, Credit and Banking* 18, 426-447.
14. Brace, A. (1996) Dual Swap and Swaption Formulae in Forward Models. *FMMA notes* working paper.
15. Brace, A. (1997) Rank-2 Swaption Formulae. *UNSW* preprint.
16. Brace, A. (1998) Simulation in the GHJM and LFM models. *FMMA notes* working paper.
17. Brace, A., Musiela, M. (1994) A Multifactor Gauss Markov Implementation of Heath, Jarrow, Morton. *Mathematical Finance* 4, 259-283.
18. Brace, A., Musiela, M. (1997) Swap Derivatives in a Gaussian HJM Framework. *Mathematics of Derivative Securities*, M.A.H. Dempster, S.R. Pliska, eds. Cambridge University Press, Cambridge, pp. 336-368.

19. Brace, A., Dun, T., and Barton, G. (1998) Towards a Central Interest Rate Model. *FMMA notes* working paper.
20. Brace, A., Gatarek D., Musiela, M. (1997) The Market Model of Interest Rate Dynamics. *Mathematical Finance* 7, 127-155.
21. Brace, A., Musiela, M., and Schlögl, E. (1998) A Simulation Algorithm Based on Measure Relationships in the Lognormal Market Models. Preprint.
22. Breeden, D.T. and Litzenberger, R.H. (1978) Prices of State-Contingent Claims Implicit in Option Prices. *Journal of Business* 51, 621-651.
23. Brenner, R.J., Harjes, R.H., Kroner, K.F. (1996) Another Look at Models of the Short-Term Interest Rate. *Journal of Financial and Quantitative Analysis* 31, 85-107.
24. Brennan, M. J., and Schwartz, E. (1979). A Continuous Time Approach to the Pricing of Bonds. *Journal of Banking and Finance* 3, 133-155.
25. Brennan, M. J., and Schwartz, E. (1982) An Equilibrium Model of Bond Prices and a Test of Market Efficiency. *Journal of Financial and Quantitative Analysis* 17, 301-329.
26. Brigo, D., Capitani, C., and Mercurio, F. (2000) On the Joint Calibration of the LIBOR Market Model to Caps and Swaptions Data. Internal Report, Banca IMI, Milan.
27. Brigo, D., Mercurio, F. (1998) On Deterministic Shift Extensions of Short-Rate Models. Internal Report, Banca IMI, Milan. Available on the internet at http://web.tiscalinet.it/damianohome and at http://web.tiscalinet.it/FabioMercurio.
28. Brigo, D., Mercurio, F. (2000a) Fitting Volatility Smiles with Analytically Tractable Asset Price Models. Internal Report, Banca IMI, Milan. Available on the internet at http://web.tiscalinet.it/damianohome and at http://web.tiscalinet.it/FabioMercurio.
29. Brigo, D., Mercurio, F. (2000b) A Mixed-up Smile. *Risk*, September, 123-126.
30. Brigo, D., Mercurio, F. (2000c) The CIR++ Model and Other Deterministic-Shift Extensions of Short-Rate Models, in: *Proceedings of the Columbia-JAFEE International Conference* held in Tokyo on December 16-17, 2000, pp. 563-584.
31. Brigo, D., Mercurio, F. (2001a) A Deterministic-Shift Extension of Analytically-Tractable and Time-Homogenous Short-Rate Models. To appear in *Finance & Stochastics*.
32. Brigo, D., Mercurio, F. (2001b) Displaced and Mixture Diffusions for Analytically-Tractable Smile Models. In *Mathematical Finance - Bachelier Congress 2000*, Geman, H., Madan, D.B., Pliska, S.R., Vorst, A.C.F., eds. *Springer Finance*, Springer, Berlin Heidelberg New York, to appear.
33. Bühler, W., Uhrig-Homburg, M., Walter, U., Weber, T. (1999) An Empirical Comparison of Forward-Rate and Spot-Rate Models for Valuing Interest-Rate Options. *The Journal of Finance* 54, 269-305.
34. Cannabero, E. (1995) Where Do One-Factor Interest Rate Models Fail? *The Journal of Fixed Income* 5, 31-52.
35. Carr, P., Yang, G. (1997). Simulating Bermudan Interest Rate Derivatives, Courant Institute at New York University, Preprint.
36. Carverhill, A. (1994) When is the Short Rate Markovian? *Mathematical Finance* 4, 305-312.
37. Castagna, A. (2001) Private Communication.
38. Chan, K.C., Karolyi, G.A., Longstaff, F.S., Sanders, A.B. (1992) An Emprical Comparison of Alternative Models of the Term Structure of Interest Rates. *The Journal of Finance* 47, 1209-1228.

39. Chapman, D.A., Long, A.B., Pearson, N.D. (1999) Using Proxies for the Short Rate: When Are There Months Like an Instant? *The Review of Financial Studies* 12, 763-806.
40. Chen, R., Scott, L. (1992) Pricing Interest Rate Options in a Two Factor Cox-Ingersoll-Ross Model of the Term Structure. *The Review of Financial Studies* 5, 613-636.
41. Chen, R., Scott, L. (1995) Interest Rate Options in Multifactor Cox-Ingersoll-Ross Models of the Term Structure. *The Journal of Derivatives* 3, 53-72.
42. Clewlow, L., Strickland, C. (1997) Monte Carlo Valuation of Interest Rate Derivatives Under Stochastic Volatility. *The Journal of Fixed Income* 7, 35-45.
43. Clewlow, L., Strickland, C. (1998) Pricing Interest Rate Exotics in Multi-Factor Gaussian Interest Rate Models, working paper.
44. Constantinides, G. (1992) A Theory of the Nominal Term Structure of Interest Rates. *The Review of Financial Studies* 5, 531-552.
45. Cox, J.C. (1975) Notes on Option Pricing I: Constant Elasticity of Variance Diffusions. Working paper. Stanford University.
46. Cox, J.C., and Ross S. (1976) The Valuation of Options for Alternative Stochastic Processes. *Journal of Financial Economics* 3, 145-166.
47. Cox, J.C., Ingersoll, J.E., and Ross, S.A. (1985) A Theory of the Term Structure of Interest Rates. *Econometrica* 53, 385-407.
48. Cox, J.C., Ross, S.A., and Rubinstein, M. (1979) Option Pricing: A Simplified Approach. *Journal of Financial Economics* 7, 229-263.
49. Davis, M. (1998) A Note on the Forward Measure. *Finance and Stochastics*, 2, pp. 19-28.
50. De Jong, F., Driessen, J., and Pelsser, A. (1999). LIBOR and Swap Market Models for the Pricing of Interest Rate Derivatives: An Empirical Analysis. Preprint.
51. Derman, E., and Kani, I. (1994) Riding on a Smile. *Risk* February, 32-39.
52. Derman, E., and Kani, I. (1998) Stochastic Implied Trees: Arbitrage Pricing with Stochastic Term and Strike Structure of Volatility. *International Journal of Theoretical and Applied Finance* 1, 61-110.
53. Dothan, L.U. (1978) On the Term Structure of Interest Rates. *Journal of Financial Economics* 6, 59-69.
54. Duffie, D. (1996) Dynamic Asset Pricing Theory, 2d. ed. Princeton: Princeton University Press.
55. Duffie, D., Kan, R. (1996) A Yield-Factor Model of Interest Rates. *Mathematical Finance* 64, 379-406.
56. Dun, T., Schlögl, E., and Barton, G. (1999) Simulated Swaption Hedging in the Lognormal Forward LIBOR Model. Preprint.
57. Dupire, B. (1994) Pricing with a Smile. *Risk* January, 18-20.
58. Dupire, B. (1997) Pricing and Hedging with Smiles. *Mathematics of Derivative Securities*, edited by M.A.H. Dempster and S.R. Pliska, Cambridge University Press, Cambridge, 103-111.
59. Dybvig, P.H. (1988) Bond and Bond Option Pricing Based on the Current Term Structure. Working Paper, Washington University.
60. Dybvig, P.H. (1997) Bond and Bond Option Pricing Based on the Current Term Structure. *Mathematics of Derivative Securities*, Michael A. H. Dempster and Stanley R. Pliska, eds. Cambridge: Cambridge University Press, pp. 271-293.
61. Flesaker, B. (1993) Testing the Heath-Jarrow-Morton/Ho-Lee Model of Interest Rate Contingent Claims Pricing. *Journal of Financial and Quantitative Analysis* 28, 483-496.
62. Flesaker, B. (1996) Exotic Interest Rate Options. *Exotic Options: The State of the Art*, L. Clewlow and C. Strickland, eds. London: Chapman and Hall, Ch. 6.

63. Flesaker, B., Hughston, L. (1996) Positive Interest. *Risk* 9, 46-49.
64. Flesaker, B., Hughston, L. (1997) Dynamic Models for Yield Curve Evolution. *Mathematics of Derivative Securities*, M.A.H. Dempster, S.R. Pliska, eds. Cambridge University Press, Cambridge, pp. 294-314.
65. Geman, H., El Karoui, N., Rochet, J.C. (1995) Changes of Numeraire, Changes of Probability Measures and Pricing of Options. *Journal of Applied Probability* 32, 443-458.
66. Geyer, A.L.J., Pichler, S. (1999) A State-Space Approach to Estimate and Test Multifactor Cox-Ingersoll-Ross Models of the Term Structure. *The Journal of Financial Research* 22, 107-130.
67. Gibbons, M., Ramaswamy, K. (1993) A Test of the Cox, Ingersoll and Ross Model of the Term Structure. *The Review of Financial Studies* 6, 619-632.
68. Glasserman, P., and Zhao, X. (1999) Fast Greeks in Forward LIBOR Models. Preprint.
69. Glasserman, P., and Zhao, X. (2000). Arbitrage-Free Discretization of Lognormal Forward LIBOR and Swap Rate Models. *Finance and Stochastics*, 4
70. Grant, D., Vora, G. (1999) Implementing No-Arbitrage Term Structure of Interest Rate Models in Discrete Time When Interest Rates are Normally Distributed. *The Journal of Fixed Income* 8, 85-98.
71. Harrison, J.M., Kreps, D.M. (1979) Martingales and Arbitrage in Multiperiod Securities Markets, *Journal of Economic Theory* 20, 381-408.
72. Harrison, J.M., Pliska, S.R. (1981) Martingales and Stochastic Integrals in the Theory of Continuous Trading, *Stochastic Processes and their Applications* 11, 215-260.
73. Harrison, J.M., Pliska, S.R. (1983) A Stochastic Calculus Model of Continuous Trading: Complete Markets, *Stochastic Processes and their Applications* 15, 313-316.
74. Heath, D., Jarrow, R., Morton, A. (1990a) Bond Pricing and the Term Structure of Interest Rates: A Discrete Time Approximation. *Journal of Financial Quantitative Analysis* 25, 419-440.
75. Heath, D., Jarrow, R., Morton, A. (1990b) Contingent Claim Valuation with a Random Evolution of Interest Rates. *Review of Futures Markets* 9, 54-76.
76. Heath, D., Jarrow, R., Morton, A. (1992) Bond Pricing and the Term Structure of Interest Rates: A New Methodology. *Econometrica* 60, 77-105.
77. Heston, S.L. (1993) A Closed-Form Solution for Options with Stochastic Volatility with Applications to Bond and Currency Options. *The Review of Financial Studies* 6, 327-343.
78. Ho, T.S.Y., Lee, S.-B. (1986) Term Structure Movements and the Pricing of Interest Rate Contingent Claims. *The Journal of Finance* 41, 1011-1029.
79. Hull, J. (1997) Options, Futures, and Other Derivatives, 3rd. edition. Upper Saddle River, New Jersey: Prentice-Hall.
80. Hull, J., White, A. (1990a) Valuing Derivative Securities Using the Explicit Finite Difference Method. *Journal of Financial and Quantitative Analysis* 25, 87-100.
81. Hull, J., White, A. (1990b) Pricing Interest Rate Derivative Securities. *The Review of Financial Studies* 3, 573-592.
82. Hull, J., White, A. (1993a) Bond Option Pricing Based on a Model for the Evolution of Bond Prices. *Advances in Futures and Options Research* 6, 1-13.
83. Hull, J., White, A. (1993b) Efficient Procedures for Valuing European and American Path-Dependent Options. *The Journal of Derivatives* 1, 21-31.
84. Hull, J., White, A. (1993c) The Pricing of Options on Interest-Rate Caps and Floors Using the Hull-White Model. *The Journal of Financial Engineering* 2, 287-296.

85. Hull, J., White, A. (1993d) One-Factor Interest Rate Models and the Valuation of Interest Rate Derivative Securities. *Journal of Financial and Quantitative Analysis* 28, 235-254.
86. Hull, J., White, A. (1994a) Branching Out. *Risk* 7, 34-37.
87. Hull, J., White, A. (1994b) Numerical Procedures for Implementing Term Structure Models I: Single-Factor Models. *The Journal of Derivatives* 2, 7-16.
88. Hull, J., White, A. (1994c) Numerical Procedures for Implementing Term Structure Models II: Two-Factor Models. *The Journal of Derivatives* 2, 37-47.
89. Hull, J., White, A. (1995a) Hull-White on Derivatives. London: Risk.
90. Hull, J., White, A. (1995b) A Note on the Models of Hull and White for Pricing Options on the Term Structure: Response. *The Journal of Fixed Income* 5 (September, 1995), 97-102.
91. Hull, J., White, A. (1996) Using Hull-White Interest Rate Trees. *The Journal of Derivatives* 3, 26-36.
92. Hull, J., White, A. (1997) Taking Rates to the Limits. *Risk* 10, 168-169.
93. Hull, J. White, A. (1999). Forward Rate Volatilities, Swap Rate Volatilities, and the Implementation of the LIBOR Market Model. Preprint.
94. Hunt, P., Kennedy, J. (2000) Financial Derivatives in Theory and Practice. Wiley. Chichester.
95. Hunt, P., Pelsser, A. (1998) Arbitrage-Free Pricing of Quanto-Swaptions *The Journal of Financial Engineering* 7, 25-33.
96. Hunt, P., Kennedy, J., Pelsser, A. (1998a) Fit and Run. *Risk* 11, pp. 65-67.
97. Hunt, P., Kennedy, J., Pelsser, A. (1998b) Markov–Functional Interest Rate Models. Working Paper, University of Warwick.
98. Inui, K., Masaaki, K. (1998) A Markovian Framework in Multi-Factor Heath-Jarrow-Morton Models. *Journal of Financial and Quantitative Analysis* 33, 423-440.
99. James, J., Webber, N. (2000) Interest Rate Modelling. Wiley. Chichester.
100. Jamshidian, F. (1988) The One-Factor Gaussian Interest Rate Model: Theory and Implementation. Working Paper, Merril Lynch Capital Markets.
101. Jamshidian, F. (1989) An Exact Bond Option Pricing Formula. *The Journal of Finance* 44, 205-209.
102. Jamshidian, F. (1991) Bond and Option Evaluation in the Gaussian Interest Rate Model. *Research in Finance* 9, 131-170.
103. Jamshidian, F. (1995) A Simple Class of Square-Root Interest Rate Models. *Applied Mathematical Finance* 2, 61-72.
104. Jamshidian, F. (1996) Sorting out Swaptions. *Risk*, March, 59-60.
105. Jamshidian, F. (1997) LIBOR and Swap Market Models and Measures. *Finance and Stochastics* 1, 293-330.
106. Jamshidian, F., and Zhu, Y. (1997) Scenario Simulation: Theory and methodology. *Finance and Stochastics* 1, 43–67.
107. Jarrow, R.A. (1996) Modelling Fixed Income Securities and Interest Rate Options. New York: McGraw-Hill.
108. Jarrow, R.A. (1997) The HJM Model: Its Past, Present, and Future. *The Journal of Financial Engineering* 6, 269-279.
109. Jarrow, R.A., Madan, D. (1995) Option Pricing Using the Term Structure of Interest Rates to Hedge Systematic Discontinuities in Asset Returns. *Mathematical Finance* 5, 311-336.
110. Jarrow, R.A., Turnbull, S.M. (1994) Delta, Gamma, and Bucket Hedging of Interest Rate Derivatives. *Applied Mathematical Finance* 1, 21-48.
111. Kennedy, D.P. (1997) Characterizing Gaussian Models of the Term Structure of Interest Rates. *Mathematical Finance* 2, 107-118.

112. Kijima, M., Nagayama, I. (1994) Efficient Numerical Procedures for the Hull-White Extended Vasicek Model. *The Journal of Financial Engineering* 3, 275-292.
113. Kijima, M., Nagayama, I. (1996) A Numerical Procedure for the General One-Factor Interest Rate Model. *The Journal of Financial Engineering* 5, 317-337.
114. Klöden, P.E., Platen, I. (1995) Numerical Solutions of Stochastic Differential Equations. Springer, Berlin Heidelberg New York.
115. Li, A., Ritchken, P., Sankarasubramanian, L. (1995a) Lattice Models for Pricing American Interest Rate Claims. *The Journal of Finance* 50, 719-737.
116. Li, A., Ritchken, P., Sankarasubramanian, L. (1995b) Lattice Works. *Risk* 8, 65-69.
117. Longstaff, F.A., Santa Clara, P., and Schwartz, E.S. (2001) Throwing Away a Billion Dollars: The Cost of Suboptimal Exercise Strategies in the Swaptions Market. *Journal of Financial Economics*, forthcoming.
118. Longstaff, F.A., Schwartz, E.S. (1992a) Interest Rate Volatility and the Term Structure: A Two-Factor General Equilibrium Model. *The Journal of Finance* 47, 1259-1282.
119. Longstaff, F.A., Schwartz, E.S. (1992b) A Two-Factor Interest Rate Model and Contingent Claims Valuation. *The Journal of Fixed Income* 3, 16-23.
120. Longstaff, F.A., Schwartz, E.S. (1993) Implementation of the Longstaff-Schwartz Interest Rate Model. *The Journal of Fixed Income* 3, 7-14.
121. Longstaff, F. A., and Schwartz, E.S. (2000). Pricing American Options by Simulation: A Simple Least-Squares Approach. *The Review of Financial Studies*, forthcoming.
122. Maghsoodi, Y. (1996) Solution of the Extended CIR Term Structure and Bond Option Valuation. *Mathematical Finance* 6, 89-109.
123. Mauri, G. (2001) Private Communication.
124. Mercurio F., Moraleda, J.M. (2000a) An Analytically Tractable Interest Rate Model with Humped Volatility. *European Journal of Operational Research* 120, 205-214.
125. Mercurio F., Moraleda, J.M. (2000b) A Family of Humped Volatility Interest Rate Models. Forthcoming in *European Journal of Finance*.
126. Merton R. (1973) Theory of Rational Option Pricing, *Bell Journal of Economics and Management Science* 4, 141-183.
127. Miltersen, K.R., Sandmann K., Sondermann D. (1997) Closed Form Solutions for Term Structure Derivatives with Log-Normal Interest Rates. *The Journal of Finance* 52, 409-430.
128. Miyazaki, K., Toshihiro, Y. (1998) Valuation Model of Yield-Spread Options in the HJM Framework. *The Journal of Financial Engineering* 7, 98-107.
129. Moraleda, J.M., Vorst, A.C.F. (1997) Pricing American Interest Rate Claims with Humped Volatility Models. *The Journal of Banking and Finance* 21, 1131-1157.
130. Musiela, M., and Rutkowski, M. (1997). Continuous-Time Term Structure Models: Forward Measure Approach. *Finance and Stochastics*, 4, pp. 261–292.
131. Musiela, M. and Rutkowski, M. (1998) Martingale Methods in Financial Modelling. Springer. Berlin.
132. Nelson, D.B., Ramaswamy, K. (1990) Simple Binomial Processes as Diffusion Approximations in Financial Models. *The Review of Financial Studies* 3, 393-430.
133. Øksendal, B. (1992). Stochastic Differential Equations: An Introduction with Applications. Springer. Berlin.
134. Pedersen, M.B. (1999) Bermudan Swaptions in the LIBOR Market Model, SimCorp Financial Research Working Paper.

135. Pelsser, A. (1996) Efficient Methods for Valuing and Managing Interest Rate and other Derivative Securities. PhD Dissertation, Erasmus University Rotterdam, The Netherlands.
136. Pelsser, A. (2000) Efficient Methods for Valuing Interest Rate Derivatives. Springer. Heidelberg.
137. Pelsser, A., Vorst, A.C.F. (1998) Pricing of Flexible and Limit Caps. Report 9809, Erasmus University Rotterdam.
138. Pugachevsky, D. (2001) Forward CMS Rate Adjustment. *Risk* 14 (March, 2001), 125-128.
139. Rapisarda, F., Silvotti, R. (2001) Implementation and Performance of Various Stochastic Models for Interest-Rate Derivatives. *Applied Stochastic Models in Business and Industry* 17, 109-120.
140. Rebonato, R. (1998) Interest Rate Option Models. Second Edition. Wiley, Chichester.
141. Rebonato, R. (1999a) Calibrating the BGM Model. *Risk* 12 (March, 1999), 74-79.
142. Rebonato, R. (1999b) On the Pricing Implications of the Joint Lognormal Assumption for the Swaption and Cap Market. *The Journal of Computational Finance* 2, 57-76.
143. Rebonato, R. (1999c) On the Simultaneous Calibration of Multifactor Lognormal Interest Rate Models to Black Volatilities and to the Correlation Matrix. *The Journal of Computational Finance* 2, 5-27.
144. Rebonato, R. (1999d) Volatility and Correlation. Wiley, Chichester.
145. Rendleman, R.J., Bartter, B.J. (1980) The Pricing of Options on Debt Securities. *Journal of Financial and Quantitative Analysis* 15, 11-24.
146. Ritchken, P., Sankarasubramanian, L. (1995) Volatility Structures of Forward Rates and the Dynamics of the Term Structure. *Mathematical Finance* 5, 55-72.
147. Rogers, L.C.G. (1995) Which Model for Term-Structure of Interest Rates Should One Use? in M. Davis, D. Duffie, W. Fleming and S. Shreve, eds. Mathematical Finance. IMA Vol. Math. Appl. 65, New York: Springer-Verlag.
148. Rogers, L.C.G. (1996) Gaussian Errors. *Risk* 9, 42-45.
149. Rogers, L.C.G. (1997) The Potential Approach to the Term Structure of Interest Rates and Foreign Exchange Rates. *Mathematical Finance* 7, 157-176.
150. Rogers, L.C.G., and Williams, D. (1987) Diffusions, Markov Processes and Martingales, Vol. II, Wiley and Sons, New York.
151. Rubinstein, M. (1983) Displaced Diffusion Option Pricing. *Journal of Finance* 38, 213-217.
152. Rutkowski, M. (1996) On Continuous-Time Models of Term Structure of Interest Rates. In Stochastic Processes and Related Topics. H. J. Englebert, H. Föllmer, and J. Zabczyk, eds. New York: Gordon and Beach.
153. Rutkowski, M. (1999) Models of Forward LIBOR and Swap Rates. Preprint.
154. Sandmann, K., Sondermann, D. (1993) A Term Structure Model and the Pricing of Interest Rate Derivatives. *The Review of Futures Markets* 12, 391-423.
155. Sandmann, K., Sondermann, D. (1997) A Note on the Stability of Lognormal Interest Rate Models and the Pricing of Eurodollar Futures. *Mathematical Finance* 7, 119-128.
156. Santa Clara, P, and Sornette, D. (2001) The Dynamics of the Forward Interest Rate Curve with Stochastic String Shocks. *The Review of Financial Studies* 14, 149-185.
157. Schlögl, E., Sommer, D. (1998) Factor Models and the Shape of the Term Structure. *The Journal of Financial Engineering* 7, 79-88.
158. Schmidt, W.M. (1997) On A General Class of One-Factor Models for the Term Structure of Interest Rates. *Finance and Stochastics* 1, 3-24.

159. Scott, L. (1995) The Valuation of Interest Rate Derivatives in a Multi-Factor Term-Structure Model with Deterministic Components. University of Georgia. Working Paper.
160. Vasicek, O. (1977) An Equilibrium Characterization of the Term Structure. *Journal of Financial Economics* 5, 177-188.

Index

Accrual swap, 262, 403
- LFM analytical approximated formula, 404
- LFM Monte Carlo price via Milstein scheme, 404

Affine models, 60, 69, 85, 88, 94
- instantaneous-forward-rate dynamics, 60
- instantaneous-forward-rate volatilities, 60
- three factor, 371
- two-factor CIR, 166
- two-factor Gaussian, 129

Andersen-Andreasen smile model, 273
Arbitrage opportunity, 25
Arbitrage-free dynamics for instantaneous forwards, 175
Arbitrage-free pricing, 23
- Harrison and Pliska, 23
- in theoretical bond market, 35
- martingale measure existence, 26

At-the-money cap (floor), 17
At-the-money caplet, 18, 204
At-the-money swaption, 20
Attainable contingent claim, 25, 35
Autocap, 381
- LFM Monte Carlo pricing with Milstein scheme, 382

Average-rate swap, 394

Backward-induction pricing with a tree, 108
Balduzzi-Das-Foresi-Sundaram model, 370
Bank-account numeraire, 2, 186
- continuously rebalanced, 202
- discretely rebalanced, 202
- domestic, 40
- foreign, 40
- in Flesaker-Hughston model, 371
- martingale measure, 27

Basis point, 401

Bermudan-style swaption, 192, 412
- Carr and Yang, 415
- Clewlow and Strickland, 416
- definition, 413
- pricing with a tree, 111
- with the LFM: Andersen, 416
- with the LFM: LSMC, 413

BGM model, 187
Black's formula
- caps, 16, 203, 267
- caps: approximate derivation, 185
- caps: rigorous derivation, 187
- floors, 17
- swaptions, 20, 224, 253

Black-Derman-Toy model, 74
Black-Karasinski model, 73
- equity derivatives and stochastic rates, 453
- example of cap calibration, 123
- examples of implied cap curves, 115
- examples of implied swaption structures, 120
- trinomial tree, 76

Bond (theoretical) market, 35
Bond option, 35
- CIR model, 58
- CIR++ model, 94
- CIR2 model, 168
- CIR2++ model, 170
- G2++ model, 145
- Hull-White model, 67
- Jamshidian decomposition, 103
- Mercurio-Moraleda HJM model, 182
- pricing under forward measure, 36
- put call parity, 47
- shifted short-rate model, 90
- shifted Vasicek model, 92
- Vasicek model, 52

Bond-price numeraire, 186
- zero-coupon bond, 3

Index

Brace's rank-r formula for swaptions, 191, 244
Brace's rank-one formula for swaptions, 190, 241
Brace-Gatarek-Musiela model (BGM), 187
Brennan-Schwartz model, 370
Brigo-Mercurio smile model, 276
Brownian motion, 471
- non differentiability, 471
- unbounded variation, 471

Calibration to caps and swaptions jointly
- G2++ model, 159
- LFM, 197, 254
- LFM desiderata, 314
- LFM one factor, 293
- LFM with Formulation 7, 295
- LFM with TABLE 5, 284
Calibration to caps/floors
- CIR++ model, 99
- CIR2++ two-factor shifted CIR, 166
- extended exponential-Vasicek model, 101
- G2++ model, 141
- LFM, 203
- market example for short-rate models, 122
- market example for the G2++ model, 156
Calibration to the swaption surface, 132
- lognormal forward model (LFM), 252
- exact fitting with TABLE 1 LFM, 303
- G2++ model, 132
- LFM with smoothed data, 312
Calibration to the zero-coupon curve
- CIR++ model, 93, 97
- CIR2++ model, 169
- G2++ model, 136
- Hull-White model, 64
- shifted short-rate model, 89
- shifted Vasicek model, 91
Cap, 16, 204
- additive decomposition in caplets, 16
- as portfolio of bond options, 36, 37
- as protection, 16
- at-the-money, 17, 205
- autocap, 381
- Black's formula, 16, 203
- CIR2++ model, 170

- G2++ model, 147
- Hull-White pricing formula, 68
- in advance, 379
- in- and out-of-the-money, 17
- market quotes, 208
- payoff, 189, 204
- with deferred caplets, 382
Cap volatility, 79, 209
- curve, 17
- one-factor short-rate models, 81
- two-factor short-rate models, 141
Cap with deferred caplets, 382
- G2++ pricing formula, 383
- LFM analytical approximated formula, 383
- LFM Monte Carlo pricing with Milstein scheme, 382
Caplet, 16, 184, 204
- and one-year-tenor swaption, 254
- as bond option, 37
- at-the-money, 18, 204
- G2++ model, 146
- in-the-money, 18, 204
- LFM, 205
- out-of-the-money, 18, 205
- prices and rates distributions, 268
- skew, 268
- smile, 268
Caplet volatility, 79, 139, 206, 209
- LFM with Formulation 6, 207
- LFM with Formulation 7, 208
- LFM with TABLE 1, 206
- LFM with TABLE 2, 206
- LFM with TABLE 3, 206
- LFM with TABLE 4, 207
- LFM with TABLE 5, 207
- stripping from cap volatility, 209
Caption, 395
- LFM Monte Carlo pricing with Milstein scheme, 395
Change of numeraire, 23
- asset with itself as numeraire, 31
- choice of a convenient numeraire, 33
- domestic/foreign measures, 41
- fundamental drift formula, 28, 30
- Girsanov's theorem, 29
- LFM forward-measure dynamics, 198
- lognormal case, 32
- Radon-Nikodym derivative, 28
- with level-proportional volatilities, 31
Cholesky decomposition, 134, 456
CIR model, 46, 56, 61, 481

- bond option, 58
- bond price, 58
- CIR2 two-factor version, 166
- forward-rate dynamics, 59
- Hull-White extension, 72
- Jamshidian extension, 73
- market price of risk, 57
- real-world dynamics, 56
- shifted two-factor version CIR2++, 130
- shifted version CIR++, 93
- T-forward dynamics, 59
- two-factor version CIR2, 130

CIR++ model, 93
- bond option, 94
- bond price, 93
- cap price, 95
- example of cap calibration, 123
- examples of implied cap curves, 116
- Jamshidian decomposition, 103
- Milstein scheme, 105
- Monte Carlo method, 105
- positivity of rates, 97
- swaption price, 95
- trinomial tree, 96
- zero-coupon swaption, 399

CIR2 two-factor CIR model, 130, 166
- bond option, 168
- bond price, 166
- bond-price dynamics, 166
- continuously compounded spot rate, 166
- T-forward dynamics, 168

CIR2++ shifted two-factor CIR, 130, 168
- bond option, 170
- bond price, 169
- cap price, 170
- instantaneous-forwards volatility, 165, 170
- positivity of rates, 165
- swaption price, 171

CMS, 385
Complete financial market, 26
- and uniqueness of the martingale measure, 26

Compounding, 5
- annual, 7
- continuous, 6
- k times per year, 7
- simple, 6

Consol bond, 369
Consol rate, 370

Constant elasticity of variance (CEV) model, 270, 273, 481
Constant-maturity swap (CMS), 385, 388
- convexity adjustment, 392
- G2++ pricing, 386
- LFM Monte Carlo pricing with Milstein scheme, 385

Contingent claim
- attainability under any numeraire, 27
- attainable, 25
- attainable in theoretical bond market, 35
- multiple payoff, 40
- pricing with deferred payoff, 38

Convexity adjustment, 387
- constant-maturity swap, 392
- general formula, 390
- quanto adjustment, 435

Correlation, 190
- curves of different currencies, 424
- equity asset/interest rate, 454
- forward rates, 189
- impact on swaptions, 189
- instantaneous for G2++, 131, 133
- instantaneous for LFM, 190, 194, 217, 230
- LFM angles constraints, 285, 288, 290, 295, 297, 300
- LFM angles parameterization, 231
- LFM instantaneous from calibration, 286, 289, 296
- LFM low rank, 230
- LFM terminal from calibration, 286, 291, 293, 294, 299, 300, 313
- LFM terminal: analytical formula, 251
- LFM terminal: MC tests, 346
- LFM terminal: Rebonato's formula, 251
- no impact on caps, 204
- short rate and consol rate, 370
- sigmoid shape, 231, 294
- terminal, 20, 128, 190, 217, 251, 353
- terminal for two-factor Vasicek, 129
- terminal: dependence on volatility, 217

Coupon-bearing bond, 14
Curvature of the zero-coupon curve, 371

Day-count convention, 4
- actual/360, 5
- actual/365, 4

- case 30/360, 5
Decorrelation, 194, 231, 294
Differential swaps, 437
- G2++ model, 438
- market-like formula, 444
- payoff, 437
Diffusion process, 470
- martingale, 473
Discount factor, 3
- deterministic in Black-Scholes, 3
- relationship with zero-coupon bond, 4
- stochastic, 3
Dothan model, 54, 61
- bond price, 55

Early exercise and path dependence together, 408
Early-exercise pricing with trees, 106
EEV, 101
Equity derivatives under stochastic rates, 453
- Black-Karasinski model, 453, 460
- general short-rate model, 459
- Hull-White model, 453
- trinomial tree construction, 460
- two-dimensional tree, 463
Equity option with stochastic rates, 458
- Hull-White model, 458
- Merton's model, 459
Equivalent martingale measure, 25
Equivalent probability measures, 483
Euler scheme, 192, 234, 478
EURIBOR rate, 1
Eurodollar futures, 200, 399
- G2++ model, 400
- LFM, 402
- LFM analytical approximated formula, 402
- LFM Monte Carlo pricing with Milstein scheme, 402
Explosion of the bank account in lognormal models, 56
Exponential-Vasicek model, 62, 101
- trinomial tree, 101
Extended exponential-Vasicek model (EEV), 101
- example of cap calibration, 123
- examples of implied cap curves, 118
- examples of implied swaption structures, 120
- trinomial tree, 101

Factor analysis, 129, 371

Feynman-Kač's theorem, 482
- link PDEs/SDEs, 482
Flesaker-Hughston framework, 371
- rational lognormal model, 372
Floater, 384
Floating-rate note, 15
- trading at par, 15
Floor, 16
- as portfolio of bond options, 37
- Hull-White pricing formula, 68
Floorlet, 16
- as bond option, 37
Floortion, 395
Forward (adjusted) measure, 51, 59, 67, 144, 168, 193
- definition, 33
- dynamics for equity derivatives with stochastic rates, 456
- general pricing formula under, 33
- martingale forward rates, 34, 193
Forward forward volatility, 209
Forward rates, 193
- absolute volatility in the G2++ model, 143
- average volatility, 190, 210
- definition, 10
- expiry and maturity, 10
- instantaneous, 12
- instantaneous correlation, 190
- instantaneous covariance in the G2++ model, 143
- instantaneous volatility, 190, 195
- risk-neutral dynamics, 200, 201
- simply compounded, 12
- terminal correlation, 217, 249, 353
Forward swap rate, 15, 221
- as average of forward LIBOR rates, 222
- as function of forward LIBOR rates, 15, 221, 222
- dynamics under forward measure, 228
- dynamics under swap measure, 223
Forward volatilities, 209
Forward-rate agreement (FRA), 10, 387
- in advance, 387
Forward-rate dynamics, 198, 228, 250
- constant elasticity of variance, 273
- drift interpolation, 262
- LFM forward measure, 198
- LFM risk-neutral measure, 202
- LFM spot LIBOR measure, 203
- mixture of lognormals, 276

- shifted mixture of lognormals, 280
- shifted-lognormal, 271
- the bridging technique, 264

Forward-swap measure, 223

G2 two-factor Vasicek model, 128
G2++ shifted two-factor Vasicek model, 130, 133
- absolute volatility of forward rates, 143
- binomial tree, 152
- bond option, 145
- bond price, 135
- calibration to caps, 156
- calibration to swaptions, 156
- cap price, 147
- cap with deferred caplets, 383
- caplet price, 146
- constant-maturity swap, 386
- continuously compounded forward rates, 143
- differential swaps, 438
- equivalence with two-factor Hull-White, 149
- Eurodollar futures, 400
- foreign-curve distribution under domestic forward measure, 431
- foreign-curve dynamics under domestic forward measure, 429
- foreign-curve dynamics under domestic measure, 427
- in-advance cap, 381
- incompatibility with market models, 187
- instantaneous covariance of forward rates, 143
- instantaneous-forwards correlation, 142
- instantaneous-forwards volatilities, 138
- joint calibration to caps/swaptions, 159
- multi currency, 421
- multi-currency correlation, 424
- probability of negative rates, 137
- swaption price, 148
- T-forward dynamics, 144
- trinomial tree, 152
- zero-coupon swaption, 399

Girsanov's theorem, 483
- example, 484

Government rates, 1

Harrison and Pliska

- continuous-time economy, 24

Hedging, 224, 316, 495

HJM framework, 13, 173
- bond-price dynamics, 176
- Harrison and Pliska, 23
- instantaneous-forward-rate dynamics, 175
- Li-Ritchken-Sankarasubramanian tree construction, 179
- Markovian higher-dimensional process, 178
- Markovian short rate, 176
- Mercurio-Moraleda HJM model, 181
- neutralizing path dependence, 178
- no-arbitrage drift condition, 175
- non-Markovian short rate, 174, 178
- non-recombining lattices, 178
- recombining (Markovian) lattice, 179
- Ritchken-Sankarasubramanian volatility, 178
- separable volatility and equivalence with HW, 177
- separable volatility structure, 176
- short-rate equation, 176
- toy-model example, 174

Ho-Lee model, 63

Hull and White's LFM swaption-volatility formula, 249

Hull-White model, 63, 92, 480
- bond option, 67
- bond price, 66
- cap price, 68
- coupon-bond option, 68
- equity derivatives and stochastic rates, 453
- equity option with stochastic rates, 458
- example of cap calibration, 123
- extended CIR, 72
- extended Vasicek, 64
- floor price, 68
- HJM framework with separable volatility, 177
- Jamshidian decomposition, 103
- swaption price, 68
- T-forward dynamics, 67
- trinomial tree, 69
- two-factor (G2++), 149

Humped volatility, 82, 116, 118, 121, 130, 139, 165, 170, 181, 210

In-advance cap, 379
- G2++ pricing formula, 381

514   Index

– LFM pricing formula, 380
In-advance forward-rate agreement, 387
In-advance swap, 378
– pricing formula, 379
In-the-money cap (floor), 17
In-the-money caplet, 18
In-the-money swaption, 20
Incompatibility of the two market
    models, 227
Instantaneous forward rates, 12, 173
Instantaneous spot rate, 2
Instantaneous-forwards correlation
– G2++ model, 138, 142
Instantaneous-forwards volatility
– affine models, 60
– CIR2++ model, 165, 170
– G2++ model, 138
– HJM formulation, 174
Interbank rates, 1
Interest-rate swap (IRS), 13, 220
– as a set of FRA's, 14
– as floating note vs coupon bond, 14
– discounted payoff, 14
– fixed leg, 13
– floating leg, 13
– in advance, 378
– payer and receiver, 14
– reset and payment dates, 13
– trigger swap, 406
– zero coupon, 395
Ito's formula, 477
– Stochastic Leibnitz's rule, 478
– second-order term, 477
Ito's stochastic integral, 471
– example, 472
– second order effects, 472

Jamshidian decomposition, 102
Jamshidian model, 187

Level of the zero-coupon curve, 370, 371
LFM, 187
Li-Ritchken-Sankarasubramanian
    recombining lattice, 179
LIBOR market model (LFM), 187
LIBOR rate, 1, 6
Lognormal forward-LIBOR model
    (LFM), 187, 200
– accrual swap, 404
– approximated (lognormal) dynamics, 250, 377, 393
– as central market model, 189

– autocap, 382
– average-rate swap, 394
– Bermudan swaptions, 413, 416
– bridging technique, 264
– Brownian motions under different
    forward measures, 200
– calibration to caps and floors, 203
– cap with deferred caplets, 383
– caplet prices, 205
– caplets vs one-year-tenor swaptions, 254
– caption and floortion, 395
– constant-maturity swap, 385
– drift interpolation, 262
– Euler scheme, 234
– Eurodollar futures, 402
– forward-measure dynamics, 198
– frozen-drift approximation, 247, 249, 250, 393
– geometric-Brownian-motion approximation, 250, 377, 393
– guided tour, 184
– in-advance cap, 380
– in-advance FRA, 387
– in-advance swap, 379
– in-between rates, 402
– incompatibility with LSM, 227
– incompatibility with short-rate
    models, 187
– instantaneous correlation, 194
– instantaneous correlation via angles, 231
– instantaneous volatility, 195
– MC tests of swaption volatility
    formulas, 318
– MC tests of terminal correlation
    formulas, 346
– Milstein scheme, 234, 263, 382
– one to one swaptions parameterization, 303
– one-factor calibration, 293
– ratchet, 384
– rates over non-standard periods, 261
– risk-neutral dynamics, 202
– spot-LIBOR-measure dynamics, 203
– swaption pricing, 189
– swaptions with Brace's formula, 241, 244
– swaptions with Hull and White's
    formula, 249
– swaptions with Rebonato's formula, 248
– terminal correlation (analytical), 251

- terminal correlation (Rebonato), 251
- trigger swap, 408
- zero-coupon swaption, 397

Lognormal swap model (LSM), 188, 220, 224
- dynamics under forward measure, 228
- dynamics under swap measure, 223
- incompatibility with LFM, 227

Longstaff-Schwartz model, 130, 165
- equivalence with two-factor CIR, 167

Longstaff-Schwartz regressed Monte Carlo (LSMC), 410
- layout of the method, 410
- path dependence and early exercise, 410

Market models
- guided tour, 184

Markov functional models, 373

Martingale, 473
- driftless diffusion, 473

Martingale measure, 25
- and bank-account numeraire, 27
- and no-arbitrage, 26
- foreign, 40
- uniqueness and completeness, 26

Maximum-likelihood estimators, 53, 265

Mean reversion, 50, 54, 62, 98, 116, 149, 157, 371

Mercurio-Moraleda HJM model, 181
- bond option, 182

Mercurio-Moraleda short-rate model, 85
- example of cap calibration, 123

Merton's model, 459

Merton's toy model, 173

Milstein scheme, 192, 234, 263, 478

Miltersen, Sandmann and Sondermann model, 187

Misalignments in the swaption matrix, 253

Modified Bessel function, 274

Money-market account, 2

Monte Carlo, 103, 317
- CIR++ model, 105
- CIR2++ model, 171
- discretization scheme for SDE, 105
- early exercise, 408
- early exercise and path dependence together, 408
- forward propagation, 104, 409
- forward-adjusted measure, 105, 145
- LFM pricing of swaptions, 192, 233
- Longstaff-Schwartz regression method, 410
- simulation, 191
- simulation of a SDE, 479
- testing LFM correlation formulas, 346
- testing LFM swaption formulas, 318
- tests for LFM formulas, 317

Moraleda-Vorst short-rate model, 84

Multi-currency derivatives, 40, 421
- correlation between different curves, 424
- differential swaps, 437, 444
- market formulas, 440
- quanto caplet/floorlet, 440
- quanto caps/floors, 443
- quanto CMS, 427
- quanto swaptions, 444

Multifactor short-rate models, 129, 370

Natural and unnatural time lag, 386
- and dependence on volatility, 387

Numeraire, 193, 200, 229, 234
- and state-price density, 371
- definition, 27
- specification in Markov functional models, 373
- specification in the potential approach, 373

One-way floater, 384
Out-of-the-money cap (floor), 17
Out-of-the-money caplet, 18
Out-of-the-money swaption, 20
Over-fitting, 64, 157

Path dependence and early exercise together, 408
Potential approach, 373
Present value for basis point numeraire, 223
Pricing kernel, 371
Pricing operator, 371
Principal-component analysis, 129
Probability space, 469

Quadratic covariation, 474
Quadratic variation, 247, 474
- Brownian motion, 474
Quanto adjustment, 435
Quanto caplet/floorlet, 440
Quanto caps/floors, 443

516   Index

Quanto CMS, 427
- G2++ model, 429
- G2++ model: a specific contract, 433
- G2++ model: Monte Carlo pricing, 431
- payoff, 427
- quanto adjustment, 435
Quanto swaps, 437, 444
Quanto swaptions, 444
- market-like formula, 448

Radon-Nikodym derivative, 456
- definition, 483
- foreign/domestic markets, 41
Ratchet, 384
- LFM Monte Carlo pricing with Milstein scheme, 384
Rational lognormal model, 372
Rebonato's LFM swaption-volatility formula, 248
Rebonato's terminal-correlation formula, 251
Risk-adjusted measure, 25
Risk-neutral measure, 25
Ritchken-Sankarasubramanian volatility and path dependence, 178

Self-financing strategy, 25
Semimartingale, 473
- model for underlying assets, 24
Shifted Dothan model, 100
- bond price, 100
- tree construction, 100
Shifted exponential-Vasicek model, 101
Shifted short-rate model
- bond option, 90
- bond price, 88
- cap price, 91
Shifted two-factor models, 130
Shifted Vasicek model, 91
- bond option, 92
- bond price, 92
Short rate, 2
- as limit of spot rates, 8
Short-rate models
- affine models, 60
- deterministic-shift extension, 88
- endogenous models, 45, 46
- example of cap calibration, 122, 156
- examples of implied cap curves, 114
- examples of implied swaption structures, 119
- exogenous models, 47, 63

- Girsanov transformation, 44
- guided tour, 43
- humped volatility, 84
- implied cap volatility, 81
- implied caplet volatility, 81
- incompatibility with market models, 187
- intrinsic caplet volatility, 80
- market price of risk, 44
- Merton's toy model, 173
- Monte Carlo method, 104
- multi-currency derivatives, 421
- number of factors, 129
- real-world dynamics, 44
- summary table, 49
- term structure of cap volatilities, 82
- term structure of caplet volatilities, 81
- time-homogeneous models, 48
- trees for pricing, 106
- two factor, 127
- volatility, 77, 138
Skew, 268
Smile, 266
- Andersen-Andreasen model, 273
- Brigo-Mercurio model, 276
- Brigo-Mercurio shifted model, 280
- G2++ model, 132
- guided tour, 266
- local-volatility model, 269
- lognormal-mixture dynamics, 276
- shifted lognormal-mixture dynamics, 280
- shifted-lognormal dynamics, 271
- stochastic volatility model, 269
Smile-shaped caplet volatility curve, 266
Spread option, 445
State-price density, 371
- Flesaker-Hughston framework, 371
- Markov functional models, 374
- potential approach, 373
Steepness of the zero-coupon curve, 370, 371
Stochastic (Lebesgue) integral, 472
Stochastic (Stieltjes) integral, 471
Stochastic differential equation (SDE), 470
- diffusion coefficient, 476
- discretization schemes, 478
- drift, 476
- drift and diffusion coefficient, 470
- drift change, 483

Index   517

- Euler and Milstein schemes, 478
- examples, 480
- existence and uniqueness of solutions, 475
- from deterministic to stochastic, 470
- geometric Brownian motion, 481
- Girsanov's theorem, 483
- in integral form, 471
- linear, 480
- lognormal linear, 480
- Monte Carlo simulation, 479
- population-growth example, 470
- square-root process, 481

Stochastic integral
- Ito, 471
- Stratonovich, 471

Stochastic Leibnitz's rule, 478
String model, 413
Submartingale, 473
Supermartingale, 473
Swap market model, 188
Swap measure, 223
Swaption, 19, 223
- at-the-money, 20, 223
- Bermudan: definition, 413
- Bermudan: Tree, 111
- Black's formula, 20, 224
- cash settled, 226
- CIR++ pricing formula, 95
- CIR2++ model, 171
- dependence on correlation of rates, 19
- G2++ model, 148
- hedging, 224
- Hull-White pricing formula, 68
- illiquid, 252
- in-the-money, 20, 223
- LFM Brace's rank-1 formula, 241
- LFM Brace's rank-r formula, 244
- LFM calibration, 252
- LFM Hull and White's formula, 249
- LFM Monte Carlo pricing, 233
- LFM one-to-one parameterization, 303
- LFM pricing, 189
- LFM Rebonato's formula, 248
- market quotes, 253
- matrix, 253
- matrix misalignments, 253
- matrix parametric form, 310
- matrix smoothing, 311
- maturity, 19
- no additive decomposition, 19

- one-year tenor vs caplets, 254
- out-of-the-money, 20, 223
- payer and receiver, 19
- payoff, 19, 21, 189, 223
- smaller than the related cap, 19
- strike, 223
- tenor, 19
- volatility surface, 20
- zero coupon, 395
- zero coupon larger than plain, 399

Swaption volatility, 246
- matrix parametric form, 310
- sensitivities, 316
- smoothing, 311

Tenor of a swaption, 19
Term structure of discount factors, 10
Term structure of interest rates, 9
Term structure of volatility, 79, 140, 210
- evolution for LFM-Formulation 6, 215
- evolution for LFM-Formulation 7, 215, 296, 299
- evolution for LFM-TABLE 1, 314
- evolution for LFM-TABLE 2, 210
- evolution for LFM-TABLE 3, 213
- evolution for LFM-TABLE 4, 213
- evolution for LFM-TABLE 5, 213, 287, 290, 291
- LFM, 210
- LFM vs short-rate models, 217
- maintaining the humped shape, 212, 215
- one-factor short-rate models, 81
- two-factor short-rate models, 141

Three-factor short-rate models, 370
Time to maturity, 4
Trading strategy, 24
- gains process, 24
- self-financing, 25
- value process, 24

Tree
- backward induction, 106, 108, 409
- binomial for G2++, 152
- for equity derivatives under stochastic rates, 460
- pricing, 107
- pricing Bermudan swaptions, 111
- pricing with two-factor models, 110
- problems with dimension > 2, 409
- trinomial for BK, 76
- trinomial for CIR++, 96

- trinomial for extended exponential-Vasicek model, 101
- trinomial for G2++, 152
- trinomial for Hull-White, 69
- two dimensional for equity derivatives with stochastic rates, 463

Trigger swap, 406
- LFM Monte Carlo price via Milstein scheme, 408

Two-curves products, 421

Two-factor short-rate models, 127
- CIR2++ model, 168
- correlation structure, 130
- deterministic shift, 130
- G2++ model, 130, 133
- implied cap volatility, 141
- implied caplet volatility, 141
- intrinsic caplet volatility, 140
- motivation, 127
- term structure of cap volatilities, 141
- term structure of caplet volatilities, 141

Vasicek model, 23, 43, 45, 50, 61, 63, 91, 103, 127, 480
- bond option price, 52
- bond price, 51
- G2 two-factor version, 128
- market price of risk, 52
- maximum likelihood estimators, 53
- real-world dynamics, 52
- T-forward dynamics, 51

Volatility
- cap, 209
- examples of cap curves for BK, 115
- examples of cap curves for CIR++, 116
- examples of cap curves for EEV, 118
- examples of swaption structures for BK, 120
- examples of swaption structures for EEV, 120
- forward, 209
- forward forward, 209
- humped shape, 82, 84, 116, 118, 121, 139, 165, 170, 181, 210
- instantaneous for LFM, 195
- instantaneous for LFM - Formulation 6, 197
- instantaneous for LFM - Formulation 7, 197
- instantaneous for LFM - TABLE 1 formulation, 195
- instantaneous for LFM - TABLE 2 formulation, 195
- instantaneous for LFM - TABLE 3 formulation, 196
- instantaneous for LFM - TABLE 4 formulation, 196
- instantaneous for LFM - TABLE 5 formulation, 197
- one-factor short-rate models, 77
- term structure, 210
- two-factor short-rate models, 130, 138, 165, 170

Volatility smile, 102
- guided tour, 266

Year fraction, 4
Yield curve, 8
- curvature, 371
- level, 370, 371
- steepness, 370, 371

Zero-coupon bond, 3
- realtionship with discount factor, 4

Zero-coupon curve, 8
Zero-coupon IRS, 395
Zero-coupon swaption, 395
- CIR++ model, 399
- G2++ model, 399
- larger than plain-vanilla swaption, 399
- LFM model, 397

Printing: Saladruck, Berlin
Binding: H. Stürtz AG, Würzburg